MATERIALS SCIENCE AND TECHNOLOGIES

FUNDAMENTALS AND ENGINEERING OF SEVERE PLASTIC DEFORMATION

MATERIALS SCIENCE AND TECHNOLOGIES

Additional books in this series can be found on Nova's website at:

https://www.novapublishers.com/catalog/index.php?cPath=23_29&seriesp=Materials+Science+and+Technologies

Additional e-books in this series can be found on Nova's website at:

https://www.novapublishers.com/catalog/index.php?cPath=23_29&seriespe=Materials+Science+and+Technologies

MATERIALS SCIENCE AND TECHNOLOGIES

FUNDAMENTALS AND ENGINEERING OF SEVERE PLASTIC DEFORMATION

V.M. SEGAL
I.J. BEYERLEIN
C.N. TOME
V.N. CHUVIL'DEEV
AND
V.I. KOPYLOV

NOVA

Nova Science Publishers, Inc.
New York

For permission to use material from this book please contact us:
Telephone 631-231-7269; Fax 631-231-8175
Web Site: http://www.novapublishers.com

LIBRARY OF CONGRESS CATALOGING-IN-PUBLICATION DATA

Fundamentals and engineering of severe plastic deformation / authors,
Vladimir M. Segal ... [et al.].
p. cm.
Includes bibliographical references and index.
ISBN 978-1-61668-190-6 (hardcover)
1. Plasticity. I. Segal, V. M. (Vladimir Mironovich)
TA418.14.F86 2009
620.1'1233--dc22
2010002981

Published by Nova Science Publishers, Inc. † New York

CONTENTS

PREFACE

Recent extensive activity on severe plastic deformation (SPD) as a new materials processing technology has generated vast experimental and analytical amounts of information. However, there is still a strong need for a multidisciplinary understanding of SPD. The present book is intended to bridge the gap between existing approaches, resolve certain controversies, and provide a united description of SPD at different scales. It consists of in-depth analyses of the fundamentals and engineering of SPD on processing mechanics (Part I), the micro-mechanics of plastic deformation (Part II) and the physics of structurally-sensitive properties (Part III). The three parts of the book were prepared by scientists from USA, Russia and Belarus, and include many original and new results. Most of the attention is paid to Equal-Channel Angular Extrusion (ECAE), which to-date has the greatest potentials for industrial applications.

Part I considers the macro-mechanics of SPD with emphasis on flow localization and deformation mode. First, the importance of simple shear versus pure shear is discussed. It is shown that simple shear provides early transition to shear band, grain boundary and rotational localizations, and is optimal for structure refinement. Different techniques for realizing simple shear mode are described: surface simple shear, simple shear at velocity discontinuities, and special processes of simple shear and combined loading. An analysis of ECAE by slip line theory demonstrates an essential influence of boundary conditions on processing parameters, and, finally, materials structure and properties. Practical engineering aspects and possible applications of ECAE are addressed.

Part II develops a multi-scale approach to the evolution of grain microstructures, crystallographic textures and material properties (yield stress, hardening, ductility, anisotropy) during SPD and subsequent loading. Central to the formalism is a polycrystal plasticity model that links the macroscopic processing path to the small-scale deformation mechanisms of crystallographic glide and twinning in individual grains. The model is extensively applied to single-phase F.C.C. and H.C.P. crystal structures. This predictive capability is vital for designing SPD processing pathways to produce ultrafine grained and nanostructured metals with mechanical properties that meet specific application needs.

Part III considers nanocrystalline and microcrystalline (NMC) materials produced by ECAE. It has been shown that their structure and properties are mainly influenced by nonequilibrium grain boundaries. Based on the theory of nonequilibrium grain boundaries, characteristics of diffusion processes and models of structure evolution have been developed, including the grain refinement limits during ECAE, recrystallization, and abnormal grain growth in NMC pure metals and alloys. Mechanical properties of NMC metals are also described in detail. In particular, deviations from the Hall-Petch law are explained and prerequisites for simultaneous increase in strength and ductility are determined. For structural superplasticity, the optimal grain size and strain hardening are analyzed.

The book is addressed to material scientists and industrial engineers involved in development of advanced materials and technologies. It should also be useful to university and college students pursuing related specialties.

INTRODUCTION

Materials play a vital role in the human society. Indeed, a classification of historical eras is based on the characteristics of materials that dominated in certain periods of human culture, such as Stone, Bronze and Iron Ages. In the modern age of new materials, metals have remained the most important industrial product. Plasticity and the ability to induce large deformation in specific conditions are included as primary characteristics of metals defining their applications and manufacturing methods. Historically, for a period more than a few millennia, the primary goal of metalworking was to change the material shape and dimensions by using forming operations such as forging, rolling, extrusion, and so forth. The scientific foundation for the description of a corresponding behavior of a metal on the macro-scale is the mathematical theory of plasticity that formulates systems of differential equations for relevant mechanical models. Together with numerical methods and advanced computer technologies, this continual approach provides realistic solutions for large and small plastic strains at adequate boundary conditions. Although the improvement of material models to include physical rather than phenomenological characteristics is an important area of academic research, the progress in traditional metal forming operations mainly relies on their optimization and engineering developments.

On the finer scales, plastic deformation depends on internal crystallographic structure and, in turn, exerts a deep influence on the structure of metals. Structural changes during plastic deformation are diverse and cannot be described by a single characteristic. Arising effects are related to deformation mechanisms on the micro scale which specify the evolution of defects and phase transformations during straining. These processes are controlled by deformation parameters such as temperature, pressure, strain and strain rate. In technology, there is always an interest to extend these parameters beyond the established limits. As for temperature, strain rate and pressure, the efforts of researchers and engineers led to the development of high temperature and cryogenic processing, creep forming, impact and hydro-extrusion. In relation to strains, the situation appears to be more complicated. A general concept to use large plastic deformation for controlling material structures and properties has long been recognized. Since the ancient age, legends of Damascus steels, Samurai swords and Chinese cold arms were linked to the secrets of multi-step and intensive forging. In the modern time, ultrahigh strains are attained for thin foils and wires after rolling and drawing with very large accumulated reductions. These products have become ordinary industrial practice. One of the first TEM analyses (Embury et. al, 1966; Armstrong et al. 1966)[1] of heavy drawn structures with total strains up to e = 7, which conforms to the present definition of "severe plastic deformation" (SPD), revealed very fine, but ordinary cell/sub-grain structures clustered into thin and long deformation bands or lamellae separated by flat high angle boundaries. Similar results were confirmed for rolling (Malin et al. 1978; Hirsch et al. 1988). A review of the early works on large plastic deformation can be found in (Gil Sevillano et al. 1980). These heavily deformed structures provide a strong hardening effect, but they cannot be classified as grain structures in accordance with Hall- Petch relation for corresponding scaling because of plain morphology in one direction and, presumably, low angle boundaries in perpendicular directions. Such products of small cross-sectional dimensions with high strength and low ductility have found restricted structural applications. Nevertheless, a few impressive examples should be mentioned: one of the strongest materials, known as "piano wire", is produced by heavy drawing of carbon steels; the "Coolidge process" used for fabrication of tungsten filaments is comprised of numerous swaging and drawing operations; "in situ" composites from two phase alloys are fabricated by rolling or drawing with extremely high area reductions (strains up to e = 10). It has also been revealed that intensive plastic deformation may enhance magnetic, superconductive, elastic and other properties of specialty alloys. However, large plastic deformation performed by ordinary forming operations in most cases resulted in quantitatively moderate effects rather than in dramatic improvements and in the discovery of fundamentally new phenomena. When deformation increases, the material characteristics quickly approximated to a plateau and further straining are not effective for modification of structures and properties.

A principally different technique for achieving practically unlimited strains in thin discs via compression with torsion was developed by P.W. Bridgman in 1930s - 1940s (Bridgman, 1952). In contrast to forming operations, the Bridgman method implements near simple shear straining under high hydrostatic pressure. The first applications of this deformation mode in Physical Chemistry in 1950s – 1970s uncovered many unusual properties such as cold polymerization and alloying, hypermobility of atoms and molecules, anomalously high diffusivities and rates of chemical reactions, and brought about the synthesis of new phases and chemicals (Enikolopov, 1981). Later, similar research into intensive simple shear of metals gave novel surprising results such as strain induced structure refinement down to submicron grained (Segal et al. 1981b; Smirnova et al. 1986) and to nano grained materials (Levit, 1989). It became apparent that both strain amount and deformation mode are equally important for severe plastic deformation and that simple shear may provide qualitatively new effects. Realization of simple shear as a production tool in materials processing for properties necessitated the elaboration of special deformation techniques. To explore these new opportunities, the research into technological applications of simple shear was initiated in the early part of 1970 at the Physical-Technical Institute, Minsk, USSR. This initiative turned into long-term developments in the areas of Plasticity, Materials Science and Engineering, including commercialization of new processing methods and technologies. Since 1990, the interest in ultra-fine grained and nano materials stimulated extensive activity in this field and an avalanche of publications. Now, severe plastic deformation is considered as one of the forefronts in the technology of advanced materials. The vast literature was recently

[1] See References to Part I

overviewed in (Meyers et al. 2006; Valiev et al. 2006). Also, proceedings of a number of conferences and collections of papers on severe plastic deformation have been published.

Owing to contributions of specialists from many countries, significant progress has been attained during last years in the accumulation of experimental data and understanding of the mechanisms of severe plastic deformation (SPD). However, there is a strong necessity for a multidisciplinary consideration of the basics of SPD. The present book is intended to bridge the gap between the existing approaches, resolve certain controversies and provide a unified description of SPD at different scales. It comprises an in-depth analysis of the fundamentals and engineering of SPD: processing mechanics, micro-mechanics of plastic deformation and physics of structure-properties relations. The corresponding parts of this book were prepared by scientists from USA, Russia and Belarus and contain many original and new results.

PART I. PROCESSING MECHANICS OF SEVERE PLASTIC DEFORMATION

V.M. Segal
Engineered Performance Materials (EPM)
11228 Lemen Road – Suite A
Whitmore Lake, MI 48843, USA

Chapter 1

CONTINUUM APPROACH TO SEVERE PLASTIC DEFORMATION (SPD)

1.1. EVOLUTION OF MICROSTRUCTURE DURING SEVERE PLASTIC DEFORMATION

Plastic flow depends on many factors. The typical situation is usually considered at room temperature for face centered cubic (f.c.c.) metals with medium to high stacking fault energy (SFE) when recovery and recrystallization are suppressed and the main micro-mechanism of deformation is a dislocation glide along certain crystallographic planes. Because of different crystallographic orientations, dispersion in grain sizes and plural interactions between of dislocations and other defects, strain distributions inside and between grains of polycrystalline metals are highly inhomogeneous. When straining proceeds, dislocations form complex structures. Some general principles of their organization were suggested to explain experimentally observed results. D. Kulhmann-Wilsdorf formulated a hypothesis of the evolution of dislocations into low energy dislocation structures (LEDS) (Kulhman-Wilsdorf, 1989). In accordance with the LEDS hypothesis, dislocations form incidental dislocation boundaries (IDBs) and geometrically necessary dislocation boundaries (GNDBs) which subdivide original grains. Based on this hypothesis, N. Hansen developed a framework for the continuous structure evolution during monotonic loading such as rolling (Hansen, 1990). The Taylor upper-bound assumption requires that at least five independent slip systems should act in each uniformly deformed grain of a grain aggregate to satisfy incompressibility and internal strain compatibility between grains (Taylor, 1938). However, the energy dissipation may be minimized when grains are subdivided on smaller areas of a uniform glide with a lower number of crystallographic planes (Figure 1.1a). Depending on the aggregate structure and loading conditions, there are optimal outcomes of the grain subdivision. Continuity of the plastic flow between subdivided areas is provided by formation of dense dislocation walls (DDWs) which are originally low angle boundaries. During monotonic straining, DDWs rotate to a direction of the principal elongation and angles of their misorientation increase. In some subdivided areas, localized crystallographic shear may appear along DDWs inducing their split and formation of the first generation of micro-bands (MB_1s). Simultaneously, trapped dislocations inside subdivided areas form structures of ordinary cells which are IDBs (Figure 1.1b). Groups of cells within subdivided areas delineated by DDWs or MB_1s are known as cell blocks (SB). At some loading stage, localized shear also arises along slip planes (111) coinciding with the maximum tangential stresses (at an angle ± 45° to the flow direction). That forms the second generation of micro-bands (MB_2s) and cut off DDWs and cell blocks into short segments.

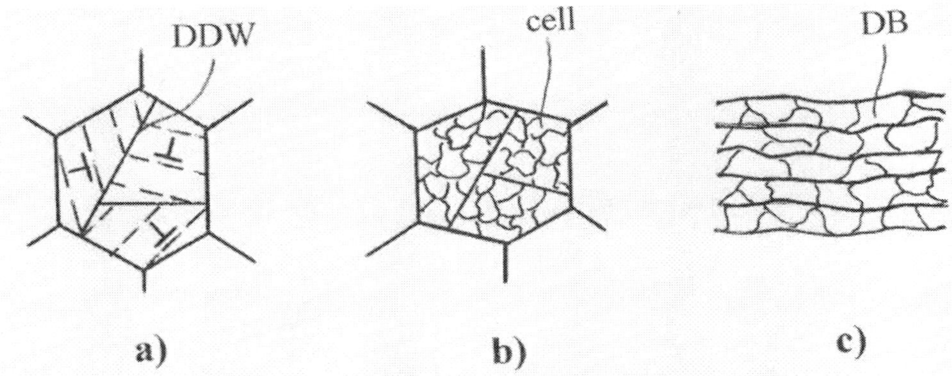

Figure 1.1.Characteristic stages of continuous structure evolution.

D.A. Hughes and N. Hansen extended this framework to large strains (Hughes et al. 1997). They found that during rolling with large reductions cell blocks are steeply transformed into flat lamellae or deformation bands (DBs) which are parallel to a rolling plane and have high angle boundaries (Figure 1.1c). Finally, lamellae comprise one row of cells and their thickness approximates to the minimum cell size. During additional rolling, very thin and long lamellae and cell blocks coalesce and annihilate each other with the increase of misorientation angles. However, boundaries of cells remain low angle configurations. Such two dimensional plane microstructures cannot be qualified as new grains. Similar long "fragments" having high angle boundaries were described by V.V. Rybin for large rolling reductions (Rybin, 1986)

The considered framework of the evolution of dislocation structures is associated with continuous plastic flows by crystallographic glide within separated grains. In materials with low SFE, under specific processing conditions micro twins arise on certain crystallographic planes and accommodate the main part of plastic deformation instead of glide. Progressively, the matrix material and twins form thin bundles aligned with the principal elongation quite similarly to flat lamellae or deformation bands. This continuous twinning mechanism may be extended into the area of large deformation.

However, experimental investigations of very intensive plastic deformation have detected qualitatively different mechanisms of discontinuous flows manifesting the strain localization inside shear bands (SBs) (Figure 1.2a). Shear bands first described by F. Adcock in 1922 were investigated in many following works (see, for example, Hatherley et al. 1984, Harren et al. 1988, and other). They are thin material layers within which strains are significantly larger than outside. A few features distinguish shear bands from micro-bands. The orientation of SBs is dictated by continual mechanics and they always follow the principal macro shear directions. Therefore, two perpendicular families of SBs are often observed. Because of the non-crystallographic nature, they can penetrate without deviation a few grains (copper-type SBs) or whole material (brass-type SBs) whereas their thickness is from only tenths of nanometers to a few microns. Strains inside SBs may exceed 10. Such intensive deformation activity is accommodated by the multiple dislocation glide or by micro twins inside shear bands. Both mechanisms provide high angle boundaries along SBs and extremely fine microstructures within SBs. Corresponding microstructures usually comprise new grains and micro twin lamellae at sub-micron and nano scales (Malin et al. 1978; Hirsch et al. 1988). Shear bands are stable structural formations that flow together with the material, penetrate the preexisting structure and intersect each other, twins and grain boundaries leaving characteristic marks. With the increase of strain, SBs join into clusters and coarse macro shear bands delineating certain fragments or whole material.

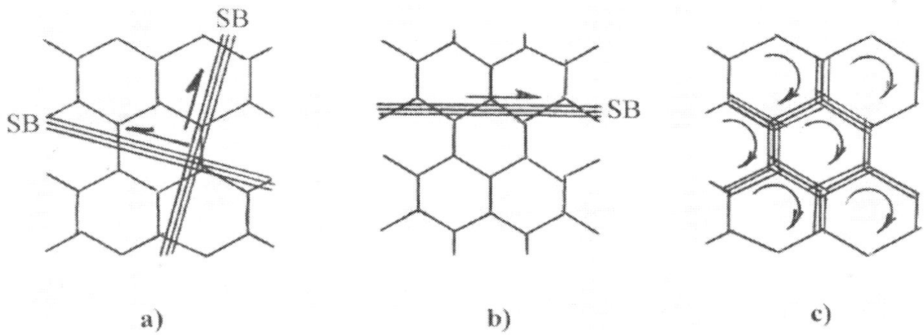

Figure 1.2. Micro localized plastic flows.

Although the mechanism of localization is not fully understood, it is generally accepted that the cause of localization is instability of the plastic flow during large deformation. The transition to localization includes a few steps. Heavily deformed microstructures and sharp textures are necessary to attain this condition. From the theory of plasticity of ideal plastic materials (Hill, 1950), it is known that the actual flow minimizing the plastic work usually includes surfaces of velocity discontinuity where the material experiences intensive shear along the principal shear directions. After large plastic deformation when hardening rate is low and the material behavior approximates ideal plasticity, localization within thin layers is accompanied by micro structural and/or textural softening that additionally reduces the energy of dissipation and provides a driving force for substitution of uniform flows by shear bands. Certain energetic thresholds during the transition to micro localization can be overcome owing to stress concentrations at grain boundaries, DDWs, twins, hard particles and other defects. All these structural irregularities and heterogeneities create specific channels for propagation of SBs through preexisting structure. Usually, they are identified as shear transition zones (STZ) (Langer, 2006) which appear at some moment of the continuous structure evolution.

When localization starts to operate, it accommodates the main part of strains. Additional loading during strain localization reorganizes the deformed structures and induces new high angle boundaries, ultra fine grains and fragments. A specific characteristic of these structures is a shift of dislocation activity from grain/fragment interiors to their boundaries. Therefore, other mechanisms of strain localization related to grain/fragment boundaries are also possible. These mechanisms are grain boundary sliding (Figure 1.2b) and grain rotation (Figure 1.2c). Depending on strain rate and temperature, they may be realized through dislocation glide or diffusion-viscous flow (Meyers et al. 2006). The grain boundary localization transforms low angle boundaries into high angle three dimensional configurations. Such modifications are usually identified as strain induced structure refinement in contrast to refinements attained by recrystallization or phase transformation. The micro localization also results in randomization of textures, refinement of second phases and precipitation, healing of volume defects, densification and bonding of particulate materials. All these effects associated with new processing techniques known as severe plastic deformation are difficult or impossible to attain with ordinary processing techniques. Therefore, the strain localization may be considered as the principal characteristic distinguishing severe plastic deformation from common large deformation. Models shown in Figure 1.1 and Figure 1.2 relate, correspondingly, to continuous plastic flows with continuous structure evolution and localized plastic flows with discontinuous structure evolution. These models describe the specific deformation steps during large and severe plastic deformation. If loading is interrupted at certain step, intermediate structures may be observed in different material areas due a to micro and macro strain non-uniformity.

1.2. Structural Scales

The primary mechanisms of plastic deformation (dislocation glide, twinning, and diffusion–viscous flow) take place at fine structural scales whereas external forces causing plastic deformation are applied to the macro bodies. An additional complication arises from the polycrystalline structure of metals with dislocation glide and twinning on different crystallographic systems of individual grains. As related characteristic scales range from whole material to grains and to dislocations, plastic deformation is usually considered at macro, meso and micro scales (Panin, 2000).

At the *micro scale*, the main strain accommodation mechanism is a dislocation glide. Movement and interaction of dislocations provide plastic flow with the increase (hardening) of the material shear flow stress τ_i on specific crystallographic planes i. The magnitude τ_i depends on a summary effect of interactions between individual dislocations and dislocation patterns. The corresponding statistics should be formulated for different loading conditions at the scale length l which is significantly larger than a dislocation core r ($l >> r$). The physical theory of dislocations is still incapable of resolving this complicated problem. Therefore, at the next scale, *mesoscopic scale,* empirical hardening functions τ_i are usually determined from experiments on single crystals. The characteristic length L of this scale corresponds to an average grain size d (L ~ d). At the *macro scale*, aggregates of grains are considered under applied (continuum) loading conditions. It is assumed that continuum parameters remain uniform within a characteristic aggregate size A (A >> d). The distribution of glide velocities v_i along slip planes i of each grain can be found from Taylor's minimization principle for dissipation of the plastic work within the aggregate (Taylor, 1938)

$$dW/dt = \min \Sigma(\tau_i v_i f_i) \qquad (1.1)$$

Here summation is performed for all grains and active glide systems i, f_i are the areas of slip planes. Minimization of (1.1) should satisfy boundary conditions applied to the aggregate. Therefore

$$\min \Sigma(\tau_i v_i f_i) = \boldsymbol{\sigma}\xi\mathbf{w} \tag{1.2}$$

where $\boldsymbol{\sigma}$ and ξ are effective von Mises macro stress and strain rate, $\mathbf{w}$ is a volume of the grain aggregate. Equation (1.2) establishes the correlation between continuum mechanics and crystal plasticity. Corresponding calculations can be performed for a number of deformation steps up to large plastic strains. This approach has been extensively used for the analysis of textures during continuous plastic flows (Wenk, 1985). It will be reviewed and applied to different problems in Part II of this book.

To simulate the microstructure evolution during SPD, different stages of the localized flow should be also considered. For continuous crystallographic glide in polycrystals, it has been known since Taylor's work that continuum mechanics can be applied to grain aggregates (see Eq. (1.2)), but not to individual grains. In polycrystals, macro stresses and strains experience significant redistributions between separated grains in accordance with their crystallographic orientations. Therefore, it may be expected that the processing mechanics has secondary effect on the structure evolution in comparison with the crystallography. However, there is a different situation during strain localization. The preexisting dislocation structure prepares the transition to localization, but further it does not play a significant role. As shear bands of a non-crystallographic nature penetrate individual grains and their aggregates into the principal shear directions, they are directly subjected to macro stress and strain rate states prescribed by continuum mechanics. The minimum scale length L for such description is an average distance 2h between shear bands (L ~ 2h). In these cases, the continuum plasticity can be extended to structural scales that usually correspond to the meso and even to lower scales. Correspondingly, the processing mechanics becomes the main factor controlling the structure evolution. During localization, dislocation activity is concentrated within shear bands which present channels for strain accommodation defining the final structure and properties. Therefore, in contrast to conventional large plastic deformation with the continuous evolution of dislocation structures, localization and shear bands can be considered as the main characteristic of severe plastic deformation. It is important to note that such approach provides the direct correlation between macro and meso scales and possibility to analyze shear bands and related phenomena by well developed methods of continuum plasticity. In comparison with a more physical "bottom-up" analysis of the statistics of dislocations still presenting an insuperable problem, the present "top - down" consideration accounting for essential micro structural features provides an effective analytical technique for the processing mechanics of SPD. Obviously, both "top - down" and "bottom - up" approaches are not contradictory and should converge at the meso scale.

1.3. Mode of Deformation

Plastic flow of metals depends on physical and mechanical properties (crystallography, stacking fault energy, microstructure and texture), parameters of deformation (temperature, strain, strain rate, hydrostatic pressure), specifics of processing mechanics and loading history. Properties of particular materials are prescribed in the initial state (before deformation). During plastic flow, metals display strong sensitivity to temperature, strain and strain rate whereas the sensitivity to the hydrostatic pressure for most properties except the ductility is rather insignificant. Temperature and strain rate define processing conditions that may be chosen within a number of broad ranges. Their summary effect is often described by the Zener-Hollomon parameter

$$Z = (\xi)^m \exp(Q/RT) \tag{1.3}$$

where ξ is the effective strain rate, *m* is a coefficient of strain rate sensitivity, Q is the activation energy, R is the gas constant and T is the absolute deformation temperature.

The effect of strains is usually introduced by the material hardening or stress – strain diagrams. For ordinary materials and monotonic loading conditions, stress-strain diagrams show a relative weak dependence on testing method and deformation history. Correspondingly, various empirical formulae were suggested to approximate experimental hardening diagrams of metals using effective von Mises strains as a single parameter. They remain true and for large monotonic plastic deformation because hardening rate quickly decreases with strains. However, effects related to plastic localization manifest a non-monotonic and singular character. In particular, plastic instability and flow localization may be observed at moderate strains when the deformation path is abruptly changed (Thuillier et al. 1994) or testing is performed by shear deformation (Gasperini et al. 1996). Therefore, a significant impact of the deformation mode should be expected for transition to localization and during severe plastic deformation.

In scientific literature "deformation mode" is often used as a specific characteristic of testing method or processing technique without a detailed description of related mechanical states and their uniformity. A more correct definition of the mode should be based on general properties of stress and strain rate tensors. Let's consider a two dimensional plain plastic flow that is the typical state during the shear band localization. We will assume that stresses and strains are uniformly distributed within a "unit" material element and elastic strains can be neglected in comparison with plastic strains. For large plastic deformation, the material inside the considered element is near ideally plastic because the ability to harden any further has been exhausted. The simplest way to describe large deformation of such materials is using the plastic flow theory of Levi-Mises. It is natural to combine the Cartesian coordinate system with directions of the principal shear stresses or α, β slip lines (Figure 1.3a) (Hill, 1950). The stress state of the small element shown on Figure 1.3a is described by a symmetrical tensor

$$\mathbf{T}_\sigma = \begin{vmatrix} \sigma & k & 0 \\ k & \sigma & 0 \\ 0 & 0 & \sigma \end{vmatrix} \tag{1.4}$$

where σ is the hydrostatic stress component, k is the material yield shear stress. When k is known, only the distribution of σ in a slip line field defines the stress state.

In accordance with conditions of isotropy and incompressibility (Hill, 1950), the principal shear strain rate components η_α and η_β are also oriented along α and β slip lines whereas the linear strain rate components along the slip lines are zero. Shear strain rate and element distortion are described by components (Figure 1.3b)

$$\eta_\alpha = \partial v_\alpha/\partial\beta,\ \eta_\beta = \partial v_\beta/\partial\alpha \tag{1.5}$$

Here v_α and v_β are velocities along α and β slip lines. For uniform strain rates η_α = const, η_β = const inside the element, and integration of (1.5) gives

$$v_\alpha = \eta_\alpha\beta + v_\alpha^{0},\ v_\beta = \eta_\beta\alpha + v_\beta^{o} \tag{1.6}$$

where v_α^0 and v_β^0 are velocities at $\alpha = \beta = 0$.

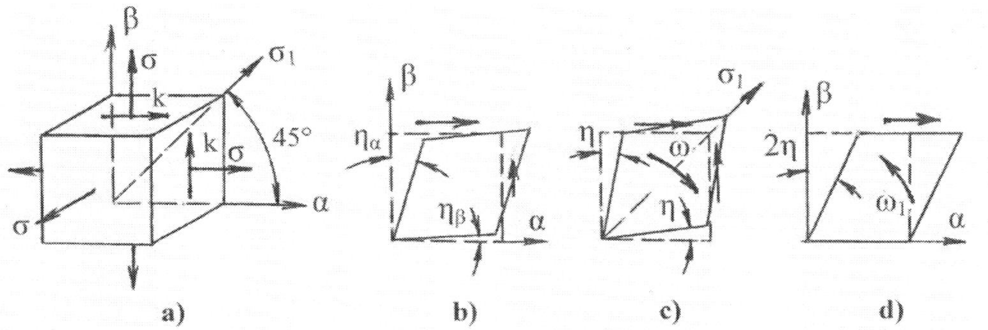

Figure 1.3. Stress (a) and strain rate (b, c, d) states along slip lines for plain plastic flow.

Corresponding strain rate tensor is

$$\mathbf{D}_\eta = \begin{vmatrix} 0 & \eta_\alpha & 0 \\ \eta_\beta & 0 & 0 \\ 0 & 0 & 0 \end{vmatrix} \tag{1.7}$$

Asymmetric tensor (1.7) describes any strain rate state. To compare different states, they should be normalized by dissipation of the plastic work within the element

$$dW/dt = k(\eta_\alpha + \eta_\beta) = 2\eta k \tag{1.8}$$

Here

$$\eta = (\eta_\alpha + \eta_\beta)/2 \tag{1.9}$$

is the effective von Mises shear strain rate. Important particular cases are pure shear defined by

$$\eta_\alpha = \eta_\beta = \eta \tag{1.10}$$

and simple shear defined by

$$\eta_\alpha = 2\eta,\ \eta_\beta = 0 \text{ or } \eta_\alpha = 0,\ \eta_\beta = 2\eta \tag{1.11}$$

For numerous intermediate states strain rates are unequal $\eta_\alpha \neq \eta_\beta$. As η_α, η_β are always positive, all possible normalized states at the plane (η_α, η_β) lay on the straight line BB_1 (Figure 1.4). Because of symmetry, consider states along a segment AB with $\eta_\alpha > \eta_\beta \geq 0$. The limit points A and B relate to pure shear and simple shear. Corresponding distortions of the material element are shown on Figure 1.5 for pure shear (a), intermediate states (b) and simple shear (c).

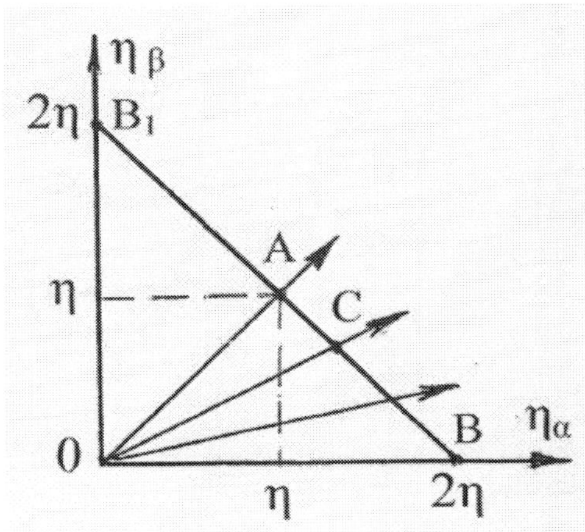

Figure 1.4. Diagram of strain rate states along slip lines.

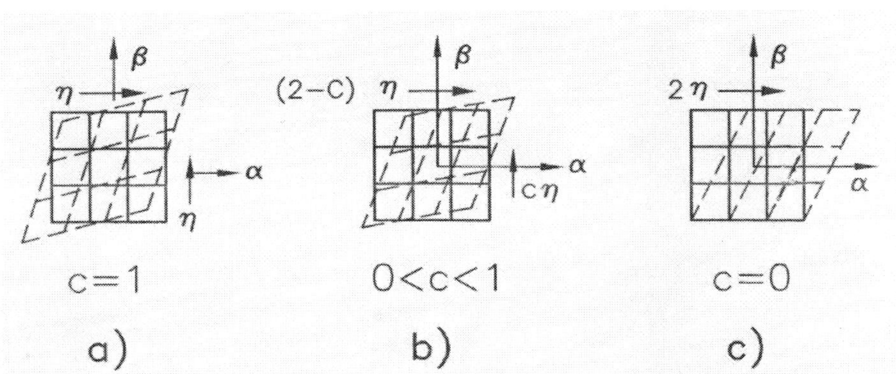

Figure 1.5. Distortions of material element for pure shear (a), intermediate states (b) and simple shear (c).

In the mechanics of solids, the Cauchy tensor (1.7) is usually divided on symmetrical and non-symmetrical parts (Cauchy-Helmholtz theorem)

$$\mathbf{D}_\eta = \begin{vmatrix} 0 & \eta & 0 \\ \eta & 0 & 0 \\ 0 & 0 & 0 \end{vmatrix} + \begin{vmatrix} 0 & \omega & 0 \\ -\omega & 0 & 0 \\ 0 & 0 & 0 \end{vmatrix} \tag{1.12}$$

The first part of (1.12) corresponds to pure shear (1.10) and the second part describes a rigid element rotation with a spin rate

$$\omega = (\eta_\alpha - \eta_\beta)/2 \tag{1.13}$$

That way, any strain rate state can be reduced to pure shear and the corresponding rigid rotation by formulae (1.9), (1.12), (1.13). In mechanical models of plastic flows, only the pure shear part of (1.12) is used to rationalize constitutive "stress-strain rate" relations whereas the rotation part (1.12) is usually skipped. However, similarly to (1.12), tensor (1.7) can be also reduced to any another state. For example, using decomposition

$$\mathbf{D}_\eta = \begin{vmatrix} 0 & 2\eta & 0 \\ 0 & 0 & 0 \\ 0 & 0 & 0 \end{vmatrix} + \begin{vmatrix} 0 & \omega_1 & 0 \\ -\omega_1 & 0 & 0 \\ 0 & 0 & 0 \end{vmatrix}$$

tensor (1.7) is presented as a sum of simple shear (1.11) and rigid rotation with angular spin $\omega_1 = -\eta_\beta$. The decomposition of tensor (1.7) into pure shear and simple shear strain rate components is shown in Figure 1.3c and Figure 1.3d, respectively. Therefore, the reduction of the strain rate tensor (1.7) to the particular deformation mode like pure shear or simple shear does not have specific physical reasons and with the disregard of the rotation strain rate component leads to the loss of its identity. From here on, only the original form (1.7) will be used.

From (1.9) one can see that the classification of different strain rate states with the same dissipation of the plastic work (1.8) depends on distribution of the effective strain rate 2η between directions of α and β slip lines. This distribution may be described by the parameter

$$C = 2(1 + \eta_\alpha/\eta_\beta)^{-1} \tag{1.14}$$

Then, strain rate components η_α and η_β are expressed through scalar quantities η and C by the formulae

$$\eta_\alpha = (2 - C)\eta,\ \eta_\beta = C\eta \tag{1.15}$$

On the other hand, formulae (1.9), (1.13), (1.15) give

$$\eta_\alpha = \eta + \omega,\ \eta_\beta = \eta - \omega$$
$$\omega = 2(1 - C)\eta$$

Therefore, different strain rate states along slip lines may be described by scalar parameters (η, C) or (η, ω). The limit cases correspond to the pure shear deformation mode when C = 1 (segment OB on Figure 1.4) or $\omega = 0$ and to the simple shear deformation mode when C = 0 (segment OA on Figure 1.4) or $\omega = \eta$. The dimensionless parameter C remains within an interval $0 \leq C \leq 1$ for all possible states shown on Figure 1.5 whereas the parameter ω depends on strain rate intensity η and C, and does not present the universal characteristic of the state. In what follows, the parameter C will be identified as a coefficient of the deformation mode reflecting specific properties of strain rate tensors.

1.4. Finite Deformation

For uniform and steady plastic flows, finite distortions and deformations can be analyzed using formulae (1.6) for velocities. Consider two limit situations of pure shear (Figure 1.5a) and simple shear (Figure 1.5c) and assume that axes (α, β) are fixed, $v_\alpha^0 = v_\beta^0 = 0$ and η_α = const, η_β = const during a time interval t. In the case of *pure shear* $\eta_\alpha = \eta_\beta = \eta$, full speed of any point along the principal direction ε_1 (Figure 1.6a) is

$$v = r\eta \tag{1.16}$$

where r is a distance from M to a center O. For square or round material elements, this direction provides maximum elongation ε_1 whereas the principal direction ε_2 provides the maximum shortening. Integration of (1.16) for the interval t gives new positions r_1 of points along the first principal direction

$$\ln(r_1/r_o) = \eta t = \gamma/2 \tag{1.17}$$

Here r_0 is the original position for t = 0 and $\gamma = 2\eta t$ is effective shear strain during t. Similarly, for the second principal direction

$$\ln (r_2/r_o) = - \gamma/2$$

Therefore, the principal von Mises strains are

$$\varepsilon_1 = - \varepsilon_2 = \gamma/2 \tag{1.18}$$

As monotonic loading directions ε_1, ε_2 are fixed, the pure shear deformation mode provides the maximum material distortion with affine transformation of square elements into rhombuses and round elements into ellipses (Figure 1.6a).

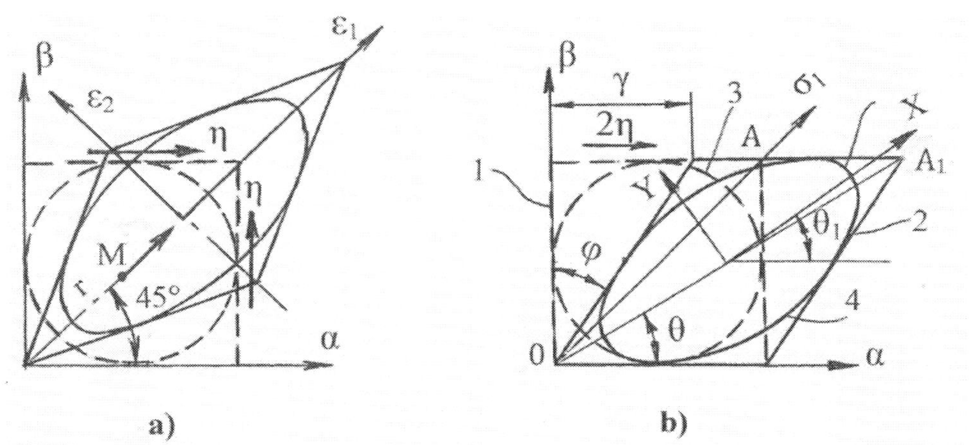

Figure 1.6. Final strains for pure shear (a) and simple shear (b).

For *simple shear*, formulae (1.6), (1.11) become

$$v_\alpha = 2\eta\beta,\ v_\beta = 0$$

After the time interval t, final displacements U_α, U_β of material points with original coordinates (α_0, β_0) are

$$U_\alpha = \gamma\alpha_o,\ U_\beta = 0$$

where $\gamma = 2\eta t$ is effective von Mises shear strain. New coordinates of material points are

$$\alpha = \alpha_o + \gamma\beta_o,\ \beta = \beta_o \quad (1.19)$$

Such deformation transforms square elements 1 into parallelograms 2 with a shear angle φ (Figure 1.6b)

$$\tan\varphi = 2\eta t = \gamma \quad (1.20)$$

Using formulae (1.19) for transition from coordinates (α, β) to characteristic coordinates (X, Y), one can find that a round element 3 is transformed into an ellipse 4

$$(X/a)^2 + (Y/b)^2 = 1$$

The ellipse axes a, b are

$$a = \sqrt{A},\ b = 1/\sqrt{A} \text{ where } A = [2 + \gamma^2 + \gamma\sqrt{(\gamma^2 + 4)}] \quad (1.21)$$

An angle θ_1 between axes X and α is

$$\text{cotan}(2\theta_1) = \gamma/2 \quad (1.22)$$

Similarly to (1.18), natural logarithmic strains corresponding to maximum elongation and shortening along ellipse axes *a* and *b* are

$$\varepsilon_1 = -\ \varepsilon_2 = -\ \ln\{[2 + \gamma^2 + \gamma\sqrt{(\gamma^2 + 4)}]/2\}/2 \qquad (1.23)$$

The angle θ_1 defines a current orientation of ε_1. At an original position $t = 0$, $\gamma = 0$, $\theta_1 = 45^\circ$ the angle θ_1 coincides with an element diagonal OA and the principal stress σ_1. When shear γ increases, the angle θ_1 rotates from the direction of σ_1 to the α-slip line and intersects different material fibers because the orientation of the diagonal OA_1 is

$$\mathrm{cotan}\theta = \gamma + 1 \qquad (1.24)$$

Calculations show (Figure 1.7) that $\theta_1 > \theta$ for any $\gamma > 0$, however the difference between both angles remains relatively small (see diagram 4 for θ_1 and diagram 5 for θ on Figure 1.7 where diagram 3 shows the orientation $\theta = 45^\circ$ of the principal elongation for pure shear). That allows one to use an approximate formula for logarithmic strains along an element diagonal and into the perpendicular direction (Figure 1.6b) instead of formula (1.23)[1]

$$\varepsilon_1 = -\ \varepsilon_2 \sim \ln\{[(\gamma + 1)^2 + 1]/2\}/2 \qquad (1.25)$$

It should be noted that the parameter γ in formulae (1.18), (1.24) defines effective shear for the spent energy and remains the same whereas strains ε_1, ε_2 relate to geometrical distortions in both cases. The comparison of strains ε_1 from formula (1.18) for pure shear (diagram 1) and from formula (1.23) for simple shear (diagram 2) in function of γ is shown on Figure 1.7. The discrepancy between maximum elongations ε_1 is relatively low for $\gamma < 1$ but increases progressively with γ and becomes large for $\gamma > 4$. In the case of pure shear ($C = 1$), parameters of final deformation (ε_1, ε_2) and γ are equivalent and give the same effective von Mises strain

$$\varepsilon = \gamma/\sqrt{3} \qquad (1.26)$$

However, for simple shear ($C = 0$), because of the rotation of axis of principal strains ε_1, ε_2, only the quantity γ (1.20) complies with the definition of equivalent strain (1.26) and can be extended to non-monotonic loading. In particular, accumulated von Mises strain for segmented loading paths is

$$\varepsilon = \sum\gamma_i/\sqrt{3} \qquad (1.27)$$

where γ_i is effective shear at each loading segment i.

[2] Finite deformation during SPD including a non-uniform strain distribution will be also considered in Part II of the book.

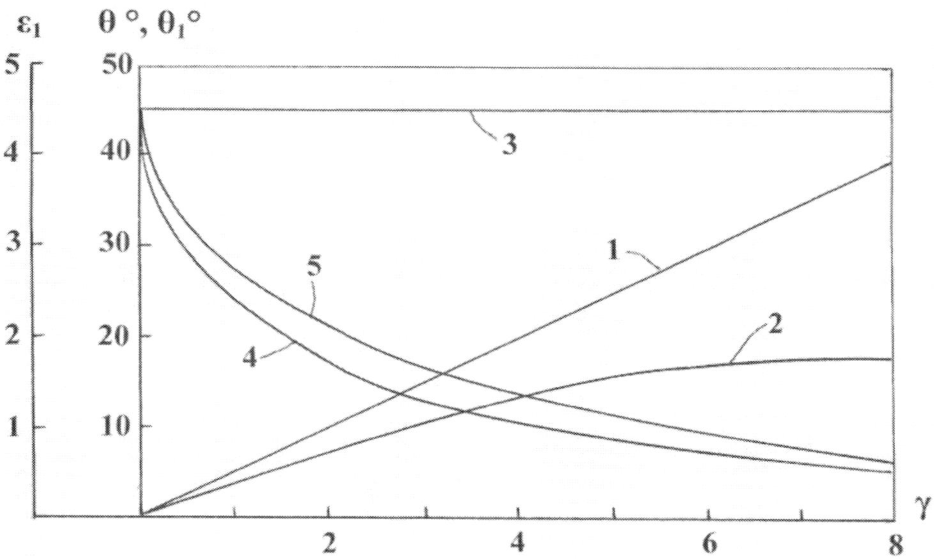

Figure 1.7. Principal elongations ε_1 and angles θ, θ_1 of their orientations in function of shear γ: 1 – ε_1 for pure shear; 2 – ε_1 for simple shear; 3 - angle θ for pure shear; 4 – angle θ for simple shear; 5 – angle θ_1 for simple shear.

For intermediate deformation modes $0 < C < 1$, parameters ε_1, θ are confined between diagrams 1, 2 and 3, 4 on Figure 1.7, correspondingly. Equation $\gamma = 2\eta t$ for effective shears in formulae (1.9), (1.17), (1.20) establishes the equivalency of the plastic work for different deformation modes. As will be shown below, this equivalency provides near similar microstructures for different deformation modes during continuous plastic flows, however for localized flows the deformation mode plays an important role on the microstructure evolution.

1.5. Transition to Localization

The localization is the way to minimize plastic work (1.1), (1.2) at meso and macro scales. Depending on material and loading conditions, the localization can be detected into continuum shear directions at different strain levels. For some metals, localization along so called Lüder's lines are observed at the beginning of the plastic flow due to the stall of piled dislocations at impurities. The general approach to plastic localization was developed in the theory of deformation instability and bifurcation (Biot, 1965; Hill et al. 1975; Sedlacek et al. 2002). Large plastic strains are accompanied by substantial changes of microstructure and texture and the monotonic decrease of hardening rate (Figure 1.8). The full exhaustion of the hardening ability in small material volumes may lead to local softening (point C, Figure 1.8) presenting new channels for strain accommodation by a localized flow in shear bands with lower dissipation of the plastic work. As the global material hardening rate at this moment may remain positive $d\sigma/d\varepsilon > 0$ (point A, Figure 1.8), a certain energetic threshold must be overcome to start shear localization. The corresponding increase of local stresses $\Delta\sigma$ between points A and B on Figure 1.8 is provided by micro stresses on elements of the micro structure.

The transition from B to C usually occurs in the avalanche manner (Lee at al, 1994). It should be noted that without even veru small primary softening BC, micro localization does not reduce the plastic work and cannot be realized. Secondary softening inside shear bands caused by adiabatic heating, development of micro cracks and pores, dynamic recovery or recrystallization also promote transition to localization. The global condition for transition to localization in polycrystalline materials is

$$d\sigma/d\varepsilon\,|_{BC} \leq 0 \qquad (1.28)$$

In terms of crystal plasticity, equation (1.28) can be rewritten in the form (Gil Sevillano et al. 1980)

$$d\sigma/d\gamma = M d\tau/d\gamma + \tau dM/d\gamma \leq 0 \qquad (1.29)$$

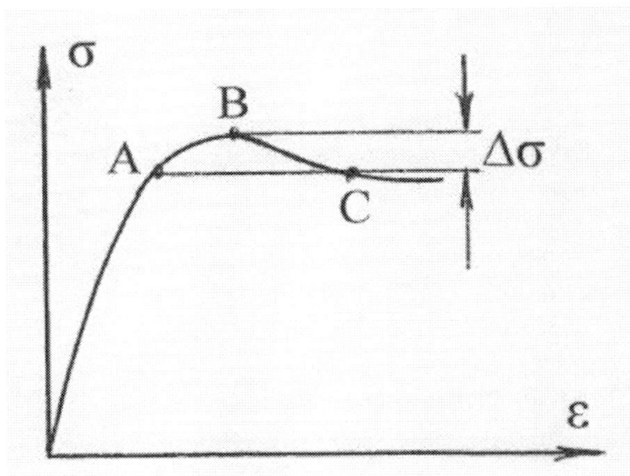

Figure 1.8. Stress-strain diagram for transition to localization.

Here M is Taylor's orientation factor; τ and γ are crystallographic stress and strain, correspondingly. The first part of (1.29) relates to dislocation hardening that plays an important role for moderate strains when a loading path changes sharply. For large monotonic plastic strains, this part remains small and positive $d\tau/d\gamma > 0$. The second part of (1.29) describes an effect of the texture evolution and should lead to textural (geometrical) softening $dM/d\gamma < 0$ to satisfy (1.29)[2]. During deformation of polycrystals, some grains with large Schmid factor and favorable orientations relative to the principal shear directions do not rotate and preserve their stable crystallographic orientations. Grains with unstable orientations experience intensive rotation accompanied by the change of Schmid factor and texture. These grains are subdivided on areas of different slip systems separated by geometrically necessary boundaries (GNBs) whose evolution strongly depends on the deformation mode. Compatibility of strains inside subdivided areas requires their reciprocal rotation while strain compatibility with a total material distortion requires rotation of GNBs into the flow direction. Figure 1.9 depicts GNBs 01, 02 and 03 at the original position (a) and after pure shear (b) and simple shear (c). During pure shear (C = 1), the unstable orientations tend to

[2] For modeling of geometrical hardening and softening see Part II, Chapter 10.

align with the first principal stress σ_1 that is at angle 45° to α and β slip lines. As tangential stresses along these orientations reduce to zero, such evolution is accompanied by a decrease in Schmid factor and hence textural hardening. Acting together with micro structural hardening, that shifts equation (1.29) in the area of large strains and delays localization. On the contrary, during simple shear (C = 0), unstable orientations rotate toward α slip lines with the increase of tangential stresses and Schmid factor resulting in textural softening and early satisfaction of the localization condition (1.29). This consideration shows that the lower the coefficient of deformation mode C (1.14), the smaller is the strain needed for transition to localization. That means that simple shear promotes early localization while pure shear delays localization until large strains. Related experimental results will be discussed later.

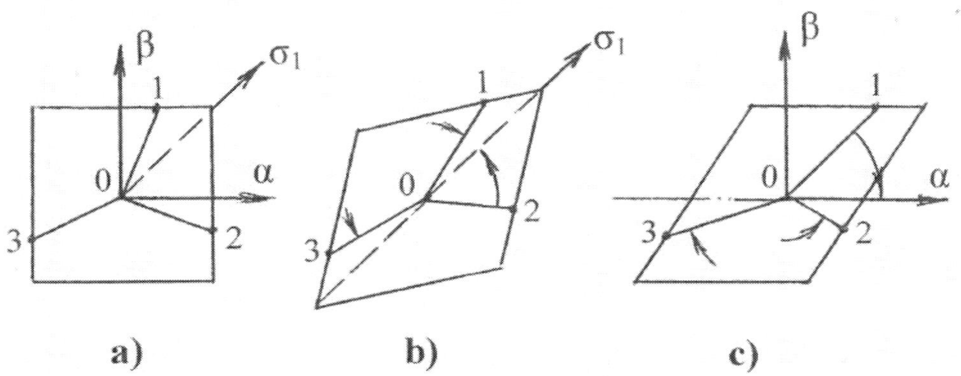

Figure 1.9. Geometrically necessary boundaries 01, 02 and 03 in the original position (a), after pure shear (b) and simple shear (c).

1.6. SHEAR BAND LOCALIZATION

Flow localization is an inevitable property of plastic deformation in most metals. For f.c.c. materials, the primary micro-bands arise on the crystallographic planes (111) concurred with boundaries of the grain subdivided areas (the first generation of micro-bands, MB_1s) or with continuum shear directions (the second generation of micro-bands, MB_2s) (Hansen, 1990). These micro-bands accommodate relatively small strains and do not play a significant role in the structure evolution remaining low angle boundaries. However, for large strains, when softening surpasses hardening, the continuous evolution is substituted by flow localization in shear bands (SBs) first arising at the meso scale. They are planar thin material layers and have non-crystallographic orientations along the continual principal shear directions. SBs penetrate grain boundaries without a noticeable deviation. With the increase of strain, they propagate from a few grains to the whole material, join into clusters and form coarse shear bands observed at the macro scale. Upon originating, SBs accommodate the main part of plastic deformation and substitute the preexisting structure.

Consider a representative material element outlined by shear bands SB_α and SB_β of thickness 2δ and spacing 2h at the moment t = 0 (Figure 1.10a). Shear bands appear in shear transition zones (STZ) and may cross grain interiors (Figure 1.2a) at the beginning of localization or grain boundaries (Figure 1.2b) at the latest stages. During deformation, they are distorted and move together with the material like a set of coordinate axes until the localization provides minimization of (1.2). Referring to Figure 1.10a and assuming that shear bands SB_α and SB_β are glide systems with the shear stress k_i = const, $h >> \delta$ and that strain rates η'_α and η'_β within SB_α, SB_β are uniformly distributed, from formulae (1.2), (1.15) one can find

$$\eta'_\alpha = \eta_\alpha h/\delta = (2 - C)\eta h/\delta,\ \eta'_\beta = \eta_\beta h/\delta = C\eta h/\delta \qquad (1.30)$$

Within a localization interval t, the accumulated shear strains within shear bands are

$$\gamma'_\alpha = (2 - C)\Delta\gamma h/\delta,\ \gamma'_\beta = C\Delta\gamma h/\delta \qquad (1.31)$$

where $\Delta\gamma = \eta t$ is the increment in the continuum shear strain during the interval t. Since $h>>\delta$, equations (1.30), (1.31) demonstrate that strain rates and strains within shear bands significantly exceed the continual values and depends on the deformation mode C. For pure shear, they are equally distributed over two shear directions (Figure 1.10b)

$$\gamma'_\alpha = \gamma'_\beta = \Delta\gamma h/\delta, \qquad (1.32)$$

whereas for simple shear strain is concentrated along one direction (Figure 1.10c)

$$\gamma'_\alpha = 2\Delta\gamma h/\delta,\ \gamma'_\beta = 0 \qquad (1.33)$$

Attainable increments of shears $\Delta\gamma$ related to any fixed shear band also show a strong dependence on the deformation mode. During pure shear, in accordance with distortions of material elements, shear bands rotate to the first principal stress direction σ_1 so that a divergence angle $\varphi/2$ between α and β slip lines and shear bands SB_α, SB_β is continuously increased with strain (Figure 1.10b). From Figure 1.6a and formula (1.17), this angle is

$$\varphi/2 = \pi/4 - \cot(e^{-\gamma}) \qquad (1.34)$$

Simultaneously, the tangential stress τ acting along considered shear bands SB_α, SB_β reduces in accordance with the formula (Hill, 1950)

$$\tau = k \cos\varphi \qquad (1.35)$$

where k is the macro shear stress along α and β slip lines. When $\tau < k_i$, the localization is transferred from SB_α, SB_β to new shear bands coinciding again with α, β slip lines. Therefore, equations (1.34), (1.35) impose a restriction on the increment of continual shear $\Delta\gamma$ related to certain shear bands SBs during pure shear

$$\sin[\cot(e^{-\Delta\gamma})] \geq k_i/k \qquad (1.36)$$

In real metals under isothermal conditions, shear stress k_i is just slightly smaller than k (Figure 1.8) (Sedlacek et al. 2002). This means that the ratio (k_i/k) remains close to 1 and continuum shear $\Delta\gamma$ from formula (1.36) is sufficiently low. Corresponding accumulated shears $\gamma'_\alpha = \gamma'_\beta$ within shear bands can be found by substitution of $\Delta\gamma$ from (1.36) into formula (1.32). Calculated diagrams of attainable strains γ' inside SBs for pure shear deformation mode are presented in Figure 1.11 as a function of the ratio (ki/k) for h/δ = 10, 20 and 30. However, adiabatic heating during high speed deformation may reduce k_i in accordance with Eq. (1.3) and extends localization along fixed shear bands to large shears $\Delta\gamma$.

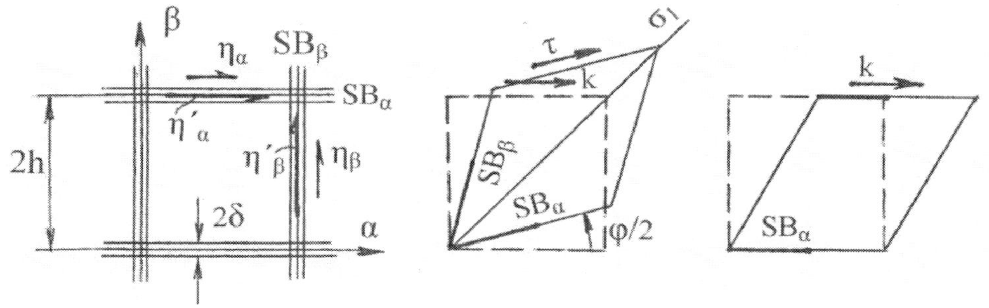

a) b) c)

Figure 1.10. Shear band localization (a) and orientations of shear bands for pure shear (b) and simple shear (c).

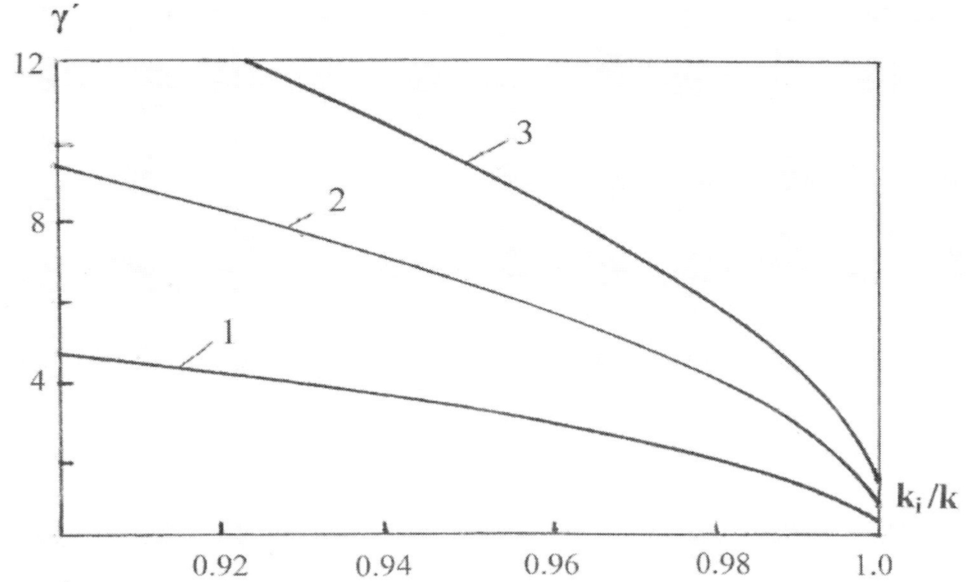

Figure 1.11. Attainable strains γ' along shear bands for pure shear deformation mode in function of local softening (k_i/k) and ratio h/δ = 10 (1), 20 (2) and 30 (3).

For simple shear, orientations of SB_α coincide with α-slip lines at any moment. Therefore, there are no restrictions on increments of continual shears $\Delta\gamma$. In this case, in accordance with equations (1.33), accumulated shear γ'_α inside one family of shear bands may be unlimited and exceed continual shear $\Delta\gamma$ by a few orders of magnitude. It is natural to suppose that crystallographic rotations within SBs and relative to a surrounding matrix are proportional to strains (1.32), (1.33). Such rotations define angles of misorientation along and within shear bands. That way, in addition to texture randomization, shear band localization induces new high angle boundaries (HABs) if strain γ'_α, γ'_β are sufficiently large. The present consideration, together with early transition to localization, demonstrates that new high angle boundaries are developed much faster for simple shear than for pure shear. This remarkable effect of the deformation mode during localization is very distinctive in comparison with its insignificant effect during continuous plastic flows that plays an important role for practical realizations of SPD.

1.7. ROTATIONAL LOCALIZATION

Shear bands appear in zones of certain structural inhomogeneity and outline material elements with strongly deformed boundaries. Because of the high concentration of vacancies and dislocations, plastic flow in SBs, similarly to grain boundaries, become sensitive to strain rates (1.30). At some stage of localization, the continuation of plastic deformation will require different tangential stresses along fragment boundaries coinciding with slip lines (Figure 1.12)

$$k_\alpha = k(\eta'_\alpha),\ k_\beta = k(\eta'_\beta)$$

When $\eta'_\alpha \neq \eta'_\beta$, moments applied to such elements are not balanced

$$\sum M_o = h^2[k(\eta'_\alpha) - k(\eta'_\beta)] \neq 0 \qquad (1.37)$$

To restore equilibrium, the elements start to rotate with an angular speed ω (Figure 2c). Consider kinematical conditions along a common boundary AA of two rotating elements 1 and 2 (Figure 12). A condition of continuity for normal $v_{\alpha 1}$, $v_{\alpha 2}$ components of a rotational velocity $\mathbf{v} = \boldsymbol{\omega} \times \mathbf{r}$ at any point M of the boundary AA gives

$$v_{\alpha 1} = \omega r \sin\varphi = v_{\alpha 2}$$

Then, a discontinuity [v] of tangential velocity components along AA is

$$[v] = v_{\beta 2} - v_{\beta 1} = 2\omega r \cos\varphi = \omega h = \text{const}$$

Here r, φ are polar coordinates of considered point M. Such rotation is kinematically admissible (Hill, 1950) and induces an additional strain rate inside the β-shear band AA

$$\eta''_\beta = \omega h/\delta$$

Using eq. (1.30), the total strain rate in the considered shear band is

$$\eta'_\beta = (C\eta + \omega)h/\delta \qquad (1.38)$$

A similar consideration gives the full strain rate inside α-shear bands

$$\eta'_\alpha = [(2 - C)\eta - \omega]h/\delta \qquad (1.39)$$

As this rotation should be accommodated by other elements, an additional moment M_i arises along boundaries of their interactions. Then, equation (1.37) gives the balance of all moments applied to the element of length l into a perpendicular direction to Figure 1.12

$$h^2 l[k(\eta'_\alpha) - k(\eta'_\beta)] = 2M_i$$

Using here $k(\eta) = k\eta^m$ from Hollomon – Zener formulae (1.3), equations (1.38), (1.39) and a plastic twist moment $M_i \sim kh^3/3$ for a square cross-section area, finally, one will have

$$[(2 - C)\eta - \omega]^m - (C\eta + \omega)^m = 2(\delta/h)^m h/3l$$

A full analysis of the last equation is complicated. A simple solution can be found for long elements when $l >> h$. In this case, the balance of moments takes place if $\eta'_\alpha = \eta'_\beta$ and

$$\omega = (1 - C)\eta \qquad (1.40)$$

This angular speed equalizes total strain rates in both families of SBs and reduces the local deformation mode to pure shear. It is interesting to note that equation (1.40) is identical with formulae (1.13). During a time interval Δt, the rotation induces an additional angle of misorientation $\Delta\chi$ along shear bands

$$\Delta\chi = (1 - C)\Delta\gamma \qquad (1.41)$$

Here $\Delta\gamma$ is the increase of effective macro shear during interval Δt. Eqs. (1.38) – (1.41) show a direct effect of deformation mode on the rotation localization. In the limiting cases of pure and simple shear, angular speeds and angles of misorientation are

$\omega = \Delta\chi = 0$ for pure shear
$\omega = \eta$, $\Delta\chi = \Delta\gamma$ for simple shear

This analysis has a simple graphical interpretation. A typical S-shape diagram $k = k(\eta)$ is shown on Figure 1.13 for simple shear (points A_α, A_β), an intermediate state (points B_α, B_β) and pure shear (point $C = C_\alpha = C_\beta$) where indexes α, β relate to corresponding SBs . The rotation shifts A_α, A_β and B_α, B_β to the point C for pure shear. Such consistent fragment rotation does not change but rotates the crystallographic texture and redistributes strains inside shear bands. That way, rotational localization forms equiaxed three dimensional material fragments and increases misorientation angles along their boundaries. This effect is strongest for simple shear and disappears for pure shear.

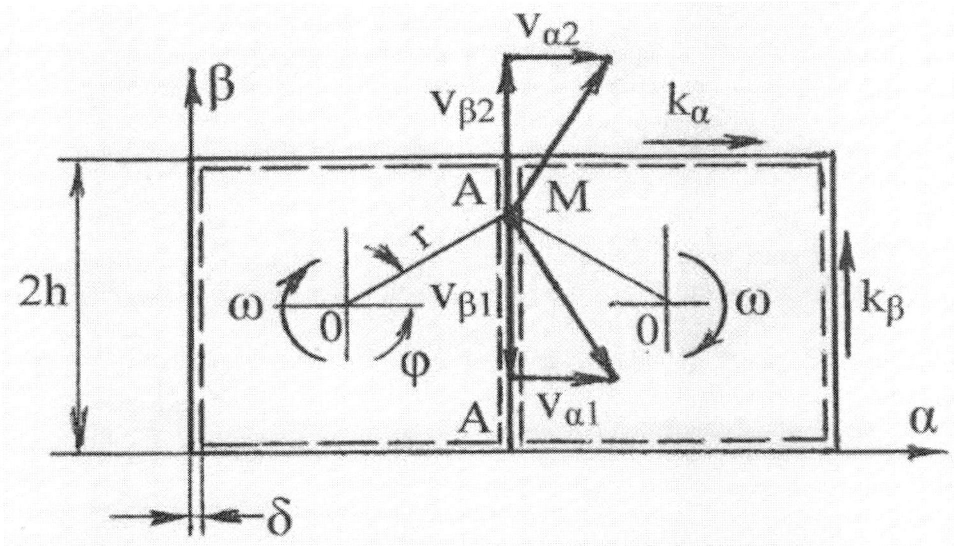

Figure 1.12. Velocity discontinuity along fragment boundaries during rotational localization.

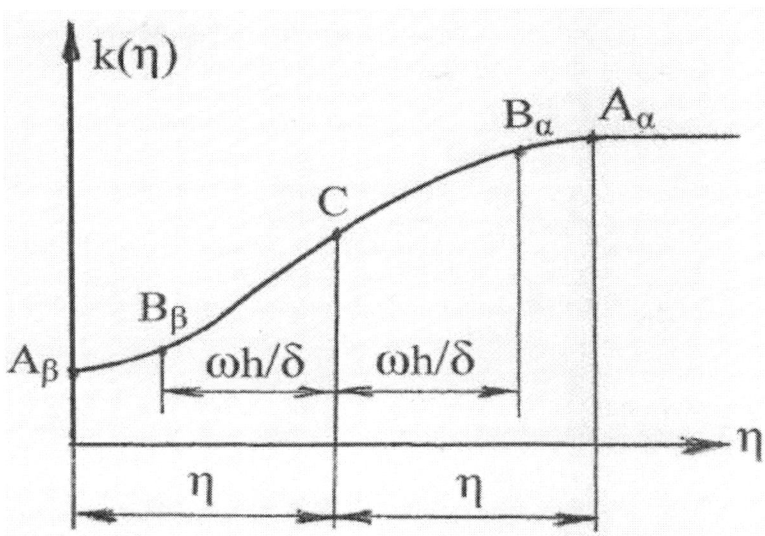

Figure 1.13. Redistribution of strain rates between shear bands during rotational localization.

1.8. Experimental Verification

In accordance with the discussion of Sections 1.6, 1.7, large plastic deformation may detect different structures induced successively by continuous evolution of dislocation structures, shear band localization and rotation localization. In contrast to continuous plastic flows which manifest an insignificant effect of a strain state, the deformation mode plays a critical role after transition to localization. Simple shear provides optimal conditions for early localization as well as the specific structure evolution by shear band and rotational localizations. This important result presents the basic idea for development of SPD techniques in materials processing for properties. To confirm some qualitative conclusions of the previous analysis, special experiments were performed on model materials at large plastic strains with pure shear and simple shear deformation modes where the biggest difference would be expected (Segal, 2002; Segal, 2005). Pure shear was realized in the central area of samples rolled with different reductions. A near uniform strain distribution through the material thickness was attained using partial reductions of less than 10% per pass and a large roll diameter of 890 mm. For an original sample thickness of H = 65 mm, an accumulated rolling reduction of 99% could be attained with a final thickness of h = 0.65mm. Simple shear was developed by equal channel angular extrusion (ECAE) (Segal et al. 1981b). Sharp corner channels with a tool angle of 90° and the careful control of processing conditions provided uniform simple shear $\gamma = 2$ ($\varepsilon = 1.15$) per pass at an angle 45° to a billet axis. Non monotonic loading was simulated by different systems of billet rotation or routes A, B, C, and D after successive passes[3]. Equivalent von Mises strains $\varepsilon = 1.15N$ between N passes of ECAE and rolling reduction

$$\psi = (1 - h/H)100\%$$

were calculated with the formula

$$\psi = (1 - e^{-\varepsilon})\ 100\ \% \tag{1.42}$$

These equivalent parameters relate to flow planes in both cases. Rolling experiments were performed without lubricant.

As localization is the specific characteristic of SPD, the appearance of shear bands depends on deformation mode. Upon generation, SBs behave as stable structural formations described by continuum mechanics. They flow together with the material and intersect each other. Separated families of shear bands remain distinguishable until angles of their intersections are sufficiently large (Gasperini et al. 1996). Typical structures of shear bands in the central area of flow planes detected by optical microscopy are shown on figures 1.14 a, b for the aluminum alloy Al0.5Cu after 4 passes ECAE via routes A and after rolling with a reduction of 95%, correspondingly. A more convenient technique for visualization of SBs during SPD is dynamic recrystallization in high purity metals. It is known that shear bands and their intersections are suitable sites for the nucleation of new grains, and the temperature of static recrystallization decreases with strain (Humphreys et al. 1995). For large strains in high purity metals, this temperature can be reduced below the room temperature. In these cases, shear bands decorated by dynamically recrystallized fine grains and their nuclei may be observed by optical microscopy directly after room temperature deformation (Segal, 2002). As an example, Figure 1.14c presents a microstructure of pure aluminum 4N5Al (99,995%) after four ECAE passes via route D. Two families of SBs can be seen at the flow plane: the first, induced during the third pass, forms an angle ~ 17° and the second, induced during the fourth pass, forms an angle 45° to the billet axis. An evolution of shear bands during straining can be detected in more pure materials. As intensity of dynamic recrystallization depends on local defects, related data, to some extent, demonstrate a mechanism of structure refinement

[3] ECAE routes define systems of billet rotation between successive passes and will be described in detail in Chapter 3.

during SPD. Figure 1.15 presents microstructures in aluminum 5N2Al (99.9992%) after ECAE with the number of passes ranging from one to four via route A when a billet orientation was not changed. All pictures relate to the flow plane. During the first pass (Figure 1.15a), the original grains are deformed and subdivided by crystallographic glide on different slip systems. For favorable oriented grains, SBs are detected into a shear direction (SD) with some nuclei of new grains observed in the polarized light as bright spots. After the second pass (Figure 1.15b, the microstructure is shown with larger magnification for more detail), the nuclei are partly transformed into recrystallized grains, and new SBs into the shear direction provide channels to extend localization in most grains regardless of their orientations. After the third pass (Figure 1.15c), SBs propagate the whole material covering the remaining grains of unstable orientations and inducing a large number of recrystallized grains. Finally, the structure is fully recrystallized after the fourth pass (Figure 1.15d). Additional passes reduce the grain size only slightly. A similar refinement mechanism by dynamic recrystallization was also observed for other ECAE routes.

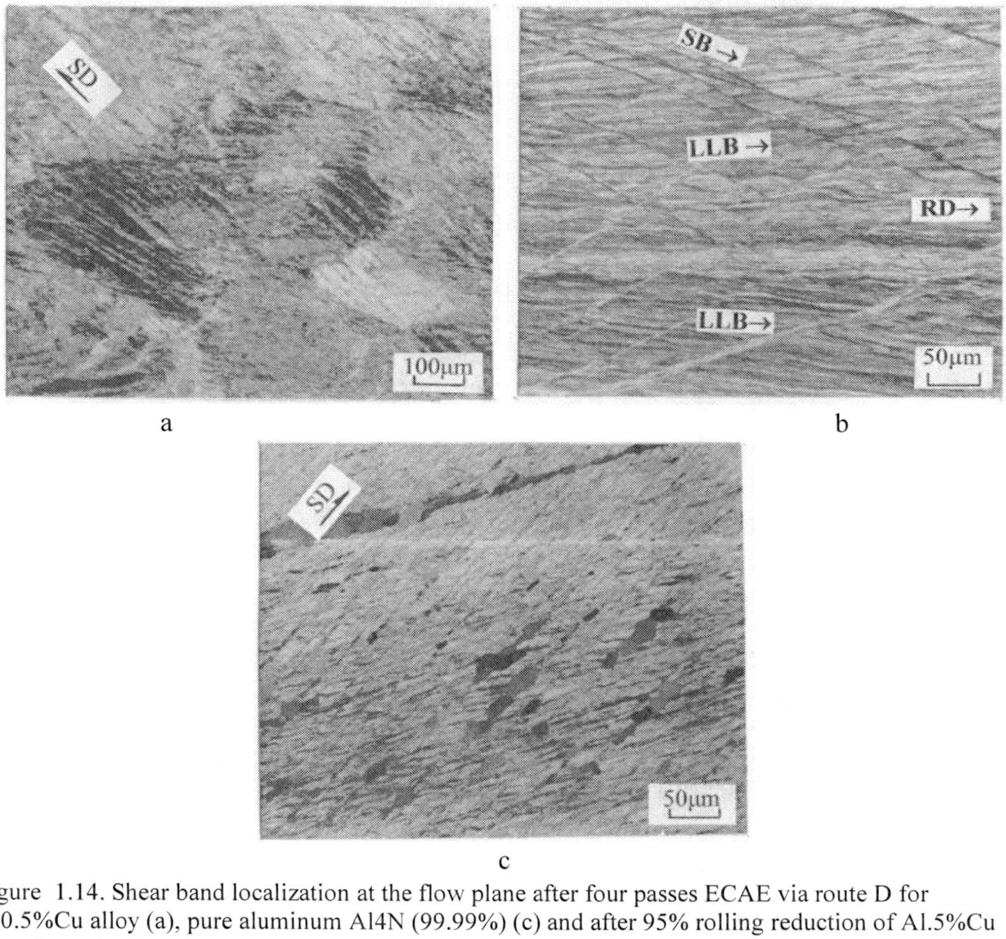

Figure 1.14. Shear band localization at the flow plane after four passes ECAE via route D for Al0.5%Cu alloy (a), pure aluminum Al4N (99.99%) (c) and after 95% rolling reduction of Al.5%Cu alloy (b, central area).

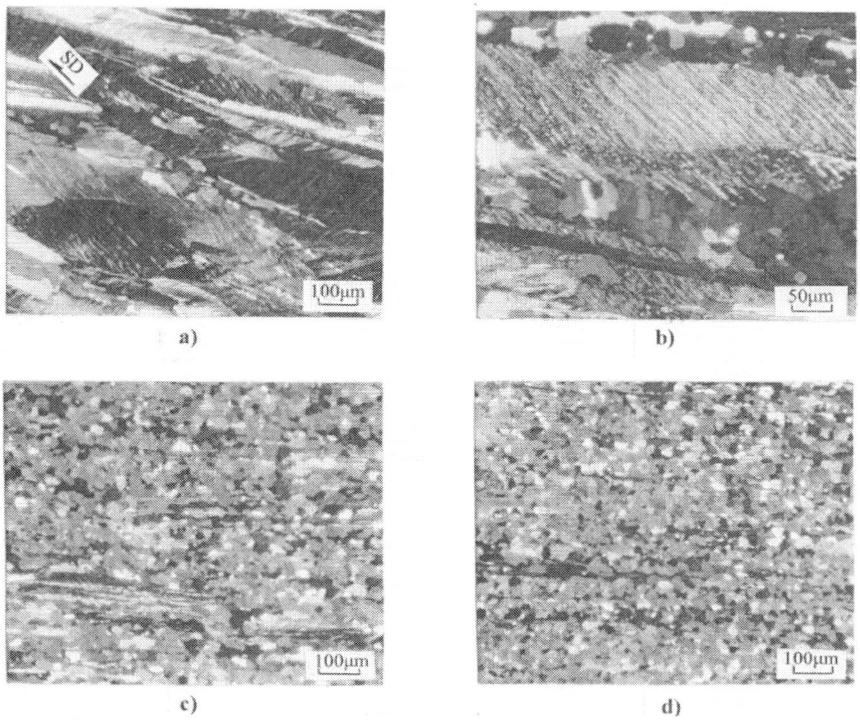

Figure 1.15. Deformation structures at flow plane for pure aluminum Al5N2 (99.9992%) after ECAE via route A: a - 1 pass: b - 2 passes; c - 3 passes; d - 4 passes.

There are remarkably different microstructures after rolling of the same 5N2Al material. Figure 1.16 demonstrates structures at the central area of the flow plane after material rolling to reductions of 68, 90, 97 and 99%. In accordance with Eq. (1.42), these reductions correspond to 1, 2, 3 and 4 ECAE passes. A reduction of 68% (Figure 1.16a) results in the typical rolling structure with grains elongated into the rolling direction (RD) by crystallographic glide on different slip systems. At 90% reduction (Figure 1.16b) grains become very elongated with the first appearance of shear bands and a few recrystallization nuclei within some grains. A reduction of 97% (Figure 1.16c) introduces deformation bands (DBs). Some of them are fully recrystallized or crossed by SBs for short segments, but a larger part of DBs remain stable orientations. Rolling with a very large reduction of 99% (Figure 1.16d, the entire billet thickness can be viewed in a single field of the microscope) demonstrates outstanding stability of stable orientations in the central area with pure shear deformation mode. However, in near surface areas the deformation mode approximates to simple shear because of intensive contact friction. It is noticeable that the microstructure in these areas is fully recrystallized and quite similar to the microstructure after a few ECAE passes.

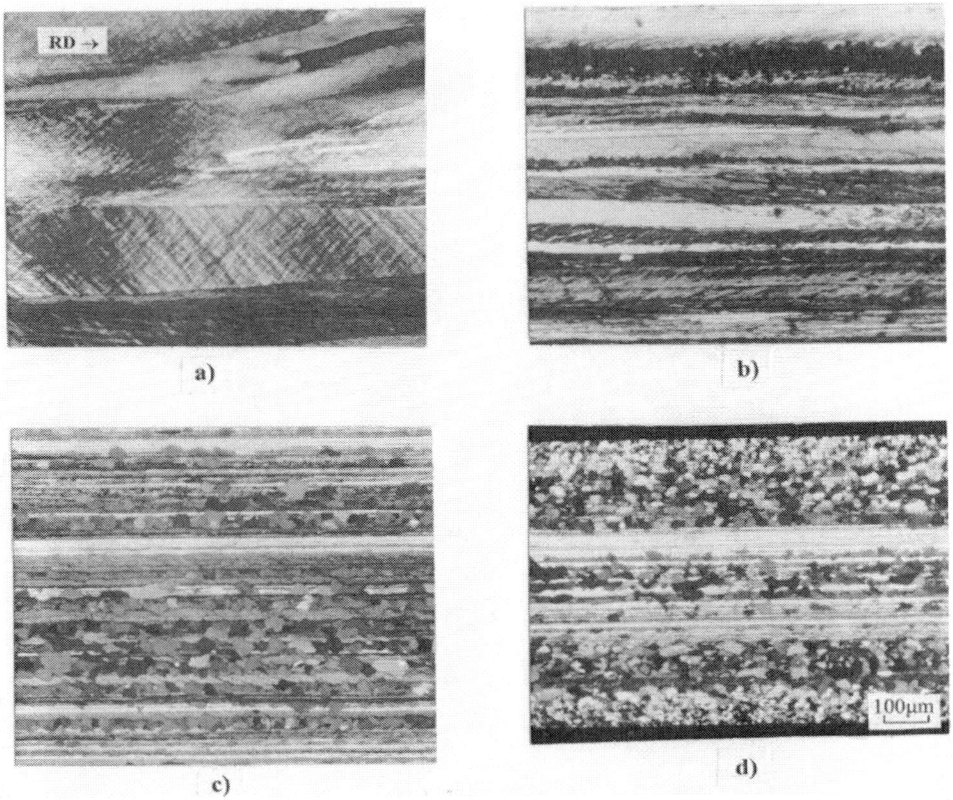

Figure 1.16. Deformation structures at flow plane for pure Al5N2 (99.9992%) after rolling reductions 68% (a), 90% (b), 97% (c) and 99% (d).

It is interesting to compare the intensity of dynamic recrystallization after ECAE and rolling. Experimental data on percentage of recrystallized areas versus accumulated strain ε for pure aluminum 5N2Al (Figure 1.17) also demonstrate high stability of the continuous plastic flow during rolling (diagram 1). In the central area of the rolled material with the pure shear deformation mode, the volume of recrystallized grains is relatively low. Even after a rolling reduction of 99%, more than 50% of the original grains continue crystallographic glide on stable slip systems. In contrast, ECAE provides a much faster increase of the recrystallized area for all processing routes and number of passes $N > 1$ (see diagrams 2, 3 and 4 for routes D, C and A, respectively). The fully recrystallized microstructure is attained after 3 passes via route D and four passes via routes C and A. Hardness (HB) after rolling (diagram 5) does not show any material softening up to a 95% rolling reduction. However, hardness after ECAE (diagram 6) reveals an evident softening effect after the first pass. These results confirm the considerable difference between simple shear and pure shear in development of localization.

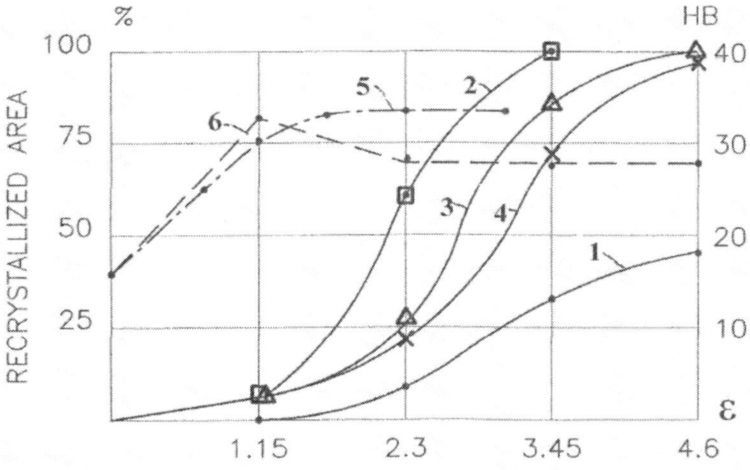

Figure 1.17. Effect of von Mises strain on percent of recrystallized area after rolling (1), ECAE via routes D (2) C (3), A (4) and material hardness HB after ECAE (6) and rolling (5) for pure aluminum Al5N2 (99.9992%).

Additional data were obtained for texture evolution during pure shear and simple shear. Since deformation textures reflect collective crystallographic glides of individual grains in polycrystalline materials, a textural analysis is very relevant to correlation between structures and processing mechanics (Hirsch et al. 1988)[4]. Crystallographic textures were measured by X-Ray irradiation with a Philips Expert Diffractometer. Partial pole figures on the {111}, {200} and {220} planes were generated. The Beartex software (Wenk et al. 1996) was used to process the raw data and calculate the orientation distribution function (ODF). Texture strength was estimated by texture index (OD index) which is the root mean square of the peak value of ODF. Special attention was paid to the correct location of measured planes. In the case of pure shear during rolling, mid-thickness sections parallel to a rolling plane were analyzed. These sections were uncovered by machining or grinding of a half material thickness followed by polishing. For simple shear during ECAE, measurements were conducted at the central areas of the flow plane. Aluminum alloy Al0.5%Cu was cast, homogenized, submitted to six passes ECAE and annealed at 250°C, 1 h. This route produces a uniform fine grained (FG) microstructure with an average grain size of 20 μm (Figure 1.18a) and random-to-weak texture with OD index 2.2. Therefore, changes in texture orientation and strength were mostly induced by crystallographic slip activity during processing. It was found that simple shear and pure shear for the FG Al0.5%Cu alloy exhibited specific modifications of the original texture with new sets of orientations. For pure shear, texture appears symmetric and a β- fiber, typical for rolling textures, is observed on all pole figures (Hirsch et al. 1988). Corresponding results for simple shear during ECAE via different routes will be considered in Chapter V. In general, the evolution of texture orientations is well described by crystal plasticity for both deformation modes. The most intriguing data relate to an effect of the deformation mode on texture strength. A diagram of OD index versus effective strain ε in the central area of the rolled fine grained material (Figure 1.19, curve 1) unveils a few steps of the texture evolution: (i) the sharp increase in

[4] See Part II for details.

texture strength up to a rolling reduction of 90%, (ii) preservation of very strong texture (OD index 37 times random) for rolling reductions from 90% to 95% and (iii) the decrease in texture strength for reductions larger than 95%. The first step confirms that continuous evolution under pure shear is accompanied by effective rotation of crystallographic orientations to the rolling direction with development of strong texture. The subsequent decrease in texture strength shows that the mechanism of the continuous evolution starts to change to localized flow along SBs resulting in texture scattering and weakening (Malin et al. 1978; Hirsch et al. 1988; Duggan et al. 1996). However, the texture was still strong for reductions $\psi > 99$ % (see Eq. (1.42)). That demonstrates that flow localization has a moderate effect even for very large strains with the pure shear deformation mode. Duggan and Lee (Duggan et al. 1996) also noted that SBs are not important for texture evolution during rolling. An example of SBs in the central area for 95% rolling reduction is shown on Figure 1.14b for Al0.5%Cu alloy. Arrows mark SBs and long lamellar deformation bands (LLBs) outside of the axis of symmetry (RD). SBs have a thickness of a few microns and a spacing of about 70 μm. So large spacing indicates a macroscopic character of localization and explains why SBs play an insignificant role for the pure shear deformation mode. That conforms to the estimated volume of SBs less than 5% during rolling (Hatherley et al. 1984). Although two families of SBs are not mutually perpendicular and intersect the axis of symmetry at an angle $\sim 30^\circ$ because of final strains $\Delta\gamma$ between their generation and detection, appearance of shear bands is similar to continuum slip line fields and resembles the Prandtl's solution for compression of thin strips (Hill, 1950).

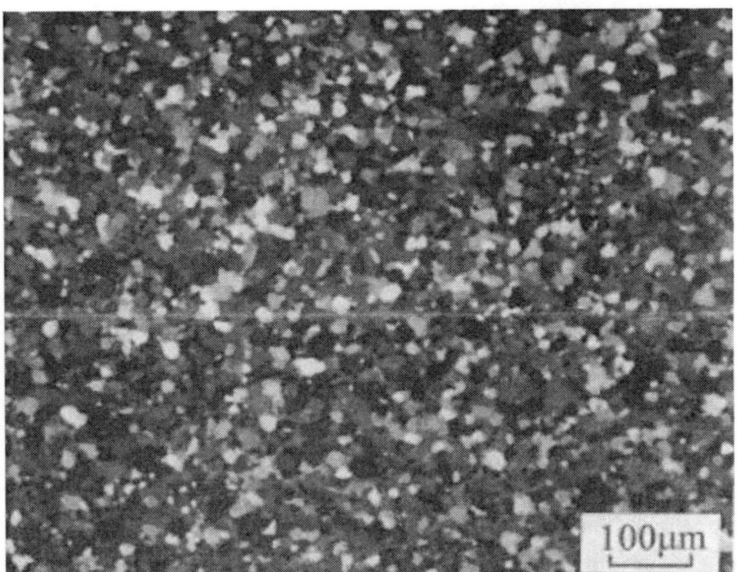

a

Figure 1.18. Continues on next page.

b

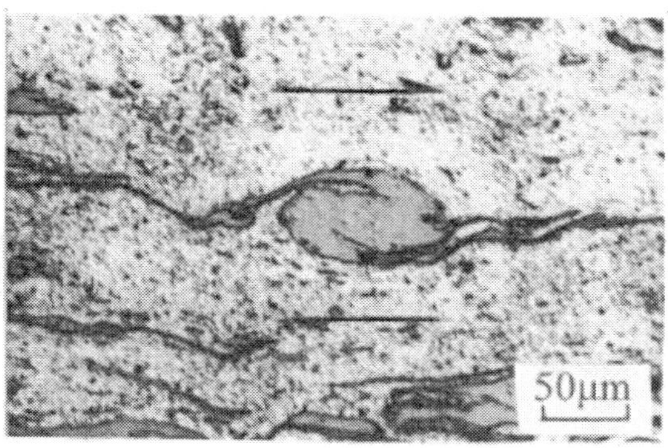

c

Figure 1.18. Microstructures of fine grained (a) and ultra-fine grained (b) Al0.5%Cu alloy; c -rotation of Nb phase during ECAE of Cu18%Nb composite.

In contrast to very strong textures after rolling, the texture strength for simple shear during ECAE remains low (Figure 1.19, curve 2). The maximum strength (a weak-to-medium texture) observed after two ECAE passes is almost an order of magnitude less than for rolling. During additional ECAE processing, OD index decreases again to a random-to-weak texture. The results on the texture strength are similar for routes A and D, although each of those routes lead to very different grain distortions and texture orientations. Such behavior is difficult to explain by the continuous evolution indicating that during simple shear the flow localization contributes significantly to the structure evolution at the earliest stages of deformation. At strains $\varepsilon > 2.3$, shear bands almost compensate for crystallographic glide and bear the major part of deformation. It can be seen from Figure 1.14a that clusters of shear bands cover all areas of the original grains and an average distance between clusters is many times finer than for rolling (Figure 1.14b). These conclusions agree with the previous observations of dynamic recrystallization for high pure aluminum 5N2Al (Figure 1.17) and with published results (Mishin et al. 1998; Wang et al. 2001).

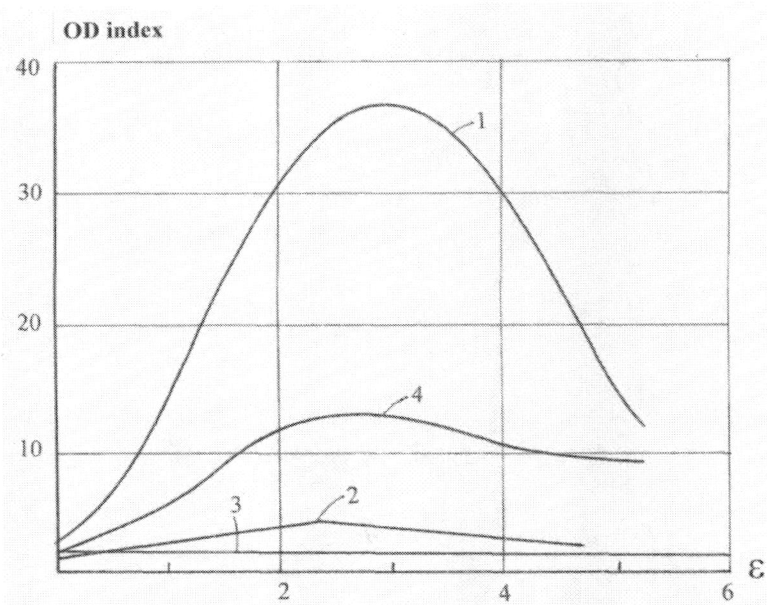

Figure 1.19. Texture strength (OD index) versus effective strains for Al0.5%Cu alloy: 1 - rolling of the fine grained material; 2 – ECAE of fine grained material; 3 – ECAE of ultra-fine grained material; 4 – rolling of ultra-fine grained material.

Severe plastic deformation of Al0.5%Cu alloy was also accompanied by significant refinement of the microstructure. Figure 1.18b presents a TEM picture of ultra-fine grained (UFG) aluminum alloy Al0.5%Cu after six passes ECAE via route D at room temperature. The strain induced microstructure is near uniform and equiaxed with an average grain size of 0.5 μm. To eliminate the possible influence of the grain size on the localization, additional experiments were performed on ECAE and rolling of the UFG alloy as described above. Despite different original textures, texture orientations after ECAE and rolling of the UFG material are quite identical to the FG material (Segal, 2005). This similarity shows that systems of the crystallographic glide depend on deformation mode irrespective of the grain size. The change of the texture strength with strain ε for ECAE (Figure 1.19, diagram 3) and rolling (Figure 1.19, diagram 4) also detects identical behaviors for FG and UFG materials. Although the maximum texture strength during rolling of the UFG material is much lower (OD index 13) than for the FG material, experimental data for both material conditions converge at large reductions. For ECAE, the OD index of the UFG material reduces continuously and steeply. It is remarkable that the alteration of deformation mode from simple shear during preparation of the UFG material by ECAE to pure shear during subsequent rolling restores the crystallographic glide and continuous evolution. However, when rolling reduction of the UFG material increases, the localization contributes more and, finally, provides the same balance with crystallographic glide as rolling of the FG material. Similar observation was reported in (Mishin et al. 1998). On the contrary, ECAE of the UFG material retains simple shear deformation mode and localization remains the primary flow mechanism resulting in the consistent texture randomization.

The presented experimental results obtained under the careful control of simple shear and pure shear demonstrate the essential influence of deformation mode on large plastic deformation. It contradicts published data reporting an insignificant role of deformation mode which were mainly obtained for the continuous evolution of dislocation structures during rolling or drawing (Hughes et al. 1997). The characteristic features of these processing techniques are near pure shear deformation mode with extended continuous flow and development of lamellar deformation bands. The material distortion under pure shear (Figure 1.6a) is accompanied by the maximum elongation of elements into one principal direction at the expense of the second principal direction. For severe plastic deformation with the preferred mode of simple shear, the stage of continuous evolution is relatively short and does not contribute significantly to development of high angle boundaries. When straining proceeds, localization induces a large number of SBs and new high angle boundaries. At later stages, shear band localization may lead to grain boundary slide and fragment rotation forming three dimensional grain structures. It is interesting to note that rotation of structural elements is often detected for simple shear, but is not observed for pure shear during rolling and drawing. An example of rotation of Nb particles in Cu matrix during ECAE of Cu18%Nb composite is shown on Figure 1.18c (Segal et al. 1997). Additionally, the material distortion during simple shear (Figure 1.6b) presents an opportunity for heavy straining of bulk products and for changing the billet orientation between successive passes. Such non-monotonic loading controls the sequence of orientations of shear planes and directions during processing and provides an additional way for the development of high angle fragmented structures by shear band intersections.

Chapter 2

PROCESSES OF SEVERE PLASTIC DEFORMATION

2.1. PROCESSING MECHANICS AND DEFORMATION MODE

The previous consideration shows that shear bands are a distinguishing characteristic of severe plastic deformation and at fixed processing conditions (temperature, strain rate) the deformation mode defines the transition to localization and structure evolution. For material processing by SPD, the control of the deformation mode is important from scientific and practical points of view and should be included in a stress-strain analysis. It is informative to evaluate this parameter in ordinary metal forming operations. A typical example of a steady 2-D plastic flow through a convergent die with friction is shown in Figure 2.1. Deformation histories of material elements at any location are determined by the continuum stress-strain states during their movement along streamlines. Because of symmetry, along the axial stream line *a-a* shear stress $\tau = 0$, α and β slip lines form angles $\pm 45^{\circ}$, velocities $v_{\alpha} = v_{\beta}$ and strain rates $\eta_{\alpha} = \eta_{\beta}$. These conditions correspond to the pure shear deformation mode. Conditions along the boundary stream-line *c–c* are defined by prescribed friction τ and a zero material flux through a tool wall (Hill, 1950)

$$\varphi_{\alpha} = [\mathrm{Arccos}(\tau/k)]/2, \ \tan\varphi_{\alpha} = v_{\beta}/v_{\alpha} \qquad (2.1)$$

where φ_{α} is an angle between α - slip line and a contact surface *c-c*. From formulae (2.1) one can find that for maximum contact friction $\tau = k$, angle $\varphi_{\alpha} = 0$, velocity $v_{\beta} = 0$ and strain rates $\eta_{\alpha} >> \eta_{\beta}$. Such a deformation mode approaches simple shear. For zero contact friction $\tau = 0$, the angle $\varphi_{\alpha} = 45^{\circ}$, velocities $v_{\alpha} = v_{\beta}$, strain rates $\eta_{\alpha} \sim \eta_{\beta}$ and the deformation mode is near pure shear. It is obvious that along streamlines confined between *a-a* and *c-c* deformation modes are within these limits. Therefore, in different metal forming operations, depending on geometry of a plastic zone and contact friction, deformation modes may vary between pure shear and simple shear.

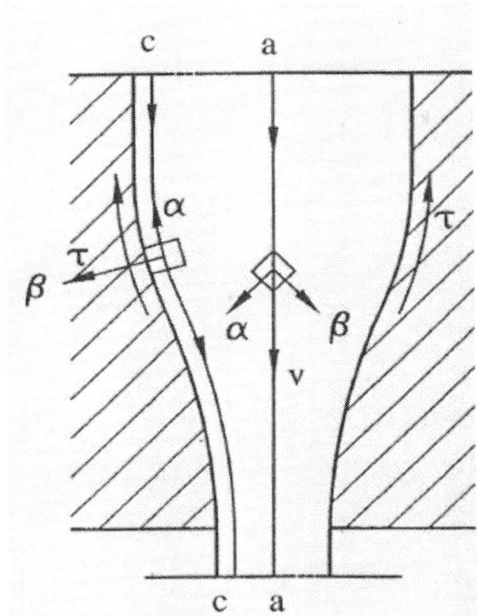

Figure 2.1. Plastic flow through curvilinear die with friction.

A few cases present some special interest. Characteristic parameters of drawing (Figure 2.2a) are low area reductions (F/f), small die angles α and insignificant contact friction $\tau << k$. In this case the deformation mode is close to pure shear along any flow line. Intensive plastic deformation during drawing requires a large number of passes and total reductions. However, each time boundary conditions and the deformation mode remain the same and continuous structure evolution without the flow localization may be extended to very large accumulated strains. In fact, effects usually associated with severe plastic deformation were not reported for drawing even with extremely large total reductions. On the contrary, drawing of fine wires with intentionally increased contact friction changes the deformation mode within a thin undersurface layer to near simple shear resulting in structure refinement and an additional strengthening effect (Kajino et al. 2006).

For multi-pass rolling of a thick plate with large accumulated reductions (Figure 2.2b), the ratio of a contact area length L to the thickness H changes from $L/H < 1$ at the beginning to $L/H >> 1$ at the final stage. Practically, this ratio may be as large as 100 or more. The compression of such thin layers develops high normal stresses on the rolls. Depending on contact friction and the ratio L/H, the deformation mode through the material thickness may vary from pure shear at a middle section *a-a* to simple shear at surfaces areas *c-c*. That explains the ambiguity of experimental results on rolling with large reductions. Although details of experimental procedures related to contact friction are usually unknown, it seems that the continuous structure evolution observed in many published works on rolling is associated with low friction and a pure shear deformation mode. However, in the cases of noticeable friction, for example, during cold rolling of steel sheets (Priester et al. 2000) or rolling of carefully degreased billets (Lee et al. 2002), an apparent strain singularity and structure refinement in subsurface areas confirm the simple shear deformation mode. These observations are quite similar with results on dynamic recrystallization of high purity 5N2Al shown in Figure 1.16d or drawing with controllable friction (Kajino et al. 2006).

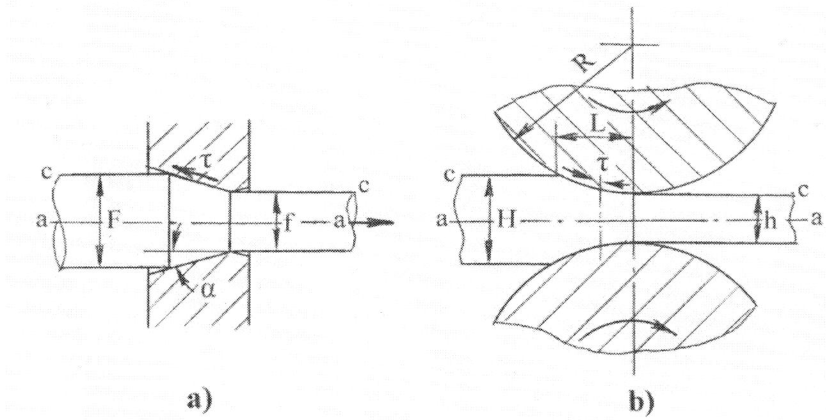

Figure 2.2. Plastic flows during drawing (a) and rolling (b).

A plastic flow through curvilinear dies (Figure 2.1) without friction $\tau \sim 0$ has a particular interest. Richmond and Devenpeck (Richmond et al. 1962) found the optimal geometry of drawing dies providing strain uniformity and minimum plastic work. Hill (Hill, 1967) extended this approach and introduced a concept of "ideal" forming operations with minimum pressure p_{if} and load P_{if} during uniform reduction of the material cross-section area from F to f:

$$p_{if} = \sqrt{3}k\varepsilon, \; P_{if} = p_{if}F \tag{2.2}$$

where $\varepsilon = \ln(F/f)$ is von Mises effective strain. In "ideal" forming operations, such as extrusion or drawing through smooth "sigmoidal dies", streamlines coincide with diagonals of continuum slip lines, deformation mode is pure shear and distortion of material elements is the maximum into the flow direction (Figure 1.6a).

In contrast to forming operations, it is difficult to formulate the same simple and unambiguous optimization principle for the structure formation because the final results may be defined by different operational mechanisms. However, when structural effects are associated with severe plastic deformation, the transition to flow localization and development of shear bands can be accepted as the optimization principle in material processing for properties. As was shown, the corresponding processing techniques should realize simple shear as the preferable deformation mode.

2.2. Simple Shear Testing

A brief overview of conventional simple shear testing techniques reveals that there are many technical problems in realization of this deformation mode for large strains and bulk samples. Torsion of cylindrical samples (Figure 2.3a) is the most common method of applying simple shear. Because strains are non-uniformly distributed in proportion with the sample radius, thin-walled tubular samples are often used. For large material volumes, linear shear (Figure 2.3b) of parallelogram-like samples was also suggested (Segal et al. 1976) where the stress-strain uniformity within a sample gage is provided by free surfaces at angles ~ 45°. In this case, ordinary tensile testing machines can be used for simple shear and alteration of a shear direction shown in Figure 2.3b changes the hydrostatic pressure σ from σ = - k to σ = + k. Both testing methods of figures 2.3a, b are restricted by moderate deformation because for large strains plastic flow localizes at some cross-section area *a-a* and further uniform straining becomes uncontrollable. For large strains, the correct shear stress state is not sufficient and should be kinematically assigned.

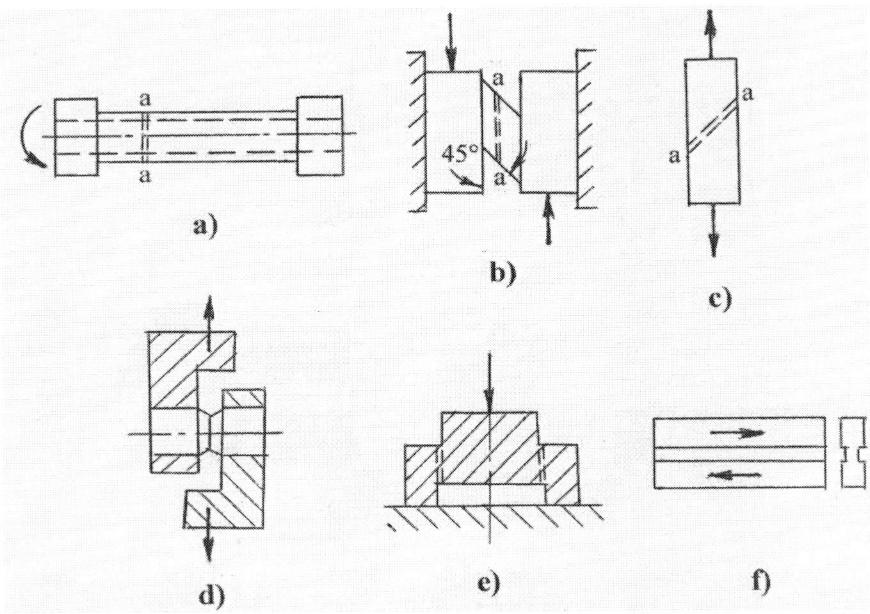

Figure 2.3. Simple shear testing: a) torsion of cylindrical/tubular samples; b) shear of parallelogram samples; c) shear of softened area; d) shear of cylindrical notched area; e) shear of hat-shaped sample; f) shear of long grooved sample.

One of the ways to resolve this problem is to constrain plastic flow and localization within a narrow zone. Typical examples are shown in figures 2.3c-f. Iosipesky suggested notched samples and special grips that prevent bending of shear zone (Figure 2.3d) (Iosipesky, 1967). Smirnov created a narrow softening area *a-a* using a lead shield and radiation-induced hardening of remaining parts of the sample (Figure 2.3c) (Smirnov, 1968). Rauch and G'Sell applied shear to sheet samples provided with long grooves into a shear direction (Figure 2.3f) (Rauch et al. 1989). Meyer et al used hat-shaped samples for dynamic testing (Figure 2.3e) (Meyer et al. 1986). In all these cases, a thickness δ of the plastic zone *a-a* is much smaller than the other sample dimensions. However, localization is transferred from a large to a small macro scale with a significant uncertainty regarding the strain distribution within δ. Such non-uniformity may be acceptable when strains are very large, the thickness δ is small, and the plastic zone spreads through the entire material.

2.3. SURFACE SIMPLE SHEAR

Many material characteristics such as friction, corrosion resistance and fatigue relate to surface properties. In these cases, material performance can be improved by severe surface plastic deformation (SSPD). Processes of SSPD have been known and found industrial applications for long time. Recently, some new techniques of SSPD were also developed. In general, the structure modification by SSPD within subsurface layers is quite similar to SPD of bulk materials. However, the processing mechanics of SSPD depends on a contact interaction between material and tool with localized plastic zones providing near simple shear deformation mode. This specificity can be traced for different methods of SSPD.

2.3.1. Simple Shear at the Plastic Contact

A few general problems of the theory of the plastic contact are relevant to different methods of SSPD. *Plastic friction* is observed in all cases of material flow along the tool at sufficiently high normal pressures. Plastic friction comprises molecular and mechanical components (Backofen, 1972, Thomsen et al. 1965). This dual nature of plastic friction includes an adhesion component τ_a acting along an ideally flat tool-material interface and a mechanical component arising between the material and micro-asperities of the ridge tool. The correct solution of the problem was presented in (Segal, 1974a) for both triangular and curved asperities[1]. The typical slip line field during a steady flow along the representative asperity ABD is shown in Figure 2.4a for the sufficiently high normal pressure with the full contact between material and tool. The additional increase of the pressure does not affect the plastic contact. For the low normal pressure, a plastic wave forms before a front asperity surface AB and defines a real contact area with adhesion friction τ_a. For moderate pressures, the plastic wave attains a back asperity surface BD. This model is in close agreement with experimental data. The biggest interest for SSPD presents an analysis of strain distribution within the plastic zone. In the case of full plastic contact, a velocity hodograph (Figure 2.4b) shows that in areas of uniform stress/strain states ABE and BFD velocities v_α, v_β are constant and strain rates $\eta_\alpha = \eta_\beta = 0$. Accumulation of strains along streamlines (like a dashed line in Figure 2.4a) takes place when material particles cross a central fan EBF and a rigid-plastic boundary AEFD. The first strain component is insignificant for all material particles as a central fan angle φ is sufficiently small and deformation histories along different streamlines are about identical. However, large shear strains ($\gamma \sim 1$) are developed when crossing AEFD. If a material displacement significantly exceeds spacing t between asperities, extremely high shear strains localized in a subsurface layer are developed during crossing of many asperities. The rigid plastic analysis shows that the subsurface layer thickness is comparable with a tool micro roughness R_a. In real situations, the strain discontinuity along the rigid plastic boundary AEFD experiences scattering over a larger layer $\delta > R_a$ due to dispersions of asperities, gradient plasticity (Fleck et al. 1993) and more complicated viscous-elastic material properties. In particular, for polishing with relatively low pressures, the very fine surface

[1] In the literature this solution is often ascribed to papers of J.M. Challen and P.I.B. Oxley, "Wear", v.53, 1979, p. 229-243 and "Wear" v. 100, 1984, p.171-193. However, these papers were published in 5 and 10 years, respectively, after (Segal, 1974a).

layer $\delta < 1$ μm with a near amorphous structure (Bilby layer) has been known for long time. For dry friction with moderate pressures, Rigney et.al (Rigney et al. 1986) observed shear strains in the range of $\gamma = 10\text{-}100$ inside the layer $\delta \sim 10$ μm with formation of shear bands and structure refinement to sub-micron scale. For friction with large pressures, Hughes et. al. (Hughes et al. 1994) found shear strains up to 100-500 and similar ultra-fine grained microstructures inside undersurface layer $\delta \sim 50\text{-}75$ μm. A practical application of the friction contact in the accumulative roll-bonding process will be considered later.

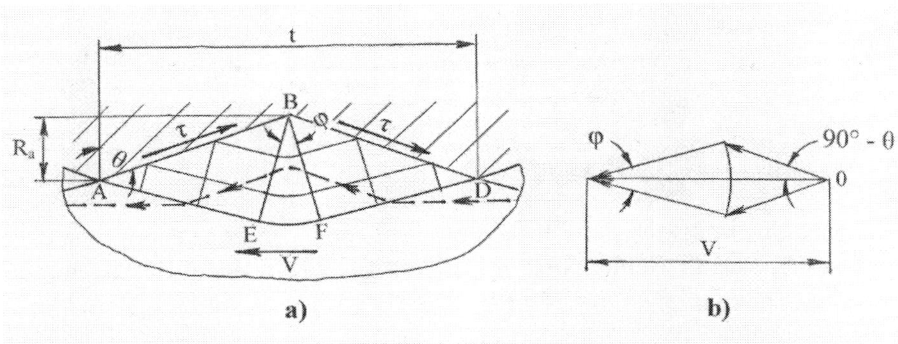

Figure 2.4. Slip line fields (a) and velocity hodographs (b) for plastic friction contact with high pressures.

Local plastic contacts may be performed in a few different ways. A slip line field for a *sliding contact* between material and a rigid wedge tool with friction τ (Figure 2.5a) is identical to considered solutions for plastic friction. Depending on the applied normal load P, the sliding contact is realized at a much larger scale than the size R_a of tool micro-asperities with the similar distribution of shear strains in the surface layer δ. For cylindrical or spherical tool (Figure 2.5b), the local sliding contact may be provided by a rigid or rotating tool. In the last case of *sliding with rolling*, a slip line solution was considered in (Segal, 1971). In general, the solution includes a plastic wave BC and areas of lagging, sticking and outstripping along AB. When the ratio of a contact length d to a tool radius R is small $d/R \ll 1$, the shear strain distribution within the subsurface layer δ is similar to the previous cases of sliding contacts. Processes of multi-pass SSPD with local sliding contacts provided by diamond or hard alloy tool (Figure 2.5a) or sliding with rolling provided by balls or rolls (Figure 2.5b) have found extensive industrial applications to improve surface finish, hardness, corrosion and fatigue resistance. The thickness δ of the subsurface layer may attain 1000 μm and more.

Figure 2.5c shows a plastic contact during the material *impression* by rolls or balls. When the contact area *d* is small in comparison with a roll radius R ($d/R \ll 1$), even low friction results in a dead metal zone AOB attached to the tool. Therefore, slip line fields and hodographs preserve near similarities, and free boundaries AB and BE are close to straight lines at any moment of non–steady impression with a speed V (Figure 2.5c). A tangential velocity discontinuity along rigid-plastic boundaries DCO and OFE is $[v] = V/\sqrt{2}$ whereas the slip line field expands with time in proportion to d. An analysis shows that an expansion speed is $\sim 2VR/d$, and strain accumulation takes place mainly during crossing of rigid-plastic boundaries DCO and OFE. Tangential and normal speeds along these boundaries are, correspondingly, v_τ = [v] and v_n ~VR/d. Then, shear strain during crossing of these boundaries is

$$\gamma = [v]/v_n \sim \sqrt{2d/2R}$$

This shear progressively spreads through the plastic zone. In contrast to sliding contacts, attainable strains during single impression remain low $\gamma << 1$, and numerous local impressions are necessary to develop severe plastic deformation. A few techniques of SSPD by multiple impressions are presented on Figure 2.6. In the multi-pass coining-forging process (MCF) (Ghosh et al. 2000), impressions in the material 1 are performed by lugs 2 of tool 3 (Figure 2.6a) alternatively into perpendicular directions by rotation of the tool about a vertical axis between successive passes. According to (Ghosh et al. 2000), accumulated strains γ = 7-10 and structure refinement to the sub-micron scale were attained in the surface layer δ = 500 μm after 40 forging steps. In the *surface mechanical attrition treatment* (SMAT) (Tao et al. 1999) numerous impressions are produced in the material 1 by balls 2 of 3-8 mm diameter agitated at a resonance frequency by ultrasonic generator 3 (Figure 2.6b). SMAT processing during 3-10 minutes results in nano structures within a layer of $\delta \sim 15$ μm. Similarly, nano scale microstructures were developed by high energy air blast peening (Altenberger et al. 1999) (Figure 2.6c) during a sufficiently long time.

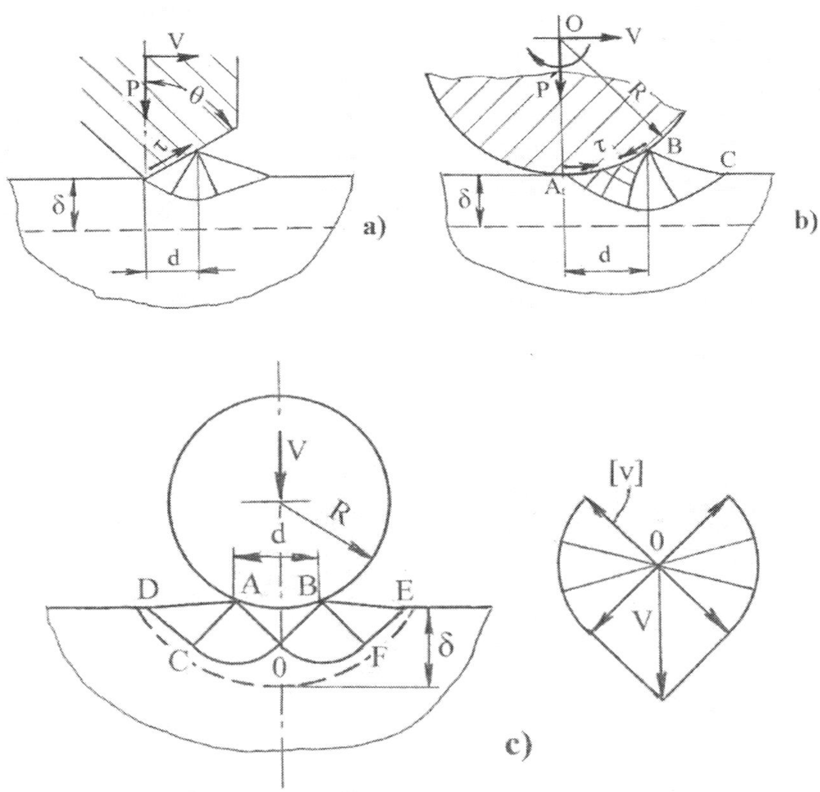

Figure 2.5. Slip line fields for linear sliding contact (a), sliding with rolling (b) and surface penetration (c).

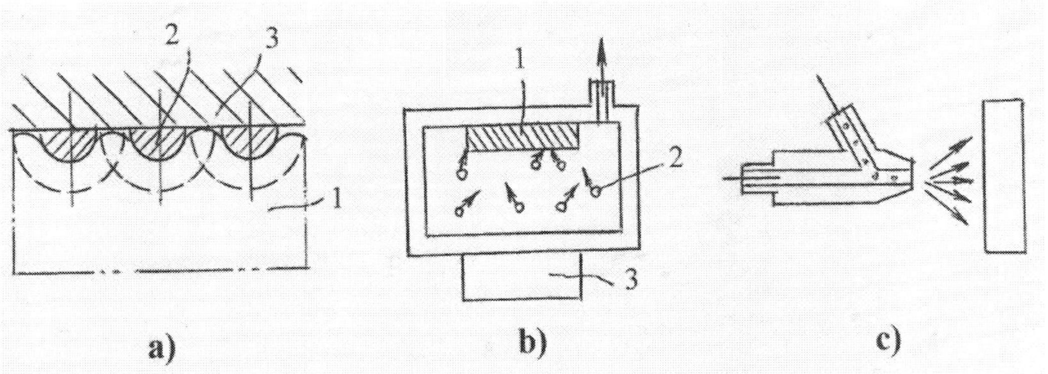

Figure 2.6. Surface impression processing: (a) multi-pass coining-forging (MCF); (b) surface mechanical attrition treatment (SMAT); (c) air blast shot peening.

2.3.2. Accumulative Roll Bonding (ARB) Process

Y. Saito et al. introduced an original method for producing ultrafine grained materials using severe surface deformation by rolling (Saito et al. 1998). When a sheet material is rolled by large rolls with moderate reductions, a contact area has an extended length L relative to the material thickness H (Figure 2.2b). For dry friction conditions, intensive shear along the contact area L produces severe deformation within an undersurface layer δ (Lee et al. 2002; Gjones et al. 1998). If rolling is continued to a small final thickness $h \sim 2\delta$, severe deformation will penetrate through all the material. Similar conditions may be also achieved for sufficiently thick sheets by co-rolling. If a stock comprising a few previously rolled and carefully cleaned sheets is rolled with reduction 50% or more, enhanced diffusivity inside surface layers δ under high contact pressures leads to diffusion bonding and embedding of layers δ into the bulk material. Practically, the accumulative roll bonding procedure includes a number of co-rolling cycles with reductions of 50% followed by material cutting into two pieces, surface cleaning and stocking after each rolling step (Figure 2.7) (Saito et al. 1998). An approximately homogeneous structure across the sheet thickness h can be reached after the number of cycles $N = \ln(h/\delta)/\ln 2$. As surface oxides and other remainder contaminations are redistributed but not eliminated during rolling, this procedure does not provide uniform bonding between different layers. It can be seen from Figure 2.8 for Al0.5Cu alloy that after 5 steps of ARB severe deformation covers the entire material; however the bonding area is minimal at the mid-thickness section O-O and increases from the center to the surface in proportion with co-rolling reductions between different layers. Potentially, ARB process can be applied for fabrication of bulk ultra-fine grained sheets from alloys that provide good diffusion bonding at relatively low temperatures.

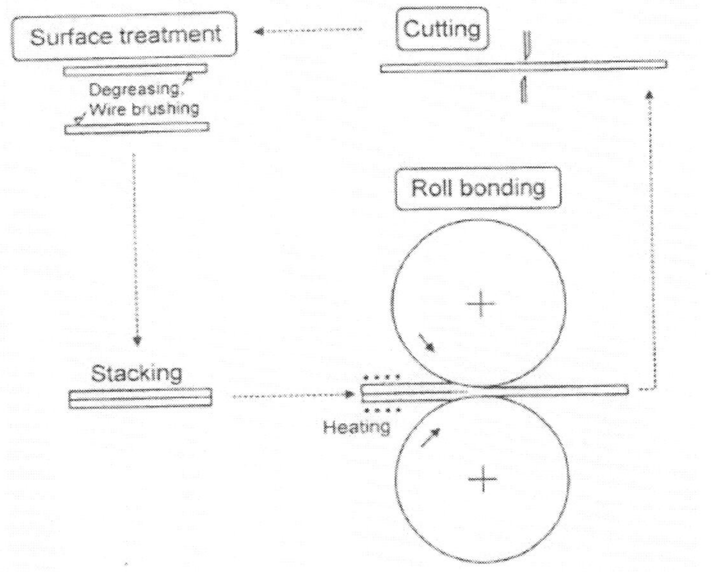

Figure 2.7. Accumulative roll-bonding process.

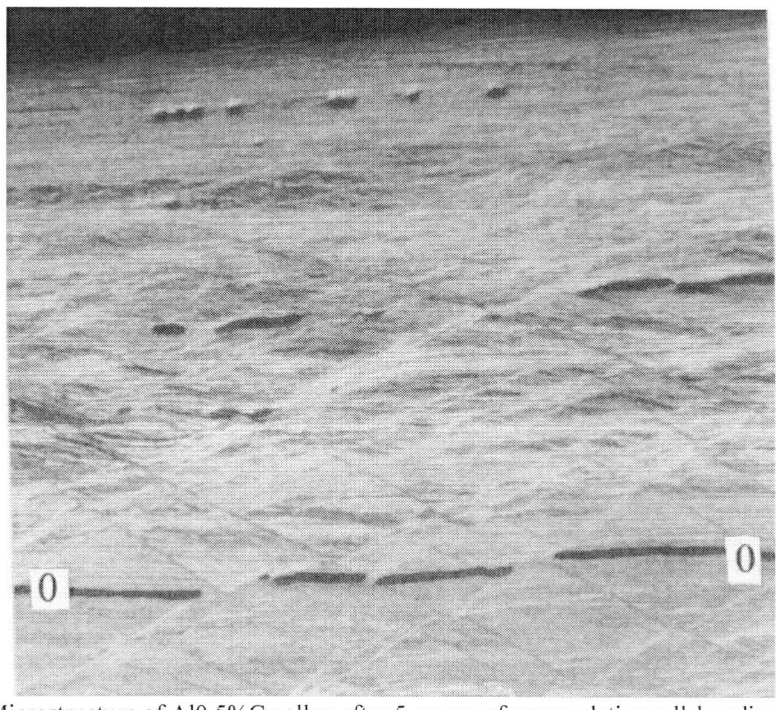

Figure 2.8. Microstructure of Al0.5%Cu alloy after 5 passes of accumulative roll-bonding.

2.4. SIMPLE SHEAR AT VELOCITY DISCONTINUITIES

It was noted that mechanical behavior of metals during large deformation approximates to the one of ideal plastic material. Correct solutions of related problems of plane plastic flows by the slip line method (Hill, 1950) reveal a significant heterogeneity in distributions of stress, strain and strain rate such as rigid and plastic zones, singularities, areas of high gradients and discontinuities. These peculiarities are general characteristics of plastic flows. For material processing, areas of large strain and strain rate gradients, play an important role. These thin and extended areas, mathematically identified as lines of velocity discontinuities along slip lines, may be observed in most technological problems. During crossing of velocity discontinuities, material particles experience simple shear with the abrupt change of flow direction. Figure 2.9 shows velocity vectors V and V' before and after crossing the line of discontinuity AO along the β slip line. Shear strain can be expressed through angles φ, φ' between vectors V, V' and a normal direction α to AO

$$\gamma = [v]/v_\alpha = \sin(\varphi' - \varphi)/\cos\varphi'\cos\varphi \qquad (2.3)$$

For slowly convergent flows like during rolling and drawing with low contact friction, the difference $\Delta\varphi = (\varphi' - \varphi)$ at any point of streamlines is insignificant and simple shear (2.3) along lines of discontinuity $\gamma << 1$ and does not contribute noticeable to accumulated strains. In "ideal" forming operations, velocities are continuous, the streamlines are smooth ($\Delta\varphi = 0$) and $\gamma = 0$. However, processes exhibiting a sharp change of flow directions should be additionally analyzed. As typical examples, consider direct extrusion through smooth ($\tau = 0$) and rough ($\tau = k$) dies with 50% area reduction and angles 30° and 45°, respectively. Related slip line solutions are presented in Figure 2.10a and Figure 2.11a (Johnson et al. 1962). Typical streamlines *a-a* were constructed using corresponding velocity hodographs (Figure 2.10b and Figure 2.11b). In the first case, a velocity discontinuity $[v] = V/\sqrt{2}$, where V is the extrusion speed, exists along rigid-plastic boundaries ABC and CO. Substituting slip line fields and hodographs by finite-linear approximations, total accumulated shear strains $\sum\gamma_i$ and the sum of simple shears γ given by (2.3) were calculated numerically at lines of discontinuity. Their distributions through an extruded material (diagram 1 for $\Sigma\gamma_i$ and diagram 2 for shears γ from Eqs. (2.3) on Figure 2.10c) demonstrates evidently that simple shear at lines of discontinuity comprises the major part of accumulated strains.

In the second case (Figure 2.11), the same velocity discontinuity $[v] = V/\sqrt{2}$ acts along boundaries AB and BO. In addition, contact friction develops infinite shear along AO. The distribution of similar accumulated strains $\Sigma\gamma_i$ (diagram 1, Figure 2.11c) and simple shears γ (diagram 2, on Figure 2.11c) across the extruded material shows again that simple shear on lines of discontinuity represents the larger part of total strains.

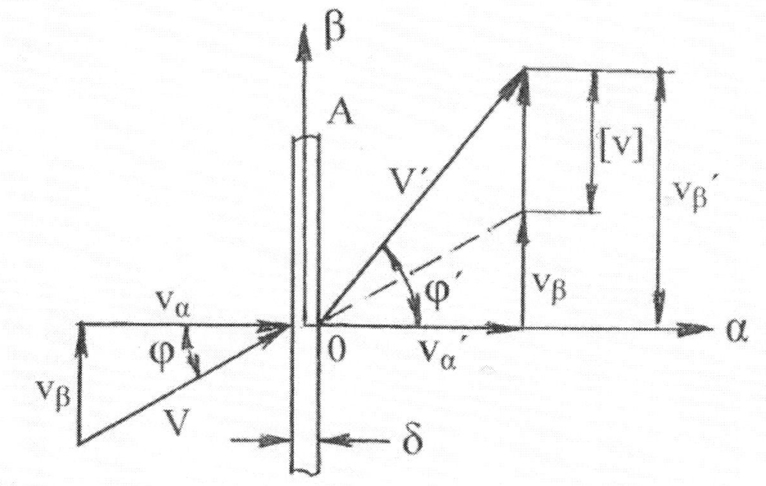

Figure 2.9. Kinematics conditions for velocity discontinuity.

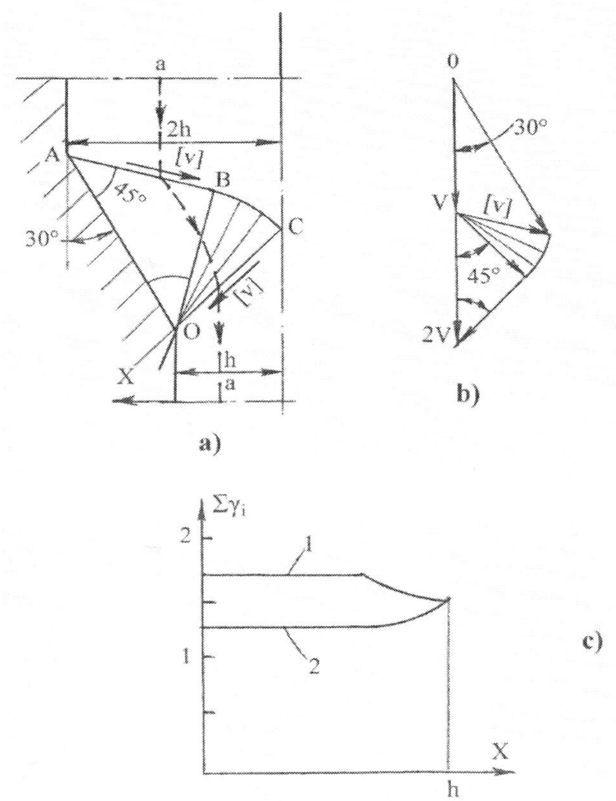

Figure 2.10. Plane extrusion through smooth die: (a) slip line field; (b) velocity hodograph; (c) distribution of accumulated strains (1) and simple shear strain (2) in the extruded product.

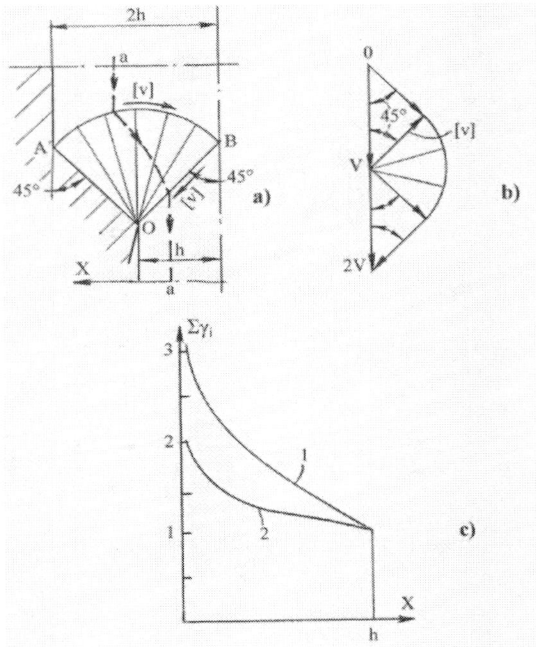

Figure 2.11. Plane extrusion through rough die: (a) slip line field; (b) velocity hodograph; (c) distribution of accumulated strains (1) and simple shear strains (2) in the extruded product.

The considered examples reveal that simple shear on lines of velocity discontinuities may be the main deformation mode in traditional forming operations. Therefore, conventional forming techniques such as extrusion and forging can be used for severe plastic deformation. In these cases, multi-step processing is necessary to attain high accumulated shears using monotonic or cyclic loading. For monotonic loading, severe deformation results in multifold reductions of the material cross-section area. With cyclic loading, the billet shape is periodically distorted and restored so that dimensions of the final product are comparative to the original billet. Both these approaches are used in some techniques of SPD. The effect of velocity discontinuities depends on the processing mechanics. Sharp tool, high friction and plane flows promote to their appearance and contribution. On the contrary, in "ideal" forming operations through smooth sigmoidal dies without friction, velocities are always continuous and the deformation mode is close to pure shear.

2.4.1. Hydro - Extrusion

Although hydro-extrusion (HE) has been known for many years, for SPD processing it was applied just recently (Kurzydlowski, 2004). Important advantages of HE are a near uniform strain distribution and processing at low temperatures under high hydrostatic pressures. Mechanically, hydro-extrusion is identical to ordinary extrusion with low friction whereas axisymmetric extrusion has many similarities with equivalent plane extrusion (Johnson et al. 1962). The considered case of a smooth low angle die (Figure 2.10) explains specific details of SPD by HE. According to Figure 2.10c, in hydro-extrusion with moderate reductions, simple shear comprises the larger part (about 75 %) of total strains. It was found in many works that minimum accumulated strain for structure refinement under simple shear is $\varepsilon \sim 4$ (see, for example, Valiev et al. 2006). That means that multi-pass hydro-extrusion should provide a total area reduction more than 100. If useful products can be fabricated with so large reductions, potentially, HE may provide finer microstructures and better strength and ductility than other SPD techniques owing to the higher hydrostatic pressures.

2.4.2. Cyclic Extrusion-Compression

Cyclic extrusion-compression (CEC) was invented and developed by M. Richert et.al. (Richert et al.1979). The process (Figure 2.12) comprises material extrusion from a chamber 1 through a die 2 by a punch 3 into the identical chamber 4 with a punch 5. During extrusion, the punch 5 has the same speed V as the punch 3 providing material upsetting in the chamber 4. After performing a full stroke, the material is extruded into the opposite direction by the punch 5. The processing cycle is repeated many times without billet ejection from the tool. As billet lubrication is not renewed after each pass, large contact friction is developed after a few passes. A related slip line field for the case of plane flow with an extrusion reduction 2 and a die angle 45° is shown on Figure 2.12. This solution in areas of extrusion and upsetting is similar to the slip line field of Figure 2.11. There are four lines AB, BO, OC and CD of velocity discontinuity [v] where flow directions are changed sharply (Figure 2.12). Intensive shears along these lines with non-monotonic loading induce numerous shear bands and their intersections observed at different scales (Richert et al. 1999; Richert et al. 2001). Accumulated strains after passing areas of extrusion and upsetting are non-uniformly distributed and twice as large as shown in Figure 2.11c. Thus, numerous passes are necessary to equalize structures in the billet center and periphery (Yuan et al. 1999). There are a few modifications of this technique. So far, CEC processing was applied at the laboratory scale to sufficiently soft materials and small samples.

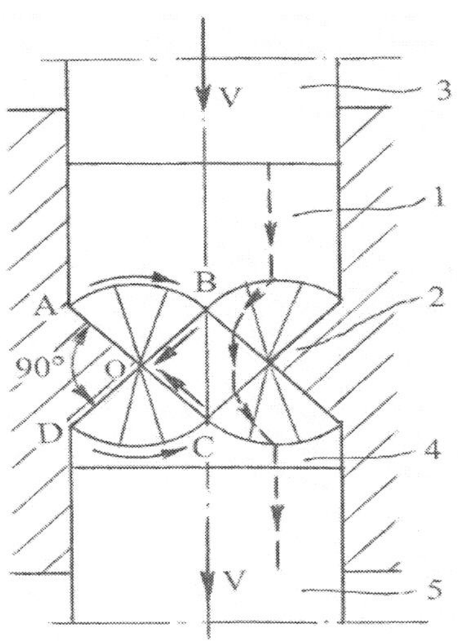

Figure 2.12. Slip line field for cyclic extrusion-compression.

2.4.3. Multi-Axis Forging

Multi-axis forging has been a well known production operation to achieve large strains and improve performance of tool and die steels, heat resistant alloys, magnets and other special materials. Since 1990, this technique was also applied for structure refinement during SPD (Imaev et al. 1992). It may be realized as two axes (plane), axisymmetric and three axes forging (Chen et al. 2000). All these approaches exploit non-monotonic, cyclic cross-loading that periodically restores a billet shape and accumulates large strains. For the plane multi-axis forging, the original billet of rectangular cross-section (a x b) (Figure 2.13a) is fixed between side plates and forged by anvils 1, 2 to an inverse ratio (b x a) (Figure 2.13c). Then the billet is rotated 90° and forging is performed in a perpendicular direction to the original ratio (a x b). Such processing may be repeated many times. The accumulated strain after N forging steps is $\varepsilon = N \ln(a/b)$. With the maximum ratio a/b = 2, strain per pass is $\varepsilon = 0.69$ or 50% of forging reduction. Distributions of strain and deformation mode are highly inhomogeneous and unsteady. At the beginning, discontinuous flow is realized along sets of coupled slip lines *aa-bb*, *a'a'-b'b'*, etc. which are confined between end rigid zones (dashed areas). The corresponding deformation mode is simple shear during short time intervals. In the middle forging position (Figure 2.13b), the billet has a square cross section divided into four rigid zones by two narrow shear bands. At the final forging position (Figure 2.13c), the rigid zones are still extensive and the deformation mode changes from simple shear along lines of velocity discontinuities AOB and A'OB' to pure shear at the axis of symmetry O'OO'. Therefore, crossed diagonal areas experience severe plastic deformation after a few forging steps whereas a large number of passes are needed for triangular peripheral areas. That remains also true for multi-pass 3D forging (Salischev et al. 2006). Nevertheless, multi-axis forging presents an opportunity for processing of bulk billets at the industrial scale because of its simplicity, realization at ordinary forging presses and without restrictions imposed by expensive tools, contact friction, high temperatures and pressures.

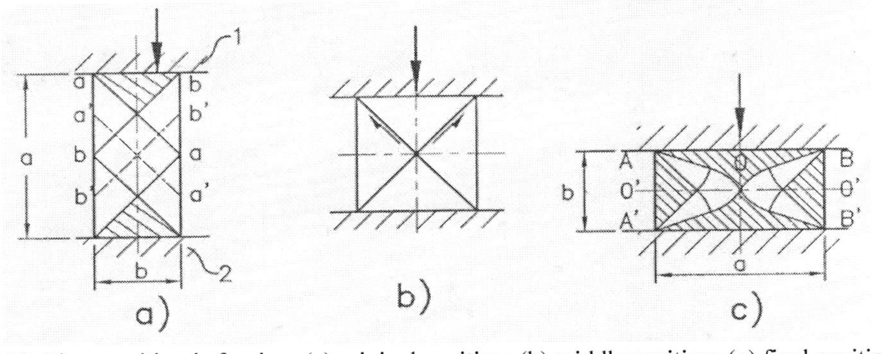

Figure 2.13. Plane multi-axis forging: (a) original position; (b) middle position; (c) final position.

2.5. PROCESSES OF SIMPLE SHEAR

A few processing techniques provide simple shear as the dominant deformation mode during SPD.

2.5.1. Torsion with Compression

This first severe deformation technique was introduced by P. Bridgman in 1930th (Bridgman, 1952) and has found numerous applications in mechanochemistry and material science. A thin disc of diameter D and thickness H is compressed between flat plates by a force P with a velocity V and twisted by a moment M with an angle speed ω (Figure 2.14a). During loading, the disc thickness is reduced and the entire material volume experiences plastic flow. Although the correct stress/strain state is unknown, it is clear that integral characteristics P, V, T and ω are interdependent. An approximate solution of the problem was found by assuming certain condition of the plastic state f(P, M) = 0 for generalized parameters P, M and V,ω (Segal et al. 1984a). The simplest quadratic form of the function f(P, M) satisfying conditions of compression (P = P*, M = 0) and torsion (P = 0, M = M*) is

$$f \equiv (P/P^*)^2 + (M/M^*)^2 - 1 = 0 \qquad (2.4)$$

Then, associated (Hill, 1950) velocities V and ω are

$$V = \Lambda \partial f/\partial P = 2\Lambda P/(P^*)^2,\ \omega = \Lambda \partial f/\partial M = 2\Lambda M/(M^*)^2\ (\Lambda > 0) \qquad (2.5)$$

Introducing a kinematical parameter m = ω/V, formulae (2.4), (2.5) give

$$P/P^* = a/(m^2 + a^2)^{0.5},\ M/M^* = m/(m^2 + a^2)^{0.5} \qquad (2.6)\backslash$$

where a = P*/M*. The total load P* for compression of thin discs (D:H) with maximum friction and the moment M* for plastic torsion of cross-section D can be found from corresponding simple solutions (Hill, 1950). Equations (2.6) show that for large *m* providing effective torsion with insignificant compression the normal load P drops sharply whereas moments M approximates to M*. However, the low normal pressure P cannot develop high shear stresses τ→ k between disc and tool necessary for torsion and slippage may arise along contact surfaces. Therefore, torsion –compression should be performed with noticeable *m* and a reduction of the disc thickness. Another shortcoming is strain non-uniformity into the radial direction.

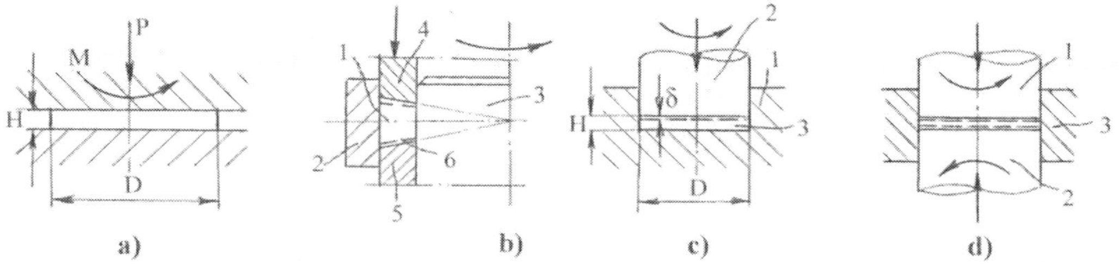

Figure 2.14. Compression with torsion (a) and high pressure torsion of conical ring sample (b), thin disc in a solid die (c) and thin disc in a floating die (d).

To overcome these problems, Saunders and Nutting developed tooling for torsion - compression of large samples (Saunders et al. 1984). Their improvements (Figure 2.14b) include a conical ring-like sample 1, a floating die 2 with a pin 3, and hollow punches 4, 5 with teeth 6. A simpler technique known as high pressure-torsion (HPT) comprises a fully constrained die cavity 1 and a punch 2 (Figure 2.14c) (Enikolopov, 1981; Smirnova et al. 1986). Under the large compressive pressure P and torsion moment M, the billet 3 gets in tight contact with the punch and the die causing radial flow to become impossible. At first, shear is initiated at the top sample surface. When intensive friction develops a sufficient hardening effect in the subsurface layer, torsion is transferred to lower layers forming parallel shear bands (Tyumentsev et al. 2000). An approximate balance of moments necessary for transferring of shear from the top billet surface with the shear yield stress k_T to a bottom billet surface with the shear yield stress k_B is

$$\pi D^3 k_T/16 = \pi D^3 k_B/16 + \pi D^2 H(k_T + k_B)/4$$

This equation gives

$$\Delta k / k = 2kH/D \tag{2.7}$$

Here $\Delta k = (k_T - k_B)$ is an increment of the shear flow stress and $k = (k_T + k_B)/2$ is the average shear flow stress through the disc thickness H. For a strain hardening condition $k = k(\gamma)$, equation (2.7) defines the difference of shears between top and bottom sample surfaces

$$\Delta\gamma = 2kH/nD \tag{2.8}$$

Here $n = dk/d\gamma$ is the material strain hardening rate at the considered moment. For large plastic deformation, n becomes small and sufficiently uniform strains ($\Delta\gamma << 1$) can be obtained only for very thin discs $H/D <<1$. For ideal plastic ($n =0$) or softening ($n < 0$) materials friction shear cannot be transferred to lower material volumes and all deformation is localized within a thin layer δ at the top billet surface Therefore, depending on the material, there are certain restrictions on the sample thickness and a thickness-to-diameter ratio H/D during HPT. Typically, $H \sim (0.15 - 0.3)$ mm and $H/D \leq 0.05$. Under these conditions, shear strain at a radius r after N rotations is

$$\gamma = 2\pi Nr/H$$

This equation shows that a distribution of the material yield stress along the radius r at any moment coincides with the diagram $k(\gamma)$ if γ is substituted by the corresponding radius r. As for most materials hardening diagrams approximate to a plateau when $\gamma >>1$, after $N >1$ the non-uniform distributions of yield stress, hardness and similar strain saturated characteristics are observed only within a small central areas $r < H$. A modification of HPT using two rotating punches 1, 2 and the floating die (Figure 2.14d) allows for doubling of the material thickness. High pressure torsion of thin samples requires large moments and loads and can be practically applied to small discs.

However, because very large strains can be induced under high hydrostatic pressures into low ductile and brittle materials, HPT has been extensively used as an effective laboratory tool.

2.5.2. Equal Channel Angular Extrusion

The optimal way to perform intensive and uniform simple shear is using velocity discontinuity along single straight slip lines. At some moment, this situation is attained during plain forging of a rectangular billet when the height–to-width ratio is 1:1 (Figure 2.13b). For homogeneous processing of large volumes, the material should flow successively through the stationary straight slip line. Such processing was first realized by the author (1970) in the "cross-extrusion" method (Figure 2.15a). In this technique, the tool is a die block 1 with two perpendicular channels 2 and 3 of the same cross-section area. A billet 4 of an identical cross-section is placed in the channel 2 and punches 5, 6 extrude it into the channel 3. After completion of the stroke, punches 5, 6 are inserted into the channel 3 and extrusion is performed in the opposite direction. This operation may be repeated many times. A plastic zone (Figure 2.15b) is formed by slip lines AOA' and BOB' along crossing planes of channels. A velocity discontinuity (see hodograph on Figure 2.15c) and shear strain along these lines are, correspondingly, $[v] = \sqrt{2}V$ and $\gamma = 2$ where V is the extrusion speed. The center O is a point of singularity where material particles such as *a* and *b* located at the left side of the central stream-line change abruptly flow directions and the distance between them and conjugate particles at the right side of the central stream-line increases dramatically. Such behavior is identical to local fracture around the center O that unveils atomically clean surfaces. Under extrusion pressure inside the channel 2, these surfaces get in touch with opposite surfaces and join together by Van der Waals forces forming specific mirror-like mechanical textures (Figure 2.15b) about a horizontal plane of symmetry. However, mechanical testing of materials after multi-pass "cross-extrusion" unveiled a noticeable weakness along the central plane. Therefore, it was the next logical step to split the plastic zone of Figure 2.15a on four separated parts. The corresponding method (Figure 2.16a) induced by the author in 1972 (Segal et al. 1994) was named equal-channel angular extrusion (ECAE). The method was first disclosed in (Segal, 1974b) and patented in the USSR in 1977 (Segal, 1977). In the former Soviet Union, information on ECAE was closed for publication and despite extensive research and development programs only a few papers appeared from 1970 to 1990. After introducing in the West at the beginning of 1990, ECAE[2] and other techniques of SPD initiated great scientific activity due to the growing interest to ultra-fine grained and nano materials.

There are several attractive characteristics that distinguish ECAE from other techniques of SPD. From the scientific point of view, this is the most clear and consistent realization of the idea of simple shear processing. Using ECAE under certain conditions, simple shear becomes the realistic technique of SPD with precision control of strain intensity, uniformity and shear orientation. With multi-pass processing, extremely high strains can be attained without any change of the billet size and cross-sectional area. A large selection of complicated deformation paths by changing the billet orientation between successive passes is also available. ECAE presents an effective tool not only for structure refinement but also for

---

2 Some authors use the name "equal-channel angular pressing" (ECAP) because a billet cross-section remains the same during processing. However, the classification of metal forming operations relies on an interaction between material and tool but not on billet shape and size. Examples of lateral extrusion with equal or larger cross section areas can be found in (Johnson et al. 1962).

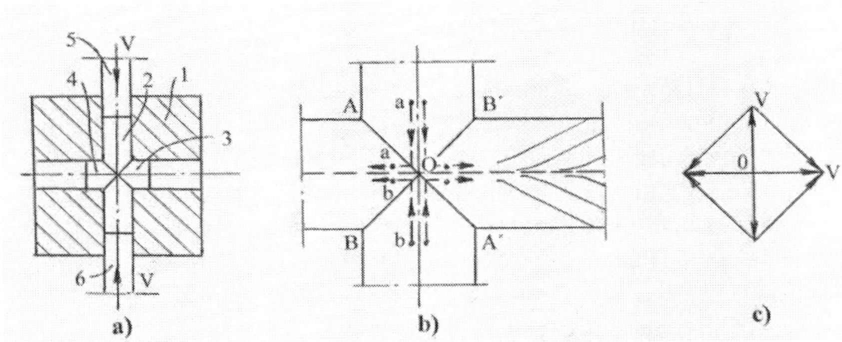

Figure 2.15. Cross-extrusion method: a - tooling; b - material flow; c - velocity hodograph.

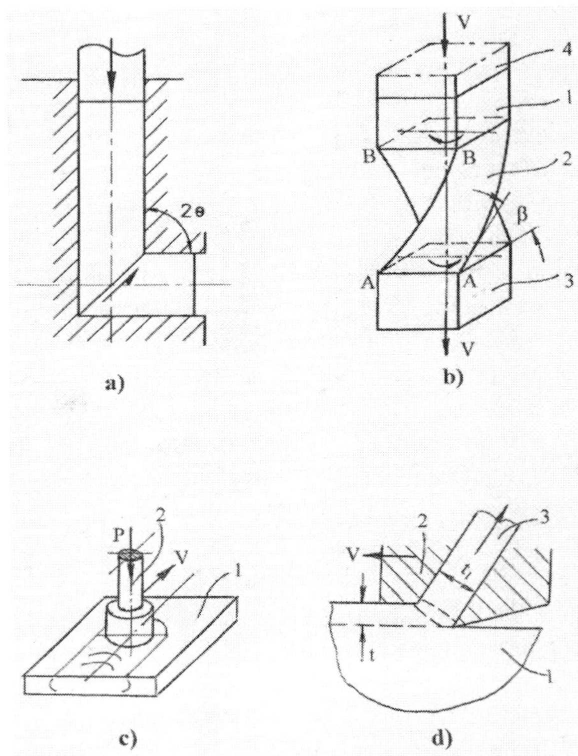

Figure 2.16. Simple shear processing: a - equal channel angular extrusion; b - twist extrusion; c - friction stir; d – extrusion- machining.

without any change of the billet size and cross-sectional area. A large selection of complicated deformation paths by changing the billet orientation between successive passes is also available. ECAE presents an effective tool not only for structure refinement but also for solving different materials science problems: control of textures, phases, particles and precipitates, consolidation and bonding of powders, fabrication of composites, materials homogenization, etc. From the practical point of view, ECAE offers the same remarkable and attractive properties. The process can be scaled up for bulk billets including large metallurgical ones. Billets may have various shapes: rods and bars, plates, sheets. Continuous semi-finished products and using of additional forming operations are also possible. The process was successfully applied to various metals and alloys including high strength and hard-to-deform materials. It does not require special equipment and may be performed at existing hydraulic presses at cold, warm or hot temperatures. By comparison with ordinary forming operations, ECAE provides noticeable reduction of pressures and loads. Certain problems related to billet lubrication, tool design, material waste and low productivity during multi-pass deformation have been a matter of concern. However, similar problems are quite typical in the early development stages of other materials processing technologies and, usually, are eliminated as the technology matures. These characteristics demonstrate significant industrial potential for ECAE in comparison with known techniques of SPD. The processing mechanics of ECAE as well as different modifications of this technology will be considered in details in Chapters 3 and 4. Finite element modeling of ECAE is the subject of Chapters 8 to 14.

2.5.3. Other Techniques

An intensive search of advanced processing techniques of SPD, which are critical for this area of materials science and engineering, is underway. However, information from patents and papers on some innovations is often incomplete and insufficiently verified. It is worthwhile to review a few recent developments that were confirmed experimentally and may present practical interest. The *twist extrusion* (TE) process (Figure 2.16b) introduced by Y. Beygelzimer (Beygelzimer et al. 2002) exploits localized twist between rigid parts of the material. A billet 1 of a rectangular cross-sectional area is hydro-extruded from a high pressure chamber 4 through a spiral die 2 into a chamber 3. The billet moves progressively in chambers 3, 4 and rotates as rigid whole in the die 2. During the crossing of planes A-A and B-B, the material particles experience simple shears into opposite directions depending on die angle β and distances from the central axis. Therefore, their loading histories conform to cyclic loading. For easy rotation inside the die, the angle β should be sufficiently small. Twist-extrusion may be repeated a number of times to accumulate large strains.

Friction stir processing (FSP) (Su et al. 2003) uses the concept of friction stir welding (Figure 2.16c). It comprises simultaneous heating and shearing of the material1by a rotating tool 2 moving at a speed V under pressure P. Intensive shear spreads over a certain material volume and provides structure refinement by dynamic recrystallization. To prevent grain growth, the material should be cooled immediately after the passage of the plastic zone. Processing of large billets is performed successively "part-by-part". The poor surface finish may require additional finish operations. Another technique is *extrusion-machining* (Figure 16d) (Moscoso et al. 2007) that explores the original idea of *extrusion-cutting* by de Chiffre (de Chiffre, 1976). The process resembles unequal-channel angular extrusion and is performed as machining of a massive blank 1 by cutting tool 2 which geometrically restrains the plastic zone and forms product 3. That way, continuous wires and strips can be produced from ductile materials with a good surface finish. Like in other cases of machining, extrusion-machining is characterized by complicated boundary conditions and inhomogeneous temperature -strain – strain rate distributions depending on the processing speed V, tool angle and a ratio t_1/t. To some extent, the processing mechanics resembles simple shear with high effective strains.

A different approach introduces intensive shear in ordinary forming operations by modification of boundary conditions. The additional tangential movement along the material – tool interfaces may change the distribution of contact friction and approximate a stress-strain state similar to simple shear. Figure 2.17a shows *differential-speed rolling* of thin sheets (Ji et al. 2007). In contrast to symmetrical rolling, rolls have different angle speeds $\omega_1 > \omega_2$. That shifts neutral points *a* and *b* on the roll surfaces and changes the distribution of contact friction τ. In the limit case $\omega_1/\omega_2 \sim H/h$, points *a* and *b* take positions near inlet and outlet cross-sections providing opposite friction along the top and bottom rolls. Further increase in the ratio of ω_1/ω_2 leads to extra slippage between rolls and material, and does not significantly affect distributions of stress and strains. For thin sheets with L/H >>1 and large contact friction $\tau \rightarrow k$, a stress-strain state is identical with shear-compression that results in lower pressure on the rolls and additional shear γ into the rolling direction. An approximate analysis gives

$$\gamma \sim [(H/h)^2 - 1]^{1/2}$$

In particular, in the case H/h = 3.3 and $\omega_1/\omega_2 \sim 3$, additional shear is $\gamma \sim 3$ which is close to the results of finite element analysis (Ji et al. 2007).

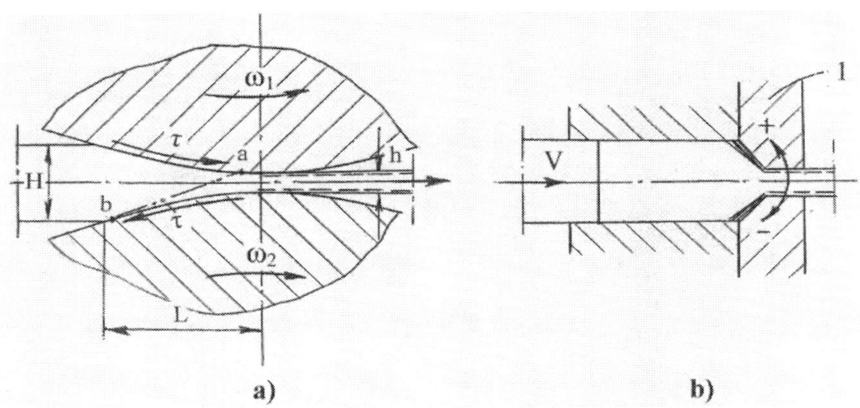

Figure 2.17.Differential-speed rolling (a) and KOBO-extrusion (b) processes.

The control of contact friction by an additional tangential movement between tool and material may affect boundary conditions for stresses but not for velocities and strains (except thin undersurface layers). Korbel and Bochniak modified the contact interaction using more complicated reversible movements (Korbel et al. 1998). They suggested a group of forming operations (KOBO-forming) with an oscillating tool to promote flow localization and structure refinement. An example of *KOBO-extrusion* through a cyclically twisted die 1 is shown in Figure 2.17b. With sufficiently large twisting amplitude and frequency, intensive shear and heating arise near the material-die contact surface. The final effect depends on the extrusion speed V. When the speed V is low, heating spreads over a material cross-section and develops a plasticized extrusion zone similarly to the known process of "friction extrusion". However, for SPD processing, heating should be insignificant to develop high friction stresses and intensive surface shear between the material and tool to refine microstructure. Conditions for transferring these surface shears to internal material volumes remain uncertain. Published experimental results (Bochniak et al. 2000) show that additional shear is localized in a thin surface layer and strain distributions through a product are almost identical for conventional extrusion and KOBO extrusion. Therefore, additional fundamental research and engineering development are needed for evaluation of KOBO-forming techniques.

Chapter 3

MECHANICS OF EQUAL-CHANNEL ANGULAR EXTRUSION (ECAE)

During multi-pass ECAE with plane plastic flow, the material experiences extensive strains from moderate to severe. Material hardening is significant for the first one or two passes and disappears during following passes. Although methods of finite element simulation are currently available in the general formulation, they are not optimal for analyzing such characteristics as deformation mode, localization, velocity discontinuity and other singularities which are essential for SPD. On the other hand, the well developed slip line method of the classic theory of plasticity (Hill, 1950) is the suitable tool to reveal the main effects of the processing mechanics SPD in a simple analytical form. In the present Chapter, this approach is consistently applied to ECAE. Polycrystalline aspects of severe plastic deformation will be considered in Part II of this book

3.1. ECAE THROUGH SMOOTH CHANNELS

3.1.1. Single Slip Line Solution

The ideal frictionless case corresponds to optimal processing conditions and presents a significant practical interest (Segal, 2003). A stress-strain state depends only on the angle between channels 2θ, the material yield stress and an applied back pressure (Figure 3.1). For $2\theta > 90°$ (Figure 3.1a), the plastic zone is the single β slip line AO. Stresses along AO are[1]

$$\tau = k,\ \sigma = -\ p_0 - k\cot\theta \qquad (3.1)$$

where p_0 is the back pressure applied to the billet end, k is the material shear flow stress.

The extrusion pressure is

$$p_e = p_0 + 2k\cot\theta \qquad (3.2)$$

---

[1] Here and further stress components σ_{ij} are positive for tensile whereas pressures p are positive for compression.

The continuation of the slip lines field in rigid zones ABO and AB′O shows that the plastic state does not reach the die walls. When $2\theta = 90°$, the plastic state contacts the tool (Figure 3.1b) and develops normal stresses $\sigma_{n1} = - p_o - 2k$ and $\sigma_{n2} = - p_0$ on die walls. For extrusion without back pressure $\sigma_{n2} = 0$. Since the tool may only compensate for compressive stresses, the slip line field of Figure 3.1b is correct when $p_o \geq 0$. For tensile pressures $p_o < 0$, other solutions with material separation from the channel wall should be considered. This restriction becomes stronger with an increase in the tool angle. For $2\theta > 90°$, the slip line field of Figure 3.1a is valid if $p_o > 0$. The extension of the plastic state to the channel wall in these cases gives an upper estimation of the back pressure

$$p_o > 2k\,(\theta - \pi/4)$$

For $2\theta < 90°$ (Figure 3.1c), a rigid area O_1AB and central fan AOB with an angle $\varphi = (90° - 2\theta)$ should be introduced to satisfy the boundary condition of zero friction. However, because the rigid area O_1AB provides the maximum shear stress $\tau = k$ along AB, ECAE with a tool angle of $2\theta < 90°$ cannot be qualified as a frictionless case.

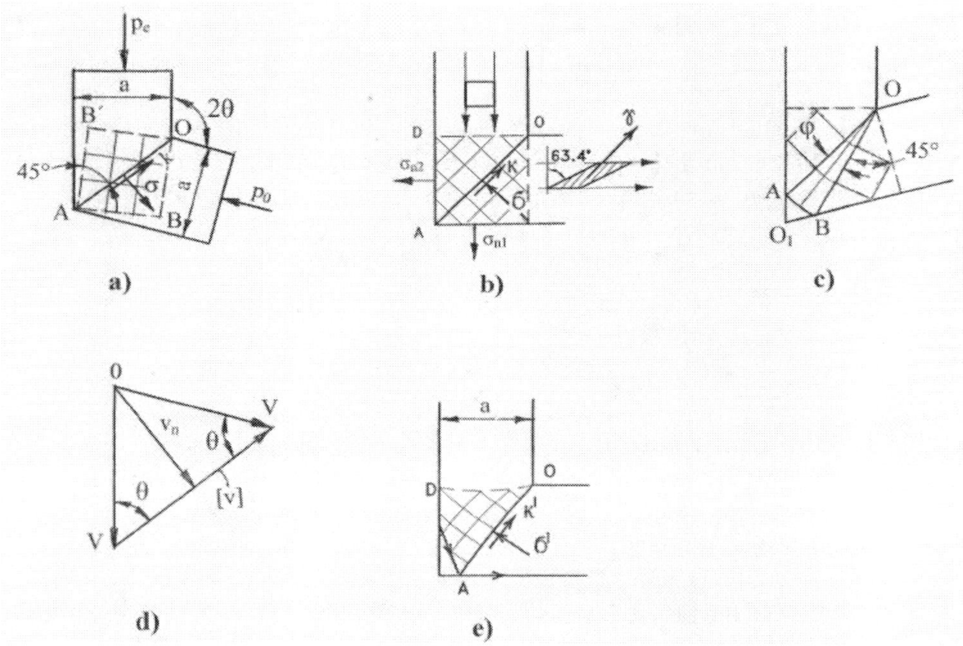

Figure 3.1. Slip line solutions for ECAE through smooth channels: a - tool angle $2\theta > 90°$; b - tool angle $2\theta = 90°$; c - tool angle $2\theta < 90°$; d - velocity hodograph for $2\theta > 90°$; e – small scale plastic flow.

From a velocity hodograph for a tool angle of $2\theta \geq 90°$ (Figure 3.1d), the discontinuity of the tangential velocity component and the normal velocity component along AO are

$$[v] = 2V\cos\theta, \; v_n = V\sin\theta$$

where V is the extrusion speed. Then, shear strain along AO is

$$\gamma = [v]/v_n = 2\cot\theta \qquad (3.3)$$

The corresponding distortion of a square material element is presented on Figure 3.2. Positions *abcd* and *a'b'c'd'* show the element before and after crossing the shear plane AO. Because of incompressibility, points *a, b, a', b'* and *c, d, c', d '* remain on the equidistant streamlines 1-1 and 2- 2 whereas *d'c'* = *a'b'*. This distortion is generated by the shear vector γ from Eq. (3.3) along AO. Similar distortion can be attained by rigid rotation of the original element *abcd* on angle (180°- 2θ) to a position *a"b"c"d"* parallel to the axis of the outlet channel followed by shear γ_1 along this axis of the same module $|\gamma_1| = |\gamma|$. Evidently, these shears without related rotations will result in different microstructures and textures. Therefore, in addition to distortions, orientations of simple shears should be also identified.

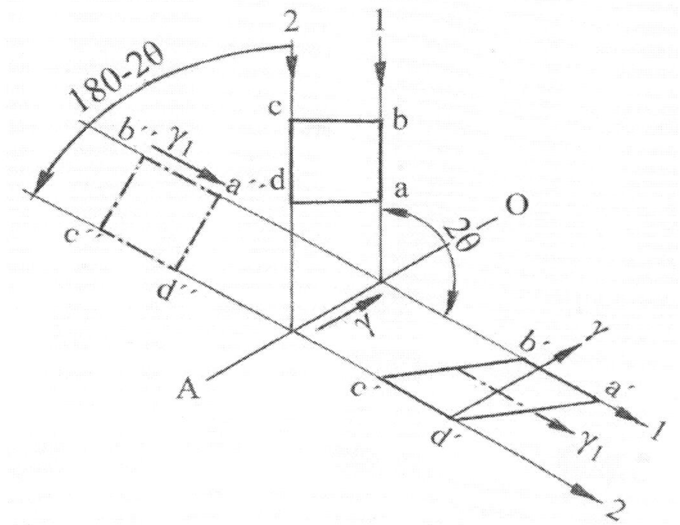

Figure 3.2. Distortion of a square material element during crossing of macro shear plane.

For verification of the considered slip line solutions, special experiments were conducted on commercially pure aluminum (99.8%) that was subjected to ECAE with tool angles of 2θ = 90°, 120° and 150°. A solid bakelite-phenol resin lubricant provided near frictionless conditions. Plastic flows were detected by coordinate grids of 1 mm spacing and longitudinal fine linear grids of 0.1 mm spacing deposited lithographically on two halves of split samples. Experiments were performed with a well annealed material and after two preliminary passes, where the material hardening ability was almost exhausted. Extrusion was extended to a steady flow with transforming of longitudinal fine grids into streamlines. This geometrical information representing all deformation history was photographed by a precision camera. The plastic zones were visualized by mechanical differentiation of two identical images of streamlines using their superposition with a small shift into a flow direction. Details of the related Moiré technique can be found in (Segal et al. 1974c). Experimental results for the three tool angles are shown on figures 3.3 – 3.5. In all cases, ECAE with back-pressure (positions "a") shows localized plastic flow with intensive shear along the crossing plane of the channels. The material outside of this plane remains rigid. The mechanical differentiation of streamlines (positions "b") reveals single shear bands whose thickness is much smaller than a billet thickness *a* (Figure 3.1a). These flow patterns conform to the slip line solution of Figure 3.1a, b. However, depending on the tool angle and material hardening, experimental data without back-pressure ($p_0 = 0$) are slightly different. When the tool angle $2\theta = 90°$, flow patterns without back-pressure are similar to Figure 3.1b for well hardened material (Figure 3.3a, b), whereas the annealed material (figures 3.3c, d) shows a more extended fan-like plastic zone. This zone is asymmetrical and comprises areas of small strains at the inlet side and localized intensive shear at the outlet side. Streamlines curve slightly in the first area and abruptly change the flow direction in the second area resembling the narrow shear zone shown in Figure 3.1e. The small-scale plastic flow before crossing of this shear zone has an imperceptible effect on strains but develops an asymmetrical "gap" between tool and material usually observed after the first pass of ECAE. With the increase of the number of passes, the material hardening ability quickly disappears and the small-scale flow becomes insignificant. Experimental billet shapes, shown in Figure 3.6 for Al0.5%Cu alloy after one (a) and three (b) passes of ECAE with tool angle 90° and low contact friction, confirm that conclusion. The first analysis of ECAE by the finite element method (Yun, 1996) also found that the "gap" vanishes when the material behavior approximates the ideal plastic one and the element size decreases.

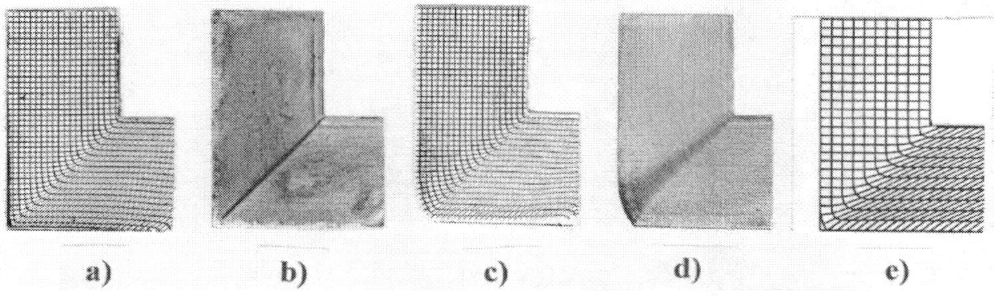

Figure 3.3. Experimental flow patterns during ECAE through smooth channels with a tool angle 90°: a - distortion of coordinate grids with back-pressure and for hardened materials; b - plastic zone with back pressure and for hardened materials; c - distortion of coordinate grids without back pressure and for annealed materials; d - plastic zone without back-pressure and for annealed materials; e - finite element modeling for ideal plastic material.

Another character of the plastic flow was detected for tool angles 120° (figures 3.4c, d) and 150° (figures 3.5c, d) both for annealed materials and for ECAE without back-pressure. Plastic zones become more extended and symmetrical with near round streamlines that do not detect shear band localization. This characteristic form of plastic zones is accompanied by lower strains at the bottom billet part. Corresponding slip line solutions are similar to round corner channels that will be considered later.

The comparison of distortions of coordinate grids after ECAE with localized simple shear (figures 3.3a,b,c, Figure 3.4a,b and Figure 3.5a,b) and distortions calculated from formulae (1.20), (3.3) show a close coincidence between theoretical and experimental results. Experiments were also performed on distortion of round elements. Two identical elements were printed on samples that were extruded in such manner that only one of them passed through the shear zone (Figure 3.4e). Experimental measurements of the principal strains ε_1, ε_2 and angle of their orientations θ_1 shown on Figure 1.6b were found to be in excellent agreement with formulae (1.22), (1.23) and (3.3).

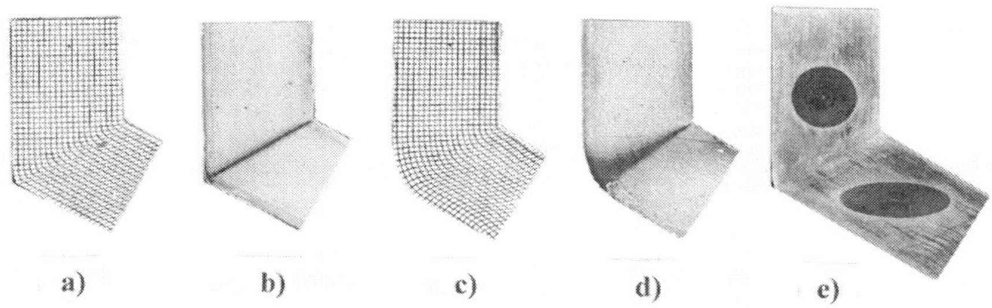

Figure 3.4. Experimental flow patterns during ECAE through smooth channels with a tool angle 120°: a - distortion of coordinate grids with back-pressure; b - plastic zone with back-pressure; c - distortion of coordinate grids without back-pressure; d - plastic zone without back-pressure; e - distortion of round elements with back-pressure.

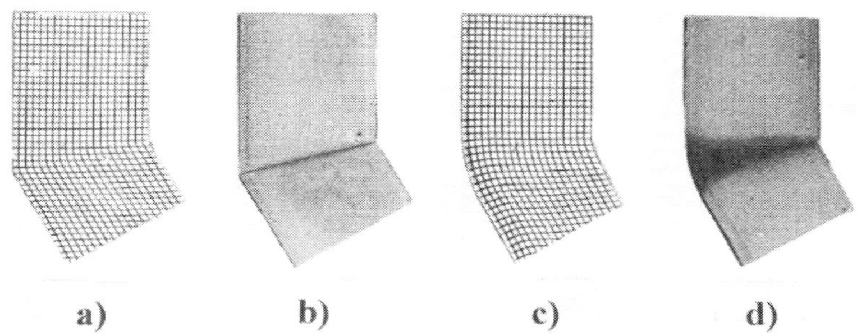

Figure 3.5. Experimental flow patterns during ECAE through smooth channels with a tool angle 150°: a - distortion of coordinate grids with back-pressure; b - plastic zone with back-pressure; c - distortion of coordinate grids without back-pressure; d - plastic zone without back-pressure.

Figure 3.6. Billet shapes after first (a) and third (b) passes of ECAE through smooth channels with tool angle $2\theta = 90°$.

3.1.2. Effects of Strain Hardening and Adiabatic Heating

Base on the slip line solution, it is possible to estimate effects induced by more complicated material properties. Referencing Figure 2.9 and assuming single slip line AO as a macro shear band with a linear distribution of the velocity v_β through thickness δ, the shear strain rate and shear strain inside AO are

$$\eta = d\gamma/dt = [v]/\delta = \text{const}, \ \gamma(\alpha) = \gamma\alpha/\delta \ (0 \leq \alpha \leq \delta) \quad (3.4)$$

For the Zener-Hollomon material (1.3) with Lüdwig strain hardening $k = k_o(1 + A\gamma^n)$, the material shear yield stress is described by formula

$$k = k_o(\eta/\eta_o)^m(1 + A\gamma^n)\exp(-a\Delta T) \quad (3.5)$$

Here *m, a* and *n, A* are coefficients of strain rate, temperature and strain sensitivity, correspondingly, k_o is the yield stress at fixed strain rate η_o, temperature T_o and $\gamma = 0$. Then, the tangential stresses $\tau = k$ along slip lines of Figure 2.9 and the distribution of shear strains through δ are described by equations (3.4) and (3.5). Consider two practically important cases. In the case of slow isothermal plastic flows when $\Delta T \sim 0$, $\eta \sim \eta_o$, eqs. (3.4), (3.5) give

$$k = k_o[1 + A(\gamma\alpha/\delta)^n] \quad (3.6)$$

At the right side of the boundary AO $\alpha = \delta$, $k = k_o(1 + A\gamma^n)$ and the hydrostatic pressure σ is prescribed by formula (3.1). This pressure cannot be extended within δ and into a rigid area at the left side of AO where $k = k_o$. Therefore, for hardening materials the stress state inside of the inlet channel does not balance stresses applied to AO and small-scale plastic flow should arise before crossing of the shear band. Also, the shear band should be slightly curved and not a straight line. Both these effects are observed experimentally. Integrating the plastic work dissipation (1.2) through δ together with (3.6), gives the effective macro shear stress ‹k› while crossing the shear band AO

$$\langle k \rangle = k_o[1 + A\gamma^n/(n + 1)] \quad (3.7)$$

This formula can be used for calculation of the extrusion pressure (3.2) for strain hardening materials with an appropriate selection of coefficients k_o, A and n at each pass.

In the second case of ECAE with a sufficiently high extrusion speed V, adiabatic heating plays a significant role. The shear band thickness δ is usually unknown and formulae (3.4) leave an uncertainty in the determination of strain rate η. However, as this thickness is much smaller than the characteristic billet size *a*, the strain rate during ECAE is much larger than in ordinary forming operations with the same processing speed V. In such cases, the strain rate effect as well as the hardening effect for strains $\gamma >> 1$ can be neglected (Johnston, 1983). Then, using $m \sim 0$, $n \sim 0$ in equation (3.5), the shear flow stress is

$$k = k_o\exp(-a\Delta T) \quad (3.8)$$

Let's define a temperature of adiabatic heating $\Delta T(\alpha)$ through δ. For adiabatic heating, the increment of plastic work dW at an interval $dt = d\alpha/v_n$ is transformed into an increase of temperature $d(\Delta T)$

$$dW = k\eta d\alpha / v_n = Jc\rho d(\Delta T) \tag{3.9}$$

where J is the mechanical equivalent of the heat, c is the material thermal capacity and ρ is the material density. Integrating (3.9) together with (3.4), (3.8) gives the temperature distribution ΔT within a shear band

$$\Delta T = T - T_o = [\ln(1 + A_1\gamma\alpha/\delta)]/a$$

Here $A_1 = k_o a/Jc\rho$ is a specific thermo-mechanical characteristic of the material. The temperature increase after crossing the shear bands is

$$\Delta T = [\ln(1 + A_1\gamma)]/a \tag{3.10}$$

Eqs. (3.8), (3.10) define the yield shear stress distribution through δ

$$k = k_o\,(1 + A_1\gamma\alpha/\delta)^{-1}$$

Similarly to (3.7), integration of the plastic work dissipation using the last equation gives an equation for the effective shear stress $\langle k \rangle$ during crossing the shear band under adiabatic conditions

$$\langle k \rangle = k_o[\ln(1 + A_1\gamma)]/A_1\gamma, \quad (\gamma > 0) \tag{3.11}$$

In material processing for properties, the adiabatic increase of temperature ΔT may play a significant role. Corresponding calculations of ΔT as a function of γ during shear band localization were performed for aluminum alloy 3003, copper 101 and magnesium alloy AZ 31 using formulae (3.10) (Figure 3.7). Related material characteristics at room temperature (Metal Handbook, 1979) are presented in Table 1.

Table 1. Material characteristics

Material	A 1/ °C	c kgm/kg°C	$\rho\times10^3$ kg/m^3	$k_o\times10^6$ kg/m^2	A_1
Al 3003	0,006	89.3	2.72	8	0.20
Cu 101	0,0025	38,5	8.99	25	0.18
Mg AZ 31	0.007	104.7	1.77	18	0.68

These results show that adiabatic heating remains relative small when strains $\gamma < 1$ and is significant when $\gamma > 2$. For the high strain rate localization such as during dynamic testing, the adiabatic temperature acts within thin shear bands for a short interval and quickly dissipates into the surrounding material volumes. However, for ECAE, practically all material successively passes through the shear zone and the temperature distribution is nearly uniform in the whole processed volume. During multi-pass cold ECAE with low speeds V, accumulated strains $\gamma >> 2$ but the heat can dissipate into the tool and surroundings. However, for warm ECAE with moderate or high extrusion speeds V, the temperature increase is sufficiently large and may affect material structures and properties. Formula (3.10) also demonstrates that SPD techniques with large shears γ during one-step processing should be performed very slowly to eliminate material overheating.

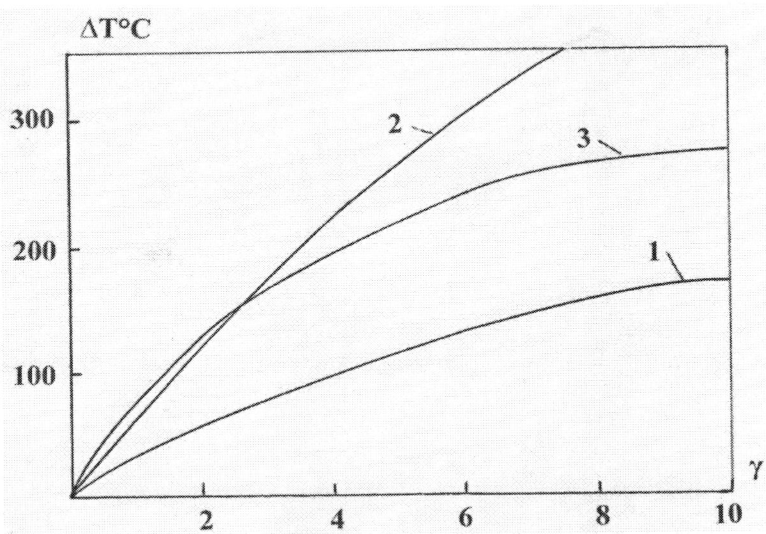

Figure 3.7. Effect of shear γ on the adiabatic increase of temperature ΔT for aluminum alloy 3003 (1), magnesium alloy AZ31 (2) and copper 101 (3).

3.2. Uniform Friction in Channels

3.2.1. Stress Analysis

A slip line solution for ECAE with a tool angle of 2θ and friction τ in both channels is shown on Figure 3.8a (Segal, 2003). It assumes that the material fills up a channel corner. The slip line field includes a central fan AOB, "dead" metal zone AO_1B and two areas OBC and OAD of uniform strain distribution. Plastic flow is concentrated inside the fan AOB while areas OBC and OAD correspond to the stress continuation into the rigid material. The angle η between OA, OB and channel walls is defined by plastic friction τ/k (Hill, 1950)

$$\eta = [\pi - \mathrm{Arccos}(\tau/k)]/2 \qquad (3.12)\backslash$$

From geometrical consideration, the fan angle is:

$$\varphi = 2(\eta - \theta) \geq 0 \qquad (3.12a)$$

Equilibrium conditions give for hydrostatic pressure on slip line BO

$$\sigma_{BO} = -\cot\eta - \tau[\sin\eta(\sin\eta + \cos\eta)]^{-1} - p_o$$

where p_o is a back-pressure applied to the outlet channel. From Hencky formulae (Hill, 1950), the hydrostatic pressure at the slip line AO is:

$$\sigma_1 = \sigma_2 + 2k\varphi$$

Hence, the axial pressure p at an entry of the inlet channel (along the slip line AO) is

$$p/2k = \cot\eta + 2(\eta - \theta) + \tau/2k[\sin\eta(\sin\eta + \cos\eta)] + p_o \tag{3.13}$$

Normal stresses along boundaries BC and AD are

$$\sigma_{n1} = - \sigma_{BO} - k\sin2\eta \quad \text{along BC}$$
$$\sigma_{n2} = - \sigma_{AO} + k\sin2\eta \quad \text{along AD}$$

Note that the positive sign in these equations corresponds to tensile stresses. Therefore, this solution is valid if:

$$\sigma_{n1} \leq 0, \sigma_{n2} \leq 0$$

An analysis shows that for $p_o \geq 0$ the normal stress σ_{n1} is negative for any contact friction τ/k (see diagram 1 on Figure 3.9 for $p_o = 0$) whereas σ_{n2} may change sign from positive to negative depending on friction τ/k and tool angle 2θ. If $\sigma_{n2} > 0$, a gap arises between the material and tool, and another slip line solution should be considered. However, there is a broad range of parameters when the slip line field of Figure 3.8a remains correct. Calculated results for σ_{n2} are presented in Figure 3.9 for $2\theta = 90°$; $105°$; $120°$ and $p_o = 0$ in function of relative friction τ/k (diagrams 2, 3, 4, correspondingly). When $\eta \leq \theta$ or $\tau/k \leq \cos2\theta$, the fan angle $\varphi = 0$, the "dead" zone AO_1B disappears and the slip line field is identical to the frictionless case. It is evident that filling a channel corner with low friction for $2\theta > 90°$ may require an additional amount of back-pressure. In the important case of the tool angle $2\theta = 90°$ the slip line field of Figure 3.8a is valid for any friction $0 < \tau < k$ without back pressure if material hardening is insignificant.

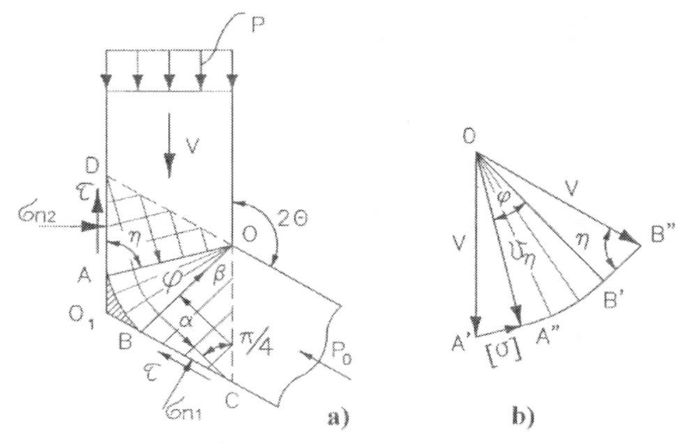

Figure 3.8. Slip line field (a) and velocity hodograph (b) for ECAE with friction.

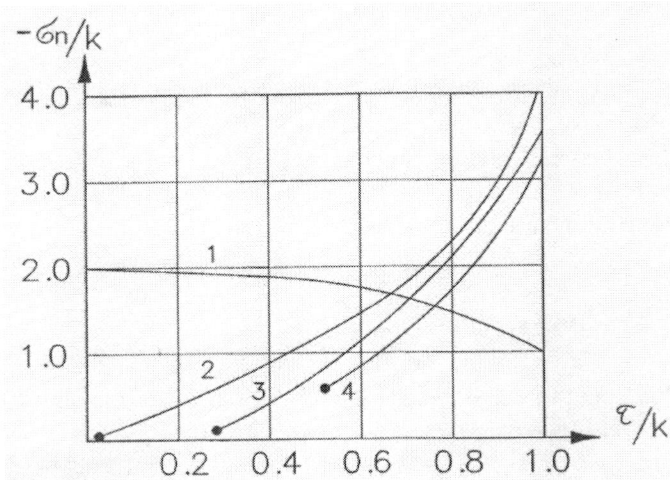

Figure 3.9. Effect of plastic friction τ/k on normal stresses at die walls ($p_o = 0$): 1 - σ_{n1}; 2 - σ_{n2} for $2\theta = 90°$; 3 - σ_{n2} for $2\theta = 105°$; 4 - σ_{n2} for $2\theta = 120°$.

3.2.2. Strain Analysis

From velocity hodograph of Figure 3.8b, the discontinuity of the tangential velocity [v] and the normal velocity component v_n on boundaries OA and BO are:

$$[v] = V\cos\eta, \quad v_n = V\sin\eta$$

During the crossing of these boundaries, stream-lines change abruptly and material elements undergo simple shears γ_1 and γ_3 along AO and BO:

$$\gamma_1 = \gamma_3 = \cot\eta \tag{3.14}$$

Inside the fan AOB velocity components on α and β slip lines are:

$$v_\alpha = V\sin\eta, \quad v_\beta = 0$$

These formulae describe a circumferential flow along the circular α slip lines with strain rate components (Segal et al. 1981b):

$$\eta_\alpha = v_\alpha/r, \quad \eta_\beta = 0 \tag{3.15}$$

where r is a distance from the center O to the considered point. Equations (3.15) conform to simple shear deformation along circular α slip lines. The increment of shear strain $d\gamma$ during an interval of time $dt = rd\varphi/v_\alpha$ and accumulated shear γ_2 after passage through the central fan area are:

$$d\gamma = \eta_\alpha dt = d\varphi, \quad \gamma_2 = \varphi \tag{3.16}$$

Simultaneously, the material elements inside the fan AOB experience a rigid rotation by an angle φ. Therefore, the deformation history of any material particle during ECAE with uniform contact friction in both channels includes three step cross-loading (Figure 3.8a):

(i) simple shear γ_1 from Eq.(3.14) along the boundary AO;
(ii) simple shear γ_2 from Eqs.(3.16) along circular slip lines and rigid rotation on the same angle inside the central fan AOB;
(iii) simple shear γ_3 from Eq. (3.14) along the boundary BO.

All shears and rotation are uniformly distributed through the material[2]. However, some strain inhomogeneity near billet surfaces may be induced by contact friction along channel walls and "dead" zone AO_1B. Figure 3.10 shows the related distortions of a small material element *abcd*. Particular positions in Figure 3.10 are:

(a) the element reaches the boundary AO of the plastic zone;
(b) the element crossed the boundary AO and experienced simple shear γ_1;
(c) the element passed through the central fan and experienced simple shear γ_2 and rotation φ. Shear bands of previous shear γ_1 change their orientation;
(d) the element crossed the boundary BO and experienced simple shear γ_3. Shear bands of previous shears γ_1, γ_2 change their orientations.

Figure 3.10 shows also dimensions of the 'unit' element after each step. Characteristic angles are:

$$\tan\varphi_1 = \sin2\eta/2,\ \tan\varphi_2 = \varphi + \sin2\eta/2,\ \tan\varphi_3 = \varphi/\sin^2\eta + 2\cot\eta \qquad (3.16a)$$

Final orientations of shear planes are described by angles:

$$\cot\varphi_4 = \varphi/\sin^2\eta + \cot\eta \text{ for } \gamma_1$$
$$\varphi_5 = 0 \text{ for } \gamma_2,\ \varphi_6 = \eta \text{ for } \gamma_3,$$

Accumulated von Mises shear strain after extrusion is (Segal et al. 1981b):

$$\gamma = \gamma_1 + \gamma_2 + \gamma_3 = 2\cot\eta + 2(\eta - \theta) \qquad (3.17a)$$

Here, in accordance with eq. (3.12), the angle η is defined by plastic friction τ/k. Calculated shears γ_1, γ_2 (diagram 1), γ (diagram 2) and the fan angle φ (diagram 4) as functions of plastic friction (τ/k) are presented in Figure 3.11 for a tool angle $2\theta = 90°$. A corresponding vector diagram can be seen in Figure 3.12 where areas OAO, OAC and ACB delineate vectors γ_1, γ_2 and γ_3 for different friction conditions. Straight lines OAB and OC relate to ECAE with zero friction $\tau = 0$ (Figure 3.1b) and maximum friction $\tau = k$ (Figure 3.13a). These limit situations provide monotonic loading with unidirectional simple shears $\gamma = 2$ at an angle of 45° to the billet axis in the first case and $\gamma = \pi/2 = 1.57$ along the billet axis in the second case.

[2] The deformation history of material elements is also considered in Chapter 11 using the tensorial kinematics formulation.

Experimental distortion of coordinate grids and plastic zone for large friction τ/k→1 (Figure 3.13c, d) are in full agreement with the theoretical solution of Figure 3.13a. In intermediate cases $0 < \tau < k$, vectors of shears γ_1, γ_2, γ_3 (Figure 3.12) have a large difference in modules and orientations conforming to numerous variants of cross-loading. Their deviations from the ideal case $\varphi = 0$ (straight line OB) are noticeable even for relative small friction τ/k or fan angles φ.

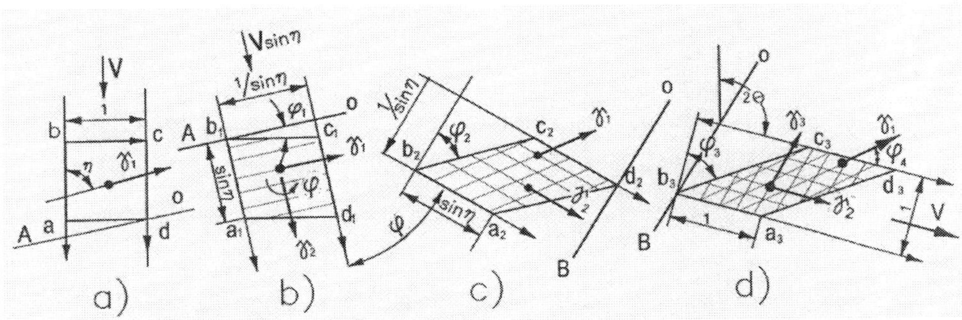

Figure 3.10. Deformation history of material elements for ECAE with friction: a - entrance of plastic zone; b - after crossing of slip line AO; c - after passage through a central fan AOB; d - after crossing of slip line BO.

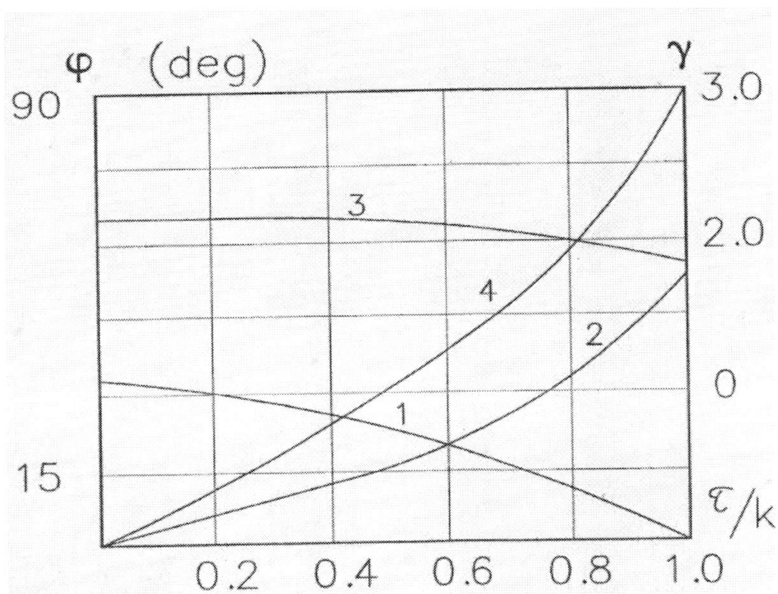

Figure 3.11. Effect of plastic friction τ/k on shears γ_1 and γ_3 (diagram 1), shear γ_2 (diagram 2), effective von Mises shear γ (diagram 3) and a fan angle φ (diagram 4) for $2\theta = 90°$.

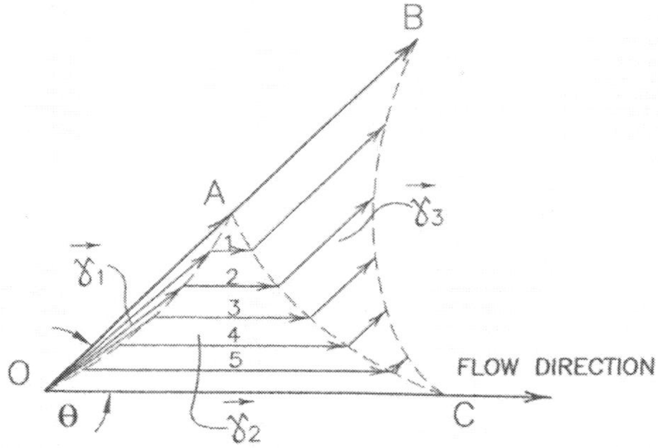

Figure 3.12. Vector diagram of successive shears γ_1, γ_2, γ_3 for sharp corner channels ($2\theta = 90°$) with friction: 1 - $\tau/k = 0.2$; 2 - $\tau/k = 0.4$; 3 - $\tau/k = 0.6$; 4 - $\tau/k = 0.8$; 5 - $\tau/k = 0.9$.

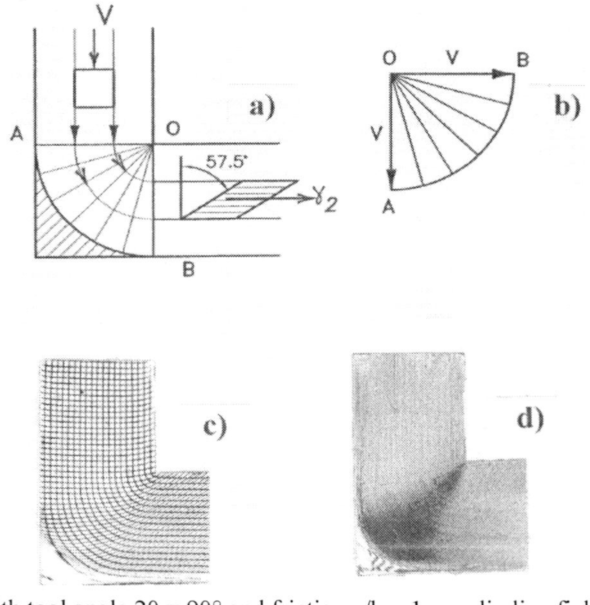

Figure 3.13. ECAE with tool angle $2\theta = 90°$ and friction $\tau/k = 1$: a - slip line field (a), velocity hodograph (b), experimental distortion of coordinate grids (c) and experimental plastic zone (d) for ECAE with tool angle $2\theta = 90°$ and friction $\tau/k = 1$.

It should be noted that many publications on ECAE used a formula for accumulated shear γ introduced in (Iwahashi et al. 1996) from a geometrical consideration for round corner channels with an angle ψ. Although this formula looks similar to (3.17), there is the principal difference in the determination of the central angles. According to formula (3.12) of the correct solution, the angle φ is solely defined by contact friction τ/k which is one of the most important characteristics of ECAE. The same angle ψ in the corresponding formulae of (Iwahashi et al. 1996) depends only on the tool design without any correlation with contact friction. The coincidence between the two formulae takes place only if (i) $\varphi = \psi = \tau/k = 0$ when both formulae are identical and (ii) for particular friction $\tau^* = k\sin\psi$ when $\varphi = \psi$. In other cases, the angle ψ should be enclosed by a "dead" metal zone $\varphi > \psi$ if $\tau > \tau^*$ or it is prescribed by round corner channels with a different slip line field and $\varphi < \psi$ if $\tau < \tau^*$. The corresponding solution will be considered in section 3.4.

3.3. Non-Uniform Friction in Channels

3.3.1. Rough and Smooth Channels

In many practical situations, contact friction in channels is different because of different processing conditions (Segal, 2003). Lubricated billets are inserted directly into the inlet channel. Following extrusion, the lubrication layer may be destroyed and pressed out causing larger friction in the outlet channel. Figure 3.14 shows the general structure of the slip line solutions for friction τ_1 in the inlet channel and τ_2 in the outlet channel when $|\tau_1| \leq \tau_2$. For $\tau_1 = \tau_2$, the slip line field COB is identical to Figure 3.8a.with a 'dead metal' zone ABC and angles $\eta_1 = \eta_2$ found from formula (3.12). For $\tau_1 < \tau_2$[3], the solutions in area BED is defined by a circular characteristic DB and friction τ_2 along EB (mixed boundary problem (Hill, 1950)). Then, it can be extended in the area DEFG using original characteristics ED, DG and a boundary condition for friction τ_1 in F. The different positions of point F between C, where $\eta_1 = \eta_2$, and A corresponds to various friction τ_1 along AC. In the particular case for a point H, $\tau_1 = 0$, the area HOBL is the plastic zone and LAH is a "dead" metal zone. Between points H and A, contact friction changes its direction from "passive friction" $\tau_2 > \tau_1 > 0$ along HC to "active friction" $0 > \tau_1 > -\cos(2\theta - \eta_2)$ along HA. That way, slip line solutions can be built for any friction $\tau_2 \geq \tau_1 \geq -\cos(2\theta - \eta_2)$.

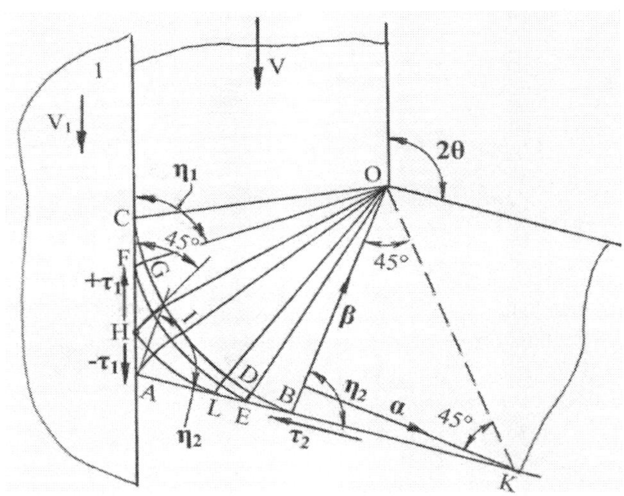

Figure 3.14. Slip line fields for different friction in channels.

To evaluate the possible effect of non-uniform friction in channels, it is worth considering the extreme cases of zero friction in the first channel and maximum friction in the second channel for the tool angle of $2\theta = 90°$. The construction of slip lines (Figure 3.15a)

[3] Cases $\tau_1 > \tau_2$ follows automatically from considered solutions by alteration of the extrusion direction.

begins with the characteristic OJ, which is perpendicular to the bottom tool wall, and circular slip line JB. An area GHJ is defined by a mixed boundary problem with friction $\tau = k$ along GJ. The initial characteristics BH and HG assign the solution in ABHG in such a manner that the slip line AG meet the boundary AK with zero friction ($\tau = 0$) at angle 45°. The area AKG is a "dead" metal zone whereas the area OJL is the continuation of the plastic state into a rigid zone with maximum contact friction along JL. The corresponding hodograph (Figure 3.15b) takes into account the velocity discontinuity $V/\sqrt{2}$ along slip lines ABO and ACEGJ thus defining two images of these characteristics at the hodograph. However, each β slip line inside the fan BDFHJO has a single image along line B"DFHJ of the hodograph. That means that velocities v_β are constant along each radial β slip line of the fan BOJ and deformation histories are identical for all material particles crossing BO. A streamline 1-1 through the point B (Figure 3.15a) separates areas of uniform OM and non-uniform MJ strain distribution in the processed material. The related distortion of transverse coordinate grids (Figure 3.15c) was found by numerical integration of displacements along streamlines. The non-uniform area below stream-line 1-1 covers about 15% of the billet cross-section. In this area shear strains increase sharply on the bottom tool wall. Important conclusions can be also made about the deformation mode. The velocity discontinuity on ABO provides simple shear during crossing ABO that changes from $\gamma_1 = 1$ at point A to $\gamma_1 = 0.84$ along BO. For maximum friction, shear along JMO $\gamma_3 = 0$. Deformation mode inside the plastic zone is not simple shear as both velocity components v_α and v_β vary along streamlines. However, while approaching slip lines JMO, the deformation mode approximates to simple shear. Therefore, in contrast with previous cases, the loading history is a complicated one because of non-uniform strain distribution and variable deformation mode.

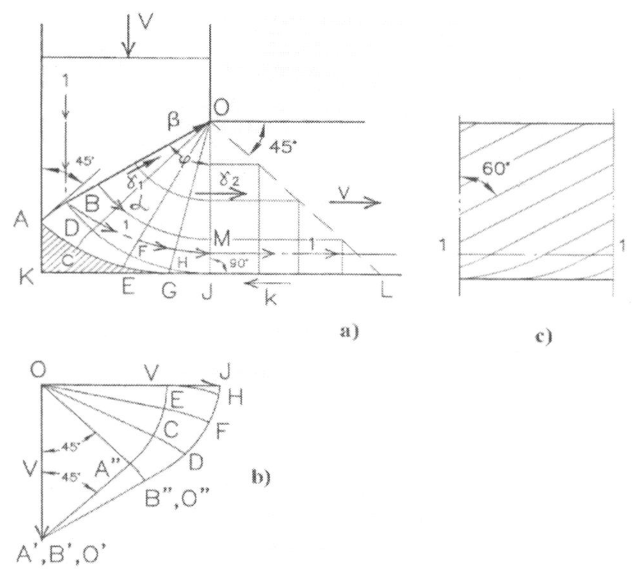

Figure 3.15. Slip line field (a), velocity hodograph (b) and material distortion (c) for ECAE with friction $\tau/k = 0$ in the inlet channel and $\tau/k = 1$ in the outlet channel.

3.3.2. Control Friction via Movable Channel Walls

Since material inside the channel moves as a rigid body with the same extrusion speed V, an effective way to eliminate friction is via movable channel walls. This concept (Segal et al. 1994) was realized in a few designs that will be considered later. Usually, the movable parts are a back wall of the inlet channel or the bottom wall of the outlet channel. Movable channel walls redistribute contact friction and may eliminate 'dead metal' zones. From Figure 3.16a for ECAE with the movable wall of the inlet channel, one can see that $\tau_1 > 0$ if $V > V_1$, $\tau_1 = 0$ if $V = V_1$ and $\tau_1 < 0$ if $V < V_1$, where V is the extrusion speed and V_1 is the wall speed. A slip line field for the case $V = V_1$, $\tau_1 = 0$ was considered in Figure 3.15a. The maximum effect of the movable channel wall takes place when $V < V_1$ and contact friction $\tau_2/k = -\tau_1/k = 1$. The corresponding slip line field (Figure 3.16a) comprises a characteristic JMO perpendicular to the channel wall, a central fan BOJ and a mixed boundary problem for an area ABJ with maximum friction $\tau_2 = k$. The fan angle φ provides the passage of slip line OBA through A at an angle 90° to AL.. From the hodograph (Figure 3.16b), one can see that the velocity discontinuity along slip line AHBO is V. Simple shear while crossing AHBO changes from $\gamma_1 = \infty$ at point A to $\gamma_1 = 1.30$ along BO. The tangential velocity along ACEJ increases from $v_\alpha = 0$ for point A to $v_\alpha = V$ for point J. The area around point A is a degenerated "dead" zone. Similarly to Figure 3.15a, the loading histories for all material elements above the streamline 1-1 are identical. This provides uniform strains along OM whereas the strain across MJ increases to infinity at the tool boundary. The calculated material distortion is also shown in Figure 3.16c. The area of non-uniform strains along JM occupies about 22% of the billet cross-sectional area. Deformation mode γ_2 inside the plastic zone ABOJ is changed from near pure shear around BO to simple shear around MO. For flow lines below 1-1, the deformation mode approximates to simple shear along the channel wall.

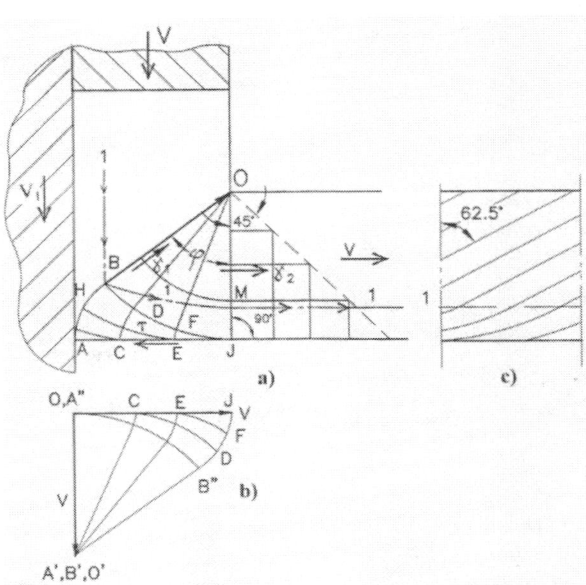

Figure 3.16. Slip line field (a), velocity hodograph (b) and material distortion (c) for ECAE with movable wall of the inlet channel and friction $\tau/k = 1$in the outlet channel.

3.4. Round Corner Channels

ECAE tool with round corner channels was introduced in (Iwahashi et al. 1996) and used in many subsequent publications. For this design, Iwahashi et.al. assumed that shear is uniform through the material and takes place along the billet axis. Using a geometrical approach, they suggested a formula for the calculation of effective strains that does not account for contact friction and real plastic flow. In particular, finite element analysis of ECAE with round corner channels (Srinivasan, 20001; Suh et al. 2001) detected a specific flow pattern with a non-uniform strain distribution. Therefore, an in-depth analysis of the problem is necessary. The structure of the slip line field for ECAE through the round corner channels of radius R and central angle ψ (Figure 3.17a) (Segal, 2003) includes the central fan BOA with a radius ρ and angle φ, and a rigid zone C_1BAC rotating about O with an angular speed ω. In contrast to (Iwahashi et al. 1996), the circular tool boundary C_1C has a center O. From the corresponding velocity hodographs (Figure 3.17b) one can see that single slip lines CA and C_1B are formed by circular arcs of radius r. The centers O_1 and O_2 of these radii are disposed on perpendiculars from O to channel walls (a correct proof can be found in (Segal et al. 1994)). Using angles ψ, θ and γ (see Figure 3.17a), other geometrical characteristics of the slip line field can be calculated from formulae:

$$R = 1/\sin(\theta + \psi),\ r = \cot(\theta + \psi)/\sin\gamma,\ \rho = r\tan(\theta + \varphi)$$
$$\cos(\theta + \varphi) = r/(1 - r\cos)\gamma,\ \theta_1 = \pi - (\gamma + \theta + \varphi)$$

Here all dimensions refer to the material thickness a (a = 1). The angles ψ and θ are prescribed by the tool whereas the angle γ should be found from additional consideration.

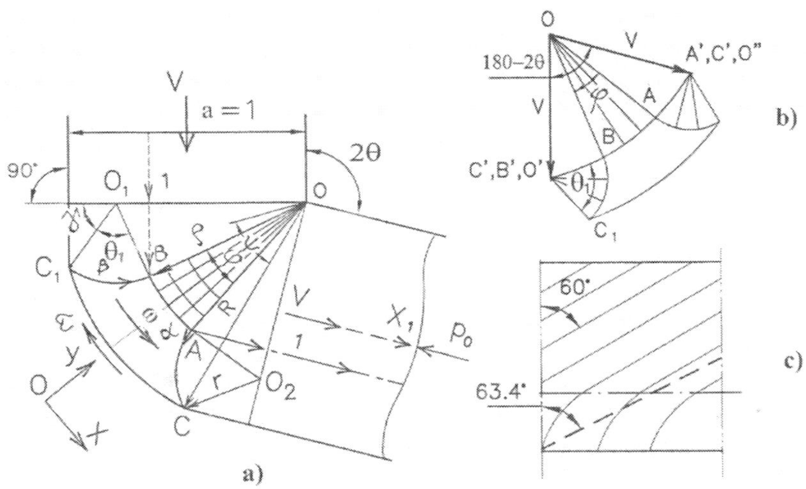

Figure 3.17. Slip line field (a), velocity hodograph (b) and material distortion (c) for ECAE with round corner channels.

From the Hencky formulae, hydrostatic stress components of any slip line point can be expressed through corresponding component σ_A of the point A. That gives for points C, B and C_1:

$$\sigma_C = \sigma_A - 2k\theta_1,\ \sigma_B = \sigma_A - 4k\varphi,\ \sigma_{C1} = \sigma_A - 2k(2\varphi - \theta_1)$$

The equilibrium conditions for stresses applied to boundaries CA, AO, OB, BC_1 and C_1C are:

$$\Sigma X = \Sigma Y = \Sigma M_O = 0$$

It is assumed that friction τ along CC_1 is also prescribed. Although the normal stresses at CC_1 are unknown, they do not contribute to the sum of moments M about the center O. Using an integration procedure developed in (MacClintock et al. 1980) for a similar problem, this condition finally gives:

$$\tau/k = [\varphi(r^2 + \rho^2) + r\rho - r(1 - r\cos\gamma)\sin\gamma - 2r(\theta_1 - \varphi)(\cos\gamma - r\cos 2\gamma + r/2)]/\psi R^2 \quad (3.18)$$

Equation (3.18) establishes a definite correlation between the angle γ and friction τ. The numerical solution of this equation was performed for a tool angle of $2\theta = 90°$ with different friction τ/k and angles ψ (diagrams 1-6, Figure 3.18). All diagrams are confined between $\gamma_{max} = (3\pi/4 - \psi)$ for maximum friction and γ_o for zero friction. When the angle γ is known, pressure p_A can be found by integrating stresses along boundary CAO into the X_1 direction using a boundary condition $\Sigma X_1 = p_0$. That way, all characteristics of the slip line field are calculated and stresses can be checked at points C and C_1. An additional restriction for the solution is $\tau/k \leq \cos 2\psi$. For sufficiently large friction τ, the rigid zone C_1BAC disappears and the slip line solution becomes identical to Figure 3.8a.

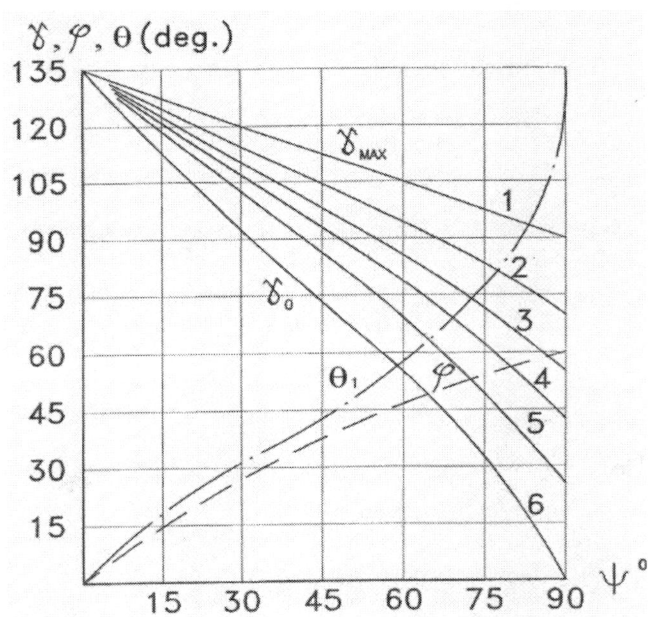

Figure 3.18. Effect of round corner angle ψ on slip line characteristics ($2\theta = 90°$): angle γ (1 - for $\tau/k = 1$, 2 - for $\tau/k = 0.95$, 3 - for $\tau/k = 0.9$, 4 - for $\tau/k = 0.75$, 5 - for $\tau/k = 0.5$, 6 - for $\tau/k = 0$) and angles θ_1 and φ.

Calculated angles φ and θ_1 for the low friction case $\tau \to 0$ are presented on Figure 3.18. When the angle ψ increases, the central fan angle φ also increases and the rigid zone C_1BAC takes a large part of the slip line field. Therefore, the angle ψ has a similar effect on the central fan as the contact friction in sharp corner channels. Owing to a more smooth flow, round corner channels improve friction in the outlet channel and lead to some reduction of the punch pressure p/2k in comparison with the identical case of sharp corner channels (compare diagrams 1 and 2 on Figure 3.19). Without back-pressure, the hydrostatic stress component σ_A changes the sign from compressive (negative) to tensile (positive) when $\psi > 62°$ (Figure 3.18, diagram 3). Therefore, contact stresses at points C and C_1 were also calculated. At point C, the contact stress σ_{n1} is always compressive but σ_{n2} at point C_1 is tensile when $\psi > 77°$ (diagram 4, Figure 3.19). In these cases, a gap around C_1 between tool and material can be observed for $p_o \leq 0$. Material elements above streamline 1-1 of the slip line field (Figure 3.17a) have the identical loading history with the slip line field of Figure 3.8a with the same fan angle φ. Thus, the former formulae (3.14), (3.16) are true after substitution of η for $(\pi/2 - \varphi - \theta)$. Similarly, the deformation mode in this area comprises a sequence of three simple shears: γ_1 along BO, γ_2 into the flow direction and γ_3 along AO. The loading history of elements below the stream-line 1-1 includes two simple shears: γ_1 along C_1B and γ_3 along CA. These shears are different for each streamline and should be determined from the hodograph (Figure 3.17b). A typical example of calculated distortions of originally transverse lines after ECAE through smooth ($\tau = 0$) round corner channels with $2\theta = 90°$, $\psi = 45°$ can be seen in Figure 3.17c. An area of non-uniform strains is extensive and covers of about 30% of the billet cross-section. The dashed line in Figure 3.17c corresponds to distortions from Eq. (3.3) for similar sharp corner channels without friction and clearly shows that the significant non-uniformity and lower strains are induced by round corner channels in comparison with sharp corner channels. This result confirmed by theoretical and finite element analysis was argued in (Xu et al. 2003; Xu et al. 2005; Valiev et al. 2006) by referring to experimental distributions of hardness through the billet cross-sectional area. Although the actual processing mechanics in these experiments was not identified, it is known that hardness attains a nearly saturated level after large plastic deformation. In particular, it remains practically unchangeable after 2-3 passes of ECAE. Therefore, in contrast to other structural characteristics, sufficiently uniform hardness distributions are quite obvious after 3 passes of ECAE. However, this conclusion cannot be transferred to strains because the correlation between both parameters becomes poorly defined for strains $\varepsilon > 2$.

The slip line solution of Figure 3.17a can be also extended for a free surface boundary CC_1. The central angle ψ in this case is unknown and should be found from an additional condition of $\Sigma Y = 0$. The geometry of the corresponding plastic zone is similar to experimental pictures shown in Figure 3.4c, d and Figure 3.5c, d. On the other hand, as shown in section 3.2, for certain friction $\tau > \tau^*$ along CC_1 a slip line field with a circular "dead" metal zone at the channel corner may provide lower load than the slip line field of Figure 3.17a. Such a solution is identical to Figure 3.8a with the same friction. In these cases, only contact friction affects the mechanics of ECAE disregarding the corner radius.

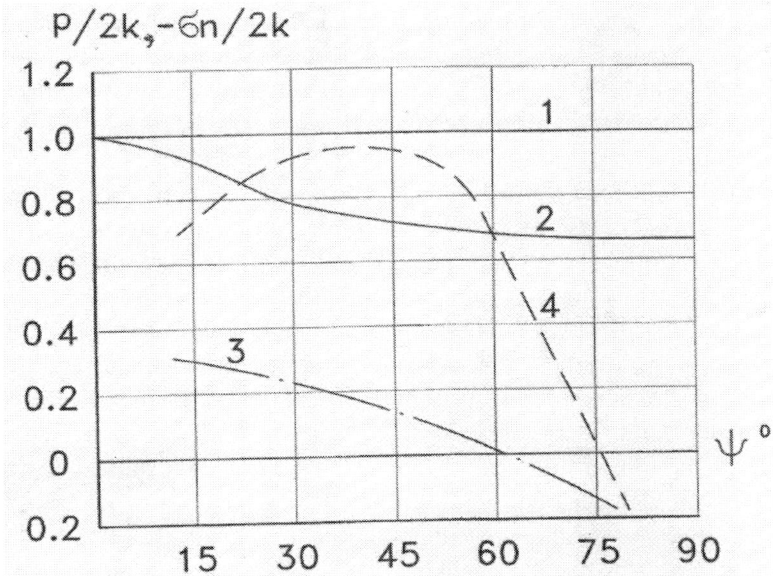

Figure 3.19. Effect of round corner angle ψ ($2\theta = 90°$, $\tau = 0$) on punch pressure (2), mean stress σ_A (3) and normal stress σ_{nC1} (4); diagram 1 shows the punch pressure for similar sharp corner channels.

The practical interest presents the comparison of strain distributions in different cases of contact friction and tool design. The final distortions of coordinate grids $\gamma = \tan\varphi_3$ through the material thickness for slip line solutions of figures 3.1b, 3.13a, 3.15a, 3.16a and 3.17a are shown in Figure 3.20 for the tool angle of $2\theta = 90°$. For a non-uniform and large friction along the bottom die wall (diagrams *c-c* and *d-d*), strains increase sharply near the bottom boundary. On the contrary, there is a sufficiently extensive area that shows a decrease of strains along the round corner boundary (diagram *e-e-e-c*). The uniform distortions are confined between the cases of zero friction ($\tau = 0$, diagram *a-a*) and the maximum friction ($\tau = k$, diagram *b-b*). The maximum difference between distortion angles $\varphi_3 = 63.4°$ and $\varphi_3 = 57.5°$, correspondingly, in these limit cases is still relatively small. Therefore, precision techniques should be used for experimental verifications of different processing conditions. Because of material incompressibility and an equal thickness of channels, the distance between stream-lines remains the same in all cases resembling simple shear into the flow direction. However, this is correct only for maximum friction $\tau = k$ in both channels. In other cases, the loading histories are more complicated and the observed angles φ_3, element distortions cannot be used as characteristics of loading histories and accumulated strains. Various orientations and intensities of real shears prescribed by the processing mechanics will define different structural effects in particular cases of ECAE.

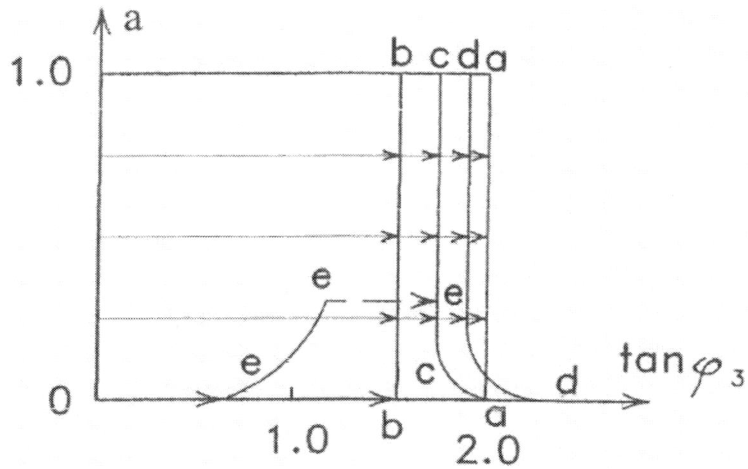

Figure 3.20. Distributions of accumulated distortions $\tan\varphi_3$ through a material thickness during ECAE ($a = 1, 2\theta = 90°$): a-a - for slip line field of Figure 37b; b-b for slip line field of Figure 49a; c-c for slip line field of Figure 51a; d-d for slip line field of Figure 52a; c-e-e-e for slip line field of Figure 53a ($\psi = 45°$, $\tau = 0$).

3.5. Multi-Pass ECAE

Multi-pass ECAE is usually necessary to attain large accumulated strains. Originally, ECAE was designated for material processing by intensive and uniform simple shear. The considered solutions demonstrate that actual stress-strain states can deviate from this ideal deformation model. The most important parameters are contact friction, the geometry of the channels and tool design. Effects of back-pressure and material characteristics on the processing mechanics are rather moderate. The plastic zone ranges from a single slip line to an enlarged central fan and may include a free surface, "dead" metal zone and area of non-uniform strains. Despite these differences, the main feature of all slip line solutions is the central fan. The plastic flow inside the central fan is near circular resulting in uniform shear into the flow direction. However, significant strain non-uniformity may exist outside the central fan. The case of round corner channels is especially deficient because of lower and non-uniform strains throughout a large part of the sample thickness. A remarkable result is that the deformation mode in most cases conforms to unidirectional simple shear or to combinations of shears into three different directions. Optimal processing corresponds to near frictionless sharp corner channels with the tool angle of $2\theta = 90°$. Related cases of single slip lines or narrow central fans outline a principal concept of simple shear processing that should be provided by certain boundary conditions and the deformation techniques during any pass of ECAE. On the other hand, multi-pass ECAE induces a large diversity of processing routes depending on the billet shape and rotations between successive passes.

3.5.1. Processing Routes

Consider the ECAE of rectangular billets *a, b, c* in smooth channels ($\tau \rightarrow 0$) with $2\theta = 90°$. The most effective billet shapes are long square billets when $a = b$; $c >> a, b$ (Figure 3.21a) and flat square billets when $c = b$; $c, b >> a$ (Figure 3.21b). The billet orientation will be described by the coordinate system (X, Y, Z) after each pass. Because of the orthogonal symmetry, the billet orientation between passes can be changed by rotations of φ_1, φ_2, φ_3 about these axes with angles of 90°, 180° and 270°. The total displacement of the material particles is the vector sum of partial displacements during each pass. For long billets, superposition of displacements after two passes rotating $\varphi_2 = 90°$ about axis Y is presented in Figure 3.22. At the first pass, simple shear γ_1 is performed in plane XOY and line 1 shows distribution of displacements through the billet thickness. Similar distribution of displacements 2 is induced by a second shear γ_2 in plane YOZ. The vector sum of two displacements is equivalent to one shear γ in a plane P which trace O_1O_1 at a transverse cross-section XOZ forms an angle ω:

$$\gamma = (\gamma_1^2 + \gamma_2^2)^{1/2}, \ \tan\omega = \gamma_1/\gamma_2 \tag{3.19}$$

However, this geometrical equivalency cannot be physically rationalized as two shears in different planes. The rotation of 180° about the same axis Y is equivalent to two shears into opposite directions restoring the material distortion. A similar consideration can be applied to any sequence of passes. There are numerous combinations of possible distortions when the number of passes increases. However, all possible situations can be reduced to four basic models of distortions (Segal, 1996; Segal, 1998):

A. The increase of distortions into the same direction.
B. The cyclic increase of distortions into two perpendicular directions.
C. The cyclic alteration of distortions into two opposite directions.
D. The cyclic alteration of distortions into four perpendicular directions.

Corresponding distortion diagrams are presented in Figure 3.23. Practical realization of these diagrams or processing routes A,B,C,D[4] are provided by special systems of billet rotations between passes. Numerous additional routes can be obtained by their combinations. Simplest systems of rotation φ_1, φ_2, φ_3 (Figure 3.21) for ECAE via routes A, B, C, D are presented in Table 2 for long and flat billets[5]. For long billets, all possible displacements are performed along the axis Y parallel to planes XY and ZY using rotations φ_2 and φ_3. In the case of flat billets, displacements are performed along axes X and Y parallel to the plane XOZ using rotation φ_3.

---

[4] In contrast to often used abbreviation A, B_A, C, B_C (Valiev et al. 2006), this designation underlines that each basic route is independent from others.

[5] Other rotations for realization of these routes are also possible. For example, route C for long billets can be performed by rotations $\varphi_1 = 180°$ or $\varphi_3 = 180°$.

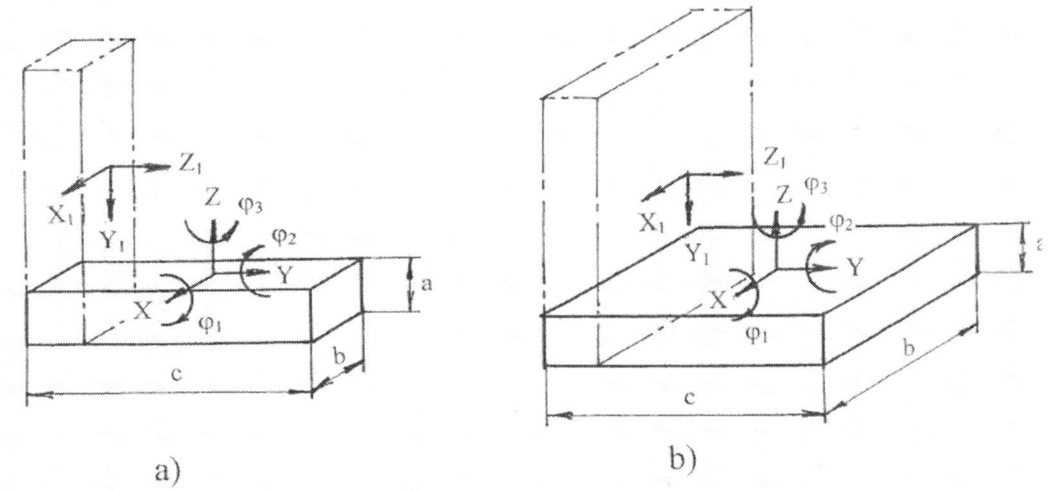

Figure 3.21. ECAE of long rectangular (a) and flat (b) billets.

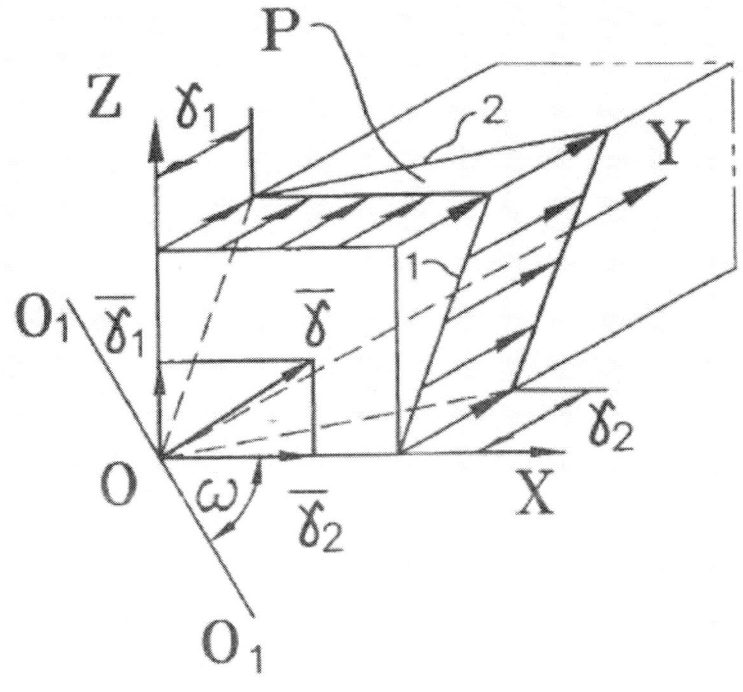

Figure 3.22. Superposition of shears.

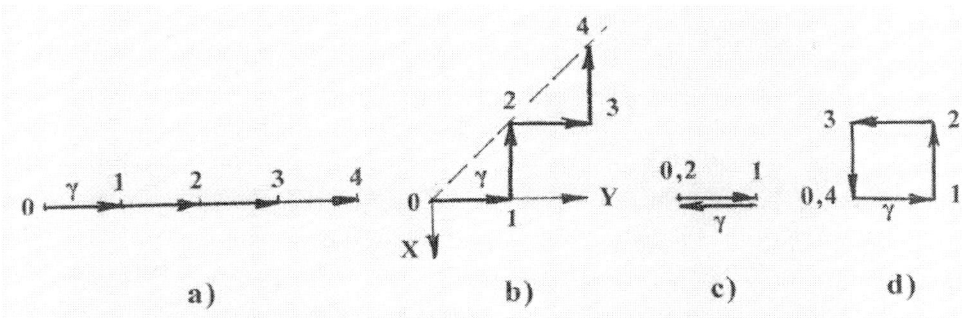

Figure 3.23. Element displacements for four passes via routes A (a), B (b), C(c) and D (d).

Table 2. Billet rotations during ECAE via different routes

Route	Long billet			Flat billet		
	φ_1	φ_2	φ_3	φ_1	φ_2	φ_3
A	0	0	0	0	0	0
B	0	± 90°	0	0	0	± 90°
C	0	0	180°	0	0	180°
D	0	90°	0	0	0	90°

Routes A and C first described in (Segal et al. 1994) provide the plane plastic flow. These routes remain identical for long and flat billets. Routes B and D introduced in (Segal, 1996) for long billets and in (Segal, 1998) for flat billets correspond to spatial flows. ECAE via routes A and B results in the maximum distortions of material elements for two dimensional and three dimensional flows, respectively. On the contrary, routes C and D restore periodically the original element shape after each second and fourth pass, respectively. Figure 3.24 presents the generalized distortion diagram for 4 passes ECAE via basic routes together with related rotations of long billets. Different combinations of these routes may be beneficial in special cases. Possible combinations and their designation for 3 passes ECAE are shown in Figure 3.25[6]. The reader is referred to Section 8.3 for modeling of crystallographic texture (grain orientation), the morphologic texture (grain shape) and material hardening during multi-pass ECAE via basic routes.

3.5.2. Processing Shear Planes and Material Shear Bands

While the processing mechanics prescribes deformation modes and orientations of macro-shear planes, processing routes develop different systems of macro-shear planes and directions during multi-pass ECAE. Correspondingly, micro-shear bands arise in the material along these planes. Each pass induces a family of numerous near parallel micro-shear bands.

---

[6] Numbers of possible combinations for more than three passes are too big for a graphical presentation.

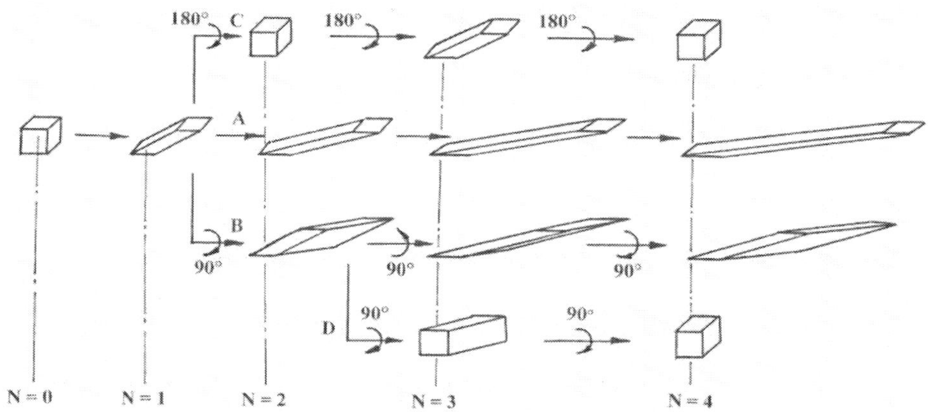

Figure 3.24. Diagram of the basic routes.

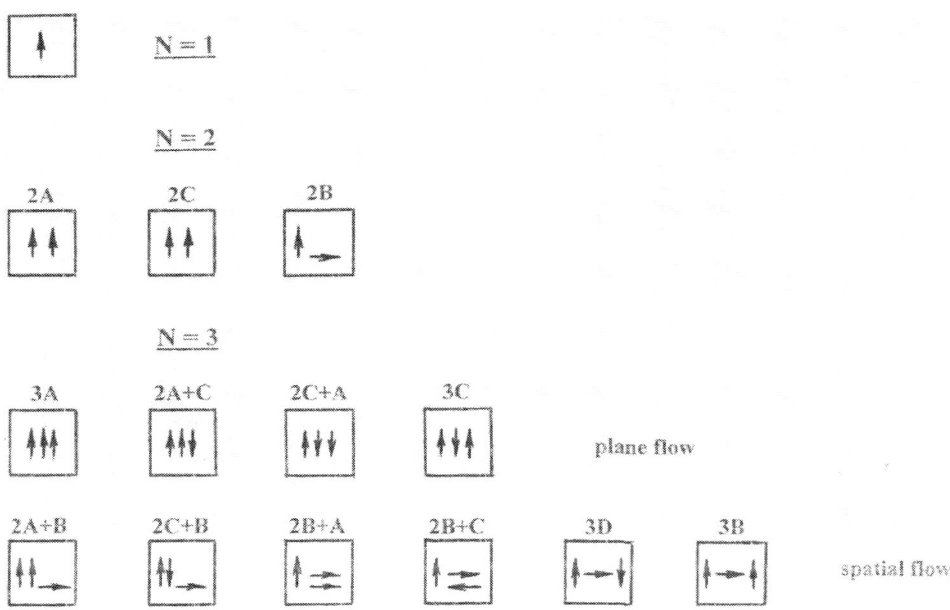

Figure 3.25. Possible combinations of the basic routes for three pass ECAE.

At subsequent passes, they flow together with the material as the accompanying coordinate system. Final orientations of shear bands induced at different passes depend on the material distortion during subsequent passes. Therefore, different routes and passes produce specific families of shear bands at the meso and macro scales. Their number, orientations and intersections affect material structures, textures and properties and present important characteristics of the processing mechanics. Calculated families of shear bands after four passes ECAE of long billets via routes A, B, C and D are shown in Figure 3.26 (positions a, b, c and d, correspondingly). Numbers 1, 2, 3 and 4 designate positions of shear bands induced at first, second third and fourth pass, respectively. For route C, there is only one family of micro-shear bands with any number of passes. After four passes ECAE via routes A, B and D, there are four different systems of shear bands. Angles between shear bands increase from route A to route B and to route D. In the last case, shear bands are also more spatially and uniformly distributed. If one supposes that spacing between shear bands induced at any pass remain identical, their intersections form smaller and more equiaxed fragments for route D than for routes B, A and C. When a number of passes increases (N > 4), new shear bands for route D coincide with previous shear bands induced during pass (N – 4). Such situation is repeated after each four passes. For routes A and B, new shear bands take positions of previous (N – 4) shear bands shown, respectively in Figure 3.26a,b, whereas the previous ones rotate to the billet axis and become quickly indistinguishable.

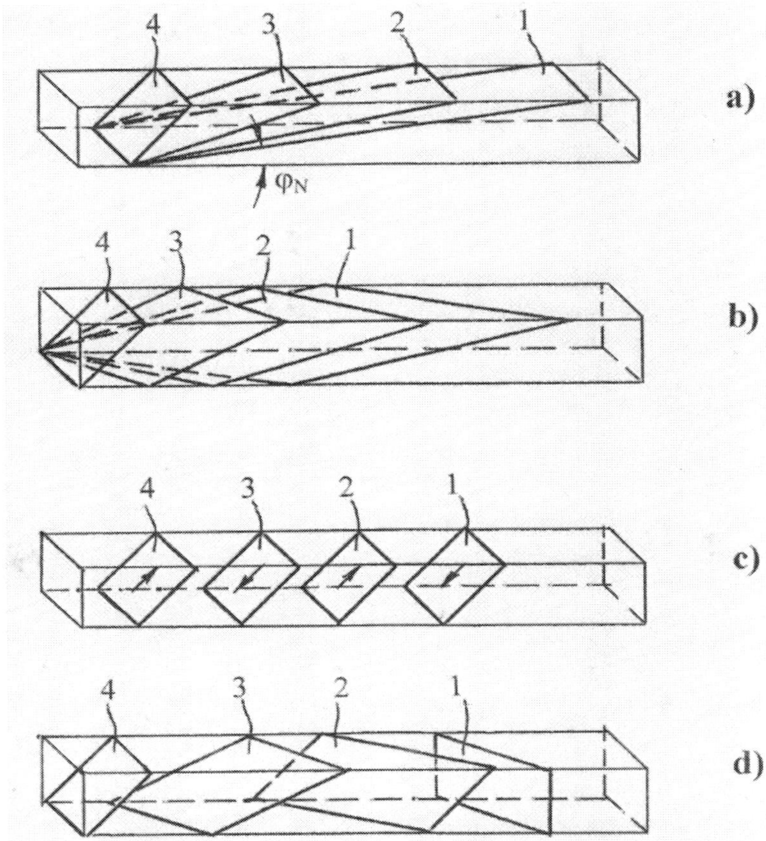

Figure 3.26. Orientations of shear bands induced at first (1), second (2), third (3) and fourth (4) passes of ECAE via routes A (a), B (b), C (c) and D (d) for long billets.

Consider the orientation of shear bands for flat billets. For two successive passes N and (N + 1) without rotation, distortions of shear bands from the Nth pass are caused by the (N + 1)th pass. However, with billet rotation 90°, shears along perpendicular planes of the (N +1) pass does not affect the orientation of shear bands induced at the Nth pass. Therefore, during multi-pass processing of flat billets there are two independent systems of shear bands into X and Y directions. Calculated orientations of shear bands after four passes via routes A, B, C and D are shown in Figure 3.27 (positions a, b, c and d, correspondingly). For routes A and C orientations of shear bands are the same as for long billets. Similarly, with the increase of the number of passes via routes A and B, shear bands progressively rotate toward the extrusion direction and soon primary shear bands become difficult to distinguish. Their intersections for route A result in thin long fragments. Route C with one family of shear bands produces plain fragments with any number of passes. After four passes via route D, the distortions are zero and two families of perpendicular shear bands coincide with two sets of processing shear planes. For route B, intersections of numerous shear bands oriented along X and Y axes are more favorable for development of three-dimensional and finer fragments than routes D, A and C. This conclusion is different than in the case of long billets

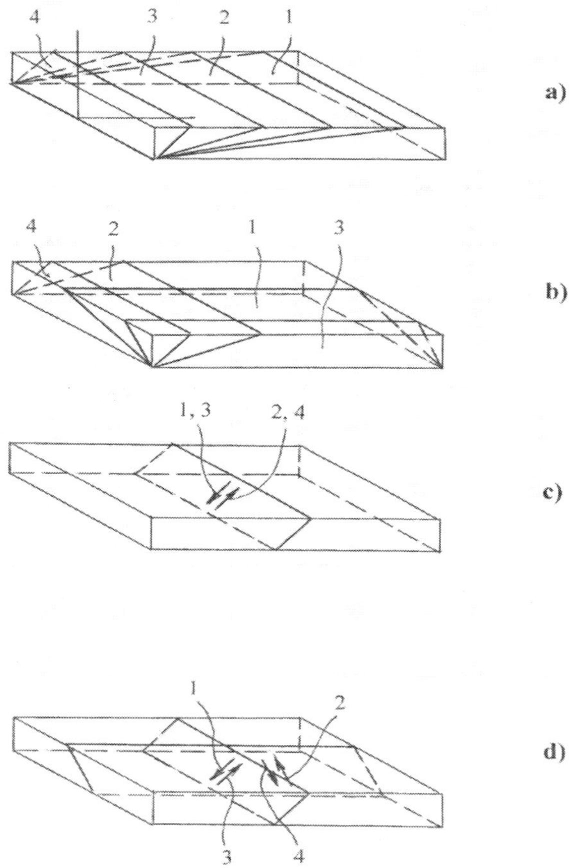

Figure 3.27. Orientations of shear bands induced at first (1), second (2), third (3) and fourth (4) passes of ECAE via routes A (a), B (b), C (c) and D (d) for flat billets.

The present analysis performed for ECAE through sharp channels with low contact friction cannot be applied to ECAE with noticeable friction ($\tau >> 0$) or with round corner channels when a plastic zone is extensive. In these cases, simple shear along circular slip lines of the central fan plays the important role and shear bands may be oriented along billet axes. An example of shear bands after four passes ECAE of long billets via routes A, B, C, D with large contact friction $\tau \rightarrow k$ and a tool angle of $2\theta = 90°$ is shown in Figure 3.28 (positions a, c, b and d, correspondingly). For all routes, intersections of shear bands produce long lamellae or fibrous structures which are not effective for structure fragmentation. Multi-pass ECAE with intermediate friction conforms to different combinations of shear planes and directions. However, regardless of processing route, the decrease of strains and their redistribution between three directions at each pass do not promote the evolution of shear bands into high angle boundaries. The slip line analysis demonstrates that structural effects of SPD are sensitive to processing mechanics and careful control of deformation characteristics is essential to attain adequate, repeatable and effective results.

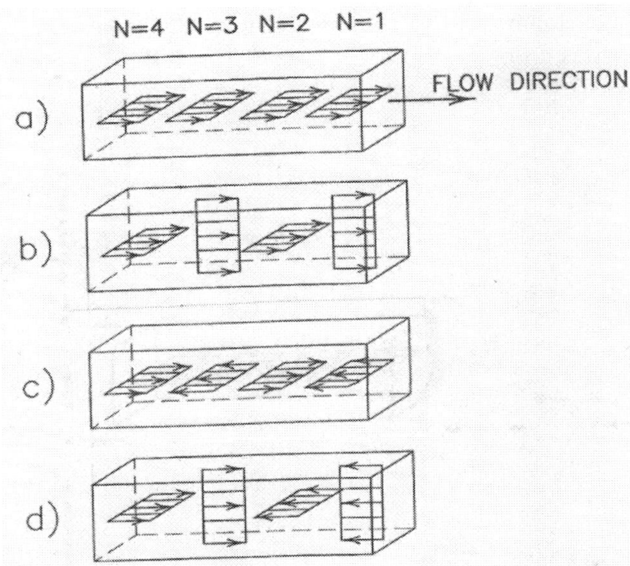

Figure 3.28. Orientations of shear bands induced at first (1), second (2), third (3) and fourth (4) passes during ECAE of long billets with maximum friction via routes A (a), B (b), C (c) and D (d).

3.5.3. 'End Effect'

During multi-pass ECAE, 'dead" metals zones on the front and back billet ends change their location and affect the strain distribution in the processed material. Positions of corresponding zones and distortions of coordinate grids were determined analytically with computer graphics. Assuming the single slip line solution ($\tau << k$), calculated results for ECAE with a tool angle of $2\theta = 90°$ via route A where the largest "end effect" should be expected are presented in Figure 3.29 for a billet with a length-to-thickness ratio of 4. Positions *a* and *b* show distortions after the second and fourth pass, respectively. For $N \geq 2$, there are three different areas of distortions induced by shears at 2N, 2(N – 1) and 2(N – 2) passes (the difference between these numbers corresponds to two passes necessary to cross end zones). The most intensive shear $\gamma = 2N$ takes place at the billet center. These theoretical results are in close agreement with experimental pictures of etched macrostructures detected for aluminum alloy 3003 after two and four passes (Figure 3.29, positions *c* and *d*). The central area of uniform and intensive strains increases in proportion with the billet length-to-thickness ratio and the number of passes N. The optimal case of a uniform strain distribution may provide ECAE of continuous or semi-continuous billets when the billet length-to-thickness ratio becomes very large.

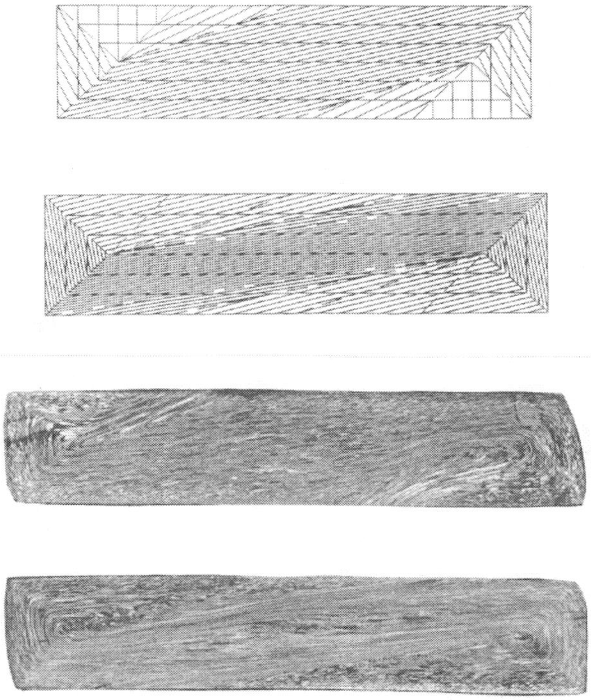

Figure 3.29. Comparison of calculated distortions of coordinate grids for ECAE of long billets (c/a = 4) via route A after second pass (a) and fourth pass (b) with macroetched flow planes of AA 3003 after second pass (c) and fourth pass (d).

For route C, the distortion of the coordinate grid periodically repeats the first pass for odd numbered passes and returns to the original grid for even numbered passes. Positions of rigid zones remain fixed at the billet ends. Their volume can be reduced by additional rotation $\varphi_2 = 180°$ about axis Y (Figure 3.21a) after even numbered passes. Route B is again similar to route A with the exception that areas of strain non-uniformity induced by the 'end effect' are turned at an angle of 45° around the Y axis and are symmetrically redistributed along opposite billet edges. However, route D is not similar to route C. For route D, rigid zones at the billet ends after four passes are only partly restored. To provide uniform processing, the 'end' effect should be considered in particular cases. Related computer modeling was developed for long billets in (Cai et al. 1998).

3.6. Stability of ECAE

For optimal ECAE, the deformation mode is simple shear inside a narrow plastic zone. This zone is thin macroscopically rather than microscopically and stable and steady macro flow provides uniform straining. However, in some cases localized shear results in an unstable flow and a non-uniform strain distribution along the billet length (Segal, 1999). From equation (3.2), the instability condition is:

$$dp_e \leq 0 \text{ or } \sin\theta dk - 2kd\theta \leq 0 \tag{3.20}$$

If $d\theta = 0$, the plastic zone and flow are stable. Because of geometrical limitations, only the case $d\theta < 0$ is possible. Then eq. (3.20) gives $dk \leq 0$. Therefore, material softening ($dk < 0$) is the necessary condition for instability ($d\theta < 0$) during ECAE. Such localization at the macro scale is similar to micro localization along fixed shear bands (Figure 1.10b) described by Eq. (1.36). Figure 3.30a shows the slip line field modification for such a situation. At the beginning, simple shear is realized along the slip line AO. The line AO is supposed to move with a speed *c* in the extrusion direction. After the interval of time Δt, the displacement is $OA_1 = ct$, the shear plane takes position AA_1 and the shear direction changes from angle θ_o to angle θ. On the hodograph (Figure 3.30b), V is the extrusion speed, a segment OC presents the velocity c and a segment CV_1 defines the angle φ of the slip line field:

$$\tan(\theta - \varphi) = (1 - c/V)\tan\theta/(1 + c/V)$$

Simple shear along AA_1 is:

$$\gamma' = 2\cot\theta/(1 - c/V) = \gamma/(1 - c/V) \qquad (3.21)$$

As $0 < c/V < 1$, shear γ' may significantly exceed shear γ from eq. (3.3) for steady flow. In accordance with eq. (3.5), for materials exhibiting dependence of the flow stress k on strain γ, strain rate η and temperature T, the correct choice of c should minimize the plastic work. The minimization condition is:

$$(\partial k/\partial\gamma)(d\gamma/dc) + (\partial k/\partial\eta)(d\eta/dc) + (\partial k/\partial T)(dT/dc) = 0 \qquad (3.22)$$

For strain hardening materials, $\partial k/\partial\gamma > 0$ and, additionally, eq. (3.21) gives $d\gamma/dc > 0$. Thus, the first part of (3.22) is positive and gradually decreasing function of γ. From eq. (3.4), strain rate inside a shear band of thickness δ ($\delta \sim$ const) is $\eta = [v]/\delta$. It can be seen from Figure 3.30 that $\theta_o > \theta$, $V < V_1$ and the velocity discontinuity [v] slightly increases with c. Together with eq. (3.5) that gives $\partial k/\partial\eta > 0$ and $d\eta/dc > 0$. So, the second part of eq. (3.22) is also positive. Further, eq. (3.5) shows that the temperature effect of the flow stress $\partial k/\partial T < 0$ whereas for $0 < c < V$ the temperature of the plastic zone increases with c and $dT/dc > 0$. Therefore, the third part of (3.22) is negative. Finally, a solution of eq. (3.22) for c should equalize temperature softening against strain and strain rate hardening. An additional restriction is

$$k\cot\theta < k_o\cot\theta_o \qquad (3.23)$$

because at any moment of unstable flow the shear plane AA_1 will return to the original position AO if that provides the lower plastic work.

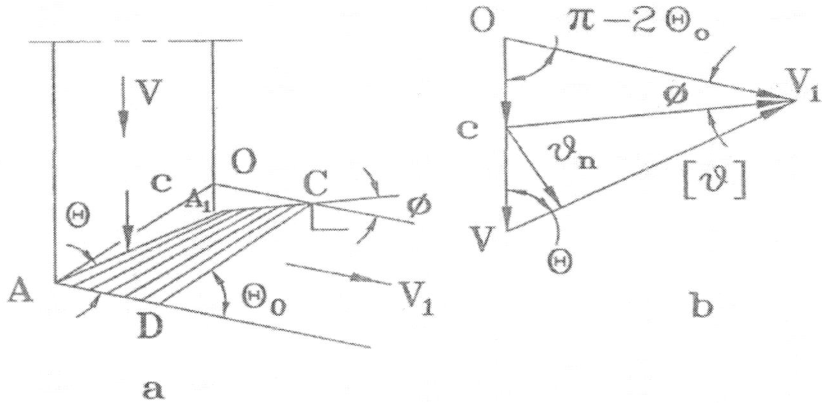

Figure 3.30. Non-steady flow during ECAE: a - slip line field; b - velocity hodograph.

A cumulative result of all these effects may be reduced into a few important cases:

1. Strain hardening is strong and/or adiabatic heating is insignificant. This is a case of the low number of passes, slow speed or sufficiently high temperature. Under these circumstances, c= 0 and flow is stable. This is the preferable case for material processing;
2. Strain rate hardening and adiabatic heating are comparative. This is the case after a sufficiently large number of passes, high speed or low temperature. Eq. (3.22) is satisfied for c << 1 while eq. (3.23) may be fulfilled only near the original position. Localization takes place mainly at the micro scale and manifests narrow separated bands of intensive shear with some roughness or waviness of the top billet surface. An example of the small scale localization is presented in Figure 3.31a for OF copper 101 after three passes of ECAE via route B with 2θ = 90°, V = 0.25 mm/sec. Macroetching of the flow plane reveals numerous fine shear bands oriented along the shear plane. The related microstructure (Figure 3.32a) demonstrates a large number of very thin micro shear bands (SBs) and their clusters. Figure 3.32a comprises shear bands SB_1 induced during the second pass and shear bands SB_2 induced during the third pass. The corresponding 'load –displacement' diagram of ECAE with small scale localization detects a slightly serrated flow (Figure 3.33, curve 1);
3. Strain rate hardening is insignificant while adiabatic heating plays a key role. This is the case after a large number of passes with sufficiently high speeds. eq. (3.22) gives values for *c* comparative with V whereas eq. (3.23) is satisfied during a finite deformation step. Under these circumstances, localized flow is realized along consecutive shear bands which form a wide deformed area. At the end of each step when eq. (3.23) is no longer satisfied, plastic flow transfers almost instantaneously to the original position. This appears in alternating areas of deformed and non deformed material with sharp teeth at the top billet surface. An example of such large scale localization is shown in Figure 3.31b for OF copper 101. All processing conditions were the same as in Figure 3.31a except that the extrusion speed was increased to V = 25 mm/sec. The macrostructure of the flow plane was detected in polarized light after etching. Direct measurements give the angle φ = 24° which corresponds to c = 0.45V, $\gamma' = 3.63$ and $\Delta\theta = (\theta_o - \theta) = 1.5°$. Details of the microstructure taken from the

transition area of Figure 3.31b are shown in Figure 3.32b. In accordance with the theoretical model, shear bands are very thin and long straight lines strictly oriented along slip lines at an angle of 45° to the billet axis. The 'load-displacement' diagram during ECAE with large scale localization (Figure 3.33, curve 2) is below the low speed diagram (curve 1) confirming significant adiabatic heating. Short amplitude waviness of curve 2 conforms to the number of shear cycles. It should be noted that even a small perturbation in the angle $\Delta\theta$ is sufficient to develop strong heterogeneity in plastic flow at the macro scale. An additional problem may present nucleation and growth of cracks along shear bands from the sharp tooth edges on the top billet surface. Usually, large scale localization is not observed for sufficiently large contact friction or back-pressure p_0 when the extrusion pressure from eq. (3.13) $p >> 2k$ and free surface A_1O (Figure 3.30a) cannot exist.

Figure 3.31. Macrostructures of copper 101 after three passes ECAE via rote B: a - steady flow, V = 0.25 mm/sec; b - non-steady flow, V = 25 mm/sec.

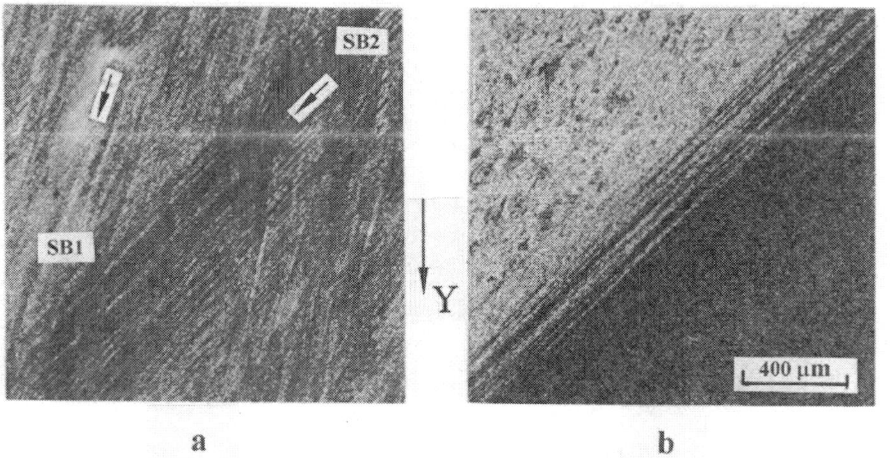

Figure 3.32. Microstructures of copper 101 after three passes via route B: a - steady flow, V = 0.25 mm/sec; b - non-steady flow, V = 25 mm/sec.

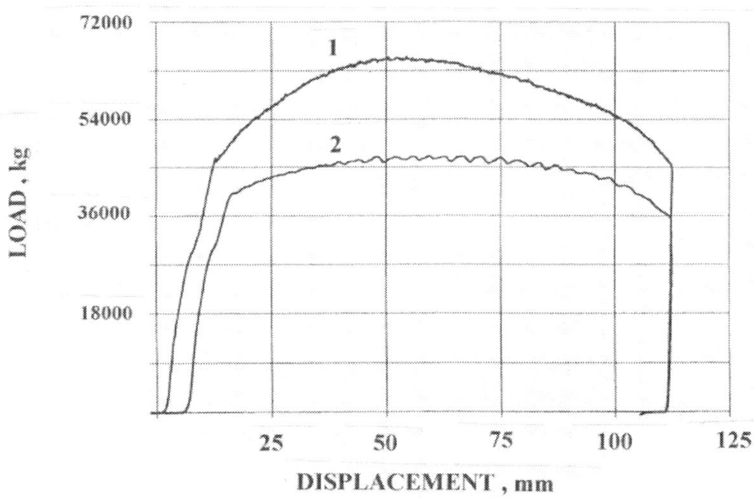

Figure 3.33. 'Load-displacement' diagrams of cooper 101 during third pass of ECAE via route B: 1 - steady flow, V = 0.25 mm/sec; 2 - non-steady flow, V = 25 mm/sec.

Chapter 4

ECAE PROCESSING

4.1. CONTACT FRICTION IN CHANNELS

It was shown in foregoing that the stress-strain state and the extension of the plastic zone depend on contact friction on the channel walls. To achieve simple shear deformation and effective processing, boundary conditions should be analyzed for inlet and outlet channels. At the beginning of extrusion, the well lubricated billets are placed into the inlet channel. Actual friction in this channel is defined by the normal pressure σ_n and a contact area between material and tool. Assuming that friction is small ($\tau << k$) and a plastic stress state inside the inlet channel approximates to linear compression, the normal pressure σ_n on the channel walls at a cross-section z = const (Figure 3.21) is:

$$\sigma_n \sim - (p_l - 2k)$$

where p_l is the axial pressure at this cross-section. Consider two specific cases. In the first, the pressure $p \sim 2k$ and the normal stress σ_n is low. For long billets with large length-to-thickness ratios c/a >> 1 (Figure 4.1), contact with the tool occurs mostly due to buckling. With good lubricants, such irregular local contact provides insignificant friction. In the second case, the axial pressure p >> 2k, the normal pressure on the tool wall σ_n>> 2k and plastic contact approximates to the entire billet surface *f*. With the same lubricants, that results in large friction force and a significant increment of the pressure Δp along the channel length. The extrusion pressure p_e is:

$$p_e = p + \Delta p \tag{4.1}$$

Here p is the axial pressure (3.13) at the channel exit and Δp is the pressure increment along the channel length. In accordance with experimental observations, after beginning of the steady extrusion, the total load and the pressure Δp decrease in near linear proportion with the channel length L. That allows one to suppose that effective plastic friction τ_l is uniformly distributed and Δp may be calculated with the formula $\Delta p \sim \tau_l f$. When τ_l is known for specific processing conditions, the maximum increment of the extrusion pressure in the stationary rectangular channel with four friction walls (Figure 4.1 a) is:

$$\Delta p/2k = [(2c' - 1)(1 + b')\tau_1/2k]/b' \qquad (4.2)$$

Here k should be found from formulae (3.7), (3.11) for each pass and parameters $c' = c/a$ and $b' = b/c$ are relative billet length and width (Figure 3.21). In particular, the case $c' >> 1$, $b' = 1$ corresponds to long square billets and the case $c' = b' >> 1$ corresponds to flat billets. Formulae (4.1), (4.2) show that extrusion pressure p_e may be significantly bigger than the material yield stress even with low friction τ_1. Therefore, the case of full plastic contact is not practical and should be eliminated. The pressure p/2k versus the coefficient of plastic friction $\tau/2k$ in both channels calculated with formula (3.13) for tool angles of $2\theta = 90°$, 120° and 150° are presented in Figure 4.2 (solid lines). This diagram demonstrates that for a tool angle $2\theta = 90°$ the local contact with $\sigma_n \sim 0$, $p \sim 2k$ can be attained only for very low friction whereas larger tool angles do not impose such strong restrictions. As both components p_1 and Δp in equation (4.1) depends strongly on contact friction in the outlet channel, low friction in this channel is especially important. Dashed lines in Figure 4.2 relate to zero friction in the outlet channel and friction $\tau_1 = \tau$ in the inlet channel providing a lower bound for the pressure p_1.

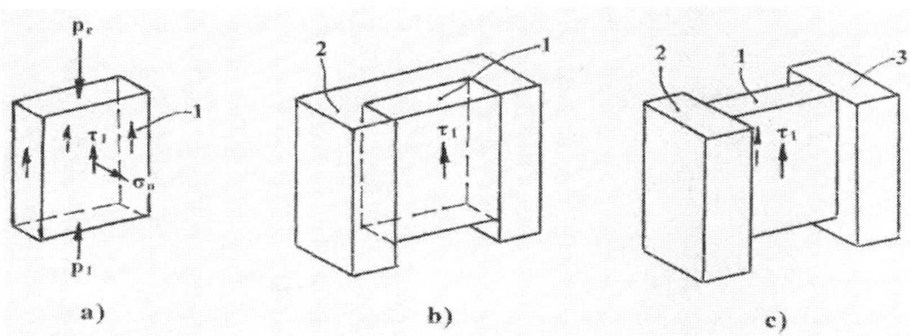

Figure 4.1. Distributions of friction in inlet channels with stationary walls (a), three movable side walls (b) and two movable side walls (c).

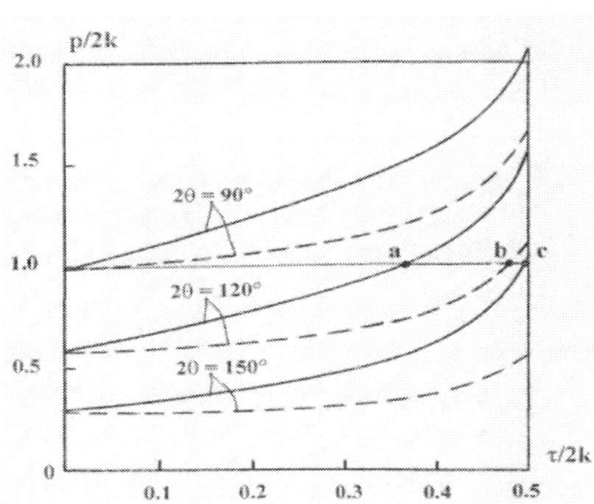

Figure 4.2. Effect of friction $\tau/2k$ on the pressure p/2k for equal friction in channels (solid lines) and zero friction in the outlet channel (dashed lines) with tool angles $2\theta = 90°$; 120° and 150°.

However, in contrast to the inlet channel, lubrication of the outlet channel is a challenging problem. Though the top and side surfaces of the billet do not experience noticeable friction, it is difficult to provide low friction along the bottom wall of the channel.

The sharp change in the extrusion direction, high contact pressure σ_{n1} (Figure 3.8a) and intensive slip squeezes out the lubricant from this surface. Even a small 'dead' metal zone at the channel corner destroys the lubrication layer. An additional complication is that, in reality, simple shear is performed in a series of small micro and macro localized steps which expose the atomically clean material along the bottom tool wall. Despite of using the best lubricants, heavy scratches, local sticking and galling are often observed at the bottom billet surface resulting in excessive pressure, tool wear and poor material surfaces. For billets with large length-to-thickness ratios this also leads to low punch stability. For applications of ECAE as a new materials processing technology, these problems should be resolved by specific techniques. One of the effective and practical ways to reduce contact friction is via movable channel walls (Segal et al. 1994; 1980; 1981; 1995).

4.2. Movable Channel Walls

4.2.1. Inlet Channel

Two versions of movable walls in the inlet channel are shown in figures 4.1b, c. In one case (Figure 4.1b), the inlet channel is formed by one stationary die wall and rectangular slot of the slider 2, which moves together with the billet 1. Therefore, friction is eliminated along three channel walls. The increment of the punch pressure at the beginning of extrusion is:

$$\Delta p/2k = (c' - 1)\tau_1/2k \qquad (4.3)$$

In another case (Figure 4.1c), two side walls of the inlet channel are formed by movable sliders 2, 3 whereas back and front die walls are stationary. Correspondingly, the increment of punch pressure is:

$$\Delta p/2k = (2c' - 1)\tau_1/2k \qquad (4.4)$$

It is informative to compare results of formulae (4.2) – (4.4). In all cases, the extrusion pressure increases with the billet length-to-thickness ratio c'. To provide better strain uniformity due to the "end effect", this ratio should be sufficiently large. Practically, c' is selected to be in the range between 5 and 8. The increment $\Delta p/2k$ is almost twice as large for Figure 4.1c than for Figure 4.1b. In the stationary channel (Figure 4.1a), the extrusion pressure also depends on the billet width-to-thickness ratio b'. However, this ratio does not affect the extrusion pressure with movable channel walls. Calculated results for typical conditions $c' = 6$, $\tau_1/2k = 0.15$ are shown in Figure 4.3 as a function of b'. Three characteristic situations are outlined: (I) long billets ($b' \sim 1$); (II) plate-like billets ($b' >> 1$); (III) strip-like billets ($b' << 1$). It is evident that ECAE of long and, especially, strip-like billets in stationary channels leads to a multifold increase of the extrusion pressure in comparison with the material yield stress. In accordance with equation (4.1), the stress increase Δp along the channel length may significantly exceed the pressure p_1 and account for the most of the punch pressure p_e. In such cases, especially for large long billets and hard materials, ECAE should be performed in dies with movable walls of the inlet channel. In contrast, for flat billets, two movable side walls provide just moderate reduction of the extrusion pressure and simple dies with stationary inlet channels can be used for many practical applications.

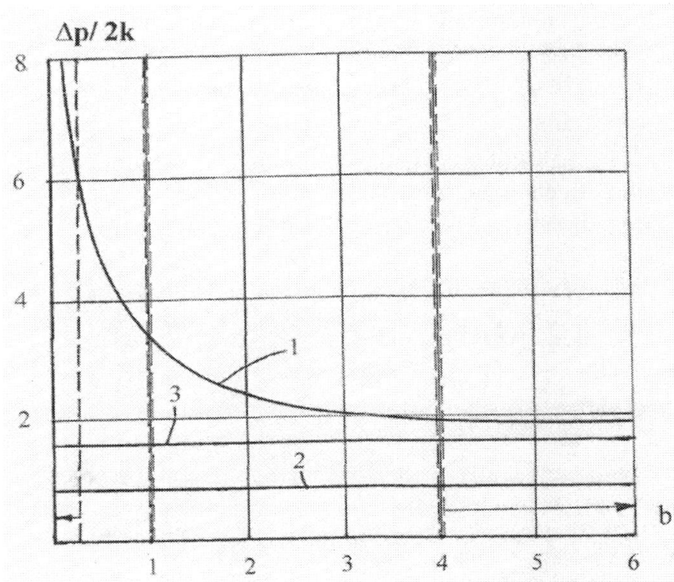

Figure 4.3. Effect of billet width-to-thickness ratio b' on the increase of pressure along inlet channel (c' = 6, $\tau/2k = 0.15$) with stationary channel walls (1), three movable walls (2) and two movable walls (3).

4.2.2. Outlet Channel

Most problems related to poor lubrication of this channel can be eliminated using a movable slider along the bottom channel wall. Two slider designs with open and closed channels are possible (Figure 4.4). In the open channel design (Figure 4.4a), billet 1 contacts the slider 2 along the bottom surface. The extrusion pressure p is transferred to the slider by the normal stresses applied at the boundary ACD of the slip line field. That develops Coulomb friction μp between slider 2 and guide plate 3. To compensate for this friction, some slippage and tangential stresses τ_2 (plastic friction) should also arise between material and slider along ACD. The balance of forces acting on the slider gives:

$$\mu p = \tau_2 l'$$

where μ is the Coulomb friction coefficient between slider and guide plate, $l' = l/a$ is the ratio of the contact area length l to a billet thickness a. Friction τ_2 in the outlet channel together with friction τ_1 in the inlet channel define the plastic zone AOC shown in Figure 4.4a.With a proper tool design, quantities $\tau_1/2k$, $\tau_2/2k$ and μ are small. Under these conditions, the slip line field AOC can be considered as a modification of the "zero solution" of Figure 3.1a, b when $\tau_1 = \tau_2 = \mu = 0$ and the plastic zone is a single slip line. Using the perturbation method for slip line fields (Spenser, 1962) and omitting intermediate calculations, the final results are:

$$\tau_2/2k \sim \mu, \quad p/2k \sim 1 + \tau_1/2k + \mu \qquad (4.5)$$

Formulae (4.5) shows that the slider with an open outlet channel does not fully eliminate plastic friction τ_2 between the billet and the tool, but replace it by Coulomb friction μ between stationary and movable die parts. When the friction coefficient μ is low, there is sufficient room for friction τ_1 to form the local plastic contact in the inlet channel and to maintain the punch pressure p_e just slightly higher than the material flow stress. As contact friction is not a stable phenomenon, a certain misalignment in velocities between the billet and slider are often observed during ECAE with this design. In some cases, a competition between plastic (τ_2) and elastic (μ) friction may result in interruption of the slider's movement and material extruding like in a stationary channel.

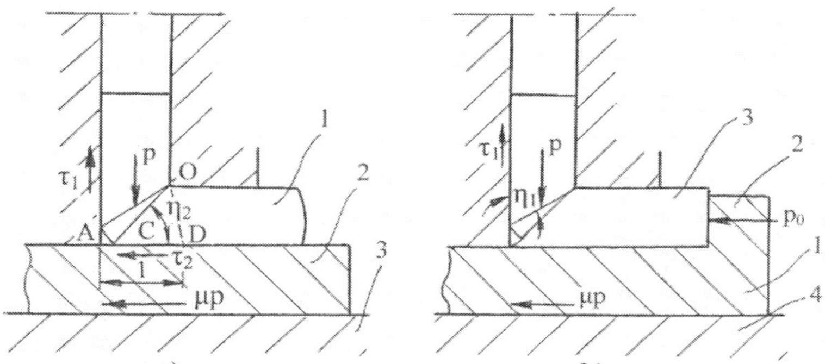

Figure 4.4. Movable bottom die walls in open (a) and closed (b) outlet channels.

A more advanced design is a slider with a closed outlet channel (Figure 4.4b). In this case, the slider 1 has a lag 2 that overlaps the channel and maintains tight contact with the front billet end 3 during extrusion. Therefore, the billet and slider always have the same speed, and contact friction on the bottom billet surface τ_2 is zero at any moment. Another advantage of this design is that an additional back pressure p_o is applied to the billet's front end to overcome Coulomb friction μ between the slider 1 and a guide plate 4 for which formula (4.5) gives:

$$p_o / 2k = \mu \qquad (4.6)$$

Full elimination of friction and development of the moderate back pressure prevent the formation of gaps and "dead" metal zones at the die corner. The billet retains a near correct rectangular shape after each pass and surface defects do not originate on the bottom surface. These improvements play an important role during multi-pass processing as that was confirmed by finite element modeling (DeLo et al. 1999).

With the movable bottom wall of the outlet channel, the inlet channel may be performed as stationary (Figure 4.1a) or with two movable side walls (Figure 4.1c).

4.3. ECAE of Batch Billets

As a materials processing technology, ECAE may be realized in different ways depending on billet shape and length-to-thickness ratio $c' = c/a$ (Figure 3.21). Billets are considered *batch* when the ratio c' is from 4 to 8 and *continuous* when this ratio is much larger than 10. The most practical batch billets are long square (a = b), long cylindrical and square flat billets (c = b) when multi-pass processing can be performed in one set of dies. From figures 4.2, 4.3 one can see that ECAE of *long square* billets (a = b << c, Figure 3.21a) in stationary channels even with low friction requires a high punch pressure p_e that may exceed significantly the material yield stress. Some experimental results (see, for example, Iwahashi et al. 1998) show that the relative punch pressure for pure aluminum is as high as $p_e/2k \sim 9$ ($p_e \sim 2$ GMPa). For most materials at cold and warm processing temperatures, such pressures are near the limit of the punch's strength and stability. Therefore, effective ECAE of long billets requires dies with a movable bottom wall in the outlet channel and two movable side walls in the inlet channel to reduce both pressure components p_l and Δp in formula (4.1). Because of friction between die parts, the punch pressure p_e for a tool angle of 90° can be practically reduced to about 1.25 -1.5 of the material ultimate tensile stress. This die design is especially effective for high strength and difficult to deform materials. After extruding, sliders are moved up to the original position and release the billet. During multi-pass ECAE, billet ejection after each pass is an important operation to renew lubricant and reshape the billet if necessary. To facilitate reinserting of long square billets without machining or rolling, both channels should have the same width b (Figure 3.21a) but a slightly different thickness (a –Δ) in the outlet channel and (a +Δ) in the inlet channel. That provides a necessary clearance Δ between channel walls and the billet after 90° rotations via routes B and D.

For *long cylindrical* billets, ECAE cannot be performed with movable walls and solid dies should be used. Therefore, contact friction may result in especially high extrusion pressures. Published data for ECAE of pure copper (Mishra et al. 2006) correspond to a punch pressure $p_e = 2700$ MPa ($p_e/2k \sim 7$) that is close to the ultimate strength of tool steels. Despite an effective MoS_2 - based lubricant, such high pressures for a relatively soft material demonstrates that in some areas contact friction is close to the material shear stress and the deformation mode is far from simple shear. Another problem is the billet ejection from the die. The ejection is usually performed by the following billet or a dummy billet with distortions of billet ends. These ends should be cut off after each pass resulting in additional operations and material waste.

Flat (or plate) square billets (c = b >> a, Figure 21b) present a few important advantages (Segal, 2008). As was previously noted, simpler dies with the solid inlet channel and a movable bottom wall in the outlet channel may be used in this case. The slider design with the closed channel (Figure 4.4b) preserves the rectangular billet shape and eliminates billet reshaping after each pass. Due to effective techniques of billet ejection and lubrication, multi-pass processing can be performed pass-by-pass without preheating. In comparison with long billets of identical length and thickness, the volume of flat billets processed is from 4 to 8 times larger. However, the basic processing routes for flat billets (Figure 3.21b, Table 2) do not induce three dimensional systems of near orthogonal shear bands which are the most effective for development of fine equiaxed structures after a low number of passes. When top and bottom billet surfaces remain the same during multi-pass processing, non-uniformity of strains and residual stresses are accumulated through the billet thickness and specific surface defects may arise at the top billet surface. These shortcomings can be eliminated by more complicated routes with additional rotations about axis X and Y (Segal, 2008). One such route, designated here as route E, comprises eight passes. The first four passes are performed as the basic route D with rotations of 90° about axis Z into the same direction inducing zero distortions and two families of shear bands into the material (Figure 3.27d). Then, the billet is rotated 180° about axis Y or axis X that changes the orientations of shear planes and alternates top and bottom billet surfaces. Next four passes via route D restore distortions again and induce two families of shear bands mutually perpendicular to the former ones. Therefore, route E conforms to an optimal route for structure refinement and a homogenization of properties throughout the material. Another processing route, route F, eliminates irregularities of the structure near the top of the billet's surface left by punch impressions at each pass. These impressions are traces of shear planes created by the sharp change of the flow direction. For route D, such abrupt changes of material distortions into perpendicular directions cause singularities of structure and residual stresses which manifest as material weakening and, sometimes, cracking. Route F includes the second rotation of 180° about axis Y together with the first rotations of 90° about axis Z performed into alternate directions after each pass. Therefore, punch impressions are located at opposite sides of the billet's top and bottom surfaces and do not intersect each other, whereas material distortions and shear band orientations remain similar to an ordinary route D schedule (Segal, 2008). Other routes can also be introduced in special cases. In particular, two modifications of the basic route B similar to routes E and F are obvious.

Modifications of ECAE were also suggested to execute multi-pass processing of batch billets in the *semi-continuous* manner without ejecting them from tool. One of these (Nakashima et al. 2000; Olejnik et al. 2005; Rosochowski, 2005) consists of extruding billet-by-billet through a few channels machined inside a solid tool (Figure 4.5a). It is supposed that simple shear takes place simultaneously in the billet along the crossing planes of the channels, which orientations correspond to a certain processing route. Similar techniques were also used in equal-channel multi-angular extrusion (Varyukhin et al. 2002) and ECAE through parallel channels (Azushima et al. 2002). In all these cases, it is difficult to provide low contact friction because several shear planes and the large channel length lead to high contact and extrusion pressures. The most likely slip line fields are central fans with 'dead metal' zones 'r' at tool corners and shears into an extrusion direction as shown in Figure 4.5a. The high extrusion pressure, material sticking to die and low tool life are a matter of concern and should be resolved. Another approach is multi-pass ECAE of a batch billet through two or three mutually perpendicular channels (Figure 4.5b). The process resembles "cross-extrusion" (Figure 2.15a) except that channels are partly blocked by stationary punches. In the original position (Figure 4.5b), punches 1, 2 are fixed, punch 3 is free and punch 4 extrudes billet 5 into the channel 6. After completion of the first pass, punches 1, 4 are fixed, punch 2 is free and punch 3 extrudes the material into the channel 2. Like ordinary ECAE, such operations can be repeated a number of times at ordinary presses using a rotary die (Nishida et al. 2001) or multi-plunger presses (Azushima, 2002). However, the processing mechanics is different. For cylindrical channels, flat punch ends do not provide the geometry of channel intersections necessary to form a thin shear zone. Arising "dead" metal areas 'r' adjacent to punches prevent lubrication of bottom channel walls. Multi-pass processing without lubricant lead to material galling to the tool, high extrusion pressure and a poor billet surface. The practical importance of these semi-continuous techniques of multi-pass ECAE is unclear and, probably, may be restricted to small samples and soft materials.

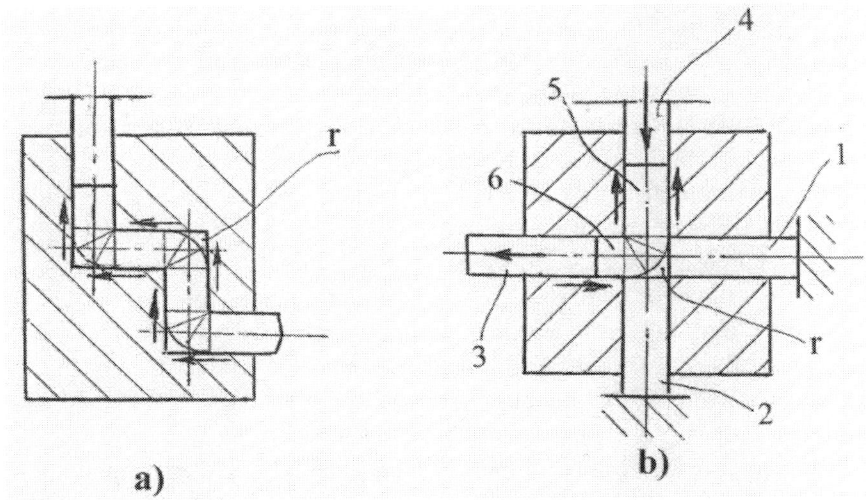

Figure 4.5. Semi-continuous multi-channel (a) and multi-pass (b) ECAE.

Modifications of the channel geometry using conjugation radii instead of sharp channel corners were also suggested. It was shown that radii at external channel corners or round corner channels (Iwahashi et al. 1996) slightly reduce the extrusion pressure and improve lubrication but lead to non-uniform stress/strain states and to a complicated deformation mode. Also, round corner channels cannot be performed with movable die walls. The effect of radii at internal channel corners was also analyzed in (Luri et al. 2006). The optimal channel geometry with increased and more uniform strains was found by the finite element method. This die configuration can be used for continuous ECAE or semi-continuous ECAE billet-by-billet. For ECAE of batch billets with their ejection after each pass, it may present a problem because of flashes forming between the punch and the internal channel corner.

4.4. Tool Design

ECAE of batch billets is realized on ordinary extrusion presses using special dies. The simplest tool for long billets (Figure 4.6a) includes a die block 1 with two intersecting channels 2, 3 and punch 4. This design is often used as a laboratory tool. Channels may have circular or rectangular cross-sections formed in split inserts and bolted or pressed together by a container. Although cylindrical channels are easier to fabricate than rectangular channels, the cylindrical billets cannot be ejected without distortion of the billet's back end (Figure 4.6a) (Azushima et al. 2002). For rectangular channels, the punch should be advanced to the top level of the second channel (a dashed line on Figure 4.6a). After that the billet can be withdrawn or ejected from the die. Correspondingly, the extrusion press should provide a control of the stroke.

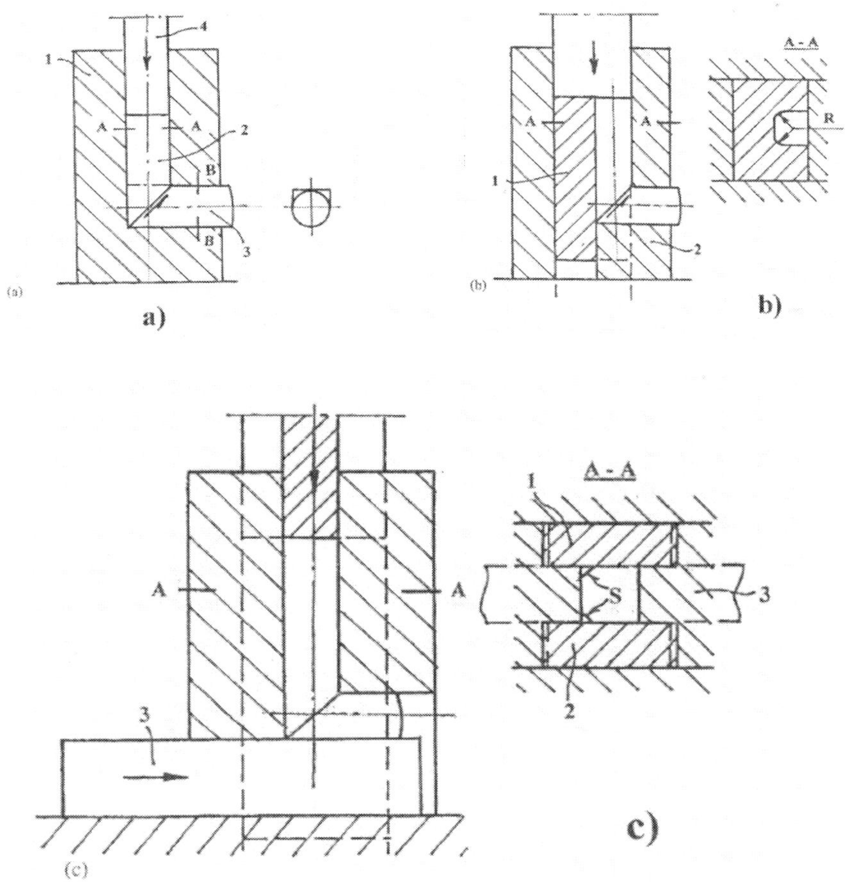

Figure 4.6. Die designs for long billets: a - solid tool block; b - three movable die walls into the inlet channel; c - two movable side wall into the inlet channel and movable bottom wall into the outlet channel.

Advanced dies include movable channel walls. The die with three movable walls of the inlet channel fabricated into the slider 1 (Figure 4.6b) allows extrusion of long billets. Therefore, friction acts only along a stationary wall and the extrusion pressure is significantly lower than with all stationary channel walls. As a punch cross-section overlaps billet and slider, it can be made sufficiently large to provide punch strength and stability necessary for processing of hard materials and long billets. Channel radii R prevent the flash formation along the bottom wall 2 of the outlet channel. The slider is connected to the punch by drafts and operated by the press to open the die, insert billet, perform a stroke and return to the original position. Working parts of the die subjected to high pressures and wear are fabricated from heat treated tool and die steels. Techniques of surface hardening and coating may be applied to the most critical parts. These dies require effective lubricants to prevent the material from sticking to the bottom wall of the outlet channel.

A movable bottom wall of the outlet channel (Figure 4.4) eliminates the sticking and reduces the extrusion pressure. A more advanced design consists of movable walls in both channels. Figure 4.6c presents a die with two movable side walls 1, 2 in the inlet channel and a movable bottom wall 3 in the outlet channel. Slider 3 is operated by a hydraulic cylinder providing the billet's ejection. This design was successfully used for cold ECAE of especially strong materials such as Ni-base Invar and Elinvar alloys, magnets, superconductors and others with ultimate tensile strength of more than 1500 MPa.

The die for ECAE of flat billets (Figure 4.7) comprises the stationary inlet channel 1 and the outlet channel 2 formed between a die block 3 and a slider 4. All parts are mounted on a bolster plate 5. The slider is operated by a cylinder 6. During extrusion by a punch 7, the cylinder 6 is open and the slider moves along with the billet by friction. At position "a-a" the punch retreats and the cylinder 6 moves the slider 4 back until opening of the inlet channel. Then, the billet is ejected from the die and the slider returns to the original position to perform the next stroke. To facilitate the ejection, the channel 1 has angle drafts α (Figure 4.7b) (Segal, 2007). In another design (Segal et al. 2004), drafts α are machined into the slider 4 (Figure 4.7c).

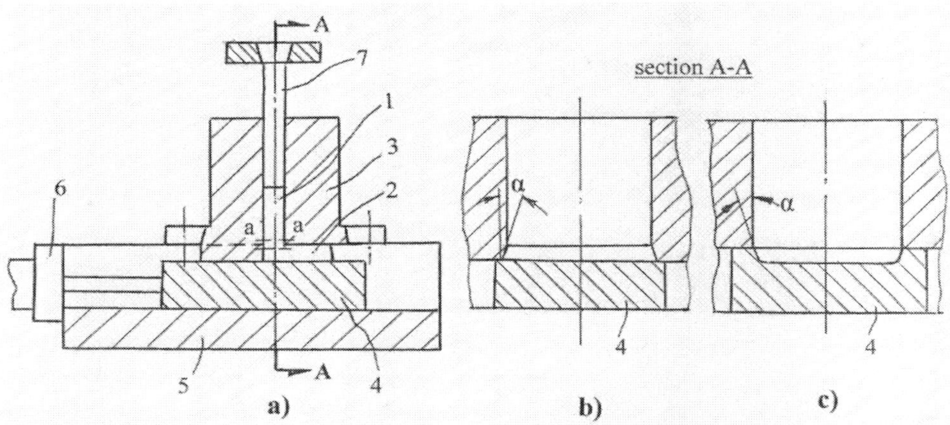

Figure 4.7. Die design (a) for flat billets with plate (b) and figured (b) sliders.

The control of the working stroke is important to preserve a flat top billet surface after each pass. Severe plastic deformation during multi-pass ECAE leads to strong material hardening. Usually, the punch pressure increases from 2 to 3 times developing a large difference in elastic strains for a press-die system. If the press is fixed for a permanent stroke, the difference in the bottom punch position may be from a few millimeters for small billets and soft materials to tenths of millimeters for large billets and hard materials. Correspondingly, punch impressions at the top billet surface form depressions resulting in laps or projections requiring machining between passes. The simple control of the press stroke by solid stops is not effective and automatic control systems were developed (Segal, 2007). These systems comprise a PLC controller, load sensor and sensors to detect press and slider positions. Depending on the extrusion load and data on press-die rigidity, elastic strains and corrected stroke are determined at each pass and introduced in the press control system. The same system controls the positions of the slider and billet ejection. It was also suggested to use a roll as a movable bottom wall of the outlet channel (Segal et al. 1981a; Park et al. 2006), but this design has not been tested experimentally so far.

At optimal processing conditions and effective lubricants, many ductile metals and alloys were successfully processed by multi-pass ECAE at cold, warm or hot temperatures. For isothermal ECAE, dies are provided with heating and temperature control systems. At temperatures above 500°C, processing may be performed using preheated dies and sufficiently high speeds to prevent material chilling. In cases of hard-to-deform and brittle materials when high processing temperatures are necessary, billets are inserted in stainless steel cans that reduces surface chilling and cracking.

4.5. Continuous ECAE

Continuous ECAE (CECAE) is an attractive alternative to batch billets because of higher productivity, low material waste, and the fabrication of long semi-finished products. Instead of the punch pressure, an extrusion load may be developed by pulling forces at the front of the billet end, pushing forces at the back of the billet end and partly pulling and pushing forces at both billet ends. The first concept was realized in equal channel angular drawing (ECAD) (Chakkingal et al. 1998). The stress-strain state during ECAD is identical to tensile bending around a sharp corner with material thinning and nearly pure shear deformation. It is notable that despite significant material hardening, typical characteristics of severe plastic deformation such as shear bands and new high angle boundaries were not observed after multi-pass ECAD. Thus, pulling forces are not practical in different techniques of continuous ECAE (Alkorta et al. 2002).

Much more effective are the pushing forces developed by contact friction at the back of the billet. Three known techniques can be used for that goal: friction forces during rolling (Figure 4.8a), friction extrusion "Conform" (Figure 4.8b) (Green, 1973) and friction extrusion "Linex" (Figure 4.8c) (Voorhes, 1975). During rolling (Figure 4.8a) (Chaudhary et al. 2005), geometrical restrictions require that $a \sim D \sim A$ and, as a consequence, the contact area between rolls and billet is significantly smaller than the contact area between billet and channels. Also, it is difficult to provide simultaneously large friction between the material and rolls, and low friction with channels, or to use multiple rolls or conveyer mechanisms. Therefore, rolling also is not practical for continuous ECAE. The Linex extrusion (Figure 4.8c) is also complicated and unreliable. In contrast, Conform extrusion (Figure 4.8b) has been shown to have significant industrial potential. The first "Conform" press for CECAE developed in 1976 at the Physical-Technical Institute (Minsk, USSR) is shown in Figure 4.9 (Segal et al. 1994). The press was used for multi-pass ECAE of long copper rods with a cross-sectional area 8 x 8 mm^2. The "Conform" principle was also applied to continuous ECAE in a number of patents and publications (Segal et al. 1981a; Stecher et al. 2003;Chung et al. 2003; Lee at al, 2002; Huang, 2005; Saito, 2000). Because of diversity of design concepts and reported results, an analysis of the mechanics of continuous ECAE is necessary for optimal processing (Segal, 2010).

Typically, the CECAE press (Figure 4.10a) comprises a main roll 1 with a groove 2, an auxiliary roll 3, a stationary sector 4 and a support 5. At the beginning, the material 6 is fed between rolls 1 and 3, pressed into the groove 2 along the roll radius R, grasped by the groove walls and advanced to the support 5. Then, slippage between the material and the roll 1 develops contact friction stress τ_1 and a pressure p into the extrusion direction. When the pressure p increases, the material fills the inlet channel formed between the groove and the sector 4 and then starts to extrude into the outlet channel 7.

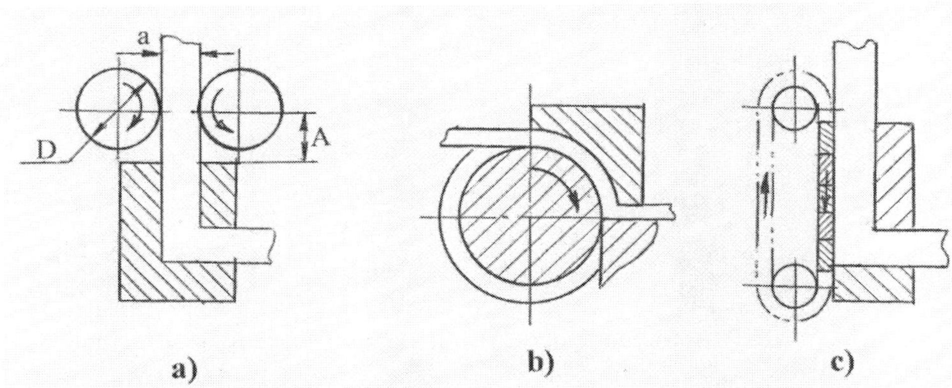

Figure 4.8. Continuous ECAE using concepts of additional rolls (a), Conform extrusion (b) and Linex extrusion (c).

Figure 4.9. Continuous ECAE mill.

During extrusion, opposing friction stresses τ_2 arises along the sector 4. The steady flow is defined by the balance of stresses inside the inlet channel. During this stage, the material confined between the roll 3 and the support 5 rotates with an angular speed ω that is just slightly less than an angular speed ω_1 of the roll 1. Since the material thickness *a* is much less than the roll radius R and $\omega/\omega_1 \sim 1$, the slip [s] between roll and material at the cross-section with an angle coordinate φ is (Figure 4.10a):

$$[s] \sim (1 - \omega/\omega_1)\varphi R \qquad (4.7)$$

This slip provides surface displacements and elastic strains which define contact friction τ_1. To account for this friction in the analysis, a related physical condition should be formulated. Figure 4.10c shows the general diagram of friction τ versus the small contact displacement [s]. For [s] < Δ, friction τ changes nearly linearly with [s]. At [s] = Δ, friction τ achieves a maximum $\tau = \tau_s$ known as static friction. With the further increase of displacements, there is a transition from static friction to sliding friction $\tau = \tau_d$. Assuming that during steady CECAE the slip [s] < Δ, formula (4.7) describes the linear distribution of friction τ_1 between material and roll

$$\tau_1 = \tau_s(1 - \omega/\omega_1)x/\Delta \qquad (4.8)$$

where $x = \varphi R$ is the length of the contact area for a considered cross-section x (φ = const) (Figure 4.10a). A whole length of the contact area $L = \varphi_{max}R$. Depending on the slip (4.7), three characteristic situations should be considered:

(i) maximum static friction τ_s is not attained along the contact area L:

$$\tau_1 < \tau_s \text{ for } 0 < x < L \text{ and } \omega/\omega_1 > 1 - \Delta/L \qquad (4.9)$$

(ii) maximum static friction takes place at some point x* and near sliding friction τ_d acts along the remaining part of the contact:

$$0 \leq \tau_1 \leq \tau_s \text{ for } 0 < x < x^*,\ \tau_1 \sim \tau_d \text{ for } x^* < x < L \text{ and } \omega/\omega_1 = 1 - \Delta/x^* \qquad (4.10)$$

(iii) sliding friction τ_d acts along the full contact length L:

$$\tau_1 = \tau_d \qquad (4.11)$$

For distributions (4.9), (4.10), ratio ω/ω_1 is definitive with low slip [s] between material and roll. These conditions describe stable processing. For friction (4.11), the ratio ω/ω_1 is indefinite and slip [s] may be large. Upon beginning, that leads to intensive surface heating, a decrease of contact friction and skidding between material and roll. Therefore, the distribution (4.11) does not provide stable processing and will not be further considered. It can be also shown that under identical circumstances, a friction moment is larger for distribution (4.10) than for distribution (4.9) if $\tau_d > \tau_s/2$. As this condition is quite realistic, the distribution of friction (4.10) may be considered as the most preferable.

Referring to figures 4.10a, b and assuming that friction τ_2 is uniformly distributed along the sector 4, an approximate equilibrium equation of the material inside the inlet channel is:

$$\tau_1 L(b + 2a) = 2\tau_2 bL + 2(p - p')ab$$

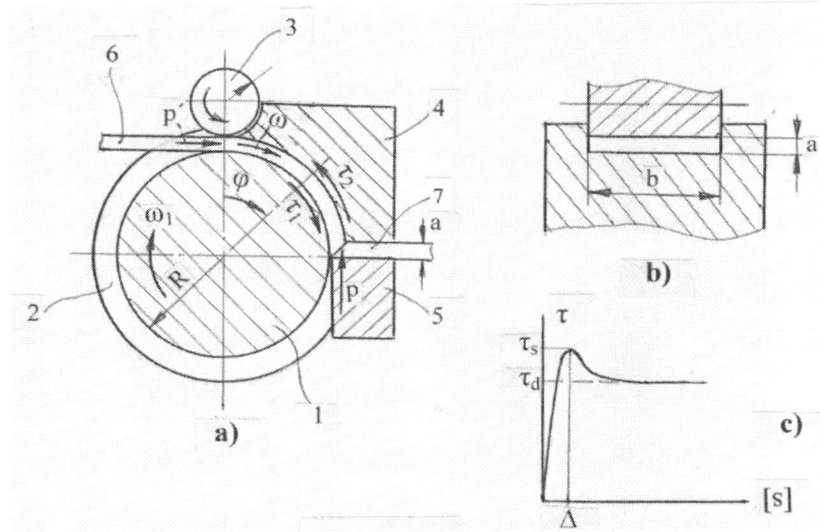

Figure 4.10. Continuous ECAE: a – principal concept, b – channel cross - section; c - friction –slip diagram.

Together with (4.8), that gives:

$$\omega/\omega_1 = 1 - 2\Delta[\tau_2 bL + (p - p')ab]/\tau_s L^2(b + 2a) \qquad (4.12)$$

Here p is the pressure at the channel exit necessary for extrusion and p′ is the pressure at the channel entrance developed by the auxiliary roll 2. The pressure p can be found from slip line fields of Figure 52 depending on τ_1. Eq. (4.12) shows that for the distribution of friction (4.9), the ratio ω/ω_1 is a self adjusting parameter at particular processing conditions. For the distribution of friction (4.10), one can find from the equilibrium equation (4.12):

$$x^* = 2[\tau_d L(b + 2a) - \tau_2 Lb - (p - p')ab]/ (2\tau_d - \tau_s)(b + 2a) \qquad (4.13)$$

The length x* is the similar self adjusting parameter for certain processing conditions.

To optimize CECAE, formulae for rolling moment M and power N at the main roll should be added to equations (4.11), (4.12):

$$M = [\tau_2 Lb + (p - p')ab]R, \; N = M\omega \qquad (4.14)$$

Here billet size (a x b), extrusion pressure p, back pressure p′, angle speed $\omega \sim \omega_1$ and roll radius R are fixed or controlled parameters within specific ranges. To minimize the contact length L, moment M and power N, the most effective distribution of contact friction is found from formula (4.10) when $x^* \to 0$. However, to prevent stalling or unstable flow near this position, the distance x* should comprise a fraction ν of L ($x^* = \nu L$, $0 < \nu < 1$) where ν is a stability factor. Then, formula (3.13) gives:

$$L/a = 2(p - p')/ [(1 + 2a/b)(2\tau_d - 2\tau_d\nu + \tau_s\nu) - 2\tau_2] \qquad (4.15)$$

For evaluation of these results, consider CECAE of square rods and thin sheets. For the former, a = b and with additional simplifications p' ~ 0, $\tau_s \sim \tau_d \sim \tau_2 \sim \tau$, eq. (4.15) becomes:

$$L/a \sim 2p/\tau(4 - 3\nu) \tag{4.16}$$

The ratio L/a as a function of contact friction $\tau/2k$ was calculated for a tool angle of 90° and ν = 0; 0.25; 0.5; 0.75 and 1 (Figure 4.11) using formulae (3.13), (4.16). One can see that this ratio remains relative low when friction $\tau/2k > 0.2$ and $\nu < 0.75$. Practically, this means that CECAE of square rods is characterized by high stability and efficiency and may be performed with effective lubricants and an idle or small auxiliary roll 3.

In the case of thin sheet materials, a/b→0 and assuming $\tau_s \sim \tau_d \sim \tau$ formula (4.15) gives:

$$L/a = 2(p - p')/[\tau(2 - \nu) - 2\tau_2] \tag{4.17}$$

As L/a > 0, it is necessary that $\tau > 2\tau_2/(2 - \nu)$. The limit situations are $\tau > 2\tau_2$ when $\nu = 1$ and $\tau > \tau_2$ when $\nu = 0$. This restriction becomes even stronger to attain acceptable ratios L/a under stable processing with $\nu >> 0$. Figure 4.12 shows calculated diagrams of L/a versus $\tau_2/2k$ for $\tau/2k = 0.5$, p' = 0, $2\theta = 90°$ and ν =0.25; 0.5 and 0.75. It is evident that CECAE of sheet products is possible with large friction τ_1 at the bottom billet surface and low friction τ_2 at the top billet surfaces ($\tau_1 >> \tau_2$), low extrusion pressure p and sufficiently large additional pressure p'. Technically, it may be achieved using driven auxiliary rolls of large diameters (Han et al. 2002) or multiple small auxiliary rolls (Saito et al. 2000), tool angles of $2\theta > 90°$ (Nam et al. 2003) and special knurls at the roll surface (Baeck et al. 2002). As lubrication of the main roll 1 cannot be used in this case, CECAE of sheets is accompanied by high contact friction along the support 5 (Figure 4.10a) with intensive tool wear and a poor surface finish on the work piece. An additional shortcoming is the low productivity of CECAE of sheets in comparison with ordinary rolling mills. It was reported in (Lapovok et al. 2007) that, at least, some of these issues were overcome by an innovative design.

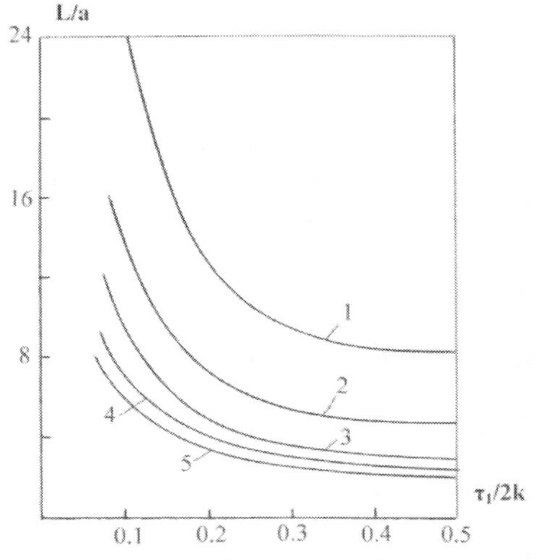

Figure 4.11. Relative contact length L/a versus friction $\tau_1/2k$ for CECAE of square rods with ν = 1(1); 0.75 (2); 0.5 (3); 0.25 (4) and 0 (5).

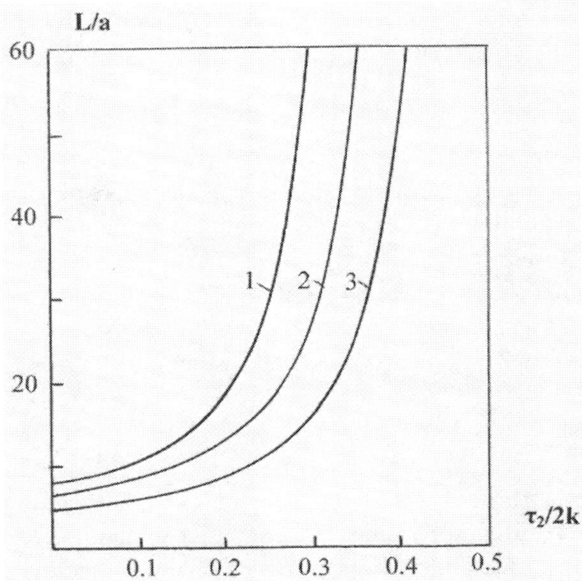

Figure 4.12. Relative contact length L/a versus friction $\tau_2/2k$ for CECAE of sheets with $\tau/2k = 0.5$ and $\nu = 0.75$ (1); 0.5 (2); 0.25 (3).

4.6. Optimization of ECAE Processing

In contrast to traditional forming operations with relatively moderate strains, SPD reveals the strong effect of processing mechanics on structure and properties. Thus, process optimization presents significant interest and, sometimes, is the key factor for commercialization. For ECAE, the optimal processing at each pass complies with intensive and uniform simple shear deformation. The actual characteristics depend on boundary conditions, tool design and temperature-strain-strain rate parameters. In this section we will discuss general guidelines for optimization of processing.

Contact friction is the most important boundary condition. Analysis and experiments show that only for zero friction the plastic zone is a thin material layer with intensive and uniform simple shear. However, the realization of frictionless conditions is not a trivial problem, and some contact friction takes place even with the best lubricants. Solid lubrication films such as a zinc stearate coating for aluminum alloys or phosphating for steels may be necessary. Friction conditions are not identical in both channels. It is especially difficult to provide low friction along the bottom wall of the outlet channel, and movable die walls are an effective solution to the problem.

Channel geometry is also important. The sharp corner channels correspond to near optimal characteristics when $\tau \rightarrow 0$, $2\theta = 90°$, $\gamma = 2$. Round corner channels lead to lower and non-uniform strains with redistribution of shears between three different directions. They do not allow for the use of movable die walls and to eject separated billets from the die.

Shear strain per pass γ depends on the tool angle 2θ and contact friction τ (see formula (3.17)). In the most effective case $2\theta = 90°$ and $\tau/k << 1$, the shear strain per pass is $\gamma \sim 2$. To attain very large strains and the required structural effects, ECAE is repeated a number of times. The accumulated von Mises strain after N passes ECAE is:

$$\varepsilon = N\gamma/\sqrt{3} \tag{4.18}$$

During multi-pass processing, the number of passes N is a factor that determines productivity, tool life and product cost. It should be minimized by optimization of processing mechanics providing intensive, uniform and controllable simple shear deformation and cross-loading history. The minimum number of ECAE passes N should be determined in each particular case depending on material, required structure and properties. Because most structural characteristics and properties evolve little past a certain strain, the practical rule is to approach the right compromise between product performance and cost.

Loading histories are realized through combinations of the basic processing routes. Optimization of routes is also performed for particular problems depending on a number of passes N.

The evaluation of the shear strain rate η is necessary to describe the material behavior during ECAE. From formulae (3.4) one can see that $\eta \to \infty$ when $\delta \to 0$. As the shear band thickness δ approaches zero, the rigid plastic analysis of velocity discontinuities is not sufficient enough to evaluate the strain rate. Experiments show that in real materials the thickness of the shear zone δ during ECAE with near zero friction composes only a small part of the billet thickness a. Usually, $\delta/a \sim (0.05\text{-}0.1)$. Then, using figures 3.1a,d, the time Δt necessary to cross this zone is:

$$\Delta t = \delta/v_n = (0.05 - 0.1)a/V\sin\theta$$

The last equation, together with formula (3.4), gives an estimation of shear strain rate in the optimal case of ECAE:

$$\eta = (10\text{-}20)\ V\cos\theta/a \tag{4.19}$$

The formula (4.19) shows that strain rate during ECAE may be much higher than in ordinary forming operations with identical processing speed V. Therefore, relatively low extrusion speeds V should be used for strain rate sensitive materials.

Punch pressure and load depend on the material shear yield stress k, friction τ_1, τ_2 in inlet and outlet channels, back-pressure p_o, die design and billet shape. Formulae (3.2), (3.13), (4.1) to (4.6) can be used for calculation of pressures in both channels for different cases of ECAE. Potentially, ECAE is an effective technique in providing low pressure and, most importantly, load. Comparison of frictionless ECAE with "ideal" forming operations (see formulae (2.2) and (3.2)) demonstrates that with similar effective strain ε and cross-section area of the final product, the extrusion pressure is the same whereas the extrusion load is $\exp(\varepsilon)$ times higher for "ideal' forming than for ECAE. For example, the effective strain $\varepsilon = 1.15$ can be attained after one pass of ECAE with the tool angle of 90° and after "ideal" extrusion with area reduction 3.18. In both cases, the punch pressure is p=2k whereas the load is 3.18 times larger for "ideal" extrusion than for ECAE. The difference in pressure and load becomes much bigger if one compares identical data for multi-pass ECAE and ordinary forming operations (Segal et al. 1981b). The control of the pressure using advanced tool design and lubricants is essential for practical realization of ECAE especially for multi-pass processing of hard materials.

The back- pressure p_o can also be used as a controllable parameter of ECAE to improve processing mechanics and formability of low-ductility and particulate materials. These effects known since Karman were also demonstrated for ECAE (Lapovok, 2005). The magnitude of the back-pressure should be comparable to the material yield stress. Existing methods for applying back-pressure during ECAE using an additional cylinder or extrusion of dump samples (Valiev et al. 2006) are not effective as they increase contact friction and extrusion pressure. To exploit possible benefits of back-pressure, significant progress should be made in the development of low cost techniques that can be used in large volume production.

The processing temperature defines the material behavior and structural changes during ECAE. Most industrial materials require processing at warm or hot temperatures. Combined effects of processing temperature and time depend on material preheating, die temperature, extrusion speed, strain rate and cooling after ECAE. Resulting thermo-mechanical histories may be complicated and non-uniform throughout the billet's volume (DeLo et al. 1999). Practically, ECAE at near isothermal temperatures, moderate extrusion speeds and without billet preheating between passes produces sufficiently uniform structures after multi-pass processing. In these cases, the optimal processing may be fixed using small scale experiments for identical thermo-mechanical conditions including similar tool, lubricant, number of passes, route and temperature. For preserving similarity of the Zener-Hollomon parameter (1.3) and the material behavior, strain rate should be identical in small scale (s) and full scale (f) experiments. Then, Eq. (4.19) gives formula for corresponding small scale V_s and full scale V_f speeds and billet thicknesses a_s and a_f:

$$a_f/a_s = V_f/V_s \tag{4.20}$$

For geometrically similar small and full scale billets with the same billet length–to- thickness ratios $c_s/a_s = c_f/a_f$ (Figure 57), strain rates are identical when the extrusion time t is equal in both cases:

$$t = c_s/V_s = c_f/V_f$$

Under conditions of similarity, identical structural effects and properties should be expected for small and large billets. With an increase of the billet size from a_s to a_f, the extrusion speed increases in proportion (a_f/a_s), the extrusion pressure p remains the same and the processing load increases in proportion to $(a_f/a_s)^2$. Extrusion temperature and speed are the main factors to control structures and produce defectless materials using ECAE.

4.7. Defects Associated with ECAE

The specific mechanics of ECAE results in characteristic defects. *Massive cracks* along the shear plane are usually observed in materials with low ductility which cannot accommodate intensive shear under certain processing conditions. In these cases, full separation of the front end of the billet 1 takes place at the beginning of extrusion along the shear plane AO (Figure 4.13a). Following extrusion, the sharp edge of the remaining material jams in the die. When the billet moves into the outlet channel, the extrusion pressure increases. At some thickness T, the extrusion pressure becomes sufficient again to separate the next material portion along plane AO. This procedure divides the billet into segments as shown in Figure 4.13b for ECAE of CP Ti at room temperature (Semiatin et al. 1999). Observations of planes AO in these cases reveal a brittle character of fracture.

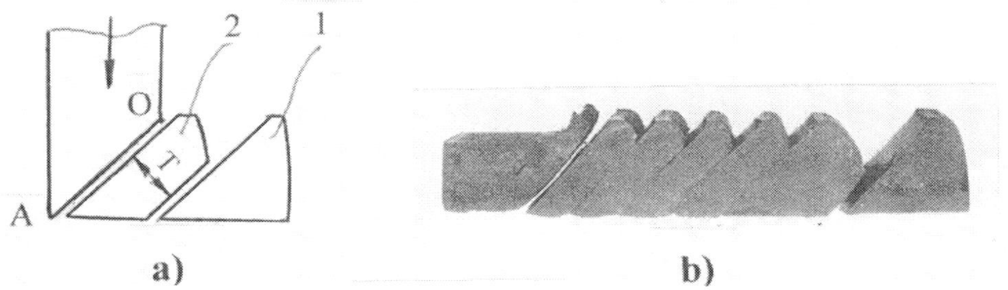

Figure 4.13. Segmented fracture during ECAE (a) and failure of CP Ti at room temperature (b).

There are some defects that lead to the *macro localization* of plastic flow in ductile materials. For very ductile material such as copper the macro localization does not detect fracture but manifests itself in areas of highly non-uniform structures with the characteristic tooth-like surface (Figure 3.31b). In less ductile materials, the macro localization results in a partial segmented fracture. Figure 4.14a shows a sample of CP Ti after ECAE at a temperature of 325°C and a strain rate of 0.2 sec^{-1} without any evidence of macro localization and fracture. However, when the strain rate was increased to 2 sec^{-1}, both macro localization and fracture were revealed (Figure 4.14b) (Semiatin et al. 1999).

Another reason for surface fracture is the development of *tensile macro stresses* during multi-pass ECAE. The considered rigid plastic analysis neglects secondary effects that may be important in some areas such as the channel vertex point O with a stress/strain singularity (Figure 4.15a). From formulae (3.15) one can see that within the central fan, the strain rate $\eta \rightarrow \infty$ when $r \rightarrow 0$. Therefore, materials with a sensitivity to strain rate (1.3) should experience significant hardening and flow non-uniformity in the vicinity of point O. Similar results were found by finite element analysis for strain hardening materials and when the friction in the channels is different (Semiatin et al. 2000; Li et al. 2004). Due to this effect, there is a tendency for the material to bend upward in the outlet channel (Figure 4.15a). The reaction N of the channel wall prevents this rotation but develops additional stresses $\Delta\sigma$. In result, tensile stresses may arise along the top surface of the billet and progress into local fracture as was observed in (Semiatin et al., 2000). Figure 4.15b shows similar surface fracture during ECAE of high purity 4N Ti at a temperature of 350°C. Surface cracks originate and propagate along shear bands. The microstructure of cracks (Figure 4.15c) reveals a typical mechanism of ductile fracture by pore growth and coalescence. For ECAE without macro localization, similar fracture may be observed with the increase of passes, as surface cracking and dilatation. For non-isothermal ECAE at warm and hot temperatures, the chilling effect of the tool reduces ductility and also promotes surface cracks. In most cases, fracture during ECAE is eliminated by increasing billet or die preheating temperatures and/or reducing processing speed to avoid macro localization and improve material ductility.

a) b)

Figure 4.14. ECAE of CP Ti at temperature 325°C and strain rate 0.2 sec^{-1}(a) and 2 sec^{-1}(b).

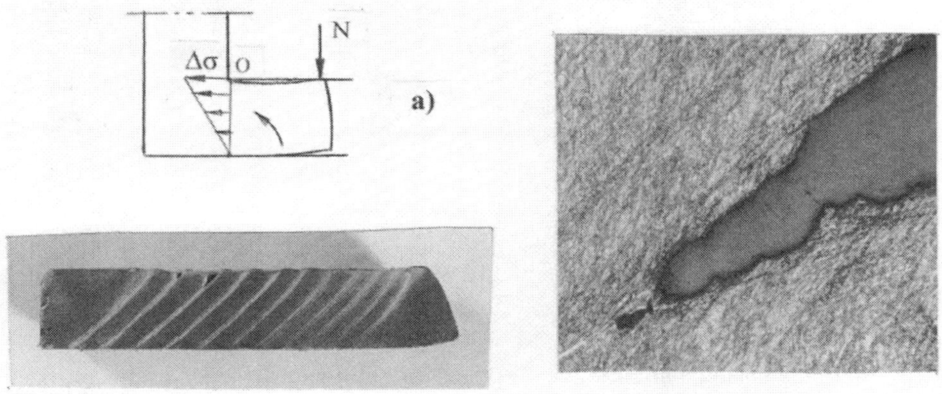

b) c)

Figure 4.15. Surface cracks at the top billet surface: a - additional tensile macro stresses; b - surface fracture of Ti4N after 2 passes ECAE via route B at temperature 350°C; c - microstructure of the crack tip.

Some specific defects of ECAE are related to irregularities of boundary conditions at the bottom surface of the billet. One of these is surface laps. At the beginning of extrusion, the originally rectangular billets change their shape and form a gap *b* at the sharp die corner (Figure 4.16a). Evolution of this area during the transition to a steady flow results in the shallow lap *c* at a distance *a* from the front of the billet's end. Similar laps may be observed due to unstable friction when small "dead" zones *d* (Figure 4.16b) periodically form, stall and release from the sharp die corner. Such instabilities are found experimentally and are also captured by finite element modeling (Semiatin et al. 2000; Li et al. 2004). The back-pressure or sufficiently large friction stabilize processing. More effective, though, is a die with a movable bottom wall and a closed outlet channel (Figure 4.4b) which eliminates laps. In most cases, because of surface defects, billets should be machined or scalped after multi-pass ECAE.

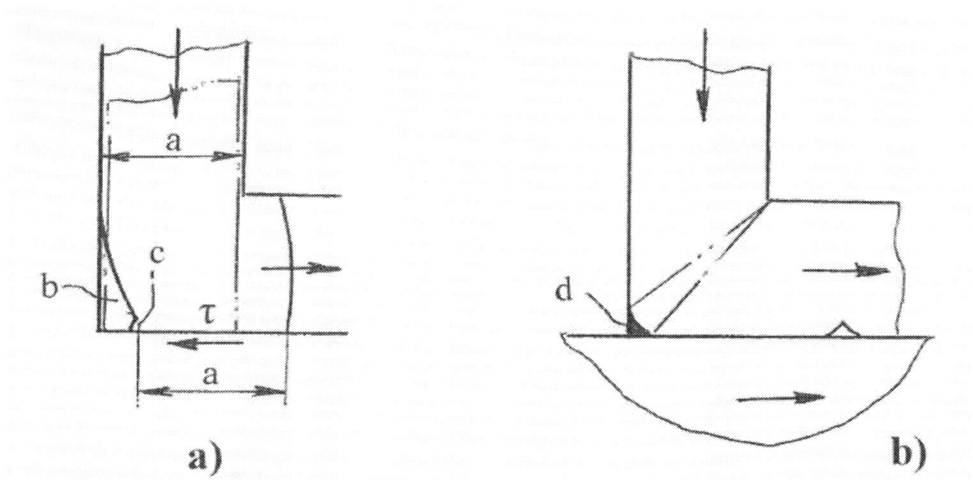

Figure 4.16. Development of laps at the beginning ECAE (a) and during steady flow (b).

Chapter 5

APPLICATIONS OF SPD

5.1. STRUCTURE MODIFICATION DURING SPD

SPD results in various modifications of structure and properties. Structure - property changes after SPD are usually strong and often lead to anomalous effects which are difficult or impossible to attain with ordinary deformation techniques. As a consequence, there is great potential for practical applications of SPD as a new material processing technology. SPD may be applied to different materials including metals, ceramics, semiconductors and polymers; polycrystals and single crystals; solids and particulate. The only precondition is to have sufficient ductility to experience large strains during processing. Presently, most activity in the field of SPD is concentrated on structure refinement into a sub-micron and sometimes nano scale. However, fabrication of ultra-fine grained materials is only one of the many possible applications of SPD. The least acknowledged applications are the control of a material texture and the consolidation and bonding of particulate materials. Other important applications relate to the breakdown and homogenization of structures, geometrical control of structural elements, "in situ "composites, phase transformations and processing of specialty materials. Each of these modifications can be used to improve materials performance for many technical problems. An additional area of interest is SPD processing to improve manufacturing characteristics (such as high strain rate /low temperature superplasticity, advanced formability and machinability) for fabrication of complicated precision parts at bulk or miniaturized scales in a productive and cost effective manner.

An overlook of structural changes induced by SPD provides background for evaluation of possible applications. Particular results published extensively during the last 10-15 years were reviewed in (Meyers et al. 2006; Valiev et al. 2006). Vast experimental information was obtained for different materials and processing conditions. We have repeatedly emphasized that the processing mechanics is a key factor of SPD that significantly affects structures and properties. During ECAE, strain intensity, uniformity, deformation mode and, finally, structure evolution depend strongly on tool design and contact friction. The control of friction is especially important because ordinary lubricants cannot provide low friction with a near simple shear deformation mode. However, in most published works, the processing mechanics was not analyzed. Experimental data on distortions of coordinate grids and hardness measurements do not show a noticeable difference in the limit cases of zero and maximum friction and do not provide reliable information on the processing mechanics. In fact, ECAE in solid dies with the best lubricants may be actually performed with large friction and an extensive fan-shaped plastic zone. In tools with round corner channels, even a small corner angle $\psi \sim 15\text{-}30°$ induces redistribution of shear planes and strain inhomogeneity that cannot be neglected merely for the sake of "simplicity". Deviations from the "ideal" case of ECAE with shear on a single plane become more significant during multi-pass processing and lead to different strain induced microstructures and textures. Processing operations can also develop uncertainty of experimental results. For example, spontaneous billet rotation may take place during ECAE in cylindrical channels causing uncontrollable processing routes especially for tool angles of more than 90° and for routes B and D with rotations of 90°. As processing conditions were not identical in different cases, it is not surprising that published results are often inconsistent and lead to controversial conclusions. Therefore, structural effects induced by SPD should be considered closely associated with the processing mechanics to understand "processing - structure - properties" relations and to define areas of applications of SPD.

5.2. Structure Refinement

5.2.1. Ultrafine Grained Structures

One remarkable effect of SPD is structure refinement into the submicron and sometimes nano scale. There is always a tendency to use materials with the finest grain size which is the most important characteristic of the microstructure. On the right side of the grain size scale of industrial metals and alloys are statically recrystallized coarse grained structures with grain sizes of more than about 100 μm. In the next range of fine grained materials from about 10 μm to 100 μm, structures are refined by thermo-mechanical treatment. On the left side of the grain size scale, there are amorphous materials (the characteristic structural size is less than 1 nm) and nano materials with grain sizes from 1 nm to 100 nm produced by techniques of rapid solidification, mechanical alloying and atomization followed consolidation. The niche from100 nm to 10 μm between fine grained and nano materials belongs to ultra fine grained (UFG) metals. This transitional position leads to many attractive characteristics of UFG materials between ordinary materials and nano materials. Severe plastic deformation provides industrial techniques such as ECAE to produce bulk UFG materials in a productive and relatively inexpensive way.

Different mechanisms of structure refinement are detected during plastic deformation and heat treatment: continuous and discontinuous recrystallization, inhomogeneous deformation with phase separation in duplex alloys, phase transformation and geometric recrystallization (Humphreys et al. 1996; Wert, 1983). For certain conditions, these mechanisms may operate and during SPD. However, the specific characteristic of SPD is strain induced new high angle boundaries organized into three dimension configurations. As was described in Chapter I, the dislocation activity during plastic deformation extends boundaries of the original grains and develops two-dimensional defects such as cell boundaries, cell blocks, sub-grains and dislocation walls outlining grain subdivided areas due to the split of grain textures. When deformation proceeds, these defects form low energy dislocation structures (LEDS) and evolve in a different manner. Cells and sub-grains remain low angle formations, whereas an intensive rotation of grain subdivided areas increases misorientation angles along their boundaries and forms dense dislocation walls (DDWs) and deformation bands (DBs). After large monotonic strains, DDWs and DBs are finally transformed into thin lamellae with high angle planar boundaries (HABs). Such extremely deformed micro-structures resulting from the continuous evolution of dislocations (Gil Sevillano et al. 1980; Kuhlmann-Wilsdorf, 1989: Hansen, 1990; Hughes et al. 1997) are usually observed after rolling or drawing with large reductions and cannot be classified as grains. Therefore, strain-induced high angle boundaries are not a new physical phenomenon. Really new effects of SPD relate to the processing mechanics that controls formation and spatial distribution of strain-induced high angle boundaries which may be identified as grain structures.

Numerous experimental works have revealed the special character of SPD structures (Mayers et al. 2006; Valiev et al. 2006; Prangnell et al. 2001). The limit of grain refinement during SPD is close to the size of cells/subgrains. New grains are almost dislocation free. The high density of dislocations and other defects is observed at grain boundaries which are in a non-equilibrium state. Refinement rate and morphology of grains are sensitive to processing mechanics and loading history. Structures are macro and micro inhomogeneous. They usually include areas of equiaxed grains and areas of elongated grains or long cell/subgrain blocks. Grain boundaries have sufficiently broad distributions of misorientation angles. Typically, misorientation angles are noticeably lower than for statistically equilibrium structures (Mackenzie, 1958). A part of high angle boundaries (HABs) is relatively small at strains $\varepsilon < 2$, then increases with strains and attains a saturation of 70-80% at $\varepsilon = 6\text{-}8$ (Prangnell et al. 2001). In some cases, a large part of boundaries remain low angle configurations (LABs). These SPD structures are different from ordinary structures of recrystallization.

5.2.2. Strain-Induced High Angle Boundaries (HABs)

In contrast to numerous publications on characterization of SPD structures, only a few works deal with mechanisms of refinement. At the micro scale, plastic deformation may be accommodated by dislocation glide, twins and diffusion flow. Consider, first, f.c.c. metals with moderate-to-high SFE at low temperatures and strain rates where the dominant deformation mechanism is crystallographic glide. Plastic deformation at the meso scale may be realized through continuous evolution of dislocation structures or discontinuous flow. Correspondingly, this leads to the formation of incidental dislocation boundaries (trapped dislocations, cells, cell blocks) and geometrically necessary boundaries (areas of subdivided grains, dense dislocation walls, micro-bands, deformation bands) in the first case and shear bands with fragments sliding or rotation in the second case. Depending on strain and deformation mode, all these successive steps contribute to structure refinement and may be observed in experiments. Therefore, different refinement models were suggested for specific conditions of SPD. A model of (Meyers et al. 2006) assumes continuous accumulation of dislocations with transformation of dislocation structures from cells to subgrains and grains. The model does not consider specific mechanisms of such transformations and takes into account accumulated strains as the only parameter irrespective of processing mechanics. This is contradictory to experimental results. For example, cell boundaries do not show the perceptible increase in misorientations and transformation into grains even at very large rolling reductions (Hughes et al. 1997; Prangnell et al. 2001). To clarify mechanisms of refinement, an analysis is necessary for successive steps of structure evolution during SPD. In the application to ECAE, such research was performed in (Xue et al. 2006). The round corner tool angles $2\theta = 90°$ and $\psi = 37°$ provided a loading history with three simple shears: along rigid plastic boundaries into radial directions and inside a central fan into a circular direction as shown in figsures 3.10b, c, d. The deformation path was traced at the flow plane along stream-lines at characteristic locations. It was found that during the first pass of ECAE ($\varepsilon \sim 1$), the loading history of material elements follows the general framework of continuous structure evolution comprising the formation of trapped dislocations, cells, long cell blocks and micro-bands. All grains are roughly grouped into two categories whose orientations are favorable for either normal or longitudinal straining relative to the streamlines. These observations show that substructures developed at small and moderate strains during SPD can be described in terms of the LEDS theory, and that the initial phase of structure refinement is generated by continuous evolution of dislocation structures. Such conclusions conform to other published results (see, for example, Wu et al. 2004; Werenskiold et al. 2006; Kaibyshev et al. 2004) and should be accounted for interpretation of experimental data together with the control of the processing mechanics.

During this phase, there is a significant effect of deformation mode on morphology of cell blocks /deformation bands. Their boundaries are oriented close to macroscopically preferable shear planes (Winther, 2003). For highly oriented simple shear like in the case of ECAE with a single slip line (Figure 3.1a), cell blocks and deformation bands form parallel to the shear plane and have extended dislocation boundaries. Such structures are observed during the first pass for suitably oriented grains and for the rest of grains during subsequent passes (Figure 1.15a, b). Iwahashi et al. (Iwahashi et al. 1998) traced cell blocks for four passes of ECAE via basic routes. They found that for the first two passes orientations of cell blocks coincide with corresponding shear planes (Figure 3.26), low angle cell blocks remain stable and their intersections outline the material fragments. For a larger number of passes, fragments are gradually transformed into three-dimensional configurations; part of their boundaries become high angle boundaries and the substructure is substituted by new grains. These changes show that the mechanism of continuous structure evolution is replaced by other mechanisms. As a physical phenomenon, such transition manifests itself in a structure refinement that is clearly detected by a comparison of the material properties after ECAE and rolling with strains $\varepsilon > 2$ (Segal et al. 1981b).

5.2.3. Grain Refinement Mechanisms

A few approaches were developed to formulate basic principles of structure refinement during SPD. Mishra et al. (Mishra et al. 2007) discussed a model of structure self organization by transformation of elongated cell blocks into long subgrains and their breaking up into new grains. This model does not take into account the processing mechanics as a crucial factor of SPD. Zhu and Lowe (Zhu et al. 2000) noted that crystal structure and mechanics should be considered together. They hypothesized that optimal refinement of f.c.c. metals takes place when the angle between the shear plane at each pass and a direction of the principal elongation of the previous pass is close to the crystallographic tetrahedron angle 70.5°. The hypothesis was not experimentally confirmed and may be justified only for the first two passes via routes A and C. Another model of structure refinement was suggested by Langdon

(Langdon, 2006) who considered evolution of bands of cell blocks (elongated subgrains) induced by multi-pass processing via different routes. As cell blocks are delineated by low angle boundaries, it was suggested that new high angle boundaries are formed by superposition of cell blocks during subsequent passes. The model does not explain mechanisms of this superposition but supposes that refinement becomes more effective when a number of bands and angles of their intersections increase. However, the transformation of low angle boundaries into high angle configurations is contradictory with strain redistribution between various directions (Wu et al. 2004) and coalescence of boundaries is impediment for large angles of intersections (Gasperini et al. 1996). On the contrary, it is known that unidirectional intensive shear during dynamic testing (adiabatic localization) produces equaxed ultrafine grain structures similar to ECAE (Meyers et al. 2006). Also, the model does not explain the special character of three dimensional structures for different routes, especially for routes A and C.

Neither of the above mentioned models associates grain refinement mechanisms with strain localization as the specific characteristic of SPD. It seems that the evolution of deformation bands or cell/subgrain blocks is not sufficient itself to develop spatially distributed high angle boundaries. On the other hand, the plastic localization provides effective and natural sources for structure refinement at the meso scale. Three refinement mechanisms are operative in this case: (i) high angle boundaries induced by shear bands, (ii) the material fragmentation by shear band intersections and (iii) fragment rotation within and between shear bands. These refinement mechanisms were described in (Hirsch et al. 1988; Korbel et al. 1985; Andrade et al. 1994). A refinement model for ECAE due to shear band localization was suggested in (Segal, 1999) and has received experimental (Prangnell et al.2001; Shin et al. 2001; Wang et al. 2004; Boczkal et al. 2007) and theoretical (Sivakumar et al. 2004; Wu et al. 2006) confirmation. Similar observations were found for cyclic extrusion-compression (Wu et al. 2006; Richert et al. 2007), multi axis forging (Belyakov et al. 2001) and high pressure torsion (Tyumentsev et al. 2000). According to the model, the first stage of the continuous structure evolution induces blocks of cells/subgrains and deformation bands mostly separated by low angle boundaries. During the successive stage of flow localization, shear bands arise along the principal macro shear directions and propagate through shear transition zones (STZ) of the preliminary substructure formed within grains or their boundaries. Suitably oriented low angle boundaries present channels for STZ. As meso strains within shear bands are much larger than continuum strains, planar high angle boundaries appear rapidly along and inside shear bands. Depending on the following loading history, two scenarios for developing spatially distributed HABs are possible. In the case of monotonic loading, the model (Hines et al. 1998) (Figure 5.1a) implies the formation of cell blocks along a shear direction SD, transformation of block boundaries into high-angle boundaries and subsequent cell rotation inducing high-angle boundaries (HABs) into transverse directions. Even for simple shear deformation mode, this model requires rather high strains. An alternative approach comprises the spatial control of HABs induced by shear band intersections (Korbel et al. 1985) into different directions during cross-loading. Figure 5.1b shows intersections of shear bands into directions SD_1 and SD_2 for two cross-loading steps with sufficiently large strains to form high-angle boundaries HAB_1 and HAB_2. After the beginning of localization, this processing route requires, at least, three cross-loading steps into mutually perpendicular directions. The present refinement model at the meso scale determines that unidirectional simple shear is the optimal deformation mode at each processing step.

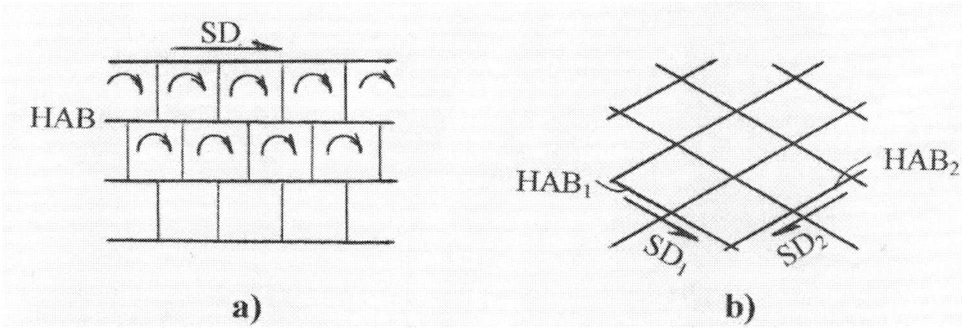

Figure 5.1. Development of three dimensional high angle boundaries by fragment rotation (a) and shear band intersection (b).

Originally, high-angle boundaries induced by shear bands are planar and diffuse. However, high density of defects and enhanced diffusivity around these boundaries provide recovery at low processing temperatures and a local boundary migration to more equilibrium configurations similar to grains. The attainable grain size should be close to the stable size of cell /subgrain for corresponding processing conditions. As shear bands do not cover the whole volume and there is certain spacing between shear bands, a part of the material retains the low-angles substructure previously induced by the continuous evolution. According to (Wang et al. 2001), the volume of shear bands in aluminum after ECAE comprises of about 50% of the material volume. For ordinary classification of HABs by misorientation angles larger than 15°, some of these boundaries are also induced during continuous evolution. Prangnell et al found that the percentage of HABs is about 15 – 20% at the beginning of localization ($\varepsilon \sim 2$) and approaches 70% after the completion of refinement ($\varepsilon > 6$) (Prangnell et al. 2001). These numbers comply with the above-mentioned volume of shear bands. It should be noted that low-angle substructures are defined by material crystallography and S.F.E. whereas refined high-angle structures depend mainly on processing mechanics. Therefore, microstructures for different materials should have more similarities after SPD than after ordinary processing techniques

5.2.4. Twinning Refinement

The shear band refinement model can be extended to materials with the twinning deformation mechanism. Twins operate globally or locally when the required shear stresses for twinning are lower than for dislocation glide. Usually, twinning is observed in h.p.c. metals with a small number of slip planes, in b.c.c.metals at low temperatures and high strain rates and in f.c.c. metals with low S.F.E. During ECAE, deformation twins were detected at ordinary processing conditions and in usually non-twinning materials such as Cu (Huang et al. 2006) and Al (Lou et al. 2008). These twins arising in grains with specific orientations with respect to the macro shear plane and in areas of high local stresses do not contribute significantly to structure refinement. However, when local shear planes take arbitrary orientations, twinning becomes the main deformation mechanism in most grains. Such a situation was attained for copper (Wang et al. 2006) and other materials (Tao et al. 2003) in the SMAT process (Figure 2.6b) owing to the plurality of local surface shears and high strain rates. Two different areas of structure evolution were traced in surface layers depending on depth and value of strains. In the subsurface layers with low and moderate strains, ordinary cell/subgrain structures are developed by dislocation glide. In the surface layer with large strains, grains are subdivided by micro twins on thin twin-matrix lamellae. When straining proceeds, lamellae are subdivided by dislocation walls and micro shear bands which, finally, form nano grains. This evolution is well understood in terms of the processing mechanics discussed in Chapter 2. Figure 2.5c shows a local plastic contact formed in the material by single ball R. During impact loading, the ball penetrates the material and successively develops shear $\gamma = 1$ through plastic zones which depth depends on the kinetic energy. For SMAT processing, a large number of such impacts at any location of the material surface induces numerous shear bands and micro twin-matrix lamellae in different directions which intersections refines the micro-structure. In contrast, the continuous structure evolution below the plastic zone retains the cell/subgrain substructure. Therefore, the twinning refinement mechanism is similar to the considered model of shear band intersections where twin boundaries play the same role as shear bands.

5.2.5. Effects of Temperature and Strain Rate

Because of enhanced diffusivity, high density of defects and latent energy, strain induced UFG and nano structures are not stable at increasing temperatures. During SPD materials may experience dynamic recovery or recrystallization. The temperatures of static recrystallization for UFG materials are noticeably lower than for ordinary heavily deformed materials. Recovery also takes place at lower temperatures extending to room temperature. It provides more equilibrium grains with almost defect free interiors, sharper boundaries and lower internal stresses. Recovery results in slightly reduced strength, improved ductility and an insignificant increase of the grain size. Therefore, warm SPD is beneficial in many cases especially for fabrication of semi-finished products. For multi-pass ECAE, static recovery and recrystallization also take place during material preheating between passes and cooling after processing. To evaluate these effects, experiments were performed on ECAE of the aluminum alloy Al0.5%Cu at temperatures cold (20°C) and warm (150°C and 225°C) deformation with a strain rate $\xi \sim 10^{-1}$ /sec. An ECAE die with a tool angle of 90° provided isothermal and near frictionless conditions. It was found that annealing at 150°C for 1h preserves the UFG structure whereas annealing at 225°C for 1h leads to full static recrystallization of the UFG structure. To trace dynamic recovery and recrystallization during ECAE, samples were water quenched after each pass of warm ECAE and then statically annealed at processing temperatures. Brinell hardness HB was measured before and after annealing. One sample was also cold extruded and annealed between passes at a temperature of 150°C, 1h. Results for 8 passes ECAE via route D are presented in Figure 5.2 as a function of number of passes N. For cold ECAE (diagram 1), the structure refinement to a grain size 0.5 micron (Figure 1.18b) was completed after four passes. Some further hardening was, probably, caused by the increase of misorientation angles of the UFG structure. Static annealing of this material at a temperature of 225°C for 0.5h produces the fine grained structure with a grain size of 7 microns (segment AB in Figure 5.2). Following heat treatment, such equiaxed and fine uniform structures show enhanced thermal stability in comparison with ordinary structures of the alloy.

For ECAE at 150°C (diagram 2), the maximum hardness was observed after four passes. Subsequent softening was induced by some coarsening of the UFG structure. Additional annealing 150C for 1h between passes reveals the insignificant decrease of hardness confirming that recovery was completed dynamically. Similar annealing after each pass of ECAE at room temperature (diagram 3) shows intensive static recovery with oscillation of diagram 3 around diagram 2. This comparison demonstrates that the extent of recovery comprises the main difference between cold and warm processing controlled by the processing temperature. For cold and warm ECAE with strains $\gamma > 1$, the coefficient m in Zener – Holomon formula (1.3) is small ((Johnston et al. 1983) whereas the processing time is sufficient to allow for complete recovery. It can be expected that in these cases strain rate has a low effect on structure evolution, if adiabatic heating does not increases the processing temperature above the temperature of dynamic recrystallization.

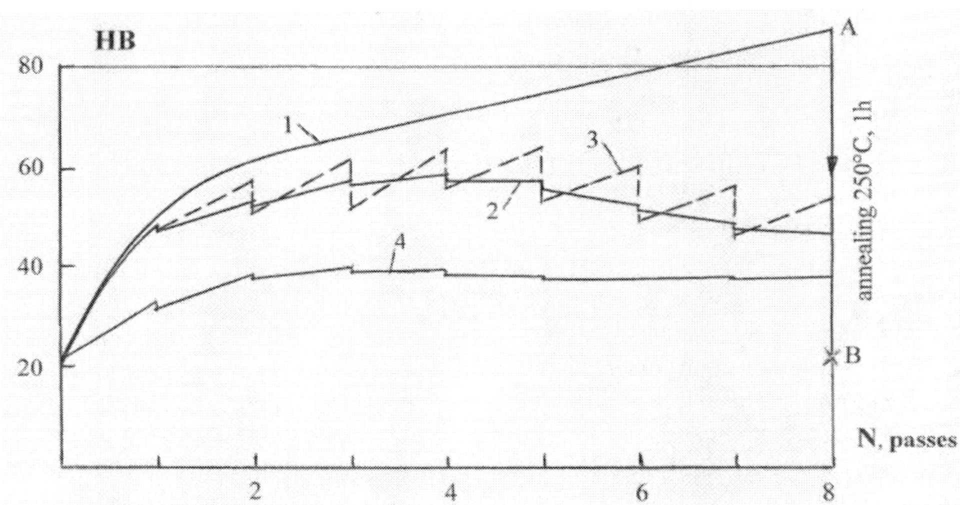

Figure 5.2. Effect of ECAE passes N on hardness HB for Al0.5%Cu alloy:1-ECAE at temperature 20°C; 2- ECAE at temperature 150°C + annealing 150°C,1h; 3- ECAE at temperature 20°C + annealing 150°C, 1h; 4 – ECAE at temperature 225°C + annealing 225°c, 1h.

For the processing temperature225°C (diagram 4), a steady state was also attained again after four passes. A very small decrease of hardness after additional annealing at 225°C for 1h between passes suggests structure stability and completion of the continuous dynamic recrystallization (CDRX). Depending on temperature and strain rate, CDRX can produce a variety of grain sizes ranging from UFG structures to ordinary structures of hot deformation. The latter structures are always observed after each pass during multi pass ECAE with billet preheating at temperatures above the temperature of static recrystallization. CDRX was described in many works at different processing conditions (Kaibyshev et al. 2004; Sitdikov et al. 2001; Goloborodko et al. 2003; Su et al. 2006; Engler, 2001). It was suggested that the mechanism of CDRX includes nucleation and growth of new grains within shear bands and their intersections (Kaibyshev et al. 2004; Goloborodko et al. 2003; Engler, 2001; Xu et al.

2001). During warm ECAE, CDRX along shear bands was also detected for Mg (Su et al. 2006). For high purity aluminum, the recrystallization temperature becomes below the room temperature after one or a few passes of ECAE, and shear bands with very small new grains can be observed by optical microscopy (figures 1.15a, b). Therefore, recovery, CDRX and static recrystallization present additional options to control strain induced microstructures of SPD. In any case, the structure morphology relates to the processing mechanics and shear band orientations. As an example, Figure 5.3 shows microstructures of Al alloy 7075 statically recrystallized at 450°C for 1h after 8 ECAE passes at 250°C via routes B and C. Montages display microstructures into planes X, Y, Z of Figure 3.21a. Lamellae like grains for route B and equaxed grains for route C follow closely with the corresponding orientations of shear bands and material distortions predicted by the processing mechanics.

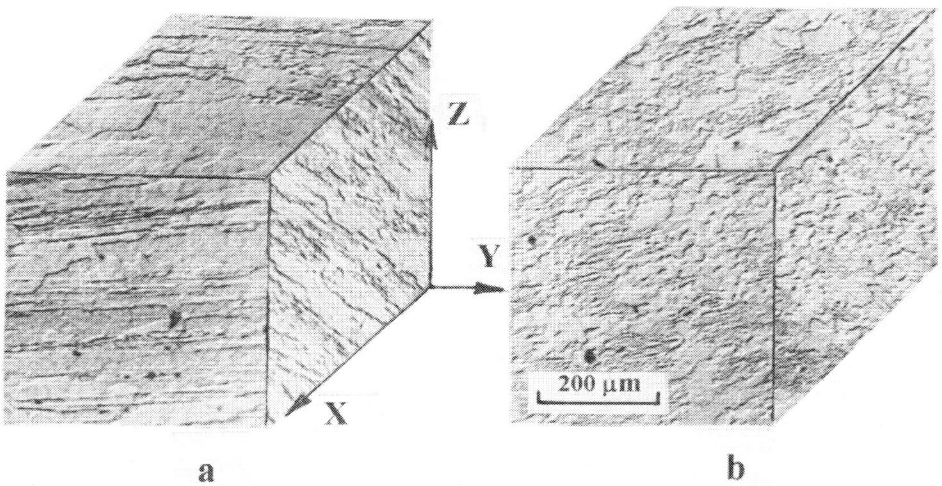

Figure 5.3. Montage microstructures of AA 7075 statically recrystallized (450°, 1h) after 8 passes ECAE (250°C, 2θ = 90°) via routes B (a) and C (b); axis X, Y, Z corresponds to Figure 57.

5.2.6. Effect of Routes

According to the model considered, the necessary conditions for structure refinement are (i) transition to localization, (ii) development of high angle boundaries along and within shear bands and (iii) their spatial distribution. The optimal deformation mode is simple shear that provides the earliest transition to localization. At the beginning of localization, intense shear should be accumulated at each processing step to develop HABs. Depending on inherent properties of the SPD techniques, the first two conditions can be satisfied by the proper tool design and the selection of optimal processing characteristics. The third condition relates to the loading paths. For monotonic loading, like in cases of simple shear tests (Figure 2.3) or high pressure torsion (Figure 2.14), unidirectional shear should be sufficiently large to initiate fragment rotation and develop three dimensional networks of HABs. Typical situations of cross–loading can be traced for basic processing routes during ECAE of long billets which were considered in the most experimental works. *For route A*, the angle φ_N between a shear band and axis Y after N passes of ECAE is (Figure 3.26a):

$$\tan\varphi_N = \cot\theta/(2N-1)$$

Here 2θ is the tool angle. The angle φ_N decreases sharply with N. For 2θ = 90°, angles between shear bands with N ≥ 3 becomes less than 12°, and shear bands start to coalesce. Two other families of shear bands with N =1 and N = 2 give φ_N = 45° and φ_N = 26.6°. Assuming that localized shear γ = 2 is sufficient for inducing HABs and that spacing between shear bands approximates to the cell size, intersections of these families of shear bands outline fragments with an aspect-ratio of about 1.5 and 3, respectively. Such elongated fragments are not suitable for rotation. For route A, UFG structures comprising similar elongated grains are observed experimentally. An estimated number of passes for structure refinement during ECAE via route A with the tool angle of 90° should include, at least, 2-3 passes to start localization and 2- 3 passes to induce shear band intersections. The resulting effective strain, ε ~ 7, is close to the experimental results of (Prangnell et.al., 2001) for route A.

For Route D, shear bands after four passes are almost mutually perpendicular and their intersections form nearly cube-like fragments (Figure 3.26d) resulting in equaxed grain structures. An important advantage of this route is that shear bands restore their orientations after every fourth pass with accumulation of angles of misorientation across the same SBs. The minimum number of ECAE passes in this case is from 4 to 5. Therefore, route D is the optimal route for structure refinement, conforming to many experimental observations.

For Route B, like in route A, shear bands (Figure 3.26b) can be distinguished for N ≤ 4. Their intersections form three dimensional elongated fragments inclined to both axis X and Y. These properties provide an intermediate position of route B between routes D and A in regards to structure refinement.

R*oute C* presents the special case of multi-pass ECAE with cyclic loading when shear planes are the same while a shear direction changes on the opposite direction after each pass (Figure 3.26c). During the continuous evolution, cyclic loading affects geometrically necessary and incidental boundaries in different ways. Alternative rotations of grain subdivided areas periodically create and annihilate geometrically necessary dislocation boundaries but accumulate incidental dislocations. As latter leads to oriented internal stresses, the material structure after each pass restores only partly and manifests certain hysteresis. It may be expected that cyclic loading during two passes ECAE results in low angle boundaries and well developed cell structures. During following passes of ECAE, these structures promote to localization and fragments rotation. Like route D, route C produces equaxed structures but delays development of HABs and is less effective for structure refinement. These conclusions also comply with experiments.

Therefore, effects of routes strongly relate to the processing mechanics of ECAE. The considered refinement model assumes unidirectional simple shear at each pass that can be attained for special boundary conditions. If the processing mechanics is not controllable, it may significantly differ from the simple shear deformation mode. Important experimental details such as contact friction, load, plastic zone, etc., are usually not disclosed whereas many published works show significant divergence of results. Inconsistency of experimental data on SPD was discussed in references (Segal, 2003; Zhu et al. 2000; Mishin et al. 2003; Zhylaev et al. 2006) and should be resolved by including actual processing mechanics in the structural analysis. For example, the controversial conclusion that for long billets route A is more preferable for structure refinement than route D (Prangnell et al. 2001) was accompanied by the observation of extensive plastic zone and shear into the extrusion direction (Gholinia et al. 2002). That presents the direct evidence of large contact friction as shown in Figure 3.13a. In this case, route A (Figure 3.28a) provides monotonic loading along one family of shear bands whereas route D (Figure 3.28d) redistributes cyclic loading between two families of shear bands and may be less effective for structure refinement.

5.3. Control of Textures

After microstructure, texture is the second important characteristic of the structure. Typically, deformation textures depend on prevalent systems of crystallographic glide in polycrystals. Different loading histories lead to characteristic final orientations and textures. Ordinary forming operations result in specific textures, such as textures attained during rolling, extrusion, etc., where texture strength increases with strains. For SPD with complex loading paths and very large strains, there are numerous combinations to induce various, sometimes uncommon textures. An experimental and theoretical analysis of the textures of SPD can be found in papers (Alexandrov et al. 1996; Gibbs et al. 1998; Li et al. 2005; Toth et al. 2004, Toth and Beyerlein, 2009) and others, and will be considered in detail in Part II. The systematic experimental investigation of the textures of ECAE was performed in (Ferrasse et al. 2004a, 2004b, 2004c) for a single phase, high stacking fault energy alloy Al0.5Cu. The starting material was subjected to two ECAE passes via route C followed by annealing at 250°C for 1h to obtain a uniform, recrystallized structure with average grain size 60 μm and a strong initial texture (O.D. index = 21.7 times of random (t.r.)) containing three major orientations (-111)<1-23>, (-102)<-28-1>, (-215)<-5-5-1>. Some samples were additionally deformed by four passes via route C and annealed at 300°C, 1h to produce a structure with grain size of 100 μm and a weak texture (O.D. index = 2.6 t.r.) of the major orientation (-113)<-8-11 1>. Both materials were subjected to ECAE with the number of passes varying from 1 to 8 via routes A, B, C, and D. An ECAE die for flat billets with a bottom slider and a tool angle of 90° provided near uniform simple shear along the intersection plane of the channels. Crystallographic textures were measured on a Phillips X'Pert diffractomer and processed by Beartex software. Textures were investigated along the mid-thickness of the billet section parallel to the X-Y plane (Figure 3.21b). Main results are described below[1].

5.3.1. Effects of Route and Passes

Partial pole figures of the {111}, {200} and {220} planes were measured for different routes and number of passes for two original textures. The raw data were processed to calculate the orientation distribution function (ODF), orientation fibers and OD indexes of the texture strength as the root mean square of the peak value of ODF. Examples of inverse (001) pole figures, which present the simplest way to compare different textures, are shown in Figure 5.4 after 2 and 4 passes via route D for weak and strong initial textures and in Figure 5.5 after 2 and 7 passes via route C for the sample with the initially strong texture. The

[1] For detail see (Ferrasse et al., 2004a, 2004b, 2004c).

following analysis of the experimental data reveals the general trends in texture evolution during ECAE.

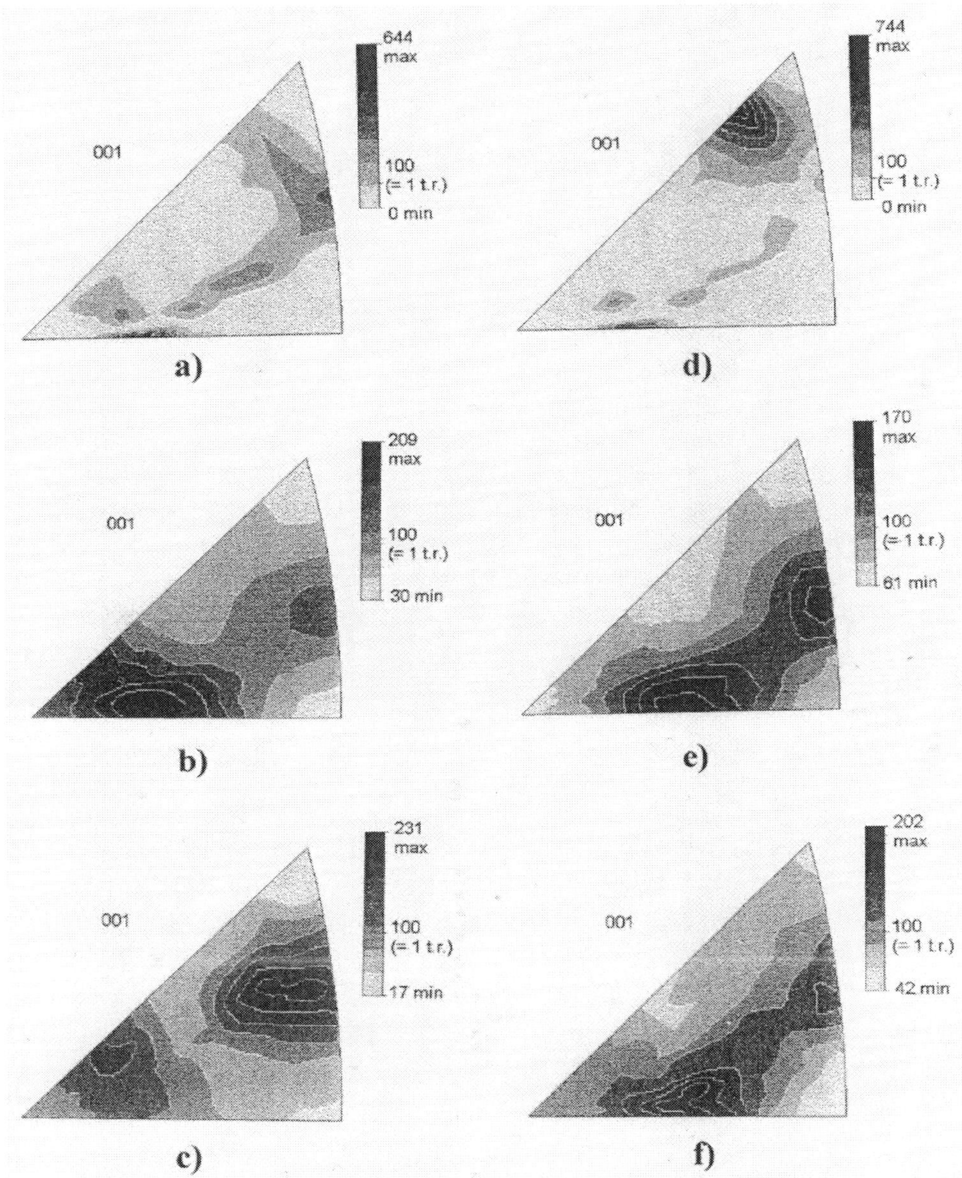

Figure 5.4. Inverse (001) pole figures for ECAE via route D: a - predicted by modeling for two passes of the original weak texture; b - experimental for two passes of the original weak texture; c - experimental for two passes of the original strong texture; d - predicted by modeling for four passes of the original weak texture; e - experimental for four passes of the original weak texture; f - experimental for four passes of the original strong texture.

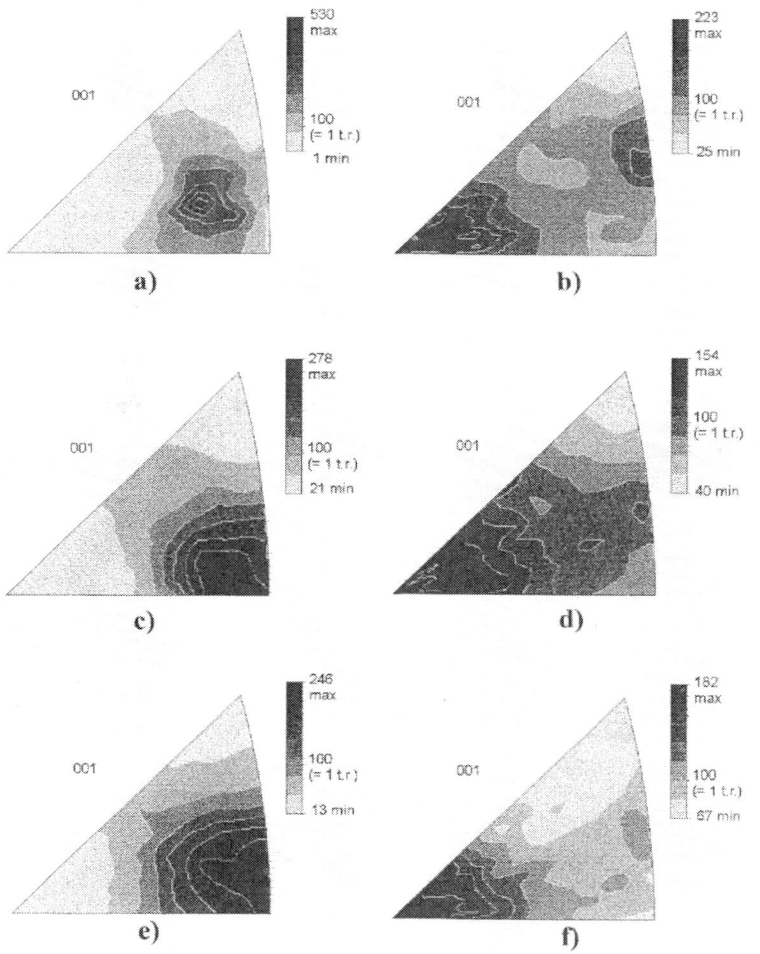

Figure 5.5. Inverse (001) pole figures for ECAE via route C for the original strong texture after 2 (a), 3 (b), 4 (c), 5 (d), 6 (e) and 7 (f) passes.

The weak initial texture exerts a limited influence during one and two passes. At higher numbers of passes, the influence of even the *strong initial texture* disappears. For example, two passes via route D shows different orientations for weak (Figure 5.4b) and strong (Figure 5.4c) initial textures whereas corresponding orientations are almost identical after three and four passes (figures 5.4 e, f). If the amount of deformation during each pass is high ($\varepsilon = 1.15$), a significant texture modification takes place after the first pass.

Routes have a strong impact on texture orientations especially at a low number of passes. Each processing route activates a unique combination of glide planes and directions resulting in different textures. A limited group of fibers containing one or several major orientations are associated with each route irrespective of the number of passes or original texture. The number of fibers may vary from two to six with one or two predominant fibers. For any route, fewer variations in orientations are seen as the number of passes increases. *Route A* results in a gradual build up of end orientations with mainly two planes (102) and (213). This type of evolution is similar to rolling. *For route D*, there are no significant changes in ODF or pole figures (Figure 5.4) after three passes in relation to an actual extrusion direction. Similarly, for *route B* the major components are stable after three passes. Because of rotations ± 90°, mirror symmetry is observed between odd and even numbered passes. *Route C* is the most atypical as the original texture evolution is reversed at each subsequent pass. However, the texture recovery is never fully realized and slightly new types of textures are permanently created due to a hysteresis phenomenon. This is evident in Figure 5.5 for even (a, c, e) and odd (b, d, f) numbered passes. Therefore, routes B, C and D produce the most variable and unique texture evolutions at each pass. Because of the multitude of combinations of processing routes, the opportunities to control texture are numerous. According to Figure 3.25, even the first three passes provide 14 different textures. They cover a significant area of the standard triangle and are fairly unusual compared to textures generated after traditional forming operations.

The evolution of the *texture strength* also unveils specific properties. Figure 5.6 presents a variation in the OD index with the number of passes via different routes and for two material conditions. The initial texture greatly impacts the texture strength at a low number of passes. For any route, the stronger the original texture the stronger the textures are after a few first passes. After three or four passes, the influence of the original texture on texture strength diminishes markedly. There is a general decrease in texture strength when the number of passes exceeds three. For a weak initial texture, a medium strength texture is observed after one and two passes for routes A, B, D and only weak textures develop afterwards. For strong textures, such weakening takes four passes. After four passes, variations in the texture strength, depending on route, decreases from medium-strong to very weak. Route C is, again, a special case which exhibits a cyclic evolution in the texture strength. Even after four passes, textures are very weak for odd numbered passes and rather strong for even numbered passes. Some oscillations are detected for routes B and D. Similar randomization/softening of textures was also found in (Alexandrov et al. 1996; Gibbs et al. 1998; Li et al. 2005).

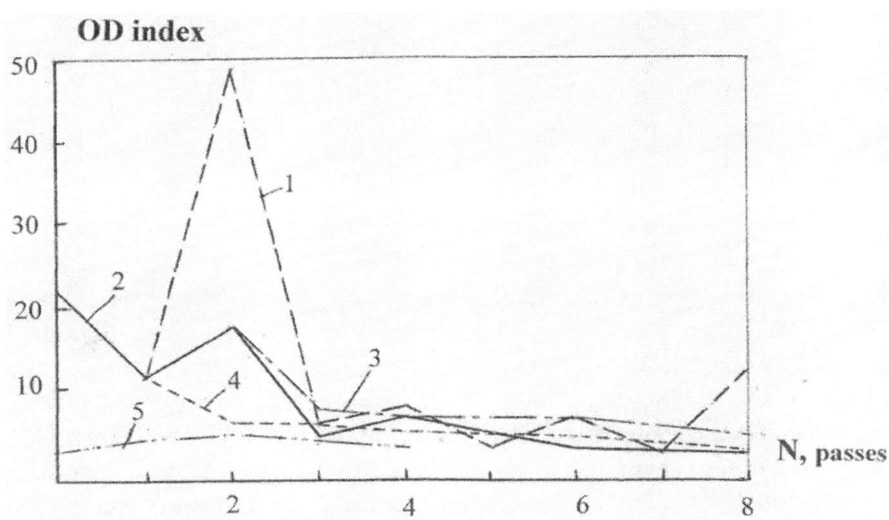

Figure 5.6. OD index of textures after ECAE via different routes in function of a number of passes: 1 – route C, strong original texture; 2 – route D, strong original texture; 3 – route B, strong original texture, 4 – route A, strong original texture, 5 – routes A and D, weak original texture.

Modeling texture evolution during different ECAE routes will be discussed in length in Part II of the book and is not going to be addressed here. It should be noted that for one and two passes, the simple Taylor polycrystalline model gives consistent predictions with experimental results (compare figures 5.4a and b) confirming that simple shear took place along the crossing plane of channels and that shear was accommodated mainly by crystallographic glide. However, agreement between simulation and experiment is poor after three passes (compare figures 5.4d and e). In particular, the model does not capture the texture hysteresis associated with route C or texture weakening observed for a large number of passes. These failures in predictions using the Taylor polycrystalline model demonstrate that other factors and deformation mechanisms start to operate after three to four passes of ECAE such as structure refinement, high internal stresses, plastic localization and rotation mechanisms that disperse X-rays and contribute to structure randomization.

Like structure refinement, textures of SPD are highly dependent on processing mechanics. When processing conditions are not controllable, there is the inconsistency between experimental and expected textures of SPD as was reported repeatedly for ECAE (Zhilyev et al. 2006; Gholinia et al. 2002; Toth et al. 2004). In these cases, characteristic textures were developed by intensive shear into the flow direction due to large contact friction similar to shown in Figure 3.28.

5.3.2. Effect of Post-Deformation Annealing

Annealing of SPD processed materials provides further control of texture strength and orientation. To analyze these opportunities , aluminum Al0.5%Cu alloy was processed with different number of passes and routes and annealed at temperatures of 150°C, 1h; 225°C, 1h and 300°C, 1h (Ferrasse et al. 2004b). These temperatures provide, respectively, recovery annealing, primary static recrystallization of the UFG material and secondary recrystallization of the fine grained material. Grain sizes after the corresponding annealing of the original UFG microstructure are 0.5μm, 7μm and 25 μm. Some examples of related inverse pole figures are shown in figures 5.7, 5.8 for 3 passes ECAE via route D and 6 and 7 passes via route C for the strong initial texture. Evolutions of the texture strength (OD index) for these routes with different number of passes and annealing temperatures are presented in figures 5.9 a, b.

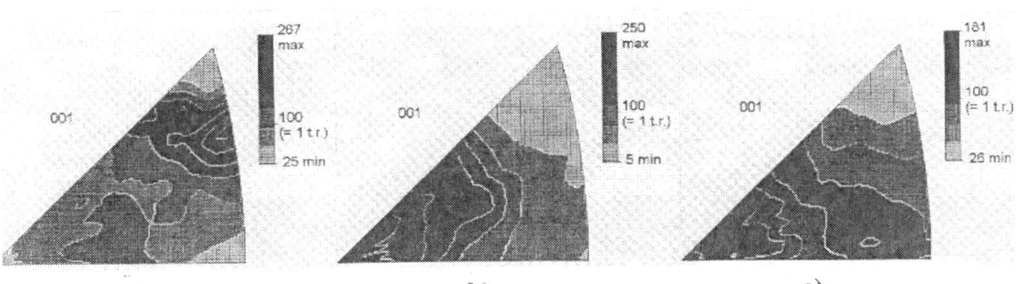

Figure 5.7. Inverse pole figures after 3 passes ECAE via route D with annealing 150°C, 1h (a), 225°C, 1h (b) and 300°C, 1h (c).

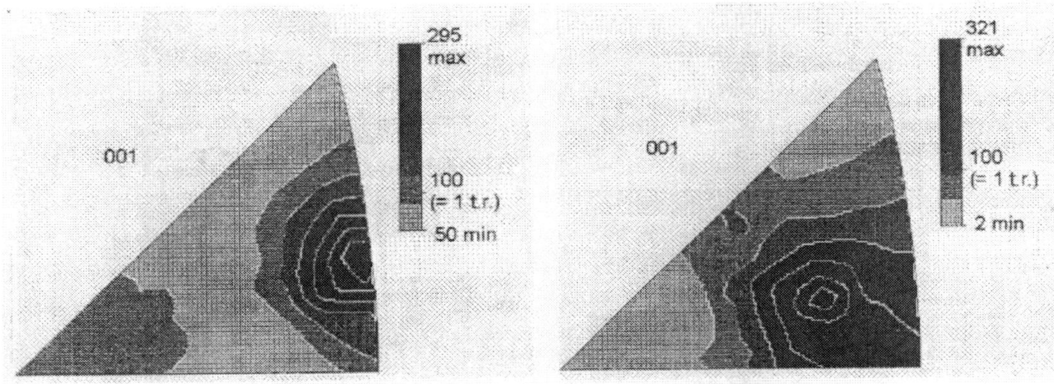

Figure 5.8. Inverse pole figures after ECAE via route C: a - 6 passes + annealing 150°C, 1h; b - 7 passes + annealing 150°C, 1h; c - 6 passes + annealing 225°C, 1h; d - 7 passes + annealing 225°C, 1h.

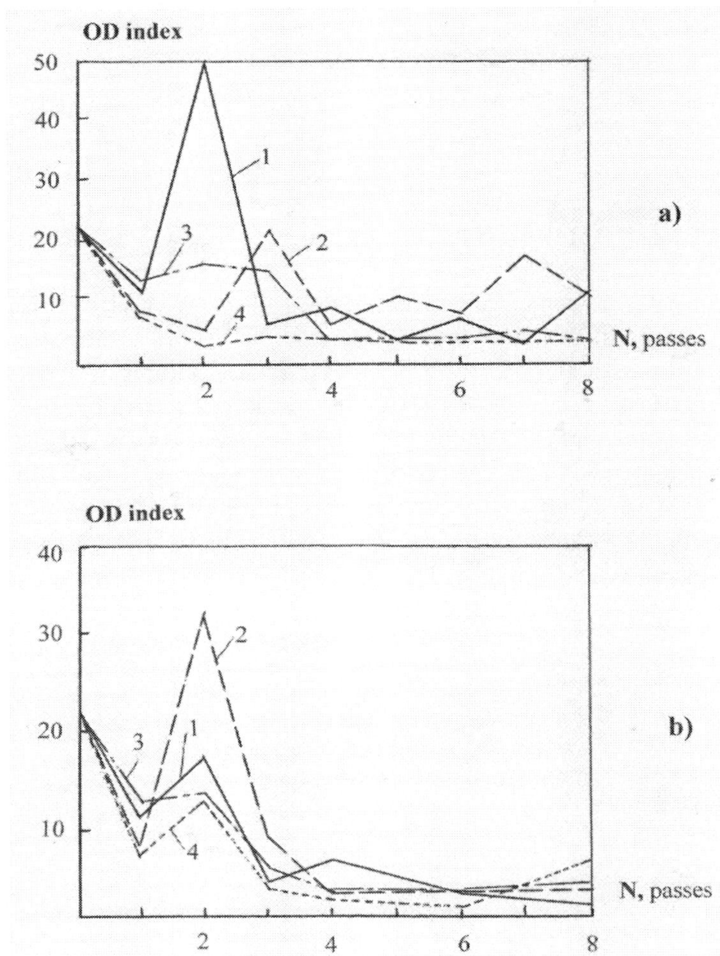

Figure 5.9. Effect number of passes on OD index of strong original texture after ECAE via routes C (a) and D (b): 1 – as deformed material; 2 – with annealing 150°C, 1h; 3 – with annealing 225°C, 1h; 4 – with annealing 300°, 1h.

Recovery annealing (150°C, 1h) reveals specific effects of deformation routes for low and large numbers of passes. *Before 4 passes*, there are obvious transformations in texture strength and orientation. The main fibers of deformed textures are visible. The most radical modifications are observed after two passes for routes C, B, D and after three to four passes for route A. Thus, for 2 passes via route C, OD index is reduced by a factor of 12 in comparison with the deformed texture. *After 4 passes*, for all routes except route C recovered textures remain quite similar to deformed textures. Routes D and B provide the most stable microstructures during recovery (Figure 5.7b). Route C exhibits a special behavior. Even after 5 and 7 passes significant changes are observed in texture strength and orientation. For odd numbered passes, texture is strong with the main orientation near (213)-(7 2 10) planes. For even numbered passes, texture is medium-strong and concentrated around planes (220) - (414). OD index (Figure 5.9a) shows the cyclic change with inverted amplitude of the deformed state. This texture evolution correlates with the evolution of dislocation structures during ECAE. For the first few passes, the continuous evolution develops heavily deformed microstructures including trapped dislocations, cells, subgrains, deformation bands and micro-bands with high internal stresses. Such low angle formations are not stable and can be partly removed, rearranged or annihilated under recovery annealing. For route C, an additional complication is strain reversal during odd and even numbered passes with alternating internal stresses. All these phenomena affect textures in a complex way. When the number of passes increases, the substructure is substituted by the UFG microstructure due to localization and fragment rotation. The new high angle boundaries are stable and do not evolve significantly during recovery annealing.

Primary static recrystallization annealing (225°C, 1h) also reveals two types of texture evolution, depending on the number of passes. With a *low number of passes*, that is less than 4 passes for routes B, C, D and 6 passes for route A, textures can be significantly modified in orientation (figures 5.7, 5.8) and strength (Figure 5.9). However, the same sets of fibers as the deformed/recovery annealed states can be identified. In this range of passes, the microstructure remains heavily deformed, non-refined and only partly recrystallized. With a *large number of passes*, where the microstructure is fully recrystallized, texture unveils most orientations of the recrystallized texture with a global decrease in texture strength. Probably, for these conditions, new grains were formed by boundary migration of the recovered UFG microstructure.

Secondary recrystallization (300°C, 1h) decreases the texture strength further regardless of the route and number of passes. OD index for routes A and C reduces continuously with passes, whereas routes B and D show cyclic changes between 1 and 2 passes. Despite a variety of orientations for routes with less than 4 passes, there is a general trend toward the orientation <100> similarly to the recrystallization of heavily rolled material. Orientations of the UFG material (after 4 or more passes) are more stable regarding recovered states especially for routes D and B. For route A, the limit of stability shifts to 6 passes. Probably, texture weakening after full recrystallization is attributed to new grains with different orientations while the stability of texture orientations after 4 passes is linked to UFG structures made up largely of high angle boundaries.

5.3.3. Effect of Additional Processing

Additional deformation processing may be necessary to fabricate a final product or modify a UFG material. The effect of post-SPD processing on texture was analyzed by comparative experiments of rolled UFG and fine grained (FG) Al0.5%Cu alloy (Ferrasse et al. 2004c). The UFG material with the grain size 0.5 μ was prepared using 6 passes ECAE via route D at room temperature. Following ECAE annealing at 225°C, 1h provided the FG microstructure with a grain size of 7 μm. Both materials were cold rolled with different reductions ranging from 50% to 99.5%. It was found that even after 50% of rolling reduction the UFG material acquires a new set of orientations, the initial texture produced by ECAE disappears and becomes very similar to a common rolling texture. With further increases in rolling reductions, textures remain stable and identical for both material conditions. However, considerable changes relate to the texture strength (Figure 1.19). Although texture strength for the UFG material (diagram 4) is noticeable lower than the FG material (diagram 1), it evolves from a medium strength texture with a reduction of 50% to a medium- strong and strong texture with a reduction of 95%.

In summary, severe plastic deformation, in particular ECAE allows the efficient control of not only microstructures but textures as well. Four independent technological parameters have a significant influence on texture: accumulated strains, number and sequence of shear planes and directions, post-deformation annealing and additional processing. SPD presents vast opportunities to control texture orientations and strength within the standard triangle ranging from near random to very strong.

5.4. Geometrical Control of Structural Elements

In single phase materials morphology of refined microstructures is defined by intersections of shear bands and, depending on the route, the aspect-ratio of new grains may vary from nearly 1 to about 3. In multi-phase materials, intensive plastic deformation of ductile phases is accompanied by structure refinement and the change of morphology of each phase in accordance with global material distortion. As it follows from Figure 3.24, aspect-ratios and orientations of phases can be controlled within broad ranges using different number of passes and routes. Under monotonic multi-pass simple shear, the originally equiaxed phases are transformed into oriented thin filaments. These types of materials, known as natural or *in situ* composites, present significant industrial interest (Verhoeven, 1975). Historically, patented steels were one of the first examples of deformation processed composites. Other applications include filament Cu-Nb-Ti superconductors, two-phase Cu-Co magnets, high-strength/high-temperature Cu-Nb, Cu-Ag conductors, etc. The main structural characteristics in such cases are the large aspect-ratio and small spacing between second phase filaments to provide strong anisotropy and an UFG structure. The critical step in fabrication of *in situ* composites is intensive cold or warm deformation used to reduce the material thickness by 2-3 orders of magnitude. With ordinary forming operations such as rolling and drawing it can be achieved only in small cross-sectional products.

An alternative method for the processing of bulk *in situ* composites, introduced in (Segal et al. 1997), comprises multi-pass ECAE via route A with a strong distortion of structural elements into the flow direction. Figure 5.10 shows the distortion of the original equiaxed element δ_0 into a ribbon-like shape induced by shear $\gamma = 2N\cot\theta$. After ECAE with a large number of passes N >>1, the element thickness δ and length l are

$$\delta \sim \delta_0/\gamma,\ l \sim \delta_0\gamma, \qquad (5.1)$$

whereas the element aspect-ratio and angle φ_n with the extrusion direction can be found from formulae

$$l/\delta \sim \gamma^2,\ \cot\varphi_N \sim \gamma \qquad (5.2)$$

Compare Eqs.(5.1), (5.2) with identical results for rolling. Assuming a uniform strain distribution during rolling from an original material thickness H_0 to a final thickness h, one will have for thickness, length and aspect-ratio of the distorted elements:

$$\delta = \delta_0/\lambda,\ l = \delta_0\lambda,\ l/\delta = \lambda^2 \qquad (5.3)$$

where $\lambda = H_0/h$ is the area reduction during rolling. Formulae (5.1) - (5.3) describe the equivalence between shear γ during N passes ECAE and rolling reduction λ that produces identical distortions of material elements

$$\gamma \sim \lambda,\ N \sim \lambda/2\cot\theta \qquad (5.4)$$

The equivalence for plastic energy or von Mises strains between ECAE and rolling is:

$$\gamma \sim \ln\lambda / 2,\ N \sim \ln\lambda / 4\cot\theta \qquad (5.5)$$

Equations (5.4), (5.5) show that simple shear is much more effective than ordinary forming operations in providing plastic energy but is less effective in providing distortions. For example, by comparison with the rolled area reduction $\lambda = 100$, only 2 passes ECAE with a tool angle of $2\theta = 90°$ induce equivalent energy in accordance with formulae (5.5) whereas 50 passes are required for equivalent distortions in accordance with formula (5.4). However, an important advantage of ECAE is the conservation of the material cross-sectional area during processing.

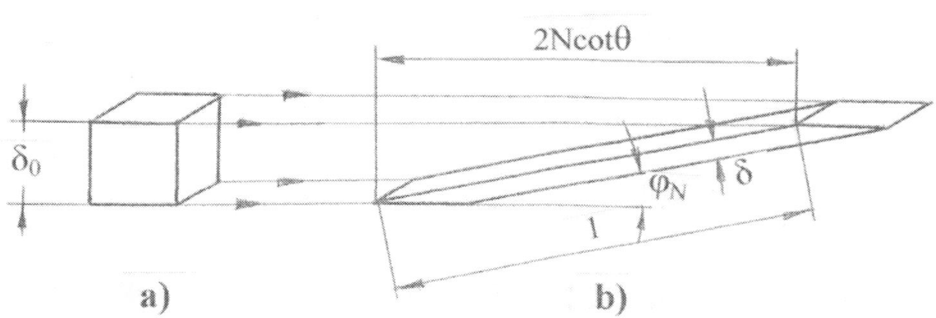

Figure 5.10. Element distortion during route A.

Consider the experimental results for ECAE of a Cu-18%Nb composite. This alloy exhibits an exceptional ductility and high strength after very large reductions (Trybus et al. 1989). In the original as-cast condition the structure is comprised of 15-25 μm diameter dendrite-like crystals of the primary Nb phase separated by a relatively pure copper matrix (Figure 5.12a). The Nb dendrites are similar in shape and size but are distributed rather non-uniformly throughout the material. Billets were SPD processed at room temperature up to 24 ECAE passes in a die having a tool angle of 90° with movable walls in both channels. The theoretical picture of plastic macro-flow after a large number of passes (Figure 5.11a) includes four uniformly deformed zones I-IV. Within these zones flow lines are straight but change direction abruptly at the boundaries SO. Such flow may be presented as simple shear in opposite directions relative to the billet axis SS. Highly elongated material elements are inclined by the small angle α to the flow lines. The experimental appearance of the flow plane after 24 passes (Figure 5.11b) closely agrees with the theoretical model. Figure 5.12 shows microstructures at the flow plane for 1 (b), 8 (c) and 24 (d) passes via route A with largest distortions. During the first pass, the flow is uniform and distortions of the Nb-phase conform to the simple shear model. Serrated flow and micro-localization was detected at the fourth pass. During the passes that followed, localization manifests itself in the finer scales (Figure 5.12c). Finally, very fine, slightly wavy filament microstructures are formed (Figure 5.12d). A comparison of tensile test results for the ECAE composite with published data for similar Cu-20%Nb composites fabricated by rolling and drawing (Trybus, 1989) as a function of equivalent strains from Eq. (5.5) shows good agreement (Figure 5.13).

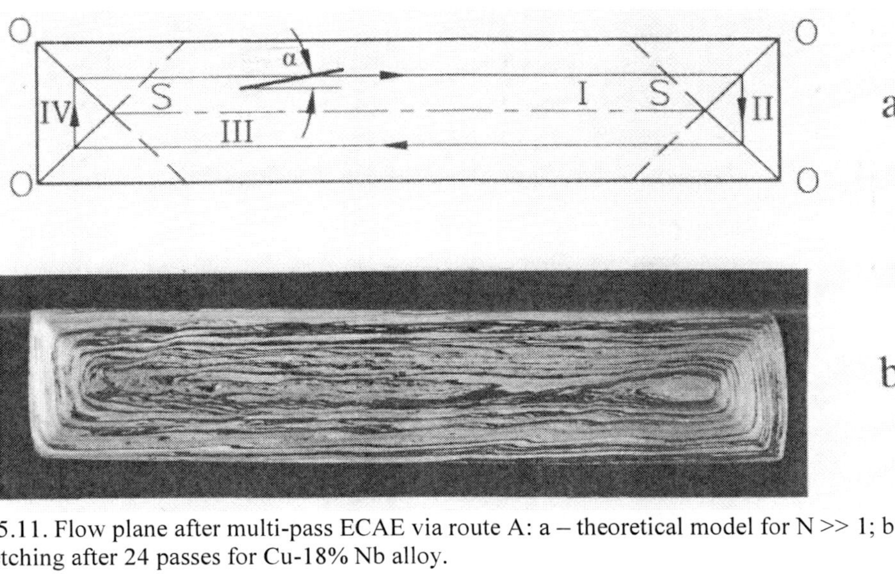

Figure 5.11. Flow plane after multi-pass ECAE via route A: a – theoretical model for N >> 1; b – macro-etching after 24 passes for Cu-18% Nb alloy.

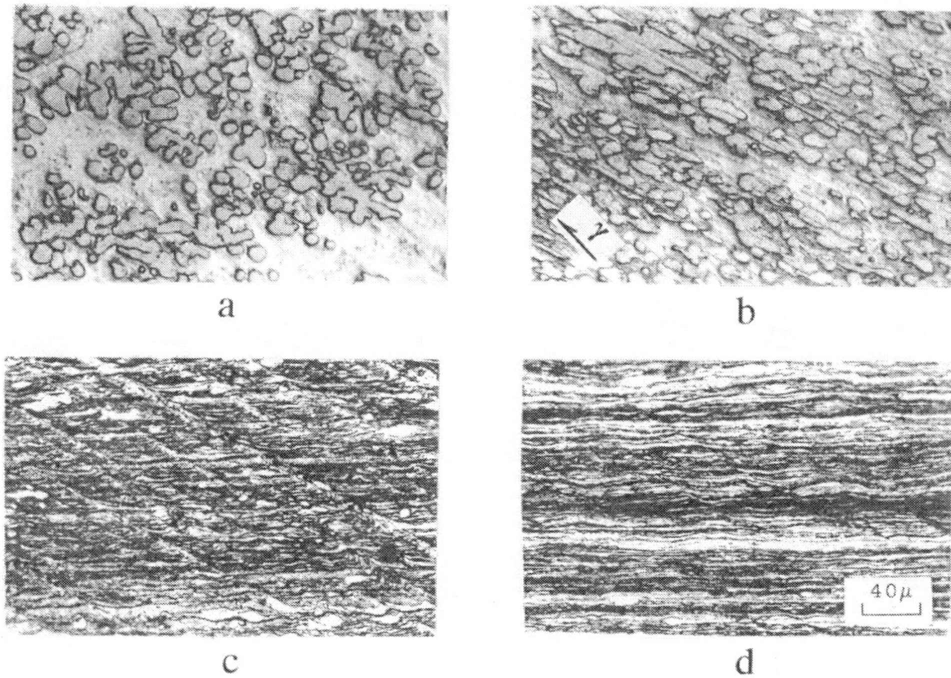

Figure 5.12. Microstructures of Cu-18%Nb alloy in the as-cast condition (a) and at the flow plane after ECAE via route A with 1 pass (b), 8 passes (c) and 24 passes (d).

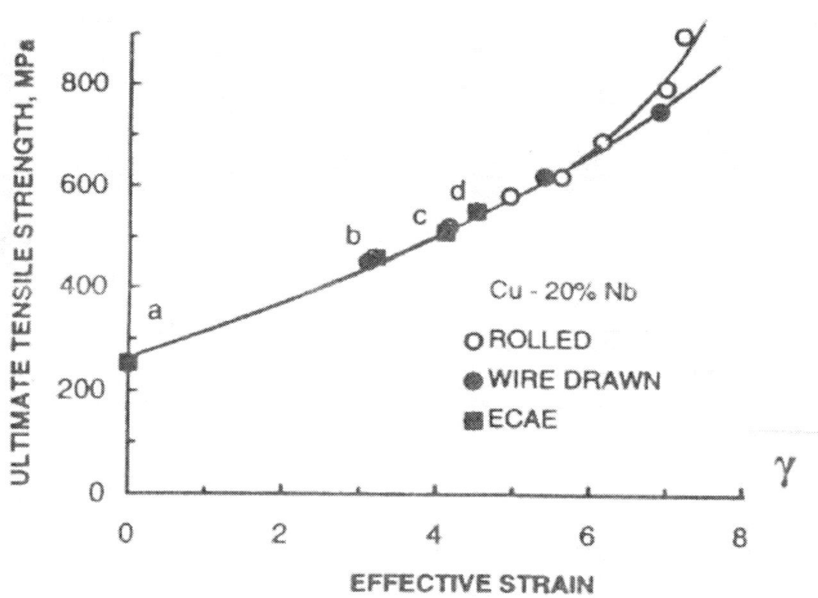

Figure 5.13. Ultimate tensile strength versus accumulated ECAE strain γ for Cu-18%Nb alloy (rolling and drawing data for Cu-20% Nb from (Trybus et al. 1989)).

These experiments demonstrate technical feasibility for fabrication of bulk *in situ* composites by SPD. As the primary goal is the reduction of the second phase thickness (δ/δ_0) by about 3 orders of magnitude, the formula (5.1) gives $\gamma \sim 10^3$ for the necessary shear. Corresponding area reduction during rolling $\lambda \sim 1000$ or number of ECAE passes ($2\theta = 90°$) $N \sim 500$ are too large and not practical for most applications. However, there is a principal difference between phase distortions during near "ideal" forming operations such as rolling and a SPD like ECAE. For "ideal" forming operations, distortions are produced by pure shear in linear proportion to strains. For ECAE, the phase distortions are produced by simple shear in sharply nonlinear proportion to strains. By differentiating equation (5.1), one can find that the rate of changing of the phase thickness δ with shear γ is

$$d\delta/d\gamma = - \delta_0/\gamma^2$$

This formula shows that the rate ($d\delta/d\gamma$) drops very quickly with the increase of shear γ. As for a tool angle 90° $\gamma \sim 2N$, only the first few passes of ECAE are effective in the fabrication of *in situ* composites. It follows from formulae (5.1), (5.2) that five ECAE passes are necessary to transform originally equaxed second phases into thin filaments with a relative thickness (δ/δ_0) $\sim$ 10 and aspect-ratio (l/δ) $\sim$ 100 orientated closely to the extrusion direction ($\alpha < 6°$). Thus, combining 5 passes of ECAE with rolling or extrusion operations, the required area reduction can be reduced or cross-sectional area of the final product can be increased by a factor 10.

Another technological opportunity is the application of spheroidization annealing after a few ECAE passes. It is known that ultrafine filaments can be transformed into stable spherical phases using the appropriate annealing (Hong et al. 1995). Therefore, the aquiaxed shape of distorted phases can be periodically restored at a much finer scale during multi-pass ECAE with intermediate spheroidizing annealing. In this case, three successive steps of 5 passes of ECAE followed by intermediate spheroidizing annealing can reduce filament thicknesses by three orders of magnitude and effectively fabricate bulk in situ composites.

5.5. Deformation of UFG Materials

Plastic behavior of UFG materials defines their strength, ductility, toughness, fatigue and other characteristics. Fabrication of useful products from bulk billets after SPD usually requires secondary forming operations with large plastic strains such as forging, rolling and extrusion. In the general context, processing and applications of UFG materials at temperatures below the temperature of static recrystallization are discontinuous loading histories with specific deformation mechanisms. They should be adequately identified and described in order to interpret results of mechanical testing and evaluate possible applications. Structural peculiarities of UFG metals include equaxed grains from a few microns to a sub-micron size with extensive non-equilibrium boundaries and practically dislocation free interiors. These structures are within a range between ordinary polycrystals and nano materials and, under particular conditions, they may exhibit mechanical characteristics similar to one or the other materials. When intragranular crystallographic glide is the dominant deformation mechanism, UFG materials follow the normal Hall-Petch relationship between the yield stress and the grain size similarly to their coarse grained counterparts. On the other hand, the plastic flow in nano structures is mainly controlled by the grain boundary activity of dislocations and other defects (Meyers et al. 2006; Gutkin et al. 2003; Dao et al. 2007). Corresponding mechanisms of shear band localization, grain boundary sliding, grain rotation and diffusion flow may operate in UFG materials. The processing conditions and, in particular, the deformation mode play a critical role in the activation of these mechanisms and their relative contribution at different stages of deformation.

5.5.1. Tensile Testing

As in ordinary materials, tensile testing of cylindrical samples is the primary experimental technique for the evaluation of UFG metals. Interpreting experimental stress-strain data does not present a problem during the stage of uniform elongation as the instantaneous sample diameter and length are easily calculated from the total elongation. Assuming that elastic and diffusion-viscous strains are small in comparison with plastic strains, one can find engineering strain ε, true strain e, area reduction ψ and diameter d_l of the cylindrical sample after uniform elongation Δl_l from formulae (Segal et al. 2006):

$$\varepsilon = \Delta l_l / l_0,\ e = \ln(1 + \varepsilon),\ \psi = \varepsilon/(1 + \varepsilon),\ d_l = d_0/(1 + \varepsilon)^{1/2} \quad (5.6)$$

True tensile stress σ and strain rate ξ are:

$$\sigma = 4P(1 + \varepsilon)/\pi d_0^2,\ \xi = 2V/l_0(1 + \varepsilon) \quad (5.7)$$

Here l_0 and d_0 are length and diameter of the original gauge region, P is the load corresponding to elongation Δl_l with $2V$ being the tensile speed. The previous equations are applicable if deformation in the gauge is uniform and the sample retains its cylindrical shape. For ductile coarse grained metals, this stage may be sufficiently long to determine characteristics of plasticity, hardening and strength. Micro- flow localization in the form of Lüder's bands and shear bands may also be observed. They reflect specific material properties and leave characteristic marks on the load-displacement diagram but do not contribute significantly to the deformation history.

When the material hardening ability is reduced to some critical level, plastic deformation is localized in the sample neck (Backofen, 1972). Subsequent interpretation of experimental results closely relates to the neck's evolution. An approximate stress analysis in the sample neck was performed in (Bridgman, 1944; Davidenkov et al. 1946; Segal, 1969). R.T. Shield (Shield, 1955) considered the kinematical state at the beginning of necking. Evolution of the neck in different materials was repeatedly analyzed by the finite element method. However, these results are difficult to apply for interpreting experimental data in particular cases, as additional measurements of the neck's shape are necessary to estimate actual stresses and strains. For UFG metals, flow localization in the sample neck is observed almost immediately after passing the yield stress. Because of experimental complexity, the stress-strain state in the neck is usually excluded from the analysis. In recently published works (Hayes et al. 2000; Wang et al. 2005; Yu et al. 2005), many specific details of tensile properties of UFG metals were described based on machine load-displacement diagrams without consideration of true stress-strain relations. Therefore, the real behavior of UFG metals remains unclear.

Practically, the biggest interest is in characterizing the change of the minimum neck cross-sectional area that is necessary to calculate the average stresses and strains. Relations between these characteristics are usually accepted as the true stress-strain diagrams during tensile tests. Slip line fields for the beginning (Figure 5.14a) and intermediate (Figure 5.15a) steps of the neck localization show that solutions in the plastic zones ABDO are defined by the geometry of free boundaries ABD. Therefore, the simplest useful estimate is to approximate these boundaries by straight lines and to reduce the problem to a quasi steady one (Segal et al. 2006). The associated plastic flow requires uniformity of the radial velocity component u along ABD. Then, the continuity of the material flux through boundaries AO, DO and ABD gives:

$$u = - V/2 \text{ along ABD} \tag{5.8}$$

This velocity u remains constant during the neck localization. The comparison of the approximation (5.8) with the correct solution at the beginning of localization (Shield, 1955) reveals a sufficiently close correlation (see curves 1, 2 on Figure 5.14b). It is reasonable to expect that the equation (5.8) provides an upper estimate for the neck diameter d at any moment of localization. This does not affect significantly the calculated true stresses because the average stresses in the neck area increase with the increase of the neck's curvature. In accordance with Eq. (5.8), the increment of neck diameter Δd during an interval t of the stage of localization with elongation $\Delta l_2 = 2Vt$ is

$$\Delta d \sim - 2ut = - \Delta l_2/2.$$

Figure 5.15b presents calculated neck shapes for two steps of localization. Plastic zones at any moment are outlined by boundaries at angles ± 45° to the sample axis whereas outer neck branches form angles arctan3 = 71.6°. For the stage of localized flow, sample elongation Δl_2 and neck diameter d are:

$$\Delta l_2 = \Delta l - \Delta l_1, \; d \sim d_1 - \Delta l_2/2 \tag{5.9}$$

Hence, the true strain, stress and strain rate in the neck during this stage are:

$$e = \ln(d_0/d), \; \sigma = 4P/\pi d^2, \; \xi = 2V/(l_0 + \Delta l) \tag{5.10}$$

In these equations Δl is the full elongation, Δl_1 and d_1 are the uniform elongation and diameter, P is the load and $2V$ is the tensile speed for the stage of localization. Poortmans et al (Poortmans et al. 2007) using finite element simulation and precision measurements of the sample neck for UFG aluminum, found that formulae (5.6) – (5.10) give a good prediction of the real behavior. Therefore, this approach provides a valid technique for reprocessing of experimental load-displacement data into true stress-strain diagrams.

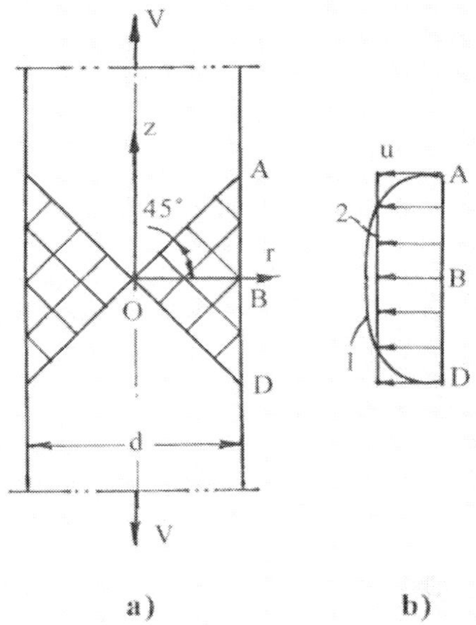

Figure 5.14. a) Slip line field at the beginning of localization during tensile testing; b) Distribution of radial velocity u at free surface ABD for Shield's solution (1) [187] and approximate solution (2) (5.8).

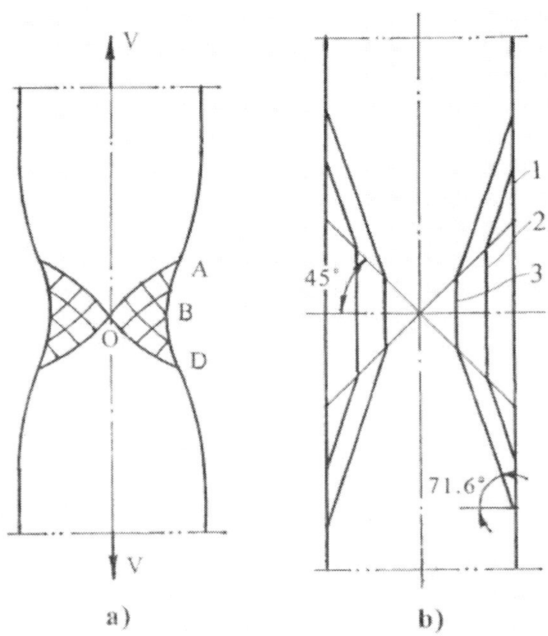

Figure 5.15. Slip line field (a) and approximations of neck shapes (b) during tensile testing.

Experiments on tensile testing of cylindrical samples Al0.5%Cu alloy were performed for three material conditions: (1) UFG (0.5 μm) structure after 6 passes ECAE via route D at room temperature; (2) UFG (0.9 μm) structure after post-ECAE recovery annealing at 175°C, 1h; (3) fine grained (FG, 7 μm) structure after post-ECAE recrystallization annealing of 225°C, 1h. The character of the load-displacement diagrams is identical for both UFG structures but show a distinctive difference with FG material (Figure 5.16). For UFG structures, diagrams 1, 2 reveals a near linear increase up to the beginning of plastic flow (point *a*), a load increase to some maximum (points *b*) and a load drop before neck localization (point *c*). During localization (segment *ce*), loads decrease systematically until point e where axisymmetric flow in the neck is substituted by planar shear within a thin material layer at an angle of 45° to the sample axis followed by fracture. The stages of localized flow are sufficiently extended whereas the stages of uniform elongation are very short. The FG material exhibits the standard diagram 3 with a continuous increase of the load until the maximum point *c*, subsequent localization in the neck and a steep decrease of the load up to failure.

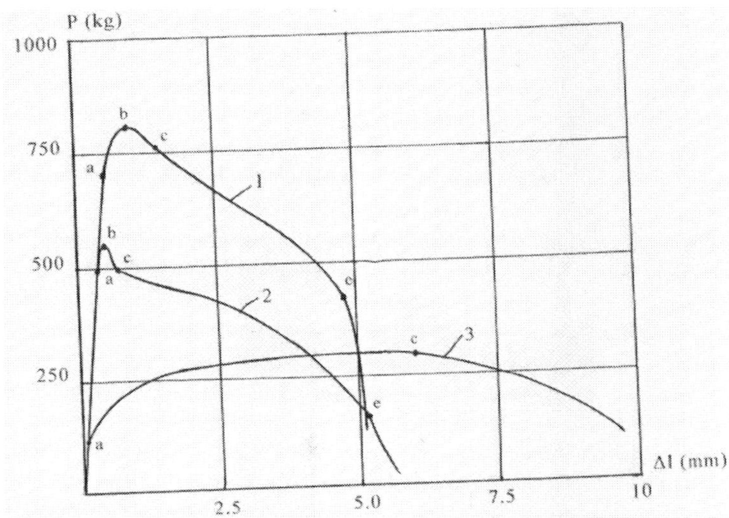

Figure 5.16. Experimental load-elongation diagrams for Al0.5%Cu UFG : 1- UFG material after 6 passes ECAE via route A; 2- UFG material + recovery annealing 175°C, 1h; 3 – FG material after recrystallization annealing 225°C, 1h.

The calculated true stress-strain diagrams (Figure 5.17) look much more similar for FG and UFG materials than load-displacement diagrams. Careful analysis of diagrams together with related modes of plastic flow (Figure 5.18) reveals important details and provides insights into mechanisms of deformation and fracture of UFG structures. Before the beginning of plasticity (Figure 5.17, points *a*), elastic deformation is homogeneous and axisymmetric (Figure 5.18a). In UFG structures (diagrams 1, 2, Figure 5.17), the yield stress (points *a*) is strongly affected by the material's grain size. The stage of uniform plastic deformation (*ab*) takes place by planar shear at an angle of 45° to the sample axis with characteristic marks at the sample surface (Yu et al. 2005). The transition to neck localization (*bc*) for UFG materials shows a specific fold which is sharper for the recovered condition than for the deformed condition. During localization (after points c), true stress-strain diagrams reveal noticeable hardening until specific points *d* where hardening turns into softening (portions *de*). Plastic flow during the macro localization in the neck is axisymmetric (Figure 5.18c). For the FG material, the true stress-strain diagram 3 (Figure 5.17) shows the ordinary behavior. The yield stress of FG material is more diffuse than that for UFG structures and the transition to localization (point *c*) is very smooth. There is an essential difference between UFG and FG materials during the phase of uniform elongation (*ac*). The FG material demonstrates an extended stage of uniform elongation with a pronounced decrease in hardening rate for higher strains. For UFG structures, this phase is very short and displays a specific transition to localization. The likely reason for these effects is the modification of deformation mechanisms after the beginning of plasticity in UFG materials prepared by SPD. During material fabrication by ECAE, the plastic flow at the meso scale conforms to shear band localization along continuum slip lines. Corresponding mechanisms of grain boundary sliding and rotation provide lower global yield stress than the competitive mechanism of intragranular crystallographic glide. As mechanisms of intragranular glide and localization are nearly balanced, trapped dislocations inside grains and some material hardening may be observed (Gutkin et al. 2003). When samples are tensile tested into the extrusion direction, macro slip lines are oriented at the same angle of 45° as the shear planes during ECAE. Therefore, plastic flow first starts along preexisting shear planes of ECAE as was observed in (Yu et. al., 2005). However, even small additional strains change the balance and transform the deformation mechanism to crystallographic glide within grains (Wang et al. 2005; Yu et al. 2005). Recovery annealing reduces the density of trapped dislocations and promotes transformation displaying a sharper drop. Testing of UFG materials at cryogenic temperatures also shows similar yield-drop effect (Wang et al. 2005; Yu et al. 2005) because the grain boundary dislocation activity becomes more difficult at low testing temperatures. During macro localization in the neck (after points *c*), UFG materials show a noticeable hardening ability despite severe straining under SPD. Hardening areas *cd* for diagrams of FG and UFG materials are quite similar confirming identical mechanisms of dislocation glide in both cases. As hardening rates $(d\sigma/d\varepsilon)_c$ at the points *c* for UFG materials are lower than for FG material, UFG structures reveals neck localization almost immediately with the modification of the deformation mode into intragranular glide.

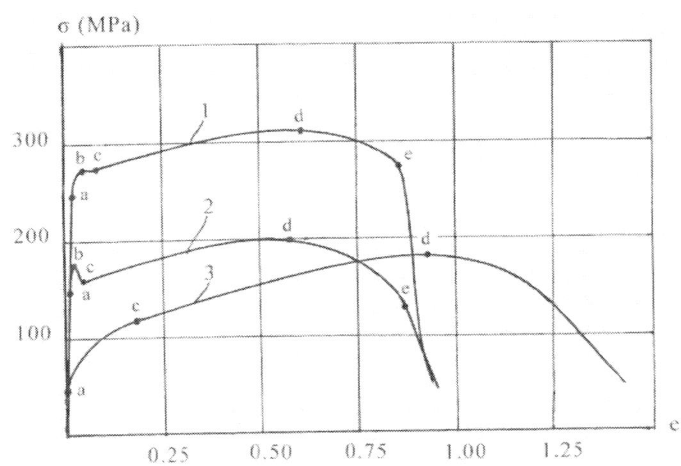

Figure 5.17. Calculated true stress-strain diagrams for Al0.5%Cu alloy: 1- UFG material after 6 passes ECAE via route A; 2 – UFG material + recovery annealing 175°C, 1h; 3 – FG material after recrystallization annealing 225°C, 1h.

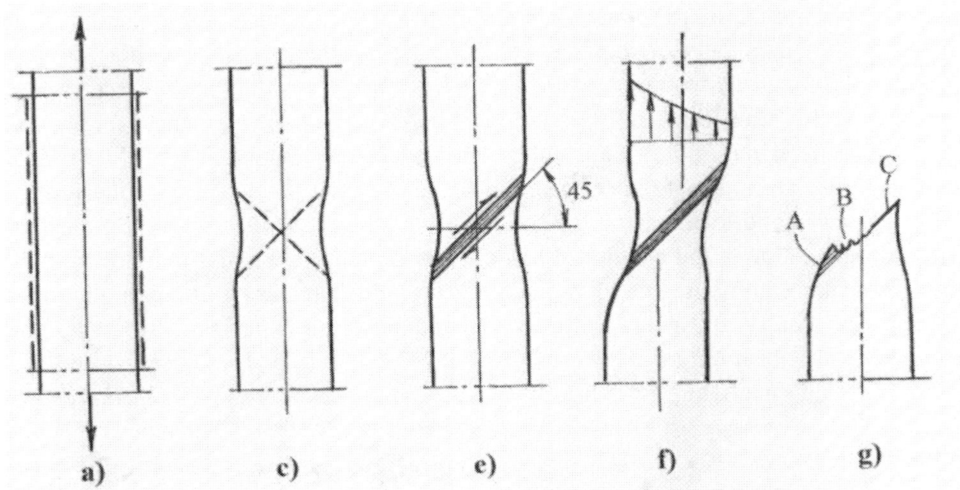

Figure 5.18. Plastic flow during tensile test of cylindrical samples: a – uniform elongation; c – macro localization in the neck; e – the beginning stage of shear band localization; f – the final stage of shear band localization; g – fracture.

The softening areas *de* of the true stress-strain diagrams are caused by accumulation of volume defects such as pores and voids. At some moment, that leads to a second transition to localization within a thin material layer at an angle of 45° to the tensile direction (Figure 5.18e). Extended shear in the layer shifts the sample ends and develops eccentric tensile stresses (Figure 5.18f) that, finally, results in a complicated fracture (Figure 5.18g). For UFG structures, the fracture surface (Figure 5.19a) includes three characteristic areas: A - free surface area of the planar shear zone; B - dimpled area of ductile tensile fracture; C - area of material shear decohesion. A micrograph of the transverse section through the neck (Figure 5.20a) confirms this observation. The planar shear zone A consists of a large number of thin shear bands. The ductile fracture area B contains many dimples uniformly distributed through the neck due to nucleation, growth and coalescence of pores and voids near hard particles. The decohesion area C manifests a similar mechanism of pore evolution under simple shear (Figure 5.20b) confirming that local stresses at hard particles in UFG structures are sufficiently high for nucleation of many pores at the beginning of localization. Their ensuing growth modifies the axisymmetric stress-strain state and, finally, substitutes it by planar shear. In contrast, the FG material manifests the usual tensile fracture mechanism (Figure 5.19b) (Backofen, 1972). During development of the neck, fracture was initiated by a single pore near the sample center where tensile stresses are at maximum and exceeds the interface strength between particles and the matrix. This pore originates in the internal neck area whose evolution together with the evolution of the external neck form a conical crater and ring like dimpled fracture area.

It is clear that true stress-strain diagrams beyond point *d* depend on fracture mechanisms rather than on deformation mechanisms. The related material behavior limits its use for most applications. In accordance with the true stress-strain diagrams (Figure 5.17), localized strains in the neck (portions *cd*) are the main component of accumulated strains whereas the contribution of uniform elongations is insignificant. Total plastic strains *ad* before softening can be accepted as an index of the material ductility. In comparison with tensile testing, even more advanced ductility of UFG materials should be expected for other stress states with compressive hydrostatic stress components preventing fracture nucleation at hard particles. In such cases, UFG materials can be successfully rolled, forged or extruded with large reductions to fabricate various products.

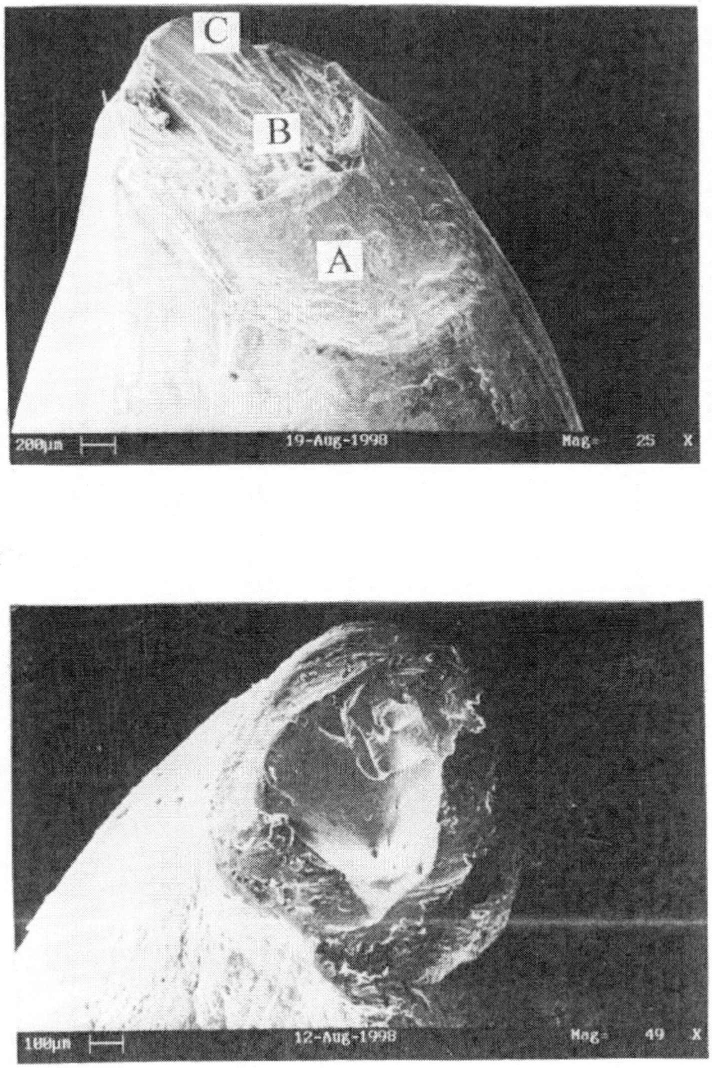

a)

b)

Figure 5.19. Tensile fracture surfaces of Al0.5%Cu alloy: a - UFG recovery annealed (175°C, 1h) material; b - FG recrystallized (225°C, 1h) material.

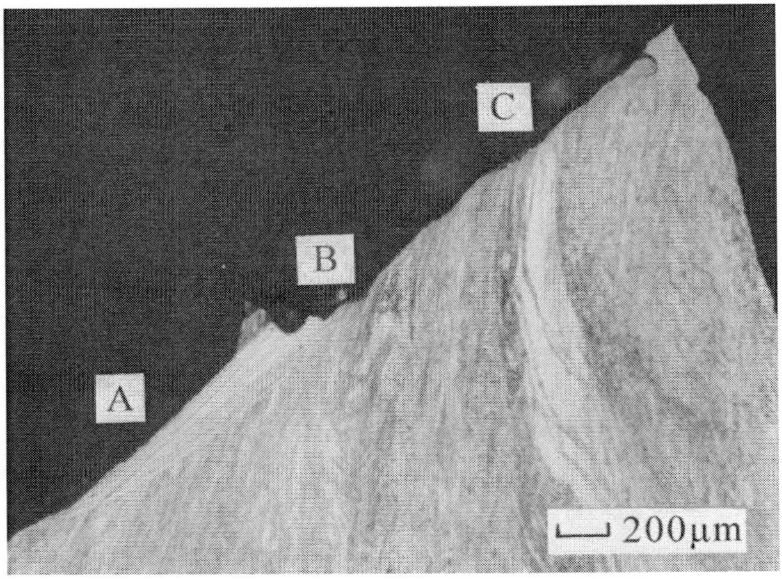

a

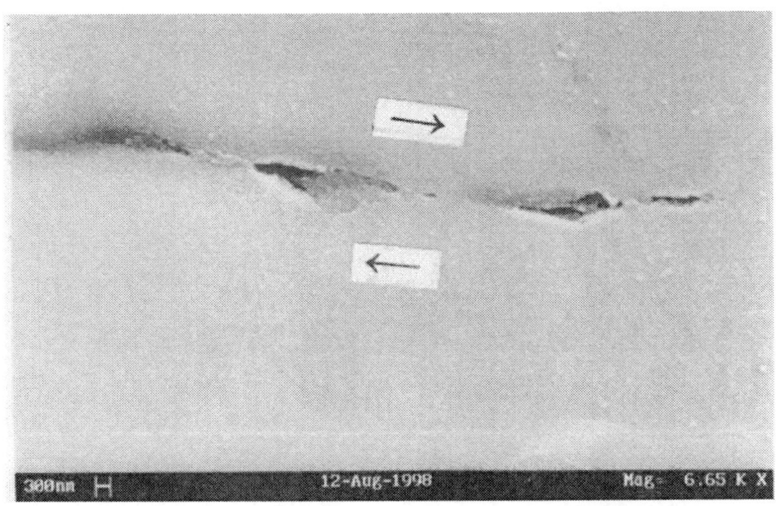

b

Figure 5.20. Transverse neck cross-section (a) after tensile fracture of UFG Al0.5%Cu alloy with recovery annealing 175°C, 1h (A – free surface of planar shear area; B – dimpled fracture area; C – shear decohesion area) and shear decohesion fracture mechanism (b).

5.5.2. Processing of UFG Materials

For large strains, there are significant problems in detection of deformation mechanisms by direct experimental methods of the structural analysis (Gutkin et al. 2001). Usually, UFG materials produced by SPD are analyzed by TEM and EBSD techniques. These methods reveal final structures after full deformation history and are not effective to manifest flow mechanisms during different steps of severe plastic deformation. Vinogradov et. al. (Vinogradov et al. 2001) demonstrated that TEM pictures are identical for ECAE processed materials and after additional loading. They applied atomic force microscopy (AFM) to separate incremental and total strains in UFG metals and found the shear localization at the fine structural scale related to grain boundary sliding during small tensile strains. Since the first stage of plastic flow during tensile testing has the same deformation mode like ECAE (areas ab of diagrams 1, 2 in figures 5.16, 5.17), it is reasonable to expect that grain boundary localization (figures 1.2b, c) is the main deformation mechanism in both cases. However, this situation is changed by modifications of deformation mode or loading conditions. Correspondingly, other mechanisms may act in UFG materials. Subsequent tensile testing of UFG structures shows the axisymmetrical plastic flow in the neck (Figure 5.18c) is quite similar to ordinary coarse grained materials until the evolution of voids stimulates the transition to the second shear band micro localization (Figure 5.18e). When the temperature of SPD is higher than the temperature of tensile tests, UFG structures demonstrate some hardening with the increasing uniform elongation (Wang et al. 2005) or with an increased stability to micro localization (Huang et al. 2003).

In most cases, various modes and conditions operate during SPD processing and during subsequent plastic deformation. The typical example is rolling of UFG materials partly considered in section 1.8. Additional results provide insight into the deformation mechanisms (Segal, 2005). Figure 5.21 presents inverse pole figures for FG and UFG Al0.5%Cu alloy after 90% rolling reductions (b, e) followed 4 passes ECAE via route A (c, f). Note that textures produced by rolling or by ECAE do not depend on original textures (a, d) and grain sizes (7μm and 0.5μm). Therefore, crystallographic glide is operational and identical in both cases irrespective of the grain size. Data on texture strength (OD index, Figure 1.19) shows that the intragranular deformation mechanism is dominant for rolling (diagrams 1, 4) and insignificant for ECAE (diagrams 2, 3). Structures of high purity aluminum Al5N5 also confirm this conclusion. The structure evolution for this material during ECAE (Figure 1.15) and rolling (Figure 1.16) were discussed in Chapter 1. During additional ECAE processing with a number of passes $N > 4$, the structure remains stable, uniform and equaxed without any traces of intragranular glide (see Figure 5.22a after 6 passes ECAE via route A) suggesting that deformation mechanisms, probably, relate to grain boundary activity. However, subsequent rolling alters the deformation mode from simple shear to pure shear and reveals a totally different microstructure evolution. Even 15% rolling reduction transforms the grain shape from equiaxed to elongated and shows a non-uniform crystallographic glide within many grains (Figure 5.22b). The developed substructure creates sites of dynamic recrystallization that was completed after 30% rolling reduction (Figure 5.22c). This coarse structure is quite similar to the original annealed material. Subsequent rolling subjects the material to repeated re-recrystallization and refinement with a gradual decrease in grain size. However, even a very large rolling reduction of 99.2% still produces a non-uniform microstructure with heavily deformed grains and a small number of recrystallized grains (Figure 5.22d). This data conforms to reports on development of substructures within elongated grains after rolling (Mishin et al. 1998) and extrusion (Stolyarov et al. 2001) of UFG materials.

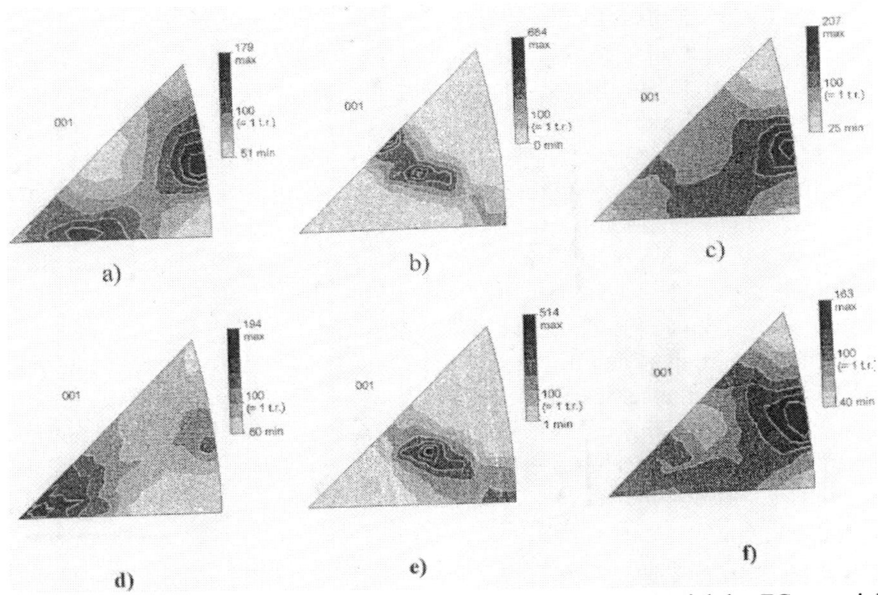

Figure 5.21. Inverse pole figures for Al0.5%Cu alloy: a - original FG material; b - FG material after 90% rolling reduction; c - FG material after 4 passes ECAE via route A; d - original UFG material; e - UFG material after 90% rolling reduction; f - UFG material after 4 passes ECAE via route A.

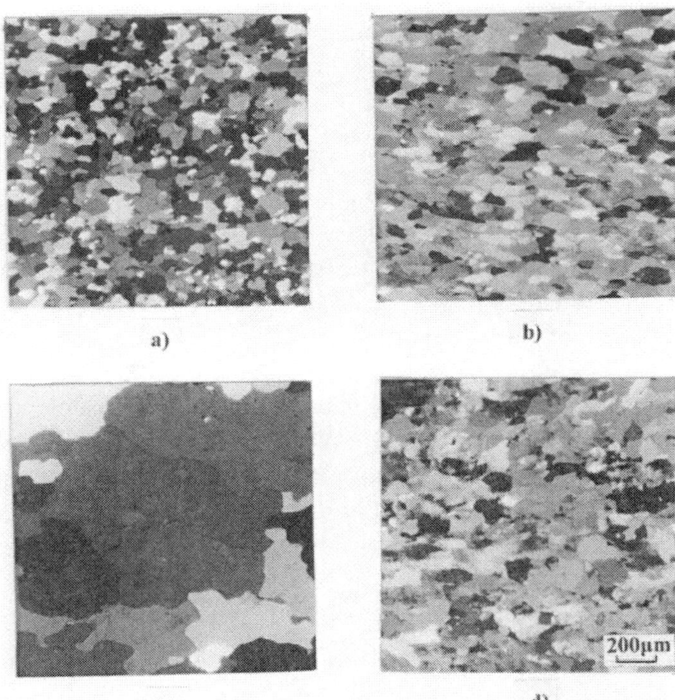

Figure 5.22. Structures of aluminum Al5N5 after 6 passes ECAE via route A (a) and additional rolling with reductions 15% (b), 30% (c) and 99.2% (d).

The transition in deformation mechanisms also manifests the specific changes of properties. Figure 5.23 presents experimental data on the ultimate tensile strength (UTS, solid lines) and relative elongation (δ, dashed lines) of FG and UFG Al0.5%Cu alloy after additional processing. Rolling of FG material (diagram 1) with large reductions provides a significant strengthening effect due to microstructural and textural hardening. ECAE of this material (diagram 2) shows noticeably lower UTS for accumulated strains $\varepsilon > 2$ presumably because of grain boundary sliding. Additional rolling of UFG material (diagram 3) reveals a low hardening for moderate reductions. However, for large rolling reductions, owing to intragranular glide with substructural hardening, UTS increases sharply to a high amount for this alloy. Relative elongation δ after rolling of the FG material (dashed diagram 1) shows the decrease at reductions of less than 75%, some increase at a reduction from 75% to 95% and, finally, the sharp drop for larger reductions. Possible reasons for such complicated behavior are the evolution of texture and the necessity to use flat tensile samples after large rolling reductions. Identical experimental results for δ after rolling of the UFG material (dashed diagram 3) are consistent with this conclusion. In contrast, ECAE of the UFG material (dashed diagram 2) shows the increase of ductility between 2 and 4 passes and near the same ductility for a number of passes N > 4.

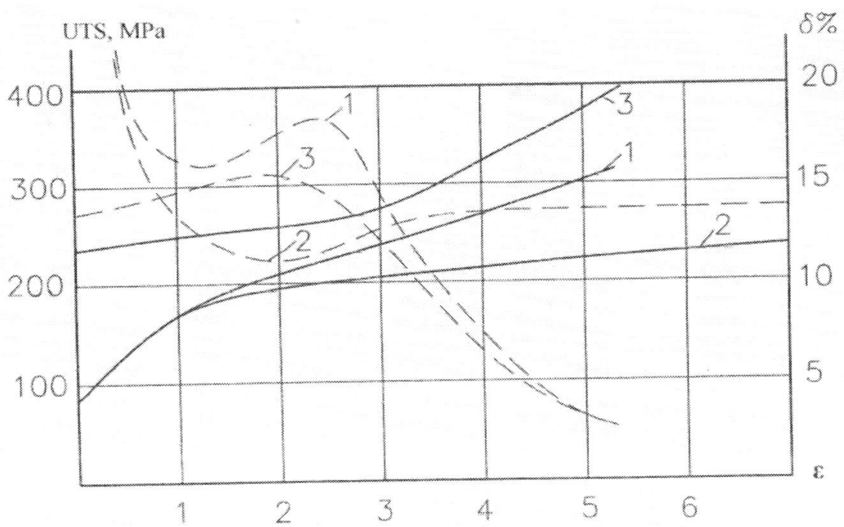

Figure 5.23. Effect of accumulated strains ε on ultimate tensile strength (UTS, solid lines) and relative elongation (δ, dashed lines) of aluminum Al5N5: 1 – rolling of FG material; 2 – ECAE of FG material; 3 – rolling of UFG material.

These results demonstrate that the deformation mechanism in UFG structures may change from grain boundary sliding/grain rotation for the simple shear deformation mode to intragranular crystallographic glide for the pure shear deformation mode. The last case is a characteristic of ordinary metal forming techniques usually used for the processing of UFG materials. As materials produced by SPD have almost dislocation free grain interiors and non-equilibrium boundaries, they may show sufficient ductility during additional processing when compressive hydrostatic stresses prevent fracture. Cold/warm rolling and extrusion with low contact friction and large reductions are favorable techniques for conversion of UFG metals and alloys into useful products.

At certain temperature–strain rate conditions, the grain boundary deformation mechanism remains irrespective of deformation mode. Related plastic flows manifest high sensitivity to strain rate and stability to localization during axial tensile testing similarly to ordinary superplasticity. Because of much finer structures, the processing temperature- strain rate window of superplasticity for UFG materials is significantly wider than for fine-grained counterparts. Corresponding effects are known as low temperature and/or high strain rate superplasticity (Valiev et al. 2006) that can be observed in different metals and alloys with UFG structures. Another important benefit is that massive UFG billets can be produced for bulk forming using SPD. In contrast to the currently available superplastic sheet forming materials, bulk superplasticity does not depend on stability to localization and allows one step forming of large intricate and precision parts providing significant potential for industrial applications of SPD.

5.6. Control of Volume Defects

Three dimensional volume defects are always detected in ordinary metals, sintered/ green powders and cast ingots. They are subdivided on micro-defects - pores, whose size is comparative with structural elements, and macro-defects - voids, cavities, observed visually. Depending on size and concentration, volume defects may significantly affect material strength and ductility and should be eliminated or minimized by thermo-mechanical processing.

5.6.1. Evolution of Voids and Pores in Shear Zone

For ordinary forming operations, healing of *macro-defects* requires high area reductions and pressures. Large voids and cavities often present a problem in the central area of billets and ingots where the deformation mode is near pure shear. They affect plastic zones with lower strains and stresses around internal free surfaces which modify plastic zones. However, volume defects may collapse and close under localized simple shear. A model for healing a cavity Δ by bifurcated shears during ECAE is shown in Figure 5.24. Two shear planes OA and A_1O_1 form angles

$$\cot\varphi = 1 - \Delta/a$$

and provide full healing of large cavities after 1 pass of ECAE as that was experimentally observed in (Segal et al. 1994).

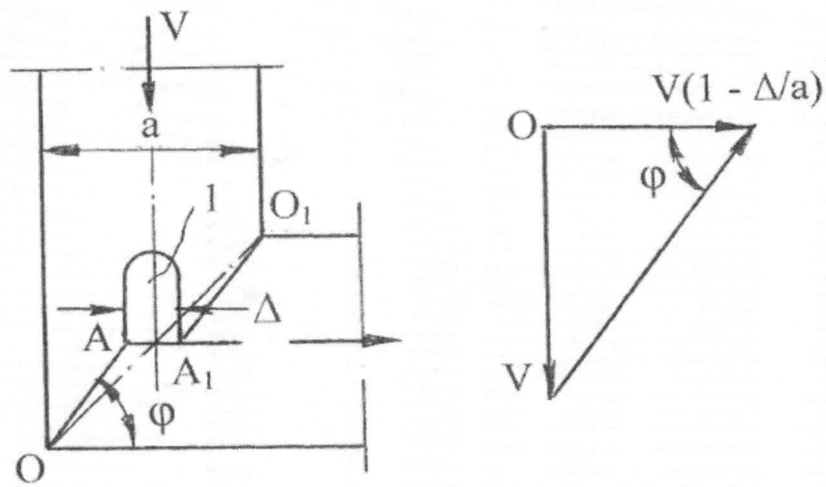

Figure 5.24. Healing of large cavities during ECAE.

In contrast, *micro – pores* are a characteristic of materials rather than plastic zones, and they experience a specific mechanism of evolution under simple shear. For a homogeneous and isotropic pore distribution, they may be described by average pore size $2a$ and distance between pores $2A$. The material porosity is usually introduced as

$$f = (a/A)^3 \tag{5.11}$$

For low porosity $f \ll 1$, $a \ll A$, interactions between pores may be neglected. Then, the representative material volume is reduced to a cubic element $2A$ oriented along the shear directions α, β with a single pore in the center (Figure 5.25a). Continuum stresses σ, k and velocities $\pm u$, v at the element boundaries develop micro shear within a layer whose thickness is comparable to the pore size 2a. It should be noted that for incompressible materials the normal speed v relates to the change of pore size. Then, a plastic analysis of the shear layer 2A x 2A x 2a (Figure 5.25b) describe the pore evolution. Because of symmetry, an approximate solution can be found by dividing the layer on three uniformly deformed areas I, II, III shown in Figure 5.25b. Characteristic stress–strain states inside of each area correspond to an identical problem for shear with tensile/compression of a thin plate $L \gg a$, $B \gg a$ (Figure 5.25c) under applied stresses

$$\sigma_x = q_x \text{ for } x = \pm L, \quad \sigma_y = q_y \text{ for } y = \pm B \tag{5.12}$$

where stresses q_x, q_y provide continuity of stresses between areas I, II, III.

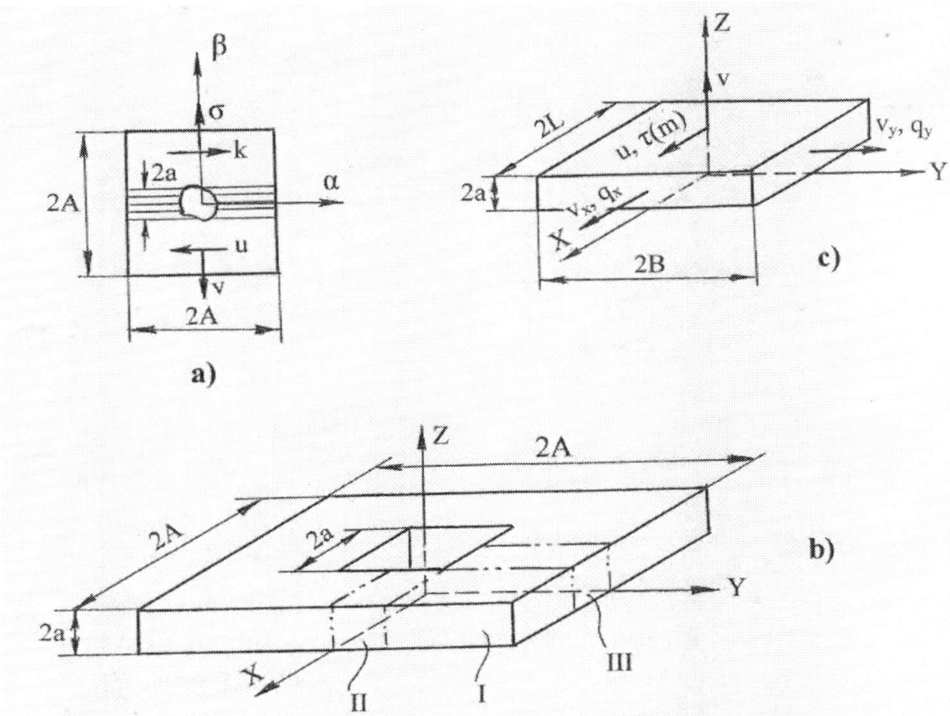

Figure 5.25. Pore evolution under simple shear: a local micro-shear through pore; b – model of micro-shear zone; c – shear-compression of thin layer.

Kinematics boundary conditions can be described by a parameter *m*:

$$m = v/u \text{ for } z = \pm a \tag{5.13}$$

For $f << 1$ the parameter m is also small ($0 < m << 1$) and the stress-strain states within areas I, II, III are slightly different from simple shear when $m = 0$. However, this parameter changes the velocity u and contact shear stresses τ at layer boundaries that may be estimated by the average value:

$$\tau = \tau(m),\ \tau(0) = k \text{ for } z = \pm a \tag{5.14}$$

where k is the shear yield stress.

Under these circumstances, stresses are nearly uniformly distributed throughout the thickness 2a:

$$\partial\sigma_{ij}/\partial z \sim 0 \tag{5.15}$$

The approximate solution of Levi-Mises equations of plastic flow (Hill, 1950) with conditions (5.12) – (5.15) was found in (Segal, 1984b). The final result, within the first order of accuracy, is:

$$p \sim (q_x + q_y)/2 + 3\ mk,\ m \sim (2p - q_x - q_y)/6k,\ \tau(m) \sim k$$

$\xi_x = u(q_x - q_y - 2mk)/8ka$, $\xi_y = u(q_y - qx - 2mk)/8ka$, $\xi_z = um/2a$, $\eta_{zy} = u/2a$

Velocities at layer boundaries x = L and y = B (Figure 5.25c) are, correspondingly:

$u_x = uL(q_x - q_y - 2mk)/8ka$, $u_y = uB(q_y - q_x - 2mk)/ 8ka$

Conditions of continuity between areas I, II, III and their boundaries give for kinematics parameter (5.13) of the shear layer:

$m = \sigma a/[k(2A + a)]$,

and velocities at pore surfaces x = y = a:

$$u_x = u_y = u\sigma(A^2 - a^2)/[4ka(2A + a)],\ v = mu_x \qquad (5.16)$$

By integrating eq. (5.16) during the time interval t, final pore width a (a << A) and thickness a_z are, respectively,

$$a = a_x = ay = a_0 + \sigma A\gamma/8k,\ a_z = \sim a_0 \qquad (5.17)$$

Here a_0 is the original (t = 0) pore size, $\gamma = ut/2a_0$ is shear strain accumulated during time t. Formulae (5.16), (5.17) show that the three dimensional pore evolution inside shear bands is non isotropic. Pores are changed uniformly along directions X, Y of the shear plane, but remain almost unchanged along the perpendicular direction Z. In contrast to near isotropic pore evolution for triaxial stress – strain states (Rice et al. 1969), simple shear provides much more effective pore growth or healing along the shear plane. A similar result was also found for elastic and viscous flows (Bandyipadhyay et al. 1979). For $\sigma > 0$, pores grow along the shear plane in proportion to the product of total shear γ and relative hydrostatic pressure σ/k. There is a similar healing process for $\sigma < 0$. Because of pore evolution to such shapes, the average porosity (5.11) cannot be used as the material characteristic under intensive simple shear. The corresponding characteristic related to the shear plane is:

$$f = (a/A)^2 \qquad (5.18)$$

Then, formulae (5.17) give:

$$f = (\sqrt{f_0} + \sigma\gamma/8k)^2 \qquad (5.19)$$

where $f_0 = (a_o/A)^2$ is the original porosity of the shear plane.

In addition to plastic flow, diffusion–viscous flow in pores should be also accounted for specific processing conditions. Plastic and diffusion -viscous mechanisms may be considered independently with their cumulative action resulting in summation of strain rates. The rate of change of a pore radius R due to the diffusion - viscous flow under applied hydrostatic stress σ and surface tension ρ is (Gegusin, 1984):

$(dR/dt)_{d\text{-}v} \sim \omega D(\sigma - 2\rho/R)/2KTR$

Here ω is the atomic volume, D is the diffusion coefficient, K is the Boltzman constant, T is the absolute temperature, ρ is the surface tension. A similar expression for plastic flow follows from formulas (5.16), (5.18):

$$(da/dt)_p = \eta\sigma a/8kf^2$$

where η is shear strain rate of the shear zone. Assuming R ~ a, the last two equations give the full rate of pore growth or healing in the shear zone:

$$da/dt \sim \eta\sigma a/8kf^2 + \omega D\sigma/2KTa - \omega D\rho/KTa^2 \quad (5.20)$$

The three terms in (5.20) account for plastic flow, diffusion flow and viscous flow, respectively. In comparison with the plastic flow mechanism, contributions of diffusion and viscous flows are insignificant for pores whose sizes, correspondingly, are:

$$a_1 >> 2f(\omega Dk/KT\eta)^{1/2},\ a_2 >> 2(\omega D\rho kf^2/KT\eta\sigma)^{1/3} \quad (5.21)$$

Calculations performed for copper at T = 1273°K (self-diffusion coefficient D ~ 10^{-10} m^2/sec)[2] show that for typical processing conditions $\eta > 10^{-2}$/sec, σ/k > 1, f < 0.2, pore radii (5.21) are a_1 ~ (1-10) μm, a_2 ~ (0.1 – 1) μm (Segal, 1982). Therefore, viscous flow induced by surface tension may be neglected in most practical cases whereas diffusion flow is effective for high temperatures and low strain rates.

5.6.2. Breakdown of Cast Structures

Breakdown of cast structures involves the elimination of coarse dendrite-like crystallites and healing of volume defects. Because of insufficient ductility of the cast materials, plastic deformation is usually performed at hot temperatures. Structure recrystallization at these conditions takes place after moderate strains while healing of cavities, voids and pores, which are primarily responsible for low ductility of cast materials, requires significantly higher strains. In ordinary operations of forging, rolling and extrusion, necessary area reductions for conversion of cast structures into wrought structures usually range from 10 to 50. Such reductions result in large original ingots, powerful machines and low productivity. Processes of SPD have significant potential for these applications because of the conservation of the billet cross-sectional area, strain uniformity and simple shear deformation mode. Neglecting the viscous-diffusion flow, an approximate condition for full healing of voids and pores by simple shear processing can be found from eq. (5.19) assuming f = 0:

$$\gamma = -\sqrt{f_0}/(\sigma/k)\ ,\ \sigma > 0 \quad (5.22)$$

Diagrams of shear strain γ as a function of the original porosity f_0 calculated from eq. (5.18) are shown in Figure 5.26 for different hydrostatic pressures σ/k = - 0.25, - 0.5, -1.0 and - 2.0.

[2] Coefficient D was increased on 2 - 3 orders of magnitude to account for enhanced diffusion during plastic deformation (Gegusin, 1984).

In accordance with formula (3.1), diagram 3 relates to ECAE with a single shear zone, a tool angle of 90° and zero back-pressure ($p_0 = 0$). In this case, full healing of cast defects with a sufficiently high porosity $f_0 = 0.2$ can be attained after two passes of ECAE.

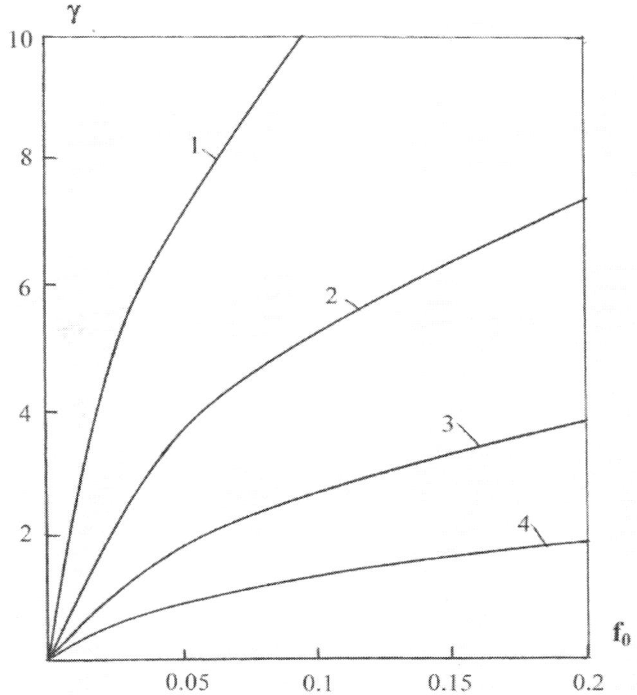

Figure 5.26. Diagrams of healing shear γ versus original porosity f_0 and hydrostatic pressure $\sigma/k = 0.25$ (1); $\sigma/k = 0.5$ (2); $\sigma/k = 1$ (3) and $\sigma/k = 2$ (4).

Experiments on the breakdown of the cast structure during ECAE were performed for steel 1045 (Segal et al. 1994). At a processing temperature of 900°C and strain rate $\eta \sim 20$ sec^{-1}, contribution of the diffusion - viscous flow was insignificant. The original micro-porosity $f_0 = 0.05$ (see eq. (5.18)) was determined by the analysis of microstructures and by direct measurements using the Archimedes method. The material after ECAE via route A with a number of passes from 1 to 4 was characterized by optical microscopy, tensile testing, impact toughness and density measurement. Structures after ECAE were fully recrystallized after the first pass without any traces of cast crystallites. Experimental data for area reduction ψ, relative elongation δ, impact toughness a_k and density ρ (Figure 5.27) shows that the transformation from cast to wrought material was completed after 1 pass of ECAE that is in a good agreement with Figure 5.26. All properties were higher than the certified characteristics of the steel 1045 after hot rolling. These results confirm that ECAE at hot processing temperatures with a low number of passes is an effective alternative to existing techniques of material conversion into wrought products in metallurgy.

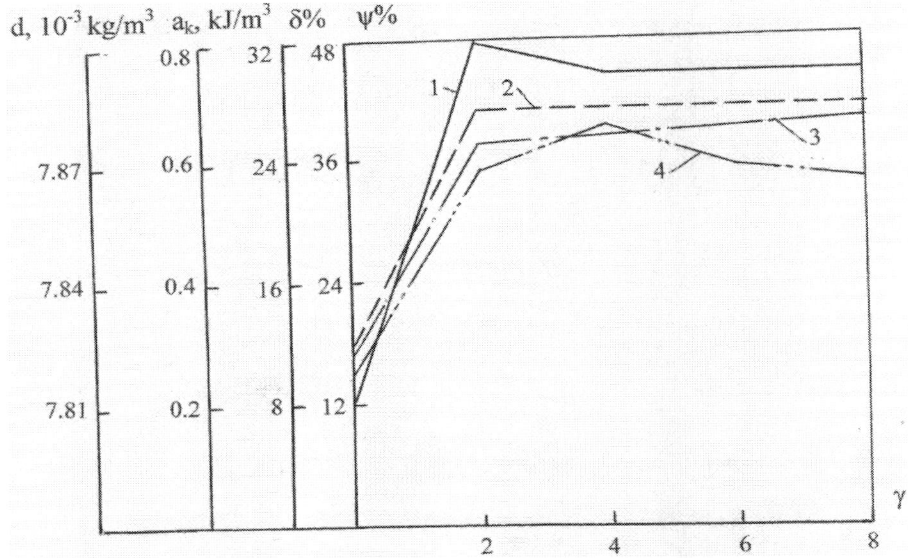

Figure 5.27. Effect of accumulated shear γ on area reduction ψ (1), relative elongation δ (2), impact toughness a_k (3) and density ρ (4) after ECAE of cast steel 1045 at temperature 900°C.

5.6.3. Materials Consolidation and Bonding

Similar processing at specific conditions can be applied to different problems of powder metallurgy. Densification of sintered billets is quite similar to healing of cast ingots. For particulate materials additional requirements are consolidation and bonding of particles. "Ideal" forming operations with the preferable pure shear deformation mode do not provide relative shear of material particles during processing. In this case, diffusion bonding of contiguous surfaces may be attained under high temperatures and hydrostatic pressures leading to material degradation, large reductions and loads. On the contrary, the simple shear deformation mode applies intensive shear upon material particles resulting in intensive heating and cleaning of their surfaces. Under these circumstances, effective densification and bonding can be attained with relatively low temperatures and pressures. Therefore, processes of SPD present promising techniques for powder metallurgy in fabrication of green briquettes, composites, rapidly solidified and mechanically alloyed materials and nano powders.

5.6.4. Simple Shear Fracture

Originally defect-free ($f_0 \sim 0$) and ductile materials may show characteristic fracture during ECAE. The mechanism of fracture under simple shear comprises generation, growth and coalescence of pores along a shear plane (Figure 4.15c) (Backofen, 1972). Pores originate at hard particles, second phases and other irregularities of the structure with blocked crystallographic glide and high micro stresses. During further growth, they experience local stress /strain rate states $\acute{\sigma}_{ij}$, $\acute{\eta}_{ij}$ that are significantly different from the macro states σ_{ij}, η_{ij}. In continuum plasticity, their effects can be accounted for by structural parameters $\langle\sigma_{ij}\rangle$, $\langle\chi\rangle$ (Segal, 1983):

$$\sigma'_{ij} = \sigma_{ij} + \langle\sigma_{ij}\rangle,\ \eta'_{ij} = \langle\chi\rangle\eta_{ij} \tag{5.23}$$

Parameters $\langle\sigma_{ij}\rangle$, $\langle\chi\rangle$ are the statistical average micro stress and coefficient of strain non-uniformity in the vicinity of pores. Actual variables (5.23) can be used in formulas (5.16) - (5.19) for evolution of micro pores. They grow independently along the shear plane until critical porosity f* when pores start to coalescence and quickly form "pore stratums" separating the material (Backofen, 1972). From (5.19), the fracture criterion during simple shear is:

$$\gamma = 8(\sqrt{f^*} - \sqrt{f_0})k/\langle\chi\rangle(\sigma + \langle\sigma\rangle) \tag{5.24}$$

where $\langle\sigma\rangle$ is the hydrostatic component of averaging micro-stresses acting on pores. Formula (5.24) gives pore growth $f > f_0$ if $(\sigma + \langle\sigma\rangle) > 0$ and pore healing $f < f_0$ if $(\sigma + \langle\sigma\rangle) < 0$. As pores originate and grow in areas of "hard" particles, the component $\langle\sigma\rangle$ is tensile ($\langle\sigma\rangle > 0$) and for many materials may approach the material yield stress (Segal, 1983). Higher processing temperatures and lower strain rates activate recovery and reduce $\langle\sigma\rangle$. On the contrary, the hydrostatic macro - stress component σ is usually compressive ($\sigma < 0$). For ECAE with single shear zone, the compressive component $\sigma \leq 0$ given by eq. (3.1) increases with the increase of back-pressure p_0 and the decrease of tool angle θ. Therefore, depending on material and processing conditions, pore growth or pore healing may be observed during ECAE. When pores grow along the shear plane, tensile stresses around a sharp tool angle O (Figure 4.15a) initiate cracks on the top billet surface propagating inside the material (Figure 4.15b). Material dilatation and local surface fracture may also be observed on the top surface during multi-pass processing.

5.7. Application to Specialty Materials

Additional structural effects of SPD become important in special cases. One of these is high *diffusivity* in UFG materials. It is known that during hot plastic deformation the coefficient of diffusion increases up to 10^3 times (Gegusin, 1984). In ultrafine grained materials produced by SPD, extended and non-equilibrium grain boundaries provide an additional increase of the diffusion more than 10^4 times against their coarse grained counterparts (Baro et al. 2001). During SPD processing of UFG materials, both factors work together and may enhance diffusivity up to 7 orders of magnitude affecting strongly many characteristics of materials. *Phase transformations* induced by diffusion and strains include allotropic modifications, phase separation or dissolution. These effects manifest the change of transition diagrams, transition rates, phase size, morphology and composition. SPD stimulates both diffusion and shear mechanisms of phase transformation because of high diffusivity and simple shear deformation mode. Similarly, severe strains and enhanced diffusivity result in *homogenization* of micro-structures during multi-pass SPD. Processing routes with an accumulated shear γ and strong distortion into one direction similar to ECAE via route A reduces the thickness of structural elements in ~ γ times (see formula (5.1)) without changing the material cross-section. Correspondingly, the diffusion time for homogenization of macro segregations (such as arms of cast dendrites) is ~ γ^2 times less than for the original material. Therefore, homogenization of bulk billets after ECAE can be performed in a short time.

Numerous publications have demonstrated the potential of SPD for the processing of materials with special mechanical, physical and functional characteristics. The first results were reported for magnetic, superconductor and special elastic properties (Segal et al. 1994; Segal, 1995a). For soft *magnet* Armco Iron, the magnetic anisotropy can be attributed to crystallographic textures developed via different routes and passes. After route C with fixed shear planes, the stable orientations of magnetization were detected in the shear plane with two axes of easy magnetization at angles ± 45° to the shear direction. Magnetic anisotropy increases with strains and after eight passes approximates to those of a single crystal. In *superconductor* 50Nb50Ti alloy, route C develops a highly oriented distribution of dislocation structures along the shear plane into a perpendicular direction to the shear direction. These structures are effective obstacles for pinning of magnetic flux resulting in a multifold increase of the critical current when the magnetic vector is oriented along dislocations. The present examples show that the control of shear orientations is an important factor in improvement of physical characteristics with strong anisotropy. It was also found that refinement of second phases and precipitates during SPD provides superior properties of dispersion strengthening materials. For *special elastic* Elinvar alloy Ni30Cr25Ti alloy, high strength, ductility and thermo-stable properties were developed after two passes of ECAE via route C followed low temperature annealing to control precipitates by dynamic and peak aging.

There are numerous options when applying intensive plastic deformation to *magnetic* alloys (Chin, 1971). However, because of processing difficulties, only some of these have been realized (see, for example, Stolyarov et al. 2001; Fujita et al. 2005). More attention has been paid to *shape memory materials* (SMM). ECAE processing of a NiTi alloy was performed at different temperatures of the austenite-martensite transformation resulting in grain sizes from sub-micron to tenths of microns (Karaman et al. 2003; Stolyarov et al. 2005; Li et al. 2006). In all cases, SPD improves mechanical properties and significantly affects functional characteristics such as transformation temperatures and recoverable deformation. With the development of industrial SPD technologies, bulk SMM may find many applications. SPD can potentially be used for *metal matrix composites* (MMC) and *semiconductors*. It was demonstrated in many publications that multi-pass ECAE of MMC refines microstructures, provides uniform distribution of strengthening phases and eliminates clusters of hard particles in cast billets. Similarly, homogenization and refinement of cast $(Bi, Sb)_mTe_n$ alloys by ECAE produces uniform fine grained polycrystal semiconductors with superior mechanical and thermo-electrical properties compared to single crystal and powder metallurgy materials (Seresara et al. 2007).

The simple shear deformation mode is especially expedient for *single crystals*. For long time, the progress in development of related testing techniques for single crystals (Figure 2.3) has been motivated by research on mechanisms of plastic deformation. Additionally, plastic processing of single crystals is often used for their strengthening and healing of defects induced by fabrication methods. In any case, the deformation technique should preserve the single crystalline structure in order to retain of many of its useful properties. However, stress-strain non-uniformity and plurality of shear planes and directions in ordinary metal forming operations transform single crystals into polycrystals with specific end textures depending on the processing method and the original crystal orientation. The same is true for ECAE of single crystals when round corner angle ψ or large contact friction τ result in extended fan-like plastic zones, distribution of shears into different directions and strain non –uniformity. In these cases, including one pass processing (Fukuda et al. 2004; Miymoto et al. 2005; Han et al. 2007), ECAE loses its unique opportunity in realization of strictly oriented simple shear that is necessary for identification of single crystals. The careful control of processing mechanics providing unidirectional shear along specific crystallographic orientations may lead to many interesting applications of ECAE for processing of single crystals with special magnetic, optical, elastic and other properties.

There are experimental evidences (Sue et al. 1998; Campbell et al. 1999) that ECAE may be also effective for reprocessing of *polymers* into highly oriented molecular structures with a dramatic increase of elastic modulus, toughness, strength and solvent resistance in large cross-sections. To reduce the effects of viscoelasticity, multi-pass processing should be performed at temperatures of semicrystallity under controllable strain rates (Zairi et al. 2006). Continuous ECAE would be the optimal processing technique for long polymer products. It should be noted that there are other possibilities of SPD such as fabrication of *biomedical* materials (Valiev et al. 2006; Niendorf et al. 2007) and *optical* materials including polycrystalline optical ceramics and materials for non-linear optics.

5.8. Commercialization of ECAE

As a new metalworking technique, ECAE leads to significant improvements of mechanical, physical and special properties that may find different applications. Presently, the main obstacles in commercialization of ECAE are engineering problems rather than fundamental problems related to the material science. From a practical point of view, ECAE should provide technical or economical effects such as superior performance and/or cost saving. These general requirements outline a few prospective directions for commercialization of ECAE.

The first direction belongs to sufficiently common industrial problems when simple shear presents the optimal deformation mode and effective processing can be performed with low number of passes. Conversion of cast ingots into wrought products may be applied to many metals and alloys at temperatures of hot deformation with a low number of passes being from 1 to 2. The existing practice of structure breakdown is via forging, rolling or extrusion with large reductions. Such operations are expensive and require heavy equipment for intensive processing of large ingots. The advantages of ECAE, especially in metallurgy of light and specialty alloys, include the reduction of ingot cross-sections, quality improvement and low cost. Similarly, ECAE technology can be used in powder metallurgy, fabrication of composites, blanks for thixoforming, thermo-mechanical treatment (TMT), and in other areas. In particular, different schemes of thermo-mechanical treatment are based on interaction of strain induced defects with phase transformations. The defect density attains near saturation at strain $\varepsilon \sim 1.6$ or area reduction ~ 80% which are optimal for TMT (Kula et al. 1978). ECAE provides similar and uniform strain after 1 pass. Examples of effective thermo-mechanical treatment of tool steels, dispersion hardened Ni- based alloy and aluminum heat-treatable alloys using ECAE with low number of passes can be found in (Segal et al. 1994; Ferrasse et al. 2008)

Another direction is multi-pass ECAE to produce ultrafine grained structures. Bulk volume production may be expected for UFG plates, sheets and strips rolled from flat ECAE billets. The large billet size reduces the cost and makes these materials available for structural applications, superplastic and precision forming of large complicated parts and panels, deep drawing, thin wall extrusions, rolling of high strength isotropic plates, and in other advanced technologies.

However, the most intriguing are materials and products with special properties.

Commercialization of the specialty materials described above may require extensive research and development of industrial technologies of SPD. Such program has been successfully completed at Honeywell International Inc. for *sputtering targets*. This high-tech product is widely used as a cathode material in electronic and semiconductor industries for fabrication of thin films by physical vapor deposition. The continuous progress of memory and logic components should be supported by corresponding improvements in target materials. To provide film uniformity, step coverage, low arcing and small numbers of particles, the target grain size should be less than a micron. As targets are bulk thin discs which work as integral parts of vacuum chambers loaded by atmospheric pressure and the pressure of cooling water, sufficiently high strength is also necessary. Existing methods of thermo-mechanical treatment cannot satisfy these requirements for high purity dilute Al and Cu alloys resulting in complicated designs with a diffusion bonded backing plate. Multi-pass ECAE of large flat billets (weighting up to 34 kg for Al targets) followed by rolling resolved these problems by reducing the grain size to 0.5 μm (Ferrasse at al, 2005). The new SPD technology allows a simple monolithic target design with extended life. The ECAE sputtering target products were commercialized in 2000 and have remained the only ultrafine grained product currently available on the market. However, other SPD products may appear soon.

The realization of these exiting new opportunities should put ECAE and other methods of SPD on the same level as presently used forming techniques, such as rolling and extrusion, and should result in products with more specific and superior properties.

REFERENCES

Alexandrov, I.V.; Wang, Y.D.; Zhang, K.; Lu, K.; Valiev, R.Z. In *Proceedings of the Conference ICITOM;* Liang, Z.; Zuo, L.; Chu, Y.; Eds.; 1996; Vol. 11, pp 929-940.

Alkorta, J.; Rombouts, M.; De Messemaeker, J.; Froyen, L.; Gil Sevillano, J. *Scr. Mater.,* 2002, *vol 47,* 13-18.

Altenberger, I; Scholtes, B.; Martin, U.; Oettel, H. *Mat. Sci. Eng.* 1999, *vol A 264,* 1-16.

Andrade, U.; Meyers, M.A.; Vecchio, K.S.; Chokshi, A.H. *Acta Metal Mater,* 1994, *vol 42,* 3183-3195.

Armstrong, R.W.; Chou, Y.T.; Fisher, R.M.; Louat, N. *Phil. Mag.*1966, *vol 14*, 943-951.

Azushima, A.; Aoki, K. *Mat. Sci. Eng.* 20002, *vol A337,* 45-49.

Backofen, W.A. *Deformation Processing,* Addison-Wesley Publishing, Massachusetts, 1972.

Baeck, S.; Seok, H.-K.; Lee, J.-C.; Kim, D.-I.; Lee, H.-C.; Oh, K.H. *Mat. Sci. Forum,* 2002, *vol 408-412,* 685-690.

Bandyopadhyay, S.N., Sigh, N. *Eng. Fract. Mech.,* 1979, *vol 11*, 635-643.

Baro, M.D.; Kolobov, Yu.R.; Ovid'ko, I.A.; Schaefer, H.-F.; Straumal. B.B.; Valiev, R.Z.; Alexandrov, I.A.; Ivanov, M.; Reinman, K.; Reizis, A.B.; Surinach, S.; Zhilyaev, A.P. *Rev. Adv. Mater.*, 2001, *vol 2,* 1-43.

Belyakov, A.; Sakai, T.; Tsuzaki, K. *Phil. Mag.* 2001, *vol 81A,* 474-483.

Beyerlein, I.J., Toth, L.S., *Progress in Materials Science,* 2009, *vol 54,* 427-510.

Beygelzimer, J.; Orlov, D.; Varyakhin V. In *Ultrafine Grained Materials –II; Zhu, Y.T.; Langdon, T.G.; Mishra, R.S.; Semiatin, S.L.; Saran, M.J.; Lowe, T.G.; Eds.;* TMS, 2002, pp 297-304.

Biot, M.A. *Mechanics of Incremental Deformations,* John Wiley, New York, 1965.

Bochniak, W.; Korbel, A. *Mat. Sci. Technol.* 2000, *vol 16*, 664-669.

Boczkal, S.; Lech-Grega, M.; Szumanski, W. *Image Anal. Steorol.* 2007, *vol 25,* 37-42.

Bridgman, P.W. *Studies in Large Plastic Flow and Fracture*; McGraw-Hill, New York, 1952.

Bridgman, P.W. *Trans. Amer. Soc. Metals,* 1944, *vol. 32,* 553-574.

Campbell, B.; Edwards, G. *Plastics, Rubbers and Composites,* 1999, *vol 28,* 467-475.

Ceresara, S.; Guinchi, G.; Bassani, P.; Cavallin, T.; Fanciulli, C. *Processings of 5th International Conference on Thermoelectrics, Ukrain,* 2007.

Chakkingal, U.; Suriadi, A.B.; Thomson, P.T. *Scr. Mater.* 1998, *vol 39,* 677-684.

Chaudhury, P.K.; Srinivasan, R.; Vismanathan, S, *U.S. Patent,* 2005, *No. 6,895,795.*

Chen, W.; Ferguson, D.; Ferguson, H. In *Ultrafine Grained Materials;* Mishra, R.S.; Semiatin, S.L.; Suryanarayana, C.; Thadhani, N.N.; Eds.; TMS, 2000, 235-245.

De Chiffre, L. *Int. J. Mach. Tool Des. Res.*, 1976, *vol. 16,* 137-144.

Chin, G.Y. *Adv. Mater. Res.*, 1971, *vol 5,* 217-280.

Chung, Y-H.; Park, H.-K.; Moon, I.-G.; Shin, M.-C. *U.S. Patent,* 2003, *No. 6,571,593.*

Cui, H.J.; Gofforth, R.E.; Hartwig, K.T. *Journal of Metals-e,* 1998, *vol 50,* No. 8.

Dao, M.; Lu, L.; Asaro, R.J.; De Hosson, J.T.M.; Ma, E. *Acta Mater.*, 2007, *vol 55,* 4041-4065.

Davidenkov, N.N.; Spiridonova, N.I. *Proc. ASTM,* 1946, *vol 46,* 1147-1158.

DeLo, D.P.; Semiatin, L.S. *Metal. Mater. Trans.*, 1999, *vol. 30A,* 1391-1402.

Duggan, B.J.; Lee, C.S. In *Processing of Conference of ICOTOM 11,* Liang, Z.; Zuo, L.; Chu, Y.; Eds.; 1996, 225-235.

Embury, J.D.; Fisher, R.M. *Acta Metallurg.*, 1966, *vol 14,* 147-159.

Engler, O. *Scripta Mater.*, 2001, *vol 44,* 229-236.

Enikolopov, N.S. In *Processing of International Symposium on Chemical Physics, Moscow,* 1981, 81-86.

Fleck, N.A.; Hutchinson, J.W. *J. Mech. Phys. Sol.*, 1993, *vol 41,* 1825-1857.

Ferrasse, S; Segal, V.M.; Kalindindi, S.R.; Alford, F. *Mat. Sci. Eng.*, 2004a, *vol A368,* 28-40.

Ferrasse, S.; Segal, V.M.; Alford, F. *Mat. Sci. Eng.*, 2004b, *vol. A372,* 235-244.

Ferrasse, S,; Segal, V.M.; Alford, F. *Mat. Sci. Eng.*, 2004c, *vol A372,* 44-55.

Ferrasse, S.; Segal, V.M.; Alford, F.; Strothers, S.; Kardokus, J.; Grabmeier, S.; Evans, J. In *Severe Plastic Deformation: Toward Bulk Production of Nanostructured Materials,* Altan, B.S., Ed.; Nova, New York, 2005, 585-601.

Ferrasse S., Segal V.M., F. Alford, J. Kardokus, S. Strothers, Mat. Sci. Eng, 2008, vol A493, 130-140.

Fujita, T.; Nishimura, S.; Kaneko, K.; Horito, Z.; Smith, D.I. *Microscopy and Microanylis,* 2005, *vol 2,* 1812-1813.

Fukuda, Y.; Oh-Ishi, K.; Horita, Z.; Langdon, T.G. *Acta Mater.*, 2004, *vol 52,* 1387-1395.

Gasperini, M.; Pinna, C.; Swiatnicki, W. *Acta Mater.*, 1996, *vol. 44,* 4195-4208.

Gegusin, I.E., *Physics of Sintering,* Nauka, Moscow, 1984 (In Russian).

Ghosh, A.K.; Huang, W. In *Investigation and Application of Severe Plastic Deformation,* Lowe, T.C.; Valiev, R.Z., Eds., 2000, Kluwer, Netherland, 29-36.

Gil Sevillano, J.; van Houtte, P.; Aernoudt, E. *Progr. Mater. Sci.*, 1980, *vol 25,* 69-134.

Gholinia, A.; Bate, P.; Prangnell, P.B. *Acta Mater.*, 2002, *vol 50,* 2121-2136.

Gibbs, M.A.; Hartwig, T.D.; Cornwell, L.R.; Goforth, R.E.; Payzant, E.A. *Scripta Mater.*, 1998, *vol. 39,* 1699-1704.

Gjones, L.; Anderson, B. *J. Mater. Sci.*, 1998, *vol 33,* 2469-2476.

Goloborodko, A.; Sitdikov, O.; Sakai, T.; Kaibyshev, R.; Miura, H., *Mater. Trans.*, 2003, *vol 44,* 766-774.

Green, D. *U.S. Patent,* 1973, No. 3,765,216.

Gutkin, M.Yu.; Ovid'ko, I.A., Pande, C.S. *Rev. Adv. Mater. Sci.*, 2001, *vol 2,* 80-102.

Gutkin, M. Yu.; Ovid'ko, I.A. *Rev. Adv. Mater. Sci.*,2003, *vol. 4,* 79-113.

Han, J.-H.; Seok, H.-K.; Chung, Y.-H.; Shin, M.-C.; Lee, J.-C. *Mat. Sci. Eng.*, 2002, *vol A323,* 342-347.

Han, W.Z.; Zhang, Z.F.; Wu, S.D.; Li, S.X. *Acta Mater.*, 2007, *vol 55,* 5881-5900.

Hansen, N. *Mater. Sci. Technol.*, 1990, *vol 11,* 1039-1047.

Harren, S.V.; Deve, H.E.; Asaro, R.J. *Acta Metallurg.*, 1988, *vol 36,* 2435-2480.

Hatherley, M.; Malin, A.S. *Scipta Metallurg.*, 1984, *vol 5,* 449-454.

Hayes, J.S.; Keyte, R.; Prangnell, P.B. *Mater. Sci. Technol.,* 2000, *vol 16,* 1259-1263.

Hill, R. *The Mathematical Theory of Plasticity,* Oxford, 1950.

Hill, R. *J. Mech. Phys. Sol.,* 1967, *vol 15,* 141-227.

Hill, R.; Hutchinson, J.W. *J. Mech. Phys. Sol.,* 1975, *vol 23,* 239-264.

Hines, J.A.; Vecchio, K.S.; Anzi, S. *Metallurg. Mater. Trans.,*1998, *vol 29A,* 1291-1303.

Hirsch, J.; Lucke, K.; Hatherley, *Acta Metallurg.,*1988, *vol 36,* 2905-2927.

Huang, C.X.; Wang, K.; Wu, S.D.; Zhang, Z.F.; Li, G.Y.; Li, S.X. *Acta Mater.,* 2006, *vol 54,* 655-665.

Hong, S.I.; Hill, M.A.; Sakai, Y.; Wood, J.T.; Embury, J.T. *Acta Metallurg. Mater.,* 1995, *vol 43,* 3313-3323.

Huang, Y.; Langdon. T.G. *Mater. Sci. Eng.,* 2003, *vol A358,* 114-121.

Huang, Y. *International Patent Application,* 2005, No. WO 2005/065856.

Hughes, D.A.; Dawson, D.B.; Korellis, J.S. *J. Mater. Eng. Perform.,* 1994, *vol 3,* 459-475.

Hughes, D.A.; Hansen, N. *Acta Mater.,* 1997, *vol 45,* 3871-3886.

Humphreys, F.J,; Hatherley. M. *Recrystallization and Related Annealing Phenomena,* Pergamon, 1995.

Imaev, R.M.; Imaev, V.M.; Salischev, G.A. *J. Mater. Sci.,* 2006, *vol. 27,* 4465-4471.

Iwahashi, Y.; Wang, J.; Horito, Z.; Nemoto, M.; Langdon, T.G. *Scipta Mater.,* 1996, *vol 36,* 143-148.

Iwahashi, Y,; Furukawa, M.; Horita, Z.; Nemoto, M.; Langdon, T.G. *Metallurg. Mater. Trans.,* 1998a, *vol 29A,* 2245-2252.

Iwahashi, Y.; Horita, Z.; Nemoto, M.; Langdon, T.G. *Acta Mater.,* 1998b, *vol 46,* 3317-3331.

Iosipesky, I. *J. of Metals,* 1967, *vol 2,* 537-566.

Ji, Y.H.; Park, J.J; Kim, W.J., *Mater. Sci. Eng.,* 2007, *vols 454-455,* 570-574.

Johnson, W.; Kudo, H. *The Mechanics of Metal Extrusion,* Manchester University Press, 1962.

Johnston, G.P.; Hogoreldt, G.M., Lindholm, U.S.; Cadgy, A. *Trans. ASME. J. Basic Eng.,* 1983, *No.1,* 51-66.

Kaibyshev, R.; Mazurina, I., *Mater. Sci. Forum,* 2004, *vols 467-470,* 1251-1260.

Kajino, S.; Asakava, M., *J. Mater. Proc. Technol.,* 2006, *vol. 177,* 704-708.

Karaman, I.; Ersin Karasa, H.; Maier, H.J.; Luo, Z.P., *Metallurg. Mater. Trans A.,* 2003, *vol. 54,* 2527-2539.

Korbel, A., Richert, M., *Acta Metallurg.,* 1985, *vol. 33*, 1971-1978.

Korbel, A.; Bochniak, W., *US Patent,* 1998, *No. 5,737,959.*

Kulhmann- Wilsdorf, D. *Mat. Sci. Eng.,* 1989, vol *A113,* 1-41.

Kurzudlowsky, K.J., *Bull. Pol. Acad. Sci., Techn. Sci.,* 2004, *vol 52*, 301-311.

Langdon, T.G., *Rev Adv. Mater. Sci.,* 2006, *vol 11,* 34-40.

Langer, J.S., *Scripta Mater.,* 2006, *vol 54,* 375-379.

Lapovok, R., *J. Mater. Sci.,* 2005, *vol 40,* 341-346.

Lapovok, R.; McKenzie, R.W.J.; Thomson, P.F.; Semiatin, S.L., *J. Mater. Sci.,* 2007, *vol 45,* 1649-1659.

Lee, S.H.; Saito, Y.; Tsuji, N.; Utsunomiya, H,; Sakai, T., *Scripta Mater.,* 2002, *vol 46,* 281-285.

Lee, C.S.; Duggan, B.J., *Acta Metallurg. Mater.,* 1994, *vol 42,* 857-860.

Lee, J.-C.;Seok, H.-K.; Park, J.-W.; Chung, U.-H.; Lee, H.-I., *U.S. Patent,* 2002, *No. 6,370,930.*

Levit, V.I. In *Physical Research. Electron Microscopy in Plasticity and Fracture of Research of Materials,* Messerschmidt, U.; Appel, F.; Heydenreich, J.; Schmidt, V., Eds., Academia-Verlag, Berlin, 1989, 59-64.

Li, S.; Bourke, M.A.M.; Beyerlein, I.J.; Alexander, D.J.; Clausen, B., *Mater. Sci. Eng.,* 2004, *vol A382,* 217-236.

Li, S.; Beyerlein, I.J.; Alexander, D.J.; Vogel, S.C., *Acta Mater.,* 2005, *vol 53,* 2111-2125.

Li, Z.; Xiang, G.; Cheng, X., *Mater. and Design,* 2006, *vol 237,* 324-328.

Liu, M.; Roven, H.J.; Yu, Y.; Werenskiold, J.C., *Mater. Sci. Eng.,* 20008, *vol A483-484,* 59-63.

Liu, Z.Y.; Liang, G.X.; Wang, E.D.; Wang, Z.R., *Mater. Sci. Eng.,* 1998, *vol A242,* 137-140.

Luri, R.; Luis, C.J.; Sebastian, M.A., *Trans. ASME,* 2006, *vol. 128,* 860-865.

MacClintock, F.A.; Clerio, M., *J. Mech. Phys. Solids,* 1980, *vol. 28,* 1-16.

Mackenzie, J.K., *Biometrica,* 1958, *vol 45,* 857-861.

Malin, A.S.; Hatherley, M., *Metals Sci.,* 1978, *No.8,* 463-472.

Metals Handbook *Properties and Selection: Nonferrous Alloys and Pure metals*, vol 2, ASM, 1979.

Meyer, L.W.; Hamworring, S. In *Metallurgical Applications of Shock-Wave and High-Strain-Rate Phenomena,* Murr, L.E.; Staudhammer, K.P.; Meyers, M.A., Eds., Dekker, New York, 1986, 657-671.

Meyers, M.A.; Mishra, M.; Benson, D.J., *Progr. Mater. Sci.,* 2006, *vol 51,* 427-556.

Mishin, O.V.; Gottstein, G., *Phil. Mag.,* 1998, *vol 78A,* 373-388.

Mishin, O.V.; Jensen, D.J.; Hansen, N., *Mater. Sci. Eng.,* 2003, *vol A342,* 320-328.

Mishra, A.; Richard, V.; Gregori, F.; Asaro, R.J.; Meyers, M.A., *Mater. Sci. Eng.,* 2005, *vols A410-411,* 290-298.

Mishra, A.; Kad, B.K.; Gregori, F,; Mayers, M.A., *Acta Mater.,* 2007, *vol 55,* 13-28.

Moscoso, W.; Shankar, M.R.; Mann, J.B.; Compton, W.D.; Chandrasekar, S., *J. Mater. Res.,* 2007, *vol 22,* 201-205.

Nakashima, K.; Horita, Z.; Nemoto, M.; Langdon, T.G., *Mater. Sci. Eng.,* 2000, *vol A281,* 82-87.

Nam, C.Y.; Han, J.H.; Chung, Y.H.; Shin, M.C., *Mater. Sci. Eng.,* 2003, *vol A347,* 253-257.

Niendorf, T.; Canandinc, D.; Maier, K.L.; Karaman, I.; Yapici, G.G., *Acta Mater.,* 2007, *vol 55,* 6596-6605.

Nishida, Y.,; Arima, H.; Kim, J.C.; Ando. T., *Scripta Mater.,* 2001, *vol. 45,* 261-266.

Niymoto, H.; Fushini, J.; Mitaki, T.; Vinogradov, A.; Hashimito, S., *Mater. Sci. Eng.,* 2005, *vol A405,* 221-232.

Olejnik, L.; Rosochovski, A., *Bullet. Pol. Acad. Sci. Tech. Sci.,* 2005, *vol. 23,* 413-423.

Panin, V.E., In *Investigations and Applications of Severe Plastic Deformation,* Lowe, T.G.; Valiev, R.Z., Eds., Kluwer, Netherland, 2000, 203-209.

Park, J.W.; Chung, Y.H., *U.S. Patent,* 2006, *No. 7,140,222.*

Poortmans, S.; Diouf, B.; Habraken, A.M,; Verlinden, B., *Scripta Mater.,* 2007, *vol 56,* 749-752.

Prangnell, P.B.; Bowen, J.B.; Gholonia, A., In Proceedings of 22th Rico International Symposium on Materials Science, Riso National Laboratory, Roshilde, 2001, 105-117.

Priester, R.; Ibrahee, A.K., *Mater. Sci. Technol.,* 2000, *vol 16,* 1273-1282.

Rauch, E.F.; G'Sell, C., *Mater. Sci. Eng.,* 1989, *vol A111,* 71-80.

Rice, J.R.; Tracey, D.M., *J. Mech. Phys. Solids,* 1969, *vol 17,* 201-217.

Richert, J.; Richert, M.; Zasadzincki, A.; Korbel, A., *Patent PRL,* 1979, *No. 123026.*

Richert, M.; Liu, Q.; Hansen, N., *Mater. Sci. Eng.,* 1999, *vol A260,* 275-283.

Richert, M.; Stuwe, H.P.; Richert, J.; Pippan, R.; Motz, C., *Mater. Sci. Eng.,* 2001, *vol A301,* 237-243.

Richert, M.; Petryk, H.; Stypkiewicz, S., *Arch. Metallurg. Mater.,* 2007, *vol 52,* 49-54.

Richmond, O.; Devenpeck, M.I., *Proceeedings of 4th U.S. Congress on Applied Mechanics,* ASME, 1962, 1053-1057.

Rigney, D.A.; Naylor, M.G.S.; Divakar, R.; Ives, L.K., *Mater. Sci. Eng.,* 1986, *vol A81,* 409-425.

Rosochovski, A., *International Patent Application*, 2005, *WO 2005/039792.*

Rybin, V.V., Large Plastic Deformation and Fracture of Metals, "Metallurgy", Moscow, 1986 (in Russian).

Saito, Y.; Tsuji, N.; Utsunomiya, H.; Sakai, T.; Hong, R.G., *Scripta Mater.,* 1998, *vol 39,* 1221-1227.

Saito, Y,; Utsynomiya, H.; Suzuki, H.; Sakai, T., *Scripta Mater.,* 2000, *vol 42,* 1139-1144.

Salischev, G.A.; Mironov, S.Yu.; Zherebtsov, S.V., *Rev. Adv. Mater. Sci.,* 2006, *vol 11,* 152-158.

Saunders, I.; Nutting, J., *Metals Sci.,* 1984, *vol 18,* 571-575.

Sedlasec, R.; Blum, W.; Kratochwil, J.; Forest, S., *Metallurg. Mater. Trans.,* 2002, *vol 33,* 319-327.

Segal, V.M., *J. Appl. Mech. Tech. Phys.,* 1969, vol 2, 319-322.

Segal, V.M., *Mechanics of Solids,* 1971, *No.3,* 184-189 (in Russian).

Segal, V.M., *Machine Sci.,* 1974a, *No. 2,* 96-102 (in Russian).

Segal, V.M., *Methods of Stress-Strain Analysis in Metal Forming*, Sc. D. Thesis, Minsk, 1974b.

Segal, V.M.; Makyshok, E.M.; Reznikov, V.I., *Investigation of Plastic Forming of Metals by Moire Method,* Metallurgia, Moscow, 1974c (in Russian).

Segal, V.M.; Reznikov, V.I.; Malyshev, V.F., *Invention Serificate of the U.S.S.R.,* 1976, *No. 690361.*

Segal, V.M., *Invention Serificate of the U.S.S.R.,* 1977, *No. 575892.*

Segal, V.M.; Kopylov, V.I., *Invention Certificate of the U.S.S.R.,*1980, *No. 780293.*

Segal, V.M.; Kopylov, V.I., *Invention certificate of the U.S.S.R.,* 1981a, *No. 902962.*

Segal, V.M.; Reznikov. V.I.; Drobyshevsky, A.E.; Kopylov, V.I., *Russian Metallurgy,* 1981b, *No. 1,* 115-123.

Segal, V.M., *Phys.-Chem. Mech. Mater.,* 1982, *vol 18*, 491-496.

Segal, V.M., *Strength of Materials,* 1983, *vol 18*, 103-106.

Segal, V.M.; Pavlik, D.A., *Russian Metallurgy,* 1984a, *No. 5,* 95-98.

Segal, V.M., *J. Appl. Mech. Tech. Phys.,* 1984b, *No.1,* 114-120.

Segal, V.M.; Reznikov, V.I.; Kopylov, V.I.; Pavlik, D.A.; Malyshev, V.F., *Processes of Plastic Structure Formation in Metals,* Science and Engineering, Minsk, 1994 (in Russian).

Segal, V.M., *Mater. Sci. Eng.*, 1995a, *vol A197*, 157-164.

Segal, V.M.; Goforth, R.E.; Hartwig, K.T., *U.S. Patent*, 1995b, *No. 5,4000,633.*

Segal, V.M., *U.S. Patent,* 1996, *No. 5,513,512.*

Segal, V.M.; Hartwig, K.T.; Goforth, R.E., *Mater. Sci. Eng.*, 1997, *vol A224*, 107-115.

Segal, *U.S. Patent,* 1998, *No. 5,850,755.*

Segal, V.M., *Mater. Sci. Eng.,* 1999, *vol A271,* 322-333.
Segal, V.M., *Mater. Sci. Eng.,* 2002, *vol A338,* 331-344.
Segal, V.M., *Mater. Sci. Eng.,* 2003, *vol A345,* 35-46.
Segal, V.M.; Willet, W.; Ferrasse, S., *U.S. Patent,* 2004, *No. 6,723,187.*
Segal, V.M., *Mater. Sci. Eng.,* 2005, *vol A406,* 205-216.
Segal, V.M.; Ferrasse, S.; Alford, F., *Mater. Sci. Eng.,* 2006a, *vol A422,* 321-326.
Segal, V.M., *Russian Metallurgy (Matally),* 2006b, *No.5,* 474-483.
Segal, V.M., *U.S. Patent,* 2007, *No. 7,191, 630.*
Segal, V.M., *Mater. Sci. Eng.,* 2008, *vol A 476,* 178-185.
Segal, V.M., *J.Mat. Proc. Technol.,*2010, *vol 210,* 542-549.
Semiatin, S.L.; Segal, V.M.; Goforth, R.E.; Frey, N.D.; DeLo, D.P., *Metallurg. Mater. Trans.,* 1999, *vol 30A,* 1425-1435.
Semiatin, S.L.; DeLo, D.P.; Shell, E.B., *Acta Mater.,* 2000, *vol 48,* 1841-1851.
Shield, R.T., *Proc. Royal Soc.,* 1955, *vol 233A,* 267-287.
Shin, D.N.; Kim, I.; Kim, J.; Park, K.-T., *Acta Mater.,* 2001, *vol 49,* 1285-1292.
Sitdikov, O.S.; Kaibyshev, R.O.; Safarov, I.M.; Mazurina, I.A., *Phys. Metal. Metallograph.,* 2001, *vol. 92,* 270-280.
Sivakumar, S.M.; Ortiz, M., *Comput. Meth. Appl. Mech. Eng.,* 2004, *vol 193,* 5177-5194.
Smirnov, B.I., *Crystallography,* 1968, *vol 13,* 721-722.
Smirnova, N.A.; Levit, V.I.; Pilyugin, V.P.; Kuznetsov, R.I.; Davydova, L.S.; Sazonova, V.A., *The Phys. Metal. Metallogr.,* 1986, *vol 61,* 127-134.
Spencer, A.J.M., *J. Mech. Phys. Solids,* 1962, *vol 10,* 17-26.
Srinivasan, R., *Scripta Mater.,* 2001, *vol 44,* 91-96.
Stecher, G.; Thomson, P., *International Patent Application,* 2003, *WO 2003/027337.*
Stolyarov, V.V.; Zhu, Y.T.; Lowe, T.C.; Valiev, R.Z., Mater. Sci. Eng., 2001a, vol A303, 82-89.
Stolyarov, V.V.; Ermolenko, A.S.; Gunderov, V.S.; Popov, A.G.; Raab, G.I.; Puzanova, T.Z.; Valiev, R.Z.; Galiko, V.S.; Belorezov, E.V., *Mater. Sci. Forum,* 2001, *vols 373-376,* 265-268.
Stolyarov, V.V.; Prokof'ev, E.A.; Prokoshkin, S.D.; Dobatkin, S.V.; Tribushina, I.V.; Khmelevskaya, I.Yu.; Pushin, V.G.; Valiev, R.Z., *Phys. Metals Metalogr.,* 2005, *vol 100,* 608-618.
Su, J.-Q.; Nelson, T.W.; Sterling, C.J., *J. Mater. Res.,* 2003, *vol 18,* 1757-1760.
Su, S.W.; Lu, l.; Lai, M.O., *Mater. Sci. Eng.,* 2006, *vol A343,* 227-236.
Sue, H.-J.; Li, C.K.-Y., *J. Mater. Sci. Letters,* 1998, *vol 17,* 853-856.
Suh, J.Y.; Kim, H.S.; Park, J.W.; Chang, J.Y., *Scripta Mater.,* 2001, *vol 44,* 677-682.
Tao, N.R.; Sui, M.L.; Ku, J.; Lu, K., *Nanostructured Materials,* 1999, *vol 11,* 433-440.
Tao, N.; Zhang, H.; Lu, J.; Lu, K., *Mater. Trans.,* 2003, *vol 44,* 1919-1925.
Taylor, G.I., *J. Inst. Metals,* 1938, *vol. 62,* 307-324.
Thuillier, S.; Rauch, E.F., *Acta Metallurg. Mater.,* 1994, *vol 42,* 1973-1983.
Thomsen, E.G.; Yang, C.T.; Kobayashi, S., *Mechanics of Plastic Deformation in Metal Processing,* McMillan Co, New York, 1965.
Toth, L.S.; Massion, R.A.; Germain, L.; Baik, S.C.; Suwas, S., *Acta Mater.,* 2004, *vol 52,* 1885-189.
Trybus, C.L.; Spitzig, W.A., *Acta Metallurg.,* 1989, vol 37, 1971-1981.

Tyumentsev, A.N.; Tret'yak, M.V.; Pinzhin, Yu.P.; Korotaev, A.D.; Valiev, R.Z.; Islamgaliev, R.K.; Korznikov, A.V., *Phys. Metals Metallograph.*, 2000, *vol 90,* 461-470.

Valiev, R.Z.; Ivanisenko, Yu.V.; Rauch, E.F.; Bandelet, B., *Acta Mater.*, 1996, *vol 44*, 4705-4712.

Valiev, R.Z.; Langdon, T.G., *Prog. Mater. Sci.,* 2006, *vol. 51*, 881-981.

Varyukhin, V.N.; Spuskanyuk, V.Z.; Matrosov, N.I.; Dugalko, A.B.; Shevchenko, B.A.; Medvedskaya, E.N.; Sennikova, L.F.; Spuskanyuk, A.V.; Pavlovskaya, E.V., *Physics and Engineering of High Pressure,* 2002, *vol 12,* 31-39.

Verhoeven, J.D., *Fundamentals of Physical Metallurgy*, Wiley, New York, 1975.

Vinogradov, A.; Hashimoto, S.; Patlan, V.; Kitagawa, K., *Mater. Sci. Eng.,* 2001, *vols 319-321*, 862-866.

Voorhes, W.G., *U.S. Patent,* 1975, No. *3,922,898.*

Wang, J.; Xu, Y.; Wang, L.; Zhao, X.; Wang, H., *Mater. Res. Soc. Symp. Proc.,* 2001, *vol 683E,* BB3.1.1-BB3.1.6

Wang, J.; Kang, S.-B.; Kim, H.-W., *Metallurg. Mater. Trans.*, 2004, *vol 35A*, 279-286.

Wang, K.; Tao, N.R.; Liu, G.; Lu, J.; Lu, K., *Acta Mater*, 2006, *vol 54*, 5281-5291.

Wang, Y.M.; Ma, E., *Acta Mater.,* 2005, *vol 52*, 1699-1709.

Wenk, H.R., Ed.,*Preferred Orientations in Deformed Metals and Rocks: Introduction to Modern Texture Analysis*, Academic Press, 1985.

Wenk, H.R.; Matthies, S., *Beartex (Berkeley Software Package),* University of California, Berkeley, 1996.

Werenskiold, J.C.; Roven, H.J., *Mater. Sci. Forum*, 2006, *vols 503-504*, 45-50.

Wert, J.A., In *Superplastic Forming of Structural alloys*, Paton, N; Hamilton, C.H., Eds., AISME, 1983, 69-83.

Winther, G*., Acta Mater.,* 2003, *vol 51*, 417-429.

Wu, P.C.; Chang, C.P.; Cao, P.W., *Mater. Sci. Eng.,* 2004, *vol A374*, 196-203.

Wu, P.D.; Huang, Y.; Lloyd, D.J., *Scripta Mater.*, 2006, *vol 54*, 2107-2111.

Xu, C.; Langdon, T.G., *Scripta Mater.,* 2003, *vol.48*, 1-4.

Xu, C.; Furukawa, M.; Horita, Z.; Langdon, T.G., *Mater. Sci. Eng.*, 2005, *vol A398*, 66-76.

Xu, Y.B.; Zhong, W.L.; Chen, Y.J.; Shen, L.T.; Liu, Q., Bai, Y.L.; Meyers, M.A., *Mat. Sci. Eng.*, 2001, *vol A299*, 287-295.

Xue, Q.; Beyerlein, I.J.; Alexander, D.J.; Grey III, G.T., *Acta Mater.,* 2006, *vol 55*, 655-668.

Yu, C.Y.; Kao, P.W.; Chang, C.P., *Acta Mater.,* 2005, vol 53, 4019-4028.

Yuan, S.Y.; Peng, C.N.; Yeh, J.W., *Mater. Sci. Technol.,* 1999, *vol 15,* 683-688.

Yun, S.-J., *An Analysis by the Finite Element Method of Material Deformation in Equal channel Angular Extrusion,* Ph.D. Thesis, TAMU, 1996.

Zairi, F.; Aour, B.; Gloaguen, J.M.; Nait-Abdelaziz, M.; Lefebre, J.M., *Comput. Mater. Sci.,* 2006, *vol 38*, 202-216.

Zhilyev, A.P.; Oh-ishi, K.; Raab, G.I.; McNelley, T.R., In *Ultrafine Grained Materials IV*, Zhu, Y.T.; Langdon, T.G.; Horita, Z.; Zehetbauer, M.J.; Semiatin, S.L.; Lowe, T.C., Eds., TMS, 2006, 1-6.

Zhu, Y.; Lowe, T.C., *Mater. Sci. Eng.,* 2000, *vol A291*, 46-53.

PART II. MATHEMATICAL MODELING OF SEVERE PLASTIC DEFORMATION

***I.J. Beyerlein* [1] *and C.N. Tomé* [2]**
[1] Theoretical Division
[2] Materials Science and Technology Division
Los Alamos National Laboratory
Los Alamos, NM 87545 , U.S.A.

PREFACE

Metals used in technological applications are usually polycrystals, that is, solid phase aggregates of many crystals. The lattice of each crystal has a specific orientation with respect to an external reference system, and the distribution of these orientations is called texture. During metal forming, permanent deformation is imparted into a workpiece, which is not recovered when the load is removed. It is called plastic deformation, as opposed to elastic deformation, which is recoverable. Plastic deformation takes place via crystallographic shear in the individual crystals, either via dislocation slip or deformation twinning. Since slip and twinning dislocations propagate on characteristic crystallographic planes and a critical stress (or activation energy) on these planes is required to activate dislocations, the orientation of the crystals with respect to the applied load (i.e., texture) will affect the macroscopic stress required to deform the aggregate.

During plastic deformation the microstructure also evolves within the crystals (or grains), usually in the form of an increased dislocation density, dislocation cells, dislocation sub-boundaries and for certain materials and conditions, deformation twins and their boundaries. Especially when large strains are involved, as in processes of severe plastic deformation (SPD), new grain boundaries may form from these dislocation patterns, a process called grain fragmentation or refinement. These internal grain structures can pose barriers to the propagation of new slip dislocations or new twins, and they can change the distribution of slip and twinning within the grain. Consequently, the applied stresses required to sustain the plastic deformation can change with straining. Strain hardening refers to an increase in the strength of the obstacle-dislocation interaction, and strain softening refers to a decrease.

It follows from the concepts expressed above that there is a direct connection between polycrystalline texture, yield stress, and anisotropic mechanical response, on one hand, and the intragranular microstructure responsible for hardening, on the other. In this context, *anisotropy* means that the response of the aggregate to external loading varies depending on the direction along which the load is applied. In turn, in the course of deformation, dislocation slip activity can rotate the crystal, and deformation twinning can reorient domains within the crystal. For large plastic deformation during SPD processes, other mechanisms such as shear band localization and grain boundary sliding also alter the evolution of microstructure and texture. Accordingly, texture and plastic anisotropy are coupled and evolve during plastic forming.

Products manufactured through severe plastic deformation will develop texture and exhibit hardening that will be substantially different that those of their initial state prior to processing. It is the connection between texture and anisotropic hardening that makes the control of texture development in metal forming processes of primary interest in achieving product quality. Modeling of texture and hardening evolution thus becomes a tool of great value in gaining fundamental understanding of a particular deformation process and the mechanical properties (strength, ductility, etc) of the final product. Modeling should also play a role when these materials are going to be designed into structural components. From a more basic perspective, polycrystal modeling allows us to bridge the gap between macroscopic response and microstructural features.

In Part II of this book we develop crystallographic-based polycrystalline plasticity models for understanding and predicting texture, anisotropy and hardening evolution in materials subjected to SPD. Part II is structured as follows. In Chapters 6 and 7, we introduce the basic concepts of polycrystal plasticity models and apply them to common bulk forming operations, such as rolling and severe simple shearing. In Chapter 8 we provide mechanics descriptions of single and multi-pass ECAE deformation for polycrystal modeling. Chapters 9 and 10 discuss texture characterization and texture evolution in ECAE for metals of different crystal structures. Chapter 11 describes mechanics models for ECAE deformation under non-ideal conditions. The remaining chapters focus on specific applications. Chapter 12 contains an in-depth, multiscale modeling analysis of texture evolution in cyclic route C, which imposes large strain reversals. Chapter 13 focuses on deformation twinning: modeling, texture evolution, and plastic anisotropy. Last, Chapter 14 presents a single crystal hardening law for modeling the effects of strain path changes (such as when going from ECAE to a mechanical test) on polycrystalline flow stress and strain hardening or softening.

Chapter 6

FUNDAMENTALS OF POLYCRYSTAL MODELING

For a comprehensive treatment of the subject of this Chapter the reader is referred to the book on "Texture & Anisotropy", by Kocks et al (1998). An excellent introductory treatment of polycrystal plasticity can be found in the book by Hosford (1993).

6.1. CRYSTAL PLASTICITY AND YIELD SURFACE

Since plasticity is mediated by crystallographic shear of single crystals, we introduce in this Section the basic relationships between crystallography and plasticity. During large plastic deformation, the contribution to strain from elasticity is negligibly small (typically $<10^{-3}$) by comparison to the plastic component (typically $>10^{-1}$). In addition, once the elasto-plastic transition is over, the stress in the grain is bounded by the threshold values associated with slip and twinning activity, and so its evolution is controlled by plastic mechanisms. As a consequence, in what follows we disregard elasticity and describe only the plastic contribution to deformation.

Here we will be concerned with crystals of cubic and hexagonal symmetry, which constitute the majority of the metallic aggregates used in technological applications. Aluminum and Al alloys, and austenitic stainless steels, have a face-centered-cubic (FCC) structure. Iron and ferritic steels have a body-centered-cubic structure (BCC), while Mg, Zr, Ti and its alloys exhibit the hexagonal-close-packed (HCP) structure. The corresponding unit cells, showing slip and twinning systems, are depicted in Figure 6.1.

Dislocations propagate by displacing atomic planes with respect to each other, thereby producing plastic shear and no volumetric dilatation. Dislocations tend to propagate in compact atomic planes and along the shortest atomic direction. Let us denote $\mathbf{n}^s$ the normal to the slip plane, and $\mathbf{b}^s$ the direction of shear (Burgers vector). If σ_{ij} is the stress tensor acting on the crystal (grain) then the resolved shear stress on the shear plane and along the shear direction is given by a linear transformation of the stress tensor [1]

---

[1] Unless stated otherwise, throughout this text we use the implicit summation convention: repeated sub-indices in equations denote summation over the range of such indices.

$$\tau^s_{res} = b^s_i n^s_j \sigma_{ij} \tag{6.1}$$

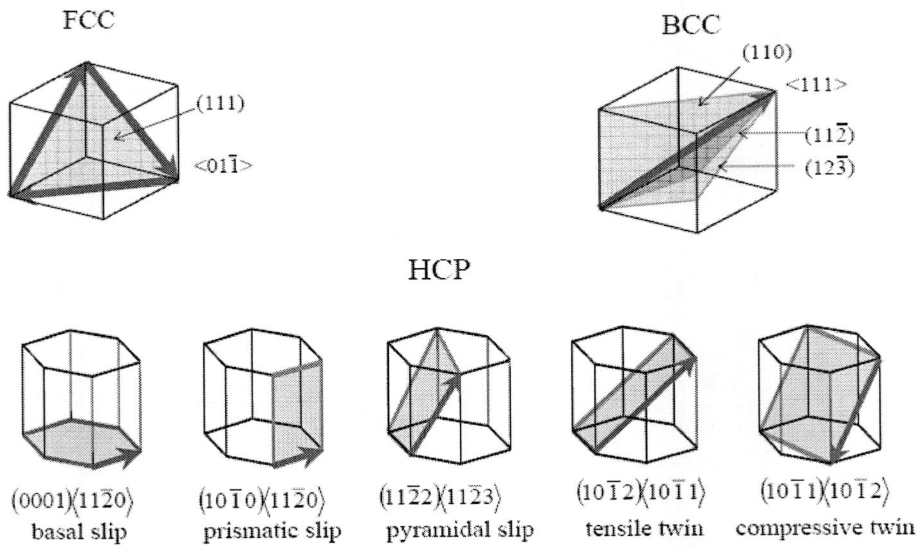

Figure 6.1. Unit cell and predominant slip planes and slip directions for: (a) FCC crystals; (b) BCC crystals; (c) HCP crystals.

The dyadic tensor $b \otimes n$, also called the Schmid tensor, provides the link between crystallographic shear and the overall stress and deformation in the grain. It will be useful in what follows to decompose the Schmid tensor into its symmetric and skew symmetric components

$$\begin{aligned} m^s_{ij} &= \tfrac{1}{2}\left(b^s_i n^s_j + b^s_j n^s_i\right) \\ q^s_{ij} &= \tfrac{1}{2}\left(b^s_i n^s_j - b^s_j n^s_i\right) \end{aligned} \tag{6.2}$$

Since $b^s_i n^s_i = 0$ it turns out that $\mathrm{trace}(m^s_{ij}) = 0$. Using Eq. (6.2) and the symmetry $\sigma_{ij} = \sigma_{ji}$ allows one to write Eq. (6.1) as:

$$\tau^s_{res} = m^s_{ij}\sigma_{ij} \tag{6.3}$$

Propagating dislocations on slip systems requires the resolved shear in the slip plane and along the slip direction to attain a threshold value τ^s. This condition, known as the Schmid law, is expressed by:

$$m^s_{ij}\sigma_{ij} = \tau^s \tag{6.4}$$

Splitting the deviatoric and the hydrostatic stress components allows one to write Eq. (6.1) as:

$$\tau^s_{res} = m^s_{ij}\sigma_{ij} = m^s_{ij}\sigma'_{ij} + p\,tr(m^s_{ij}) = m^s_{ij}\sigma'_{ij} \tag{6.5}$$

which expresses the important fact that activation of plastic shear is independent of the hydrostatic stress and depends only on the deviatoric stress component. In addition, because the stress tensor is symmetric, only five deviatoric components are independent.

A shear rate $\dot{\gamma}^S$ in system *s* has associated a strain rate tensor **D** when expressed in an arbitrary reference system (i.e.: crystal axes). Once again, the Schmid tensor mediates the transformation

$$m^s_{ij}\dot{\gamma}^s = D^c_{ij} \tag{6.6}$$

Because tr(**m**)=0 it turns out that, as is to be expected from shear deformation, the strain rate **D** induced by $\dot{\gamma}^S$ is also traceless, and as a consequence, the rate of volume change, i.e., (tr(**D**)=$\Delta\dot{V}/V$) is zero. This result provides another important property of plastic deformation--plastic deformation is non-dilatational. Finally, if several systems are active simultaneously in a grain, the total strain rate is given by the sum of the individual slip rates:

$$\sum_s m^s_{ij}\dot{\gamma}^s = D^c_{ij} \tag{6.7}$$

Since the tensor $\mathbf{D}^c$ is traceless and symmetric, it only has five independent components. An important consequence of the linear system of equations in Eq. (6.7) is that five linearly independent slip systems need to be active if an arbitrary strain rate is to be accommodated by the grain. According to Eq. (6.4), this can be accomplished if the resolved shear stress equals the threshold value in five systems simultaneously.

The FCC, BCC, and HCP structures studied here differ in the type and number of slip modes available for shear. In FCC crystals slip takes place in {111} planes along [110] directions, and there are 12 crystallographically equivalent systems plus the opposites (Figure 6.1a). In BCC crystals slip takes place along the [111] direction, on the {110} and {112} planes, and each contributes twelve crystallographically equivalent systems (Figure 6.1b). Additional slip on other planes may be present at high temperature (pencil glide). In HCP crystals several modes may be active depending on temperature or rate, and each mode contributes a particular number of crystallographically equivalent systems. Prismatic and basal slip contribute three each, pyramidal <c+a> slip contributes 12, and twinning modes contribute six each. However, not every combination of five systems is independent, and not all combinations of independent ones are allowable. In what follows we discuss how the active systems are chosen.

The linear form of Eq. (6.4), written in terms of the five independent deviatoric components, represents in 5-dimensional deviatoric stress space a plane with normal $\mathbf{m}^s$ located a distance τ^S from the origin. The Schmid condition indicates, then, that the stress state has to be on this plane for slip to take place. A convention frequently used to define the five independent components (i.e.: Kocks et al. 1998, Chapter 7) is:

$$(\sigma_1,\sigma_2,\sigma_3,\sigma_4,\sigma_5)=\left[(\sigma_{22}-\sigma_{11})/\sqrt{2},(2\sigma_{33}-\sigma_{22}-\sigma_{11})/\sqrt{6},\sqrt{2}\sigma_{23},\sqrt{2}\sigma_{13},\sqrt{2}\sigma_{12}\right] \quad (6.8)$$

The same convention is used for representing the five independent components of the plastic strain rate. Stress and strain rate vectors so defined are work conjugates. The set of hyper-planes associated with all the slip systems in the grain (12 plus the opposites for FCC) defines an inner envelope called the single crystal yield surface (SCYS). The SCYS bounds the stress states attainable in the crystal because the resolved shear stress cannot exceed the threshold by definition. Figure 6.3 depicts sections of the SCYS of FCC, BCC and HCP crystals, corresponding to the 2-dimensional subspace defined by the first two components in Eq. (6.8). This subspace comprises the diagonal stress components and is usually called the π-plane representation of the yield locus. Observe that all the SCYSs associated with the Schmid criterion are faceted, and that the equipotential surfaces associated with the rate sensitive response to be discussed next (Eq. 6.11) are close to the faceted surfaces and have slightly rounded vertices. In the case of BCC the inner surface shown is calculated using the 48 slip systems depicted in Figure 6.1 (pencil glide), which provide a nearly isotropic response. In the case of HCP, easy prismatic slip, intermediate tensile twinning and hard pyramidal <c+a> slip were considered. Observe that twinning being directional breaks the centro-symmetric character of the SCYS.

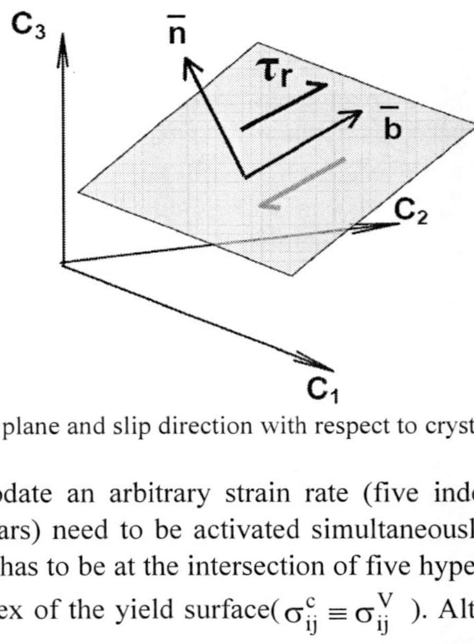

Figure 6.2. Schematic of slip plane and slip direction with respect to crystal axes.

In order to accommodate an arbitrary strain rate (five independent components) five independent systems (shears) need to be activated simultaneously. This requirement means that the stress in the grain has to be at the intersection of five hyper-planes; that is to say, they must coincide with a vertex of the yield surface($\sigma_{ij}^{c} \equiv \sigma_{ij}^{V}$). Although there are many such vertices (for instance, 28 plus the opposites in a FCC crystal; Kocks et al. 1983), the maximum work principle of Bishop and Hill (1951a,b) usually removes the ambiguity associated with choosing the value of the stress. The maximum work principle states that the active vertex V^* is the one that maximizes the projection of the strain rate on the stress

$$D_{ij}^{c}\sigma_{ij}^{V^*} = \max_{V} D_{ij}^{c}\sigma_{ij}^{V} \quad (6.9)$$

The geometric interpretation of Bishop and Hill's principle is that the strain rate vector $\mathbf{D}^c$ has to be contained ~~by~~ within the cone formed in five dimensional space by the normal vectors to the SCYS at the vertex. Such a situation is depicted in Figure 6.3a for the 2-dimensional section of the SCYS of a FCC crystal. Anisotropy is highlighted by the fact that the directions of deformation rate and stress are not necessarily co-linear.

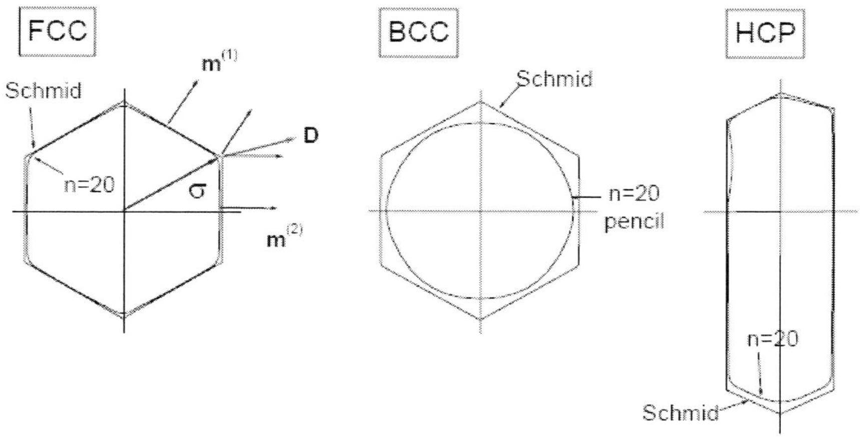

Figure 6.3. Single Crystal Yield Surface of FCC, BCC and HCP crystals, using the Schmid criterion of Eq. (6.4) and the rate sensitivity law of Eq. (6.11) with n=20. Shown in the Figure is the π-plane projection, corresponding to the subspace (σ_1, σ_2) as defined by Eq. (6.8).

Plastic deformation of HCP crystals is more complex than that for FCC or BCC crystals because shear on the compact planes - either basal or prismatic depending on the c/a ratio – lacks enough independent systems for accommodating an arbitrary imposed strain rate. As a consequence, energetically less favorable pyramidal <c+a> slip and/or tensile twinning and/or compressive twinning needs to be active if the aggregate is to deform without fracturing. The slip systems observed in hcp Zr deforming at quasi-static rates in the low temperature regime are depicted in Figure 6.1c. They are: $\{10\bar{1}0\}<11\bar{2}0>$ prismatic slip (3 systems), $\{10\bar{1}1\}<11\bar{2}3>$ pyramidal <c+a> slip (12 systems), $\{10\bar{1}2\}<10\bar{1}1>$ tensile twins (6 systems) and $\{11\bar{2}2\}<11\bar{2}\bar{3}>$ compressive twins (6 systems). Basal $\{0001\}<11\bar{2}0>$ slip (3 systems), plus other types of pyramidal and twinning systems may be active, depending on the material, the deformation temperature and the deformation rate. The SCYS of an HCP crystal with easy prismatic slip, medium hard tensile twinning, and hard pyramidal is more anisotropic than those of FCC and BCC crystals (Figure 6.3c), and may adopt several morphologies depending on the relative threshold between modes (Tomé and Kocks, 1985).

While the Bishop and Hill principle removes the ambiguity in choosing the yield stress, another type of ambiguity arises when more than five planes intersect at one vertex, a usual situation in cubic systems (Kocks et al. 1983). Depending on the five independent active systems chosen to solve Eq. (6.7), different shears, and so a different reorientation of the grain, will result. This ambiguity is a consequence of the faceted character of the SCYS, and the fact that the normal is not defined at a vertex. Asaro and Needleman (1985) proposed a rate-sensitive visco-plastic response which removes the ambiguity and facilitates the numerical implementation of slip in polycrystal models. The shear rate in a given system is given by a power of the resolved shear stress τ_{res}

$$\dot{\gamma}^s = \dot{\gamma}_o \left| \frac{\tau_{res}}{\tau^s} \right|^n sign(\tau_{res}) \tag{6.10}$$

While the power n can be interpreted as the inverse of the rate sensitivity, it may also be interpreted in a numerical way as "rounding" the sharp vertices of the SCYS. There are several advantages in using the constitutive law represented by Eq. (6.10). First, it provides shear rates directly. Second, it indirectly accounts for the fact that dislocation glide is a thermally activated mechanism: a higher resolved shear stress is required for attaining higher shear rates. Finally, for large values of n (i.e.: $n > 20$) Eq. (6.10) gives a response which is close to the Schmid criterion, but which is not burdened by the ambiguity in the choice of active slip systems. Every system with a non-zero resolved shear contributes to deformation, but only those where the resolved shear is close to the threshold will contribute significantly. The strain rate in the grain is still given by the linear superposition of the shear rates on all systems:

$$D_{ij}^c = \sum_s m_{ij}^s \dot{\gamma}^s = \dot{\gamma}_o \sum_s m_{ij}^s \left(\frac{m_{kl}^s \sigma_{kl}}{\tau^s} \right)^n \tag{6.11}$$

The implications of the rate sensitivity law for the SCYS are the following. First, since any stress will induce some amount of plastic yield, the 'Schmid' concept of yield locus is lost. Instead, the SCYS is identified with an equal-dissipation surface, defined as the locus $\boldsymbol{\sigma}$:**D** = cte. The SCYSs given by the Schmid law (Eq. 6.4) are plotted in Figure 6.3. The inner surfaces are equal-dissipation surfaces obtained using $n = 20$: they give a locus close to the previous one but with rounded corners. A positive consequence of identifying the yield surface with the locus of equal-dissipation states is the normality rule: according to Eq. (6.11) the gradient $\partial(D_{ij}^c \sigma_{ij}) / \partial \sigma_{kl}$ of the plastic potential $\boldsymbol{\sigma}$:**D** is proportional to the strain rate **D**, and since the gradient is normal to an equipotential surface, it turns out that **D** is normal to the yield locus defined by an equipotential. In practice, for the algorithmic calculations one uses Eq. (6.11) to relate stress and strain-rate instead of using an explicit form of the SCYS.

6.2. Texture Evolution and the Kinematics of Lattice Rotation

Section 6.1 outlines the connection between crystallography and plasticity. In this Section we define the basic kinematic tensors, report their relations, and give expressions for calculating the change in crystallographic orientation associated with plastic deformation (texture evolution). For a more detailed treatment of the subject the reader is referred to Gurtin (1981) and Bonet and Wood (1997).

For each crystal (grain) we are going to consider an *initial* and a *current* configuration. In the initial (undeformed) configuration the coordinates of a material point are denoted with **X**. In the final (current) configuration, where the crystal has deformed by crystallographic shear and has rotated, the coordinates of the same point are denoted with **x**. The associated displacement is denoted **u** = **x** – **X**. In addition, we define a linear transformation matrix $\mathbf{R}^c$ that allows us to express tensors from initial to current crystal axes.

The deformation in the grain is characterized by the velocity gradient tensor $\mathbf{L}^c$ and the deformation gradient tensor $\mathbf{F}^c$, defined as:

$$L_{ij}^c = \frac{\partial \dot{u}_i^c}{\partial x_j} \tag{6.12}$$

$$F_{ij}^c = \frac{\partial x_i}{\partial X_j} \tag{6.13}$$

With the property:

$$x_i = F_{ij}^c X_j \tag{6.14}$$

Using the definitions of the tensors above it can be shown that :

$$\dot{F}_{ij}^c = \frac{\partial F_{ij}^c}{\partial t} = L_{ik}^c F_{jk}^c \tag{6.15}$$

In addition, since plastic deformation is accommodated by shear, and since shear deforms the crystal but preserves its orientation, it is useful to utilize what is called the 'multiplicative decomposition' of the deformation gradient. Such procedure, which is illustrated in Figure 6.4, decomposes the displacements into two sequential steps: a 'plastic stretch' $\mathbf{F}_o^c$ which shears the crystal without reorienting it, followed by a rigid crystal rotation $\mathbf{R}^c$ that transforms from initial to final crystal axes orientation:

$$\mathbf{F}^c = \mathbf{R}^c\,\mathbf{F}_o^c \tag{6.16}$$

In crystal axes the stretch obeys a relation like Eq. (6.15):

$$\dot{\mathbf{F}}_o^c = \mathbf{L}_o^c \mathbf{F}_o^c \tag{6.17}$$

where

$$L_{o_{ij}}^c = \frac{\partial \dot{u}_i^c}{\partial X_j} \tag{6.18}$$

is the velocity gradient in the reference frame attached to the crystal axes and is given by the linear superposition of shear rates on all active slip and twinning systems:

$$L^c_{o_{ij}} = \sum_s \dot{\gamma}^s b^s_i n^s_j \tag{6.19}$$

The advantage of the multiplicative decomposition is that the vectors **n** and **b** remain invariant in crystal axes. Decomposing the dyadic $b^s \otimes n^s$ into the symmetric and skew symmetric components:

$$\begin{aligned} m^s_{ij} &= \tfrac{1}{2}\left(b^s_i n^s_j + b^s_j n^s_i\right) \\ q^s_{ij} &= \tfrac{1}{2}\left(b^s_i n^s_j - b^s_j n^s_i\right) \end{aligned} \tag{6.20}$$

allows us to decompose $\mathrm{L}^{\mathrm{c}}_{\mathrm{o}}$ into a strain rate and a rotation rate (spin):

$$L^c_{o_{ij}} = D^c_{o_{ij}} + W^c_{o_{ij}} \tag{6.21a}$$

where

$$D^c_{o_{ij}} = \sum_s \dot{\gamma}^s m^s_{ij} \tag{6.21b}$$

$$W^c_{o_{ij}} = \sum_s \dot{\gamma}^s q^s_{ij} \tag{6.21c}$$

Substituting Eq. (6.16) into Eq. (6.15) and using Eq. (6.17):

$$\mathbf{L}^c = \dot{\mathbf{F}}^c \mathbf{F}^{c^{-1}} = \left(\dot{\mathbf{R}}^c \mathbf{F}^c_o + \mathbf{R}^c \dot{\mathbf{F}}^c_o\right)\left(\mathbf{F}^{c^{-1}}_o \mathbf{R}^{c^{\mathrm{T}}}\right) = \dot{\mathbf{R}}^c \mathbf{R}^{c^{\mathrm{T}}} + \mathbf{R}^c \mathbf{L}^c_o \mathbf{R}^{c^{\mathrm{T}}} \tag{6.22}$$

$\mathbf{L}^{\mathbf{c}}$ in the current frame can be decomposed, using Eq. (6.21), into a strain rate $\mathbf{D}^{\mathbf{c}}$ and a rotation rate $\mathbf{W}^{\mathbf{c}}$:

$$\mathbf{L}^c = \mathbf{D}^c + \mathbf{W}^c \tag{6.23}$$

where

$$\begin{cases} \mathbf{D}^c = \mathbf{R}^c \mathbf{D}^c_0 \mathbf{R}^{c^{\mathrm{T}}} = \mathbf{D}^{c,\mathrm{R}}_0 \\ \mathbf{W}^c = \mathbf{R}^c \mathbf{W}^c_0 \mathbf{R}^{c^{\mathrm{T}}} + \dot{\mathbf{R}}^c \mathbf{R}^{c^{\mathrm{T}}} = \mathbf{W}^{c,\mathrm{R}}_0 + \dot{\mathbf{R}}^c \mathbf{R}^{c^{\mathrm{T}}} \end{cases} \tag{6.24}$$

Eq. (6.24) shows that determining $\mathbf{D}^c$ simply involves a transformation from crystal frame into current frame through $\mathbf{R}^c$, but determining the rotation rate $\mathbf{W}^c$ involves a transformation and an extra contribution due to the time rate of change of $\mathbf{R}^c$.

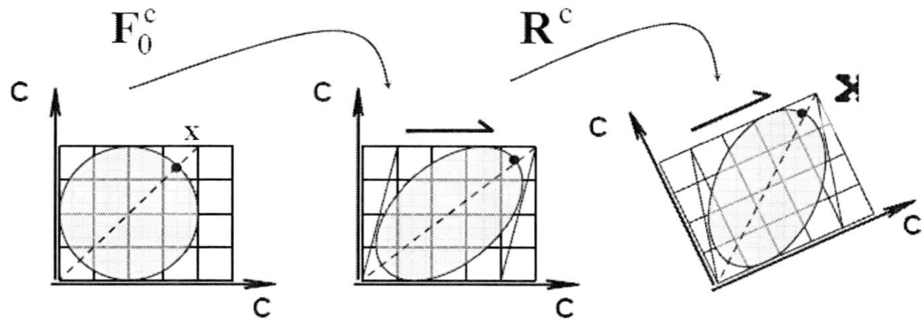

Figure 6.4. Schematic showing the stretch and rigid rotation sequence associated with the multiplicative decomposition.

The kinematic expressions derived above are completely general and applicable to any polycrystal model, which neglects elastic deformation. Specifically, the polycrystal model will provide a value for the velocity gradient in each grain $\mathbf{L}^c = \mathbf{D}^c + \mathbf{W}^c$. The strain rate $\mathbf{D}^c$ determines the crystallographic shear (Eqs. 6.21b and 6.24). The rotation rate $\mathbf{W}^c$, in combination with Eq. 6.24, determines the rate of change of the crystal orientation matrix $\mathbf{R}^c$ as

$$\dot{\mathbf{R}}^c = (\mathbf{W}^c - \mathbf{W}_o^{c,R})\mathbf{R}^c \tag{6.25}$$

which is used to update incrementally the orientation of the crystal and, as a consequence, to follow the texture evolution.

In addition, the deformation gradient of the grain is updated using an incremental form of Eq. 6.15:

$$\mathbf{F}^c_{(t+\Delta t)} = (\mathbf{I} + \mathbf{L}^c \Delta t)\mathbf{F}^c_{(t)} \tag{6.26}$$

The deformation gradient admits a geometrical interpretation which provides a physical basis to the Effective Medium polycrystal model, which will be discussed in Section 7.3. Consider the locus of points in the crystal, which, before deformation, are on the surface of a sphere, characterized by the equation $X_i X_i = X^T X = 1$ (see Figure 6.4). According to Eq. (6.14), upon deformation those points will evolve into the quadratic form $\mathbf{x}^T (\mathbf{F}^{c^{-1}})^T (\mathbf{F}^{c^{-1}})\mathbf{x} = 1$, which is the equation of an ellipsoid. The eigenvalues of $(\mathbf{F}^c)^2$ are the square of the ellipsoid axes lengths, and the eigenvectors describe the ellipsoid orientation (Gurtin, 1981). This property of the deformation gradient is relevant to the self consistent polycrystal model discussed in Section 7.3, where grains are regarded as ellipsoidal inclusions embedded in a Homogeneous Effective Medium (HEM) and shape effects play a role in the interaction between the grain and the surrounding medium.

6.3. Hardening

The earlier polycrystal models were only concerned with capturing the texture evolution associated with certain deformation modes (such as tension, rolling, torsion, etc.), and usually assumed that the threshold shear stress (CRSS) required to activate slip remained fixed during deformation. However, for a constitutive equation to be realistic, it not only has to describe the evolution of texture but also the evolution of yield stress along an arbitrary deformation path. As a consequence, the CRSS in each system has to evolve as the grain structure evolves with deformation. Its evolution is not easy to determine, even for the least complicated situations, and becomes extremely challenging when the grains evolve directional dislocation patterns or twins. Furthermore, in lower symmetry metals, such as the HCP, different deformation modes characterized by different CRSS values are present. In addition, these modes exhibit different hardening evolution depending on how they interact among themselves. For the interested reader, the following references represent a few examples of hardening models implemented into polycrystal formulations: Karaman et al. (2000) treat a Hadfield steel with twinning effects, Kok et al. (2002) propose a rate and temperature dependent law for FCC, Beyerlein and Tomé (2007) account for dislocation structure effects in Cu subjected to ECAE, Tomé et al. (2001) consider interaction between slip and twin modes in Zr, Beyerlein and Tomé (2008) model dislocation density evolution for various slip modes in Zr, and Proust et al. (2007, 2008) and Capolungo et al. (2008a,b) account for twinning effects in Zr and Mg using a composite grain model.

In Chapters 13 and 14 of this book we provide more details about some of the hardening models quoted above, in connection with ECAE deformation. In what follows we will only describe a basic and general hardening formulation that has proven useful in developing constitutive laws and simple for implementing into polycrystal models.

6.3.1. Voce Hardening

The Kocks-Mecking equation (Mecking and Kocks, 1981; Kocks and Mecking, 2003) expresses the time evolution of dislocation density as a balance between the dislocation generated and the ones that are annihilated (recombined). In its original form, Kocks-Mecking describes an average dislocation density in the whole aggregate without individual grain considerations. Karaman et al (2000) applied the concept of dislocation evolution to individual orientations and coupled it with crystallographic shearing. Their expression is:

$$\dot{\rho} = \left(k_1\sqrt{\rho} - k_2\rho\right)\dot{\Gamma} \tag{6.27}$$

Here $\dot{\Gamma} = \sum_s \dot{\gamma}^s$ is the sum of the shear rates over all systems in the grain. In addition, the Taylor law, coupling macroscopic yield stress with average dislocation density, is rewritten for single grains, in terms of the threshold stress τ^s as:

$$\tau = \tau_0 + \alpha\mu b\sqrt{\rho} \tag{6.28}$$

Using Eqs. (6.27) and (6.28) it is possible to show that

$$\dot{\tau} = \left(k_1 \frac{\alpha \mu b}{2} - k_2 (\tau - \tau_0) \right) \dot{\Gamma} = \left(\frac{\theta_0}{\tau_1} (\tau_0 + \tau_1) - \frac{\theta_0}{\tau_1} \tau \right) \dot{\Gamma} \tag{6.29a}$$

where:

$$\theta_0 = k_1 \frac{\alpha \mu b}{2} \quad \textit{and} \quad \frac{\theta_0}{\tau_1} = \frac{k_2}{2} \tag{6.29b}$$

Integrating Eq. (6.29a) leads to a form of the Voce law that describes the evolution of the threshold stress with shear as:

$$\tau = \tau_0 + \tau_1 \left(1 - \exp\left(-\Gamma \frac{\theta_0}{\tau_1}\right)\right) \tag{6.30}$$

The law (6.30) predicts that the threshold stress in the grain saturates with accumulated shear at a value $\tau^s = \tau_0 + \tau_1$. Many materials, however, do not saturate at large strain, but show instead a constant rate of hardening known as Stage IV. As a consequence, Tomé et al (1984) proposed an empirical extension of the Voce law, with the intent of capturing this hardening stage, while retaining the monotonic dependence of the threshold stress with accumulated shear in the grain. The 'extended Voce law' is of the form

$$\hat{\tau}^s = \tau_0^s + (\tau_1^s + \theta_1^s \Gamma)\left(1 - \exp\left(-\Gamma \frac{\theta_0^s}{\tau_1^s}\right)\right) \tag{6.31}$$

where $\Gamma = \sum_s \gamma^s$ is the accumulated shear in the grain; τ_0, θ_0, θ_1, $(\tau_0+\tau_1)$ are the initial CRSS, the initial hardening rate, the asymptotic hardening rate, and the back-extrapolated CRSS, respectively. Equation (6.31) is illustrated in Figure 6.5 for some arbitrary parameters. As shown, the condition $\theta_0 \geq \theta_1 \geq 0$, $\tau_1 \geq 0$ corresponds to increasing yield stress and decreasing hardening rate tending to linear saturation. Linear hardening (constant hardening rate) is a limit case of this law corresponding to $\tau_1^s = 0$. The case of rigid-perfectly-plastic hardening corresponds to $\theta_0 = \theta_1 = \tau_1 = 0$ (Figure 6.5). Observe that when $\theta_1 = 0$ the classical Voce law is recovered.

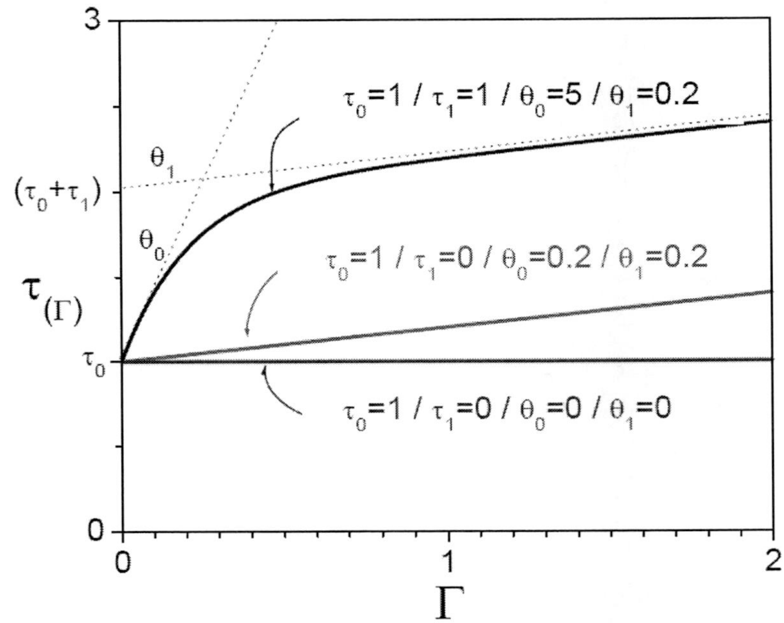

Figure 6.5. Extended Voce law (Eq. 6.31) and some limit cases. The hardening parameters used for producing the plots are indicated in each case.

In addition, one may allow for the possibility of 'self' and 'latent' hardening by defining coupling coefficients $h^{ss'}$ which empirically account for the obstacles that new dislocations associated with s' activity represent for the propagation of system s. The latent hardening parameters couple shear activity $\Delta\gamma^{s'}$ in system s' with the increase in threshold stress of system s. When combined with the Voce law, the increase in the threshold stress of a system due to shear activity in the grain is calculated as:

$$\Delta\tau^s = \frac{d\hat{\tau}^s}{d\Gamma}\sum_{s'} h^{ss'}\Delta\gamma^{s'} \tag{6.32}$$

It is customary to use self-hardening as a reference and set $h^{ss} = 1$. 'Self' and 'latent' hardening are indistinguishable when $h^{ss'} = 1,\ \forall\, s, s'$, in which case the evolution of the threshold stress is given by only the Voce hardening function via:

$$\Delta\tau^s = \frac{d\hat{\tau}^s}{d\Gamma}\Delta\Gamma \tag{6.33}$$

6.4. Description of Texture

The mathematical description of texture and the calculation of texture from experimental pole figures are subjects beyond the scope of this Chapter. The interested reader is referred to the book by Kocks et al (1998) and the references therein. In this Section we only provide the basic concepts required to understand polycrystal models and texture simulation.

Texture refers to the distribution of crystal orientations in a sample and is mathematically represented by the Orientation Distribution Function (ODF). The ODF describes the relative fraction (density) of material having a certain orientation inside the aggregate. The relative orientation between the sample and the crystal axes can be expressed in different ways: in terms of three sequential rotations (Euler angles), in terms of an axis and a rotation around it (Rodrigues representation), in terms of a linear transformation (orthogonal matrix $\mathbf{R}^c$), etc. As a consequence, the ODF is a function defined in the space of parameters associated with the representation used. In addition, the ODF may be represented as a continuum or as a discrete function. A frequently used continuum representation due to Bunge (1982) is a series expansion in symmetrized spherical harmonics, function of the three Euler angles, giving the intensity of the distribution at every point in Euler space. In the discrete representation, volume fractions are assigned to a finite number of crystal orientations in order to reproduce a given texture.

The continuum representation may find some use in calculations of average polycrystal properties, such as elastic constants, but the discrete representation is more suitable in polycrystal models of texture development and is the method adopted for the simulations presented in this book. In applications involving non-textured polycrystals it is common to start with a set of grains of equal volume fraction, whose orientations are given by randomly generated Euler angles. Otherwise, when the starting aggregate is textured, it is typical to assign volume fractions to a set of regularly spaced orientations by integrating the experimental ODF in the region of Euler space surrounding each orientation.

Textures are represented graphically by means of 2-dimensional sections of the three-dimensional Euler space or pole figures. We will mostly use the latter in this book. Pole figures are built by first locating the crystal in the center of a hypothetical sphere, then finding the intersection point between the sphere and the vector perpendicular to a chosen crystallographic plane, and making a polar projection of the points on the sphere onto the equatorial plane.

As an example, the (111) poles of the typical rolling and shear texture components for FCC are represented in Figure 6.6. Because there are four crystallographically equivalent (111) poles per grain, and because of the 2-mirror and 1-mirror symmetry associated with rolling and shear, respectively, there can be a maximum of 16 and 8 poles, respectively, associated with each component.

These components are often referred to as ideal components because they represent relatively 'stable' orientations that are achieved during the corresponding deformation mode. They are frequently used as a benchmark to characterize deformation textures and to compare predictions from polycrystal models. The ideal rolling component denoted {hkl}<uvw> indicates that the crystal plane {hkl} is parallel to the rolling plane (ND) and the crystal direction <uvw> is parallel to the rolling direction (RD). For simple shear, {hkl}<uvw> defines the crystal plane parallel to the shear plane (SP) and the crystal direction parallel to the shear direction (SD).

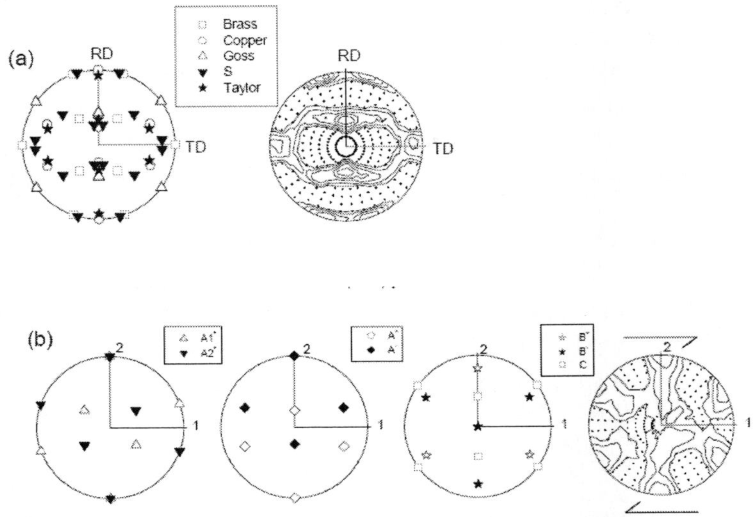

Figure 6.6. (111) pole figures of the ideal texture components associated with rolling (plane strain) and torsion (simple shear) for an FCC aggregate. The plot includes symmetrically equivalent orientations. Contours levels are 0.8/1.2/1.8/2.6/3.6 mrd.

Ideal components associated with simple shear will be relevant to the discussion of SPD textures presented in Chapters 8-14. These have been identified for FCC metals by Canova et al. (1984), for BCC metals by Montheillet et al. (1984), and recently for Mg, an HCP metal, by (Beausir et al 2007, Li 2008). In the FCC metals, for instance, the preferred orientations either lie on the A fiber ({111}||SP) or the B fiber (<110>||SD). The A fiber contains the A, $\bar{A}$, A_1^* and A_2^* components, while the B fiber contains the A, $\bar{A}$, B, $\bar{B}$ and C components. A *two-fold symmetry* about the 3-axis occurs when A and $\bar{A}$ have the same intensity and B and $\bar{B}$ have the same intensity. The components A_1^* and A_2^* can still have different intensities in two-fold symmetry, and C is self symmetric. Analogous fibers, components and symmetries for BCC metals are related to those for an FCC metal by exchanging the indices of slip planes {hkl} and slip directions <uvw>.

Chapter 7

MODELS FOR TEXTURE DEVELOPMENT

7.1. INTRODUCTION

Chapter 6 explains how to calculate the reorientation of individual grains when their velocity gradient and slip activity, induced by plastic deformation, are known. Texture development of an aggregate can be predicted if one knows the velocity gradient of each individual grain in the aggregate as a function of deformation. The way in which macroscopic deformation is partitioned between the grains is a characteristic of the polycrystal model used. In the following, two polycrystal formulations frequently used for the simulation of plastic deformation and the prediction of texture and hardening evolution are outlined. They are the upper-bound Full Constraints model (also known as the Taylor model) and the Effective Medium approach (also known as the Self Consistent formulation). We discuss the merits and ranges of applicability of these formulations and give comparative examples for the cases of plane strain and simple shear.

For comprehensive treatments of the subject of this Chapter the reader is referred to Kocks et al (1998) and Hosford (1993). In addition, a comprehensive review by Dawson et al (2003) covers the subject of sheet forming simulations using polycrystal based models. We do not address recrystallization issues or recrystallization modeling in this Part II. For a discussion of recrystallization simulations applied to plastic forming the reader is referred to the article by Beaudoin & Engler (2001).

At the polycrystal level, the imposed macroscopic velocity gradient can be decomposed into a stretch rate and a spin rate component, as is done for an individual crystal in Eq. (6.23):

$$\mathbf{L} = \mathbf{D} + \mathbf{W} \tag{7.1}$$

Numerically, the simulation of plastic deformation and texture evolution is done according to the following sequence. For a particular crystallographic orientation 'c':

- A velocity gradient $\mathbf{L}^c$ (Eq. 6.12) is imposed to each grain. It is related to--but not necessarily equal to--the macroscopic velocity gradient $\mathbf{L}$ given by Eq. (7.1).
- The symmetric component of $\mathbf{L}^c$, $\mathbf{D}^c$, is accommodated by plastic shear, and one can find the grain stress σ_{ij}^c by solving the non-linear system given by Eq. (6.11).

• Once the stress is known, the shear rates in the slip systems can be calculated using Eq. (6.10).
• The skew-symmetric component of $\mathbf{L}^c$, $\mathbf{W}^c$, plus the shear rates, give the rate of rotation for the crystallographic orientation (Eq. 6.25).
• Crystal orientation (Eq 6.25), the deformation tensor of the crystal (Eq. 6.26), and hardening of the slip systems (Eq. 6.32), are updated incrementally by assuming that the rates remain constant through a small time interval Δt.
• The macroscopic stress, strain rate and rigid rotation rate are given by the weighted averages over all grains in the aggregate:

$$\sigma = \left\langle \sigma^c \right\rangle \tag{7.2a}$$

$$\mathbf{D} = \left\langle \mathbf{D}^c \right\rangle \tag{7.2b}$$

$$\mathbf{W} = \left\langle \mathbf{W}^c \right\rangle \tag{7.2c}$$

The incremental steps described above are applied sequentially until the total prescribed deformation is accumulated. The simulation of texture and stress-strain evolution for a deforming aggregate results from carrying out the above procedure for many orientations (grains) at each step.

While all polycrystal models have to verify Eqs. (7.2), they differ in their treatment of compatibility and equilibrium between the grain and the surrounding medium. Specifically, the assumptions used to decide how much of an imposed macroscopic deformation is going to be accommodated by each grain depending on its orientation and shape are what differentiates the various polycrystal models.

7.2. Full Constraints (Taylor) Model

If one pictures the aggregate as a collection of space filling polyhedra (grains), the simpler and more intuitive picture of aggregate deformation is one where all polyhedra deform proportionally at the same rate $\mathbf{L}^c = \mathbf{L}$. Equation (7.2b) tells us that the strain rate of the grains has to be equal to the macroscopic one:

$$\mathbf{D} = \left\langle \mathbf{D}^c \right\rangle = \mathbf{D}^c \tag{7.3}$$

This renders the Full Constraints (FC) model. Compatibility between the grains is automatically fulfilled but the stress, which follows from solving Eq. (6.11), is different for different orientations, and equilibrium conditions among crystals are not satisfied. Bishop and Hill (1951a,b) prove that the FC model provides an absolute upper bound on the plastic dissipation energy. And because the strain rate is uniform, it follows that the FC model provides an upper bound on the macroscopic stress.

A consequence of enforcing all strain components $\mathbf{D}^c$ is that (at least) five independent slip systems need to be active in each grain if Eq. (6.7) is to be fulfilled. In other words, the stress state has to coincide with a vertex of the SCYS. This is usually a very demanding condition and at large strains the textures predicted with the FC approach tend to differ from the measured textures. If the visco-plastic constitutive expression Eq. (6.11) is used, in combination with a large rate exponent *n*, the stress state and system activity that results is still very close to the one given by the Schmid law, as we show for the crystal yield surfaces in Figure 6.3. Numerical examples are given in Section 7.4 for the case of rolling and shear of an FCC aggregate.

7.3. VISCOPLASTIC SELF-CONSISTENT MODEL

The visco-plastic self-consistent (VPSC) approach regards each grain as a viscoplastic inclusion embedded in a visco-plastic homogeneous effective medium (HEM) represented by the polycrystal. Within the VPSC model one solves the interaction problem between the inclusion and the medium. In this Section we only discuss the main assumptions and results and present the relevant equations. For a complete description of the self-consistent method as it applies to visco-plastic aggregates the reader is referred to (Molinari et al. 1987; Lebensohn and Tomé, 1993; Kocks et al. 1998, Chapter 11; Lebensohn et al. 2007).

The aim of the self-consistent approach is two-fold: first, to find the stress and strain rate in the individual grains, and second, to find the effective viscoplastic properties of the medium. Since the constitutive response of the grains and of the medium are non-linear, the first approximation made involves linearizing them. For a grain, Eq. (6.11) is written in a pseudo-linear form

$$D_{ij}^{c} = P_{ijpq}^{c,\mathrm{sec}} \sigma_{pq} \tag{7.4}$$

by defining the *secant* visco-plastic compliance

$$P_{ijpq}^{c,\mathrm{sec}} = \dot{\gamma}_o \sum_s \frac{m_{ij}^s m_{pq}^s}{\tau^s} \left(\frac{m_{kl}^s \sigma_{kl}}{\tau^s} \right)^{n-1} . \tag{7.5}$$

In order to solve the equilibrium equation $\nabla \sigma_{ij,j} = 0$ it is assumed that:

$$\sigma_{(x)}^{c} = (\mathbf{P}^{c,\mathrm{sec}})^{-1} : \mathbf{D}_{(x)}^{c} \qquad \text{for } x \text{ in the domain of the grain} \tag{7.6a}$$

For the effective properties of the medium, one assumes a similar relationship, and defines an effective visco-plastic compliance $\mathbf{P}^{\mathrm{sec}}$ which relates the overall stress to the overall strain rate:

$$\sigma = (\mathbf{P}^{\mathrm{sec}})^{-1} : \mathbf{D} \qquad \text{on the boundary of the medium} \tag{7.6b}$$

Finally, if one assumes that the constitutive response in the vicinity of the inclusion is represented by the effective visco-plastic compliance, one can write the local relationship between stress and strain rate as

$$\sigma_{(x)} = (\mathbf{P}^{\sec})^{-1} : \mathbf{D}_{(x)} \qquad \text{for } x \text{ outside the grain domain} \tag{7.6c}$$

If each grain is considered as an ellipsoidal inclusion, the boundary problem represented by Eq. (7.6) is formally identical to the Eshelby problem for an elastic inclusion (Eshelby, 1957) when the linear relationship between stress and strain is replaced by a linear relationship between stress and strain rate. The solution of the equilibrium equation for this problem finds that the velocity gradient and the stress are uniform within the domain of the inclusion. In addition, it is found that the deviations between the grain and effective medium stress and rate quantities are linearly related through the interaction equations

$$(\mathbf{D}^c - \mathbf{D}) = -n^{eff}\,(\mathbf{I}-\mathbf{S})^{-1} : \mathbf{S} : \mathbf{P}^{\sec} : (\sigma^c - \sigma) \tag{7.7a}$$

$$(\mathbf{W}^c - \mathbf{W}) = \mathbf{\Pi} : \mathbf{S}^{-1} : (\mathbf{D}^c - \mathbf{D}) \tag{7.7b}$$

In Eq. (7.7a) we introduce for the first time an effective interaction coefficient n^{eff}. Up to now, the secant approach has been described, in which $n^{eff} = 1$. However, the constitutive law Eq. (7.4) can also be posed in terms of tangent moduli by making a first-order expansion around the average stress. In this case, for a grain $\mathbf{P}^{c,tg} = n\mathbf{P}^{c,\sec}$. Likewise, the effective tangent and secant compliance are related through $\mathbf{P}^{tg} = n\mathbf{P}^{\sec}$ (Hutchinson, 1976). As a consequence, $n^{eff} = n$ for the tangent approach (we use $n = 20$ in our calculations).

The Eshelby tensors **S** and **Π** are complex functions of the visco-plastic tensor $\mathbf{P}^{\sec}$ and the ellipsoid axes. **S** varies from null to identity and measures the capability of the HEM to accommodate deviations in strain rate between itself and the inclusion. A medium much stiffer than the inclusion will not accommodate extra strain and will drive the deformation of the inclusion. Such a stiff medium is characterized by an Eshelby tensor close to zero and, as a consequence, Eq. (7.7a) predicts that $\mathbf{D}^c \cong \mathbf{D}$, which is precisely the condition associated with the Full Constraints (upper bound) approach. Conversely, a medium that is much more compliant than the inclusion will easily accommodate extra strain, and its Eshelby tensor is close to unity. The inverse of Eq. (7.7a) indicates that in this case $\sigma^c \cong \sigma$, which is the condition associated with a lower bound approach. Finally, the specific dependence of the Eshelby tensor components with the ellipsoid axes may help accommodate strain-rate deviations in some components, while it may constrain others. For example, in the case of a flat ellipsoid associated with rolling (with axes $a_1 >> 1, a_2 = 1, a_3 << 1$), the associated components $S_{ij33} \cong 1$ tend to enforce $\sigma_{13}^c \cong \sigma_{13} = 0$ and $\sigma_{23}^c \cong \sigma_{23} = 0$, while relaxing the corresponding rate components. These conditions were the basis of the Relaxed Constraints approach proposed originally by Honneff and Mecking (1981) for explaining the rolling textures after large reductions of FCC aggregates. We will see in the example of Section 7.4.1

that the SC approach induces a transition in these conditions as strain increases and the grains become flatter.

The relevance of the interaction equation (7.7) associated with the Effective Medium approach cannot be overemphasized. First of all, it explains the simpler upper and lower bound polycrystal models (Taylor and Sachs, respectively) in terms of the relative stiffness of the HEM and the inclusion. Second, in the case of highly anisotropic crystals, the interaction equation applies to the relative *directional* stiffness between the medium and the grain. For example, in the case of HCP crystals, grains are plastically much stiffer along the c-axis than along the a-axis. As a consequence, the inclusion formalism allows the HEM to accommodate more strain and the grain less strain along the grains' c-axis direction. Accommodating each crystal's plastic anisotropy is an important feature when simulating the plastic deformation of low symmetry materials, and makes the Self Consistent schemes invaluable for modeling the plastic forming of highly anisotropic materials.

With all the elements above we can now formulate the Self Consistent polycrystal approach. The interaction equations (7.7) provide the strain rate and spin rate of the grains, $\mathbf{D}^c$ and $\mathbf{W}^c$, as a function of the secant visco-plastic modulus $\mathbf{P}^{sec}$ of the HEM, which is not known *a priori.* The moduli $\mathbf{P}^{sec}$ has to be found by means of an implicit *self-consistent* scheme and depends on the response of the individual grains. Enforcing the condition in Eq. (7.2a), which states that the average stress over all grains has to equal the macroscopic stress, and combining Eqs. (7.6) and (7.7), lead to:

$$\mathbf{P}^{sec} = \left\langle \mathbf{P}^{c,sec}\mathbf{B}^{c} : \right\rangle \tag{7.8a}$$

where

$$\mathbf{B}^{c} = \left(\mathbf{P}^{c,sec} .+ \widetilde{\mathbf{P}}\right)^{-1}\left(\mathbf{P}^{sec} .+ \widetilde{\mathbf{P}}\right) \tag{7.8b}$$

and

$$\widetilde{\mathbf{P}} = \left(\mathbf{I} + \mathbf{S}\right)^{-1}\mathbf{S}\overline{\mathbf{P}} . \tag{7.8c}$$

Here $\mathbf{B}^c$ is the localization tensor and $\mathbf{S}$ is the Eshelby tensor, which is a function of $\mathbf{P}^{sec}$ and the ellipsoid aspect ratios (Lebensohn et al. 2007). Since the localization tensor $\mathbf{B}^c$ depends in a complex way on the unknown $\mathbf{P}^{sec}$, Eq. (7.8a) has to be solved iteratively. A guess for $\mathbf{P}^{sec}$ is entered to evaluate the right hand term of the equation, and then the resulting new value of $\mathbf{P}^{sec}$ is re-entered into the equation until the input and output coincide (within a given tolerance). At this point the solution is said to be *self-consistent.*

7.4. APPLICATIONS

In this section we will illustrate how the polycrystal models described above are applied to two large plastic deformation cases, namely, plane strain and simple shear of FCC aggregates. The former is relevant to rolling processes, while the latter is relevant to ECAE.

In practice both, rolling and ECAE, exhibit non-homogeneous deformation across the volume of the plastically formed material as a consequence of friction between the sample and the tools. Rolling introduces a shear gradient across the plate, and Engler et al (2000) discuss a polycrystal modeling approach to this problem. The deformation conditions associated with ECAE in connection with polycrystal modeling have been addressed by Beyerlein and Tomé (2004) and are discussed in detail in Chapters 9 through 12 of this book. In this Section we regard homogeneous plane strain and homogeneous simple shear as introductory examples of polycrystal modeling, which can also be considered to approximate the more complex rolling and ECAE processes. The associated reference axes are shown for each mode in Figure 7.1.

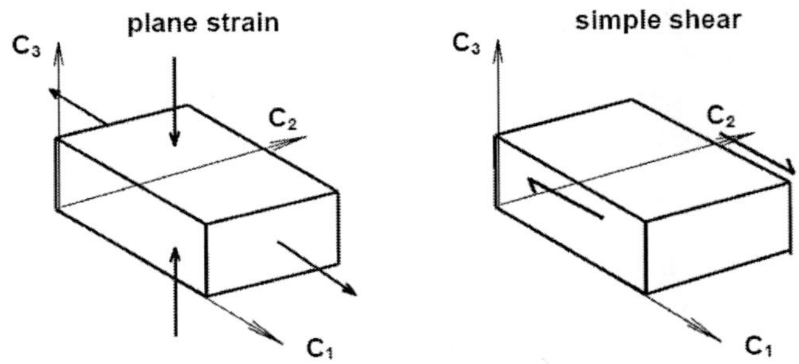

Figure 7.1. Sample reference axes associated with plane strain and simple shear deformation.

7.4.1. Plane Strain Deformation

For rolling we adopt the usual reference frame where the axis 3 is the compression axis (ND), axis 1 the extension axis (RD), and axis 2 the transverse direction (TD). The macroscopic velocity gradient tensor adopts the form

$$\mathbf{L}=\begin{bmatrix}1 & 0 & 0\\ 0 & 0 & 0\\ 0 & 0 & -1\end{bmatrix}\sec^{-1} \tag{7.9}$$

It is clear that there is no macroscopic spin in this case: $\mathbf{D} = \mathbf{L}$ and $\mathbf{W} = 0$. For the rolling simulations, the velocity gradient $\mathbf{L}$ given by Eq. (7.9) is imposed to an aggregate composed by 1000 initially random orientations during time increments $\Delta t = 2\text{sec}$ to enforce steps of 2% reduction. We simulate 63% and 86% rolling reduction, which correspond to an accumulated strain $\varepsilon_{33} = -1$ and -2, respectively.

In what follows we compare the rolling texture, intergranular stress and strain-rate fluctuations, and slip system activity predicted by the Full Constraints and the Self Consistent models. A system is considered to be active if its shear rate $\dot{\gamma}^s$ is at least 5% of the maximum shear rate in the grain. If n^c denotes the number of active systems in grain 'c', the Average Active Systems per grain (AVACS) can be defined as:

$$AVACS = \frac{\sum_{c=1}^{ngrain} n^c}{ngrain} \tag{7.10}$$

As for the fluctuations, they are reported as the standard deviation of each tensor component, normalized by the norm of the average stress (or strain-rate) tensor:

$$\delta\sigma_{ij} = \sqrt{\frac{\sum_c (\sigma_{ij}^c - <\sigma_{ij}^c>)^2}{\sum_{kl} <\sigma_{kl}^c>^2}} \tag{7.11}$$

The predicted distribution of (111) poles after 63% and 86% rolling reduction is shown in Figure 7.2, while the average number of active systems per grain and the deviation of the shear components $\sigma_{13,}\sigma_{23}$ and $D_{13,}D_{23}$ with respect to their average value (zero in this case) are shown in Figure 7.3.

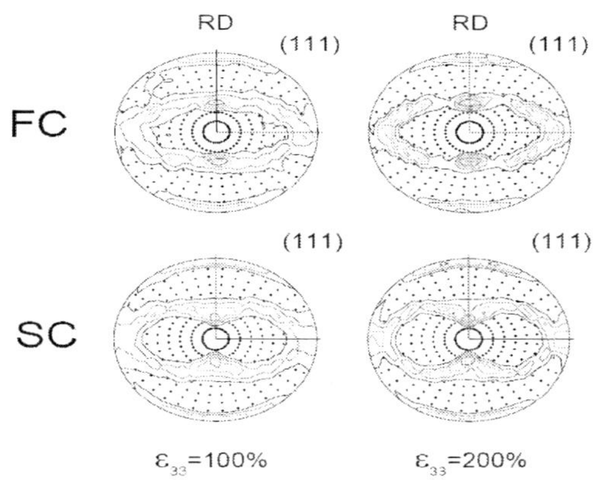

Figure 7.2. (111) pole figures corresponding to simulated rolling of $\varepsilon_{33}=-1$ and $\varepsilon_{33}=-2$ (rolling reductions of 63% and 86% respectively) using the Full Constraints (FC) and the Self Consistent (SC) approach. Contour levels are 0.6/1.5/3.0/5.3/9.0/15 mrd.

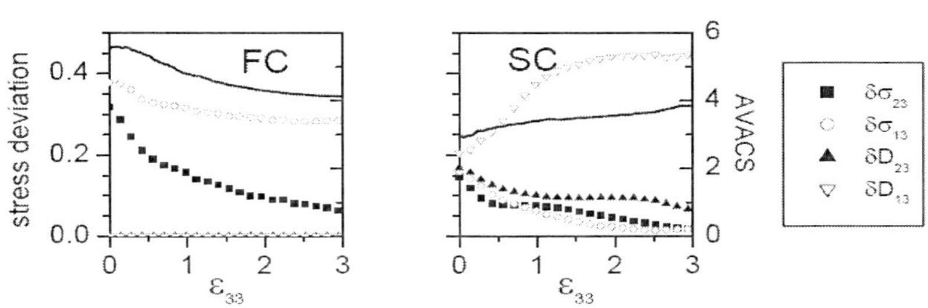

Figure 7.3. Normalized standard deviation of grain shear stresses σ_{13}, σ_{23} and strain rates D_{13}, D_{23} as a function of rolling strain predicted by: (a) Full Constraints (FC) approach; (b) Self Consistent (SC) approach (symbols; left axis; see Eq. (7.11) for definition). Also shown is the average number of active systems per grain (solid line; right axis; see Eq. (7.10) for definition).

The FCC rolling texture is characterized by the development of the β fiber, with <110> tilted 60° to the rolling direction. Three individual components distinguish the fiber: the copper component, with Miller indices {112}<111> (the first set indicates the crystallographic plane perpendicular to the ND, the second set the crystallographic direction parallel to the RD); the S component, near {123}<634>; and the brass, or B component, with indices {011}<211>. These components are represented in Figure 6.6, together with the Goss {110}<001> and Taylor $\{4\ 4\ 11\} < 11\ 11\ \bar{8} >$ components. A comparison of the predicted textures with the ideal components shows that the FC model tends to reorient the grains towards the Taylor and the Goss components, while the SC approach favors the S and Brass component.

The purpose of this exercise is to illustrate the differences between the predictions of both models and we will avoid here a discussion of how they compare with experimental evidence. As a matter of fact, both models tend to predict similar textures for cubic materials. However, it is interesting to note that in the case of FC the average number of active systems per grain starts at about six and, as the grains become more favorably oriented for accommodating plane strain, reduces to about four (see Figure 7.3). The fact that this number may exceed the value 5 is a consequence of having either six or eight planes associated with each vertex of the SCYS, and our specific procedure for counting active systems. The SC approach, on the other hand, accommodates deformation with a lesser number of systems per grain, namely, between three and four (see Figure 7.3). An explanation for the difference between FC and SC can be found in the intergranular stress and strain-rate deviations and their evolution with deformation (also shown in Figure 7.3). In the case of FC, forcing plane strain, that is, $\varepsilon_{13}^{c} = \varepsilon_{23}^{c} = 0$, induces a large deviation in σ_{13}^{c} *and* σ_{23}^{c} In the SC case, on the other hand, as the ellipsoid representing the grain becomes flatter, the intergranular deviations in σ_{13} *and* σ_{23} asymptotically approach zero, as required by continuity conditions for grains that are infinitely flat. The latter is the condition empirically enforced by the Relaxed Constraints model of Honneff and Mecking (1981). In addition, the increasing spread in the shear rate D_{13} indicates that grains tend to deform by shear rather than plane strain. While the average strain over all grains is still plane strain, individual grains opt to deform locally according to a mode that involves a lesser plastic dissipation rate and less slip activity. Clearly, the different slip activities between FC and SC are responsible for the relative strengths of their various rolling texture components.

7.4.2. Simple Shear Deformation

Simple shear is usually difficult to impose experimentally. If torsion of solid bars is used the presence of a radial shear gradient complicates the analysis of the results; in the case of a thin-walled tube the gradient is negligible, but large deformation is difficult to achieve because of instabilities and buckling; short-gauge thin-wall tubes subjected to torsion, or shear of a narrow groove inscribed in a plate eliminates instabilities but the close proximity of the shoulders makes it difficult to guarantee a large enough region of homogeneous simple shear. Obviously, none of these issues arises when *simulating* simple shear up to large strain. The macroscopic velocity gradient for the case sketched in Figure 7.1 is:

$$L = \begin{bmatrix} 0 & 1 & 0 \\ 0 & 0 & 0 \\ 0 & 0 & 0 \end{bmatrix} \sec^{-1} = \begin{bmatrix} 0 & 0.5 & 0 \\ 0.5 & 0 & 0 \\ 0 & 0 & 0 \end{bmatrix} \sec^{-1} + \begin{bmatrix} 0 & 0.5 & 0 \\ -0.5 & 0 & 0 \\ 0 & 0 & 0 \end{bmatrix} \sec^{-1} \quad (7.12)$$

and can be decomposed into a stretch component (first term) and a rigid rotation component (second term). For the simulations we consider the same initial aggregate as for rolling, consisting of 1000 randomly generated orientations. We also considered that the CRSS in the (111)<110> systems obeys a rigid plastic law (Voce parameters $\tau_0 = 1$ and $\tau_1=\theta_0=\theta_1 = 0$ in Eq. 6.31), although this does not affect the texture and activity results presented below. The velocity gradient given by Eq. (7.12) was applied repeatedly in time increments $\Delta t = 4\text{sec}$ ($\Delta\varepsilon_{12} = 0.02$), until strains of $\varepsilon_{12} = 1$ and $\varepsilon_{12} = 2$ were accumulated. The (111) pole figures corresponding to the predictions of the FC and the SC approach are shown in Figure 7.

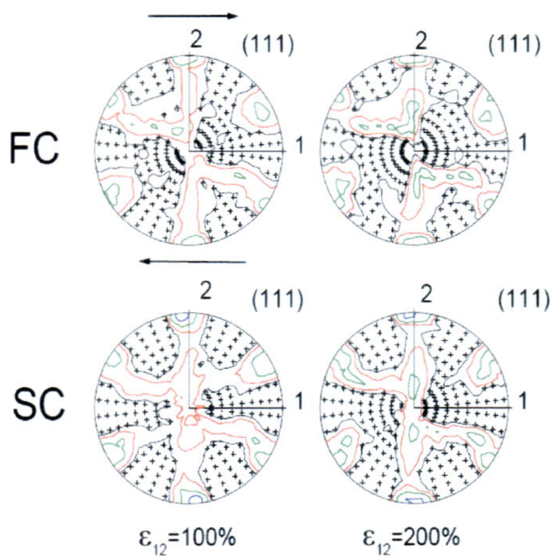

Figure 7.4. (111) pole figures corresponding to simulated shear of $\varepsilon_{12}=1$ and 2 ($\gamma=2$ and 4 respectively) using the Full Constraints (FC) and the Self Consistent (SC) approach. Contour levels are 0.7/1.4/2.8/5.6 mrd.

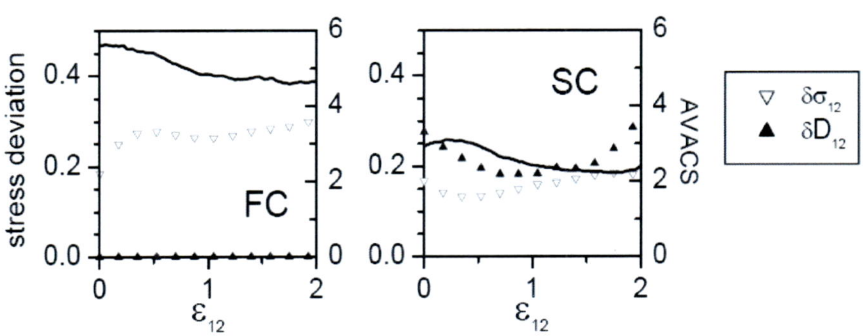

Figure 7.5. Normalized standard deviation of grain shear stress σ_{12} and strain rate D_{12} as a function of shear strain predicted by: (a) the Full Constraints (FC) approach; (b) the Self Consistent (SC) approach (symbols; left axis; see Eq. (7.11) for definition). Also shown is the average number of active systems per grain (solid line; right axis; see Eq. (7.10) for definition).

The textures predicted with FC and SC are similar, although the relative intensity of the shear texture components varies between the two cases for the same strain (see Figure 6.6). In the FC case the average number of active systems per grain is similar as for rolling (between 5 and 6), but substantially smaller in the case of SC (between 2.5 and 3) as is to be expected for an overall shear deformation mode (see Figure 7.5). The reason is to be found in the relatively large intergranular dispersion of the rate component D_{12} (Figure 7.5), which allows some orientations (grains) to accommodate more shear than others.

7.4.3. Grain Shape, Grain Splitting, and Simple Shear

In this Section we discuss the role of grain shape in texture evolution. This gives us the opportunity to show how experimental evidence can be tied to modeling. In doing so the model is improved and some of its parameters can be related to physical features of the microstructure. Under severe plastic shearing or rolling, grains tend to adopt distorted shapes and extreme aspect ratios. Grains with extreme shapes may tend to rotate with the flow field at a faster rate than equiaxed grains, and in addition, grain shape can affect slip activity. Both effects will affect crystal reorientation and so texture evolution. An example in this respect is provided in Figure 7.6, corresponding to annealed OFHC Copper subjected to one pass Equal Channel Angular Extrusion (ECAE) in a 90° die. Observe the elongation and inclination of the grain shape and also the changes in color along the shape, indicative of misorientation development and eventual grain fragmentation. The VPSC model, which explicitly accounts for individual grain shape and its evolution with strain, provides a tool to analyze how these effects influence texture.

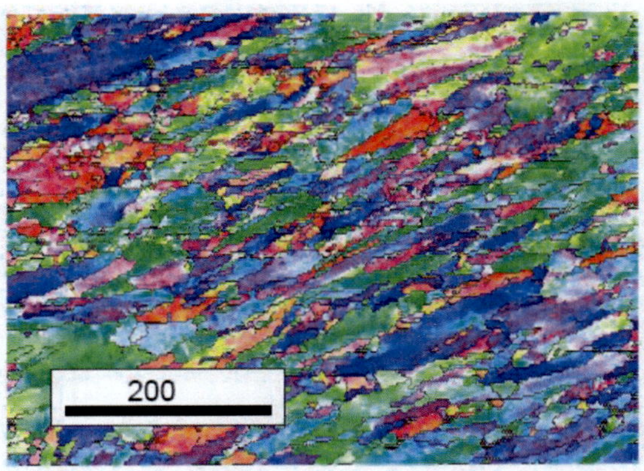

Figure 7.6. Electron Back Scattering Diffraction (EBSD) image of OFHC copper after 1 pass ECAE in a 90° die (image provided by C. T. Necker).

VPSC, as most polycrystal models do, assumes that the stress state is uniform in the ellipsoidal grain. However, there is a physical and a numerical limit to how distorted a grain can become. From a numerical point of view the evaluation of the integrals associated with the Eshelby tensors (Eq. 7.7) becomes less accurate. From a physical point of view, when grains become severely distorted, widely separated regions within the grain will experience different neighborhoods and are likely to develop a non-uniform stress state and to build enough internal misorientation as to split into sub-grains with different orientations. Such a situation is likely to arise with large shear strain or with ECAE, although it should not be identified with the grain fragmentation induced by the formation of dislocation structures inside the grain. The latter occurs at a much smaller scale. To address this issue a *grain splitting scheme* was introduced by Beyerlein et al (2003), implemented in VPSC, and applied to the specific case of ECAE. The criterion and the associated model, though, are applicable to any process where grains deform to extreme shapes.

In the grain splitting scheme, subdivision of grains is determined according to the length ratios of the long (d_l), medium (d_m) and short (d_s) axes of the ellipsoidal grain in comparison with a critical value R. As illustrated in Figure 7.7, a grain is divided into two grains when $d_l/d_s \geq R$ and $d_m/d_s < R/2$ are satisfied or into four grains when $d_l/d_s \geq R$ and $d_m/d_s \geq R/2$. The crystallographic orientation immediately before and after the split remains the same. To demonstrate the influence of the grain-splitting scheme, we repeat here the VPSC simulations of simple shear presented in Section 7.4.2, changing some of the run parameters as follows. We update the shape of each individual grain independently, rather than using the average grain shape to calculate the Eshelby tensor and the associated grain-matrix interaction (Eq. 7.7). We consider three cases for grain shape evolution: 1) a spherical grain shape is enforced throughout deformation; 2) grain splitting is allowed to take place at a critical ratio $R = 5$; and 3) grain splitting is allowed to take place at a critical ratio $R = 20$. In cases 2 and 3, the grains evolve individually depending on their orientation, and therefore each grain will reach R at different moments in deformation.

Figure 7.8 compares the (111) pole figures predicted for each of the three cases, at three values of applied shear strain $\gamma = 2$, 4, and 6. Between $\gamma = 2$ and 4 the effects of grain shape begin to impact texture evolution and after $\gamma = 4$, the texture 'saturates' to a distribution dependent on R. For all values of R, the shear textures up to $\gamma = 2$ contain A, B, and C shear components (see (Canova et al., 1984) and discussion in Chapter 6 for review of shear components). For case 1, the A components strengthen and the C components weaken with straining beyond $\gamma = 2$. A different result occurs when the grain aspect ratio is allowed to deform to larger values beyond $\gamma = 2$. For $R = 5$, both A and C strengthen. For $R = 20$, the C component dominates after large strain.

The critical aspect ratio R is likely a material-dependent parameter, as some materials tend to develop highly distorted grains and others less so after severe plastic deformation. Dynamic recrystallization tends to produce equiaxed grains and will put a limit on the maximum distortion. Based on torsion textures reported in the literature for a variety of other FCC metals, a reasonable choice for R can range from 2 to 20. After large strain, (ε_{vM} = $\gamma / \sqrt{3 \approx 4}$), Ni – 60%Co produces room temperature textures resembling the $R = 5$ textures in Figure 7.8, whereas pure Ni produces those resembling the $R = 20$ case (Hughes et al., 1996). The ability of Ni to deform to extreme shapes in route A has been reported by Segal (1995). At half the strain level ($\gamma = 3.63$ or $\varepsilon_{vM} = 2.1$), but higher temperatures (537K), both Ni – 60%Co and pure Ni develop textures close to those predicted with $R = 5$ for $\gamma = 4$. Similar trends are observed after torsion testing of Cu and Al. After $\varepsilon_{vM} \approx$ 4-6, both metals develop room temperature textures similar to the $R = 5$ and $R = 20$ cases, respectively (Montheillet et al., 1984). Under higher temperatures and similar strain levels, Cu exhibits a texture closer to an $R = 20$-like texture. Apparently in the higher stacking fault energy materials, like Ni and

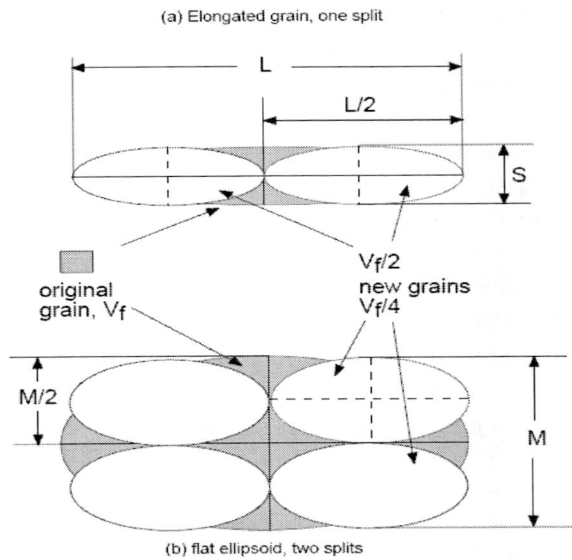

Figure 7.7. Grain splitting criteria for (a) elongated grain and (b) flat grain. Taken from (Beyerlein, Lebensohn, and Tomé, 2003).

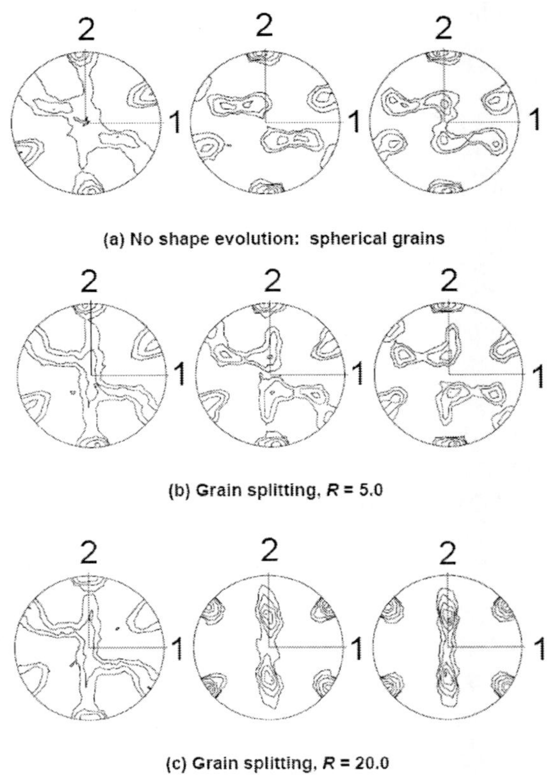

Figure 7.8. (111) pole figures showing the effect of grain shape evolution on texture evolution in a FCC metal during simple shearing for $\gamma = 2$, 4, and 6. In (a) grains are forced to remain spherical. In (b) and (c) the critical aspect ratios are $R = 5$ and 20, respectively. Contour levels are 1.0/2.0/4.0/8.0/11.3/16.0 mrd.

Al, or under high temperature conditions, grains are more likely to stretch to higher R under SPD than lower stacking fault energy materials, like Ni-60%Co and Cu, or under room temperature. Some features of their large strain shear textures, therefore, can be attributed to differences in grain shape.

Grain splitting is not intended to model grain refinement by the formation of subgrain dislocation structures and the evolution of misorientations between them. In an actual metal, under large strains, subgrains develop that are one to two orders of magnitude smaller than the parent. This high rate of reduction in grain size with strain cannot be modeled by the grain splitting scheme. It is also tempting to relate the grain shape in the VPSC model to the *subgrain* morphologies (cellular versus lamella sub-boundaries) rather than *grain* morphologies found in real metals. Although deformation mode and thermally-activated dislocation motion both contribute to microstructural evolution, grain shape is controlled more so by the former and grain refinement by the latter.

In this Section we have tried to keep the analysis qualitative and simple. We emphasize the fact that even in FCC aggregates, some orientations are 'harder' than others when a given deformation mode is macroscopically enforced. It is intuitively clear that, whenever possible, the aggregate will tend to accommodate plastic deformation by shifting the task to those grains that are more efficiently oriented. In the case of plastically anisotropic grains, this means that grains will tend to use the 'easy' deformation path. As a consequence, less than five independent systems per grain will be active in general. The local variations in strain are compensated by the surrounding medium such as to preserve, on average, the imposed macroscopic deformation. From the perspective of a polycrystal constitutive law, this interplay is captured by self-consistent models by regarding the grain as an inclusion embedded in an effective medium with the average properties of the aggregate. By enforcing the macroscopic strain on all grains, the Full Constraints approach requires at least five active systems per grain and especially for low symmetry materials (but not so much for cubics), leads to very different predictions of mechanical response and deformation texture than VPSC.

Chapter 8

MODELING SINGLE AND MULTIPLE PASSES IN ECAE

8.1. BASIC DEFINITIONS IN MECHANICS

In this chapter we develop the mechanics descriptions of deformation in single and multiple pass ECAE as needed for polycrystal models. First, it is important that the reader be familiar with some basic concepts in continuum mechanics. We begin by recalling the deformation gradient and velocity gradient defined in Section 6.3. The former expresses the deformation required to distort a material from one configuration to another. Let $x = f(X)$ be a one-to-one mapping of a point X in the previous configuration to a point x in the distorted one, where X and x are position vectors. The gradient of f is called the deformation gradient F:

$$F = \nabla f \text{ or } F_{ij} = \frac{\partial x_i}{\partial X_j} \qquad \text{and} \qquad x_i = F_{ij} X_j \tag{8.1}$$

The velocity gradient L defined as

$$L_{ij} = \partial \dot{u}_i / \partial x_j$$

is related to F via:

$$L = \dot{F} F^{-1} \tag{8.2}$$

where $i,j = 1,3$, $u = x - X$ is the displacement, and $\dot{u}$ is the displacement rate. F is generally composed of an elastic and plastic part, $F = F^e F^p$. In modeling SPD processes, we neglect the elastic contribution to F, as it plays a negligible role in large strain processes and microstructural development, i.e., $F^p = F$. The determinant of F, det$F$, represents the dilatation. If the deformation is incompressible, which will be the case when $F^p = F$, then det$F = 1$.

The displacement gradient $\partial u / \partial x$ can be determined from F as:

$$\frac{\partial u}{\partial x} = (F - I)F^{-1} = I - F^{-1}$$

where I is the identity matrix.

With the deformation expressed in terms of F and L, one can determine several outcomes of the deformation, such as the stretches and rotations of a material element, texture evolution, and the strains and strain rates experienced by a material point. In particular, the evolution of the velocity gradient L is required by the VPSC polycrystalline model to calculate texture and flow response associated with a loading history. VPSC also ignores the elastic effects and represents only incompressible plastic deformation.

Next F, L, and the associated material distortions are presented for the ideal case (simple shear model). Later, in Chapter 11, analytical and numerical methods for calculating F and L in non-ideal situations are described.

8.2. Single Pass in ECAE

8.2.1. Simple Shear Model

Consider ECAE tooling with a die angle Φ, a sharp bend $\Psi = 0°$, the entry channel parallel to the 2-direction, and the exit channel at an angle $(\Phi - \pi/2)$ clockwise from the 1-direction, as illustrated in Figure 8.1. The imposed ECAE deformation is simple shear concentrated in the plane of intersection between the two channels, which in our convention lies at an angle $(\pi/2 - \Phi/2)$ from the 1-direction, as indicated in Figure 8.2.

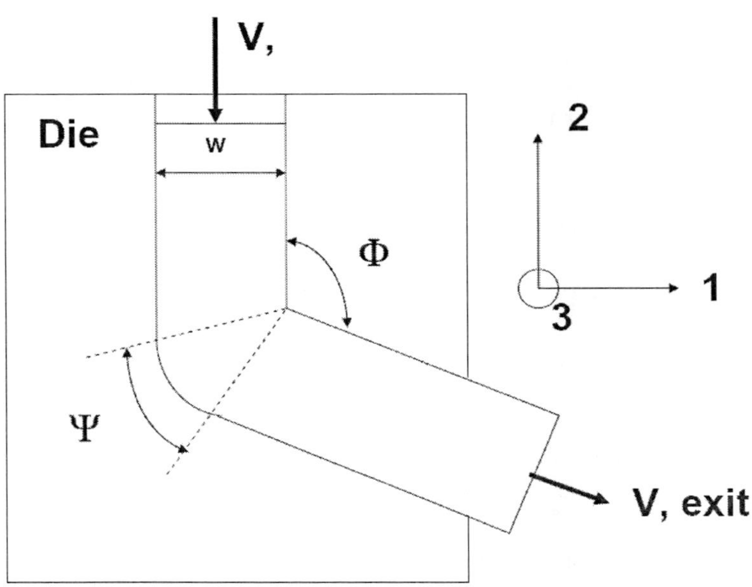

Figure 8.1. Schematic of ECAE die.

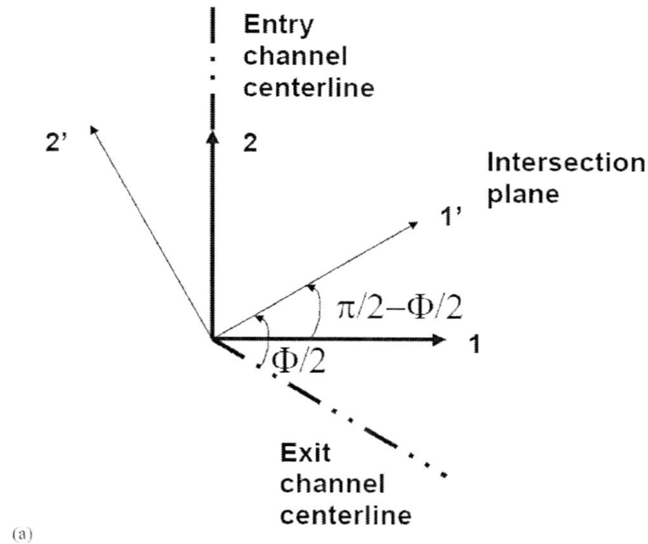

Figure 8.2. Laboratory coordinate system, 1-2-3, relative to the coordinate system of the intersection plane 1'-2'-3 of an ECAE die with die angle Φ.

When expressed in a coordinate system x'-y'-z' lying parallel to the intersection plane (See Figure 8.2), the final deformation gradient is

$$F = \begin{bmatrix} 1 & -2\cot(\Phi/2) & 0 \\ 0 & 1 & 0 \\ 0 & 0 & 1 \end{bmatrix} \tag{8.3}$$

The only non-zero component of the velocity gradient is $L_{1'2'} = -\dot{\gamma}$, where $\dot{\gamma}>0$ is the applied shear rate in the shear plane. Thus,

$$L = \begin{bmatrix} 0 & -\dot{\gamma} & 0 \\ 0 & 0 & 0 \\ 0 & 0 & 0 \end{bmatrix} \tag{8.4}$$

After a single pass, the accumulated shear $\gamma = \int \dot{\gamma} dt$ is

$$\gamma = 2\cot(\Phi/2) \text{ and } \varepsilon_{vM} = \gamma/\sqrt{3} \tag{8.5}$$

as presented earlier in Eq. 3.3 (see Part I). According to Eq. 8.5, the lower Φ the higher the accumulated shear strain per pass.

For the purposes of multiple-pass modeling in ECAE, in which the direction of shearing with respect to the sample changes from pass to pass, it is best to express the deformation in a fixed laboratory coordinate system *x-y-z,* as shown in figures 8.1 and 8.2. In this coordinate system, F and L in Eqs. (8.3) and (8.4) transform to

$$F = \begin{bmatrix} 1+\cot\left(\frac{\Phi}{2}\right)\sin\Phi & \cot\left(\frac{\Phi}{2}\right)(\cos\Phi-1) & 0 \\ \cot\left(\frac{\Phi}{2}\right)(\cos\Phi+1) & 1-\cot\left(\frac{\Phi}{2}\right)\sin\Phi & 0 \\ 0 & 0 & 1 \end{bmatrix} \quad (8.6)$$

$$L = \dot{\varepsilon}\begin{bmatrix} \sin\Phi & \cos\Phi-1 & 0 \\ \cos\Phi+1 & -\sin\Phi & 0 \\ 0 & 0 & 0 \end{bmatrix} \quad (8.7)$$

where $\dot{\varepsilon}$ is a proportionality constant chosen to enforce the strain rate[1]. Note that when expressing ECAE deformation in this fixed system, all four in-plane components of L are non-zero and depend explicitly on Φ.

8.2.2. Grain Distortions

One advantage of having the deformation described by F is that the distortion of any element with an initial arbitrary shape can be determined. In this section we apply F from Eq. (8.6) to calculate the distortion of an element during a single pass of ECAE. The same calculation for multiple passes will be discussed shortly.

The eigenvalues and eigenvectors of $U^2 = F^TF$ or $V^2 = FF^T$ correspond to the principal stretches and stretch directions of a volume element as a result of the deformation F. As an example, suppose the initial shape is a sphere. After one ECAE pass, an initially spherical element distorts into an ellipsoid. Let N and S be the lengths of the major and minor axes of this ellipsoid, respectively. Lengths N and S are the square root of the largest and smallest eigenvalues of U or V (i.e., principal stretches of the deformation F). For F given by Eq. (8.6) they are (Beyerlein and Tomé, 2004):

$$N^2 = 1 + 2\cot^2\left(\frac{\Phi}{2}\right) + 2\cos\left(\frac{\Phi}{2}\right)\csc^2\left(\frac{\Phi}{2}\right) \quad (8.8)$$

$$S^2 = 1 + 2\cot^2\left(\frac{\Phi}{2}\right) - 2\cos\left(\frac{\Phi}{2}\right)\csc^2\left(\frac{\Phi}{2}\right) \quad (8.9)$$

---

[1] For the simulation we choose a component of Eq. (8.7), e.g. ε_{22}, increment this strain component, $\Delta\varepsilon_{22}$, and impose a time increment Δt and n steps, such that $\Delta\varepsilon_{22} = \dot{\varepsilon}\,\Delta t$ and $n\Delta\varepsilon_{22} = \varepsilon_{22}$.

Lengths N and S are dimensionless quantities, representing the ratio of the newly deformed ellipsoidal axis lengths to the original lengths along the largest and smallest principal stretch directions. Due to volume preservation and the planar deformation of a single pass of ECAE, $NS = 1$ for all Φ.

As shown in Figure 8.3, the aspect ratio N/S decreases with Φ because the imposed strain decreases with Φ. The angle of orientation of the long axis of this ellipsoid θ_s corresponds to one of the principal stretch directions. W*ith respect to the horizontal 1-direction* it will be

$$\theta_s = \tan^{-1}\left(\csc\left(\frac{\Phi}{2}\right) - \cot\left(\frac{\Phi}{2}\right)\right) = \frac{\Phi}{4} \tag{8.10}$$

and *with respect to the axis of the exit channel* it is

$$\theta_s^* = \theta_s + \left(\Phi - \frac{\pi}{2}\right) = \frac{5}{4}\Phi - \frac{\pi}{2} \tag{8.11}$$

The angle θ_s increases with Φ due to (i) the change in the shear plane orientation with Φ, and (ii) decrease in the imposed shear strain with Φ. In the impractical case of $\Phi = 180°$, $\theta_s = 45°$ according to Eq. (8.10), and $N/S = 1.0$ according to Eqs. (8.8) and (8.9). Thus, the initial sphere remains a sphere, which is to be expected.

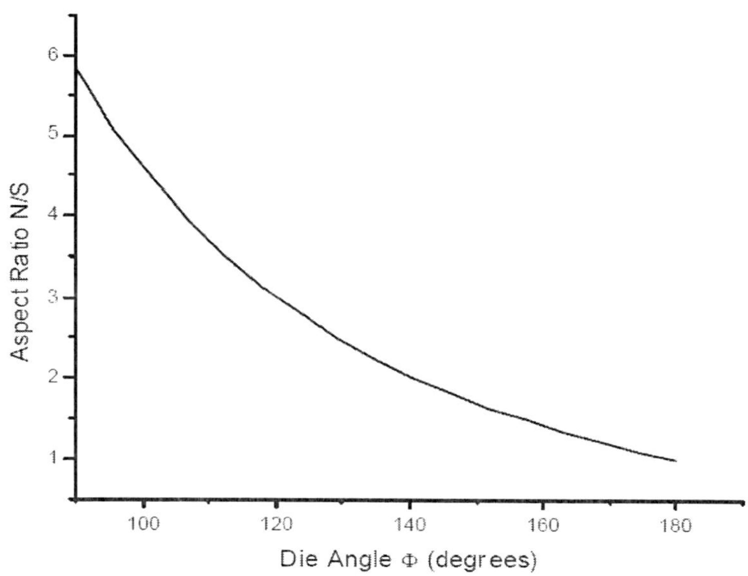

Figure 8.3. Aspect ratio N/S of an original sphere after being distorted into an ellipsoid by an ECAE pass as a function of die angle Φ. (Beyerlein and Tomé, 2004)

As mentioned above, any initial shape can be treated. For instance, rather than a sphere, consider the deformation of a grid element (or cube), as was done in Part I (see figures 1.6b and 1.7). Applying F to an initial square element, the angle of orientation of a grid element θ_r *with respect to the axis of the exit channel* will be

$$\theta_r = \cot^{-1}\left(2\cot\left(\frac{\Phi}{2}\right)\right) \tag{8.12}$$

An important point to realize is that $\theta_r \neq \theta_s$ (and $\theta_r \neq \theta_s^*$). Figure 8.4 compares a circle and a square that have distorted after one ECAE pass for $\Phi = 90°$ into an ellipse and a parallelogram, respectively. As shown, the long axis of the ellipsoid is not aligned with the inclination angle of the parallelogram. Also the square element has rotated significantly: its vertical sides (circles) become horizontal and its horizontal sides (pluses) distort into lateral inclined segments. This rotation is contributed by the skew part of F, which is an essential part of ECAE deformation. Figure 8.5 compares the variation of θ_r and θ_s^* with Φ according to Eqs. (8.11) and (8.12). Both angles are defined with respect to the axis of the exit channel. The billet axis is aligned with the axis of the exit channel, and therefore, θ_r and θ_s^* correspond to the angles of elements (e.g., scribed grids, grains) that would be observed in the billet. As shown in Figure 8.5, the larger Φ, the bigger the difference between θ_r and θ_s^*. For $\Phi = 90°$ (120°) die, the θ_r of a cube 'grain' deformed after one pass is 26.6° (40.9°) but for a spherical 'grain', θ_s is 22.5° (30°). The angle $\theta_s^* = \theta_s$ for $\Phi = 90°$ but $\theta_s^* = 60 \neq \theta_s = 30°$ for $\Phi = 120°$.

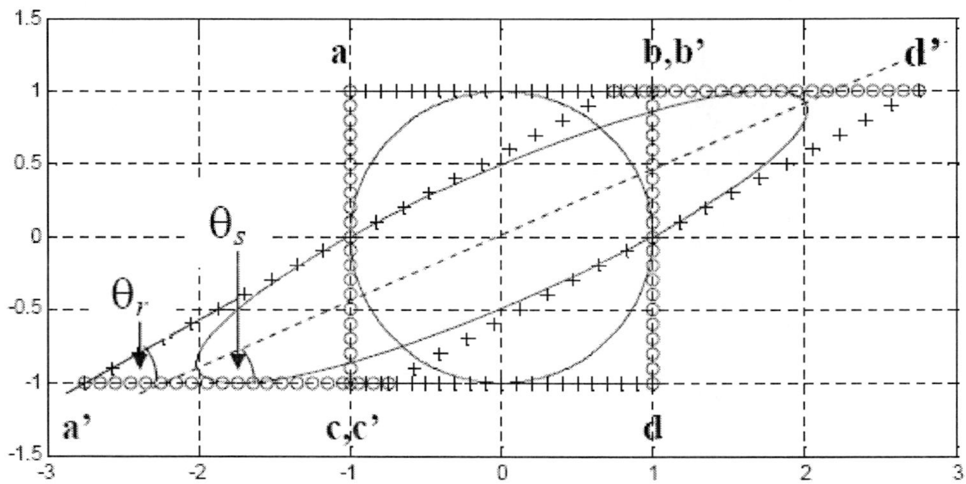

Figure 8.4. Deformation of an original circle inscribed in a square under the ideal case of simple shear along the intersection plane after one pass deformation with $\Phi = 90°$. The original configuration has distorted into an ellipse inside a parallelogram. Angles θ_r and $\theta_s = \theta_s^*$ are also defined. (Beyerlein and Tomé, 2004)

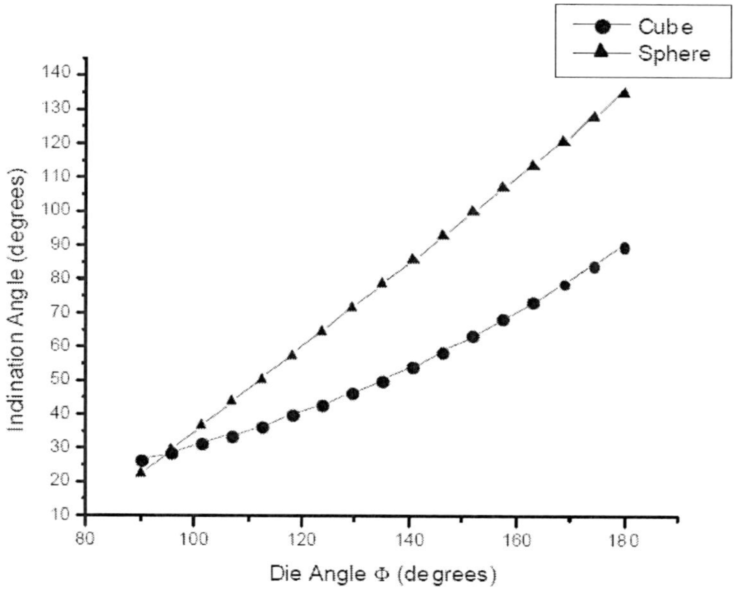

Figure 8.5. Final inclination angle for a distorted cube (or grid element) θ_r (Eq. 8.12) and distorted sphere (or model grain) θ_s^* (Eq. 8.11) with respect to the axis of the exit channel as a function of ECAE die angle Φ. (Beyerlein and Tomé, 2004)

Element distortion calculations are useful in estimating grain morphologies after ECAE. Grains in an actual polycrystal are not, of course, perfect cubes or spheres and do not deform as homogeneous continuum elements, as assumed above. Nonetheless θ_s^* and θ_r appear to be good predictors of the average inclination angles expected in grains after one ECAE pass, regardless of crystal structure (FCC, BCC, and HCP) (Fang et al., 2006; Kamachi et al., 2003; Shan et al., 1999; Han et al., 2006; Han et al., 2007; Shin et al., 2003; Suzuki et al., 2004; Segal, 1995; Li et al. 2006b; Beyerlein et al., 2005a).

In the past, the average grain inclination measured after an ECAE extrusion has been mistakenly interpreted as the orientation of the shear plane, implying, in other words, that $\Phi/2$ is not the plane of shearing but that θ_r or θ_s (or θ_s^*) is. Under ideal conditions, simple shearing takes place along a plane rotated $\Phi/2$ from either channel. With this model, the above mechanical analysis clearly shows that the grain inclination angles θ_r and θ_s that result from assuming simple shearing along the $\Phi/2$ intersection plane do not coincide with $\Phi/2$.

8.3. Multiple Passes in ECAE

8.3.1. ECAE Routes

In ECAE, strains accumulate in the sample following multiple passes through the die. After each pass, the sample must be rotated to re-align the billet axis with the axis of the entry channel. Re-insertion into the die for each subsequent pass unavoidably introduces a strain path change between passes. As a result, strain accumulation in ECAE is non-monotonic as in a torsion test. The strain path change sequence can be altered by rotating the sample about its long axis by various angles. The second rotation and its angle for a given pass number define the ECAE *route*. To date, four standard routes—routes A, Bc (or D), Ba (or B), and C (see footnote 6 in Part I)—have been introduced and studied intensively in the literature. These four routes have already been described in Part I (Section 3.5.1) and so we only briefly recall them here as it is relevant for simulating multi-pass ECAE. In route C, the billet is rotated 180° between each pass.

8.3.2. Rigid Rotations between Passes

The standard routes can be described, in general, by two rotations, R_{I} and R_{II}. Suppose the entry channel is aligned with the vertical axis y and the exit channel varies according to Φ as in Figure 8.1. The first rotation R_{I} is a CW $\theta = -(\pi - \Phi)$ about the z-axis for re-insertion of the 'head' of the sample into the entry die. The second rotation R_{II} is about the y-axis and depends on the ECAE route Z through the rotation angle $\phi(Z,N)$. The rotation tensors for the first and second rotation are

$$R_I = \begin{bmatrix} \cos\theta & -\sin\theta & 0 \\ \sin\theta & \cos\theta & 0 \\ 0 & 0 & 1 \end{bmatrix} \quad R_{II}(Z,N) = \begin{bmatrix} \cos\phi(Z,N) & 0 & -\sin\phi(Z,N) \\ 0 & 1 & 0 \\ \sin\phi(Z,N) & 0 & \cos\phi(Z,N) \end{bmatrix} \tag{8.13}$$

Assuming counter-clockwise positive, for route A, $\phi(A,N) = 0$; for route B_C, $\phi(B_C,N) = -\pi/2$; and for route C, $\phi(C,N) = \pi$ for all N. For route B_A, $\phi(B_A,N) = -\pi/2$ for odd N and $\pi/2$ for even N. When multi-pass ECAE is formulated in this way, the model can be applied to any number of rotations between passes and any arbitrary sequence from pass to pass.

There are two ways in which one can account for these rigid rotations when simulating deformation in multi-pass ECAE. One can either use the sample axes at the exit channel as a reference and rotate the die for reinsertion, or as we have described, keep the die fixed and rotate the sample for reinsertion. In both schemes, it is necessary to first select the orientation of ECAE die and the order in which these rotations are applied. The selection does not matter, just as long as it is known and fixed. Both methods should lead to the same result.

The first scheme involves rotating the die, not the sample, between passes, and can be envisioned as extruding the billet through a network of connecting dies. In this method, F and L change from pass to pass because they have to be expressed in the axes of the sample. The deformation gradient of the previous pass F_1 is related to that of the current pass F_2 by $F_2 = R_{II} R_I F_1 R_I^T R_{II}^T$ and their velocity gradients by $L_2 = R_{II} R_I L_1 R_I^T R_{II}^T$. This procedure is then repeated for the third pass, i.e., $F_3 = R_{II} R_I F_2 R_I^T R_{II}^T$ and $L_3 = R_{II} R_I L_2 R_I^T R_{II}^T$, and so on. The F_N and L_N for pass N and for any possible route (not just the four standard ones) can, therefore, be determined.

Because of the rotation R_{I} for re-insertion between passes, no ECAE route can apply deformation continuously in the same way from pass to pass. For route A, with no second rotation R_{II} = I (identity), a frequent misunderstanding is that this route applies monotonic strain. From the first to the second pass of route A, the ellipsoidal grain does indeed elongate further, appearing to react in the same way as if the deformation was monotonic. For the first two passes of A, $F_2(A) = R_1 F_1 R_1^T$, whereas for continuous $F_{\text{cont}} = F_1 F_1 = F_1^2$. The shapes of the initial configuration of a circle inscribed in a square element after (a) $F_2(A)$ and (b) F_{cont} are compared in Figure 8.6. The final distorted shapes are strikingly different, although both of them have been stretched. In particular, with F_{cont}, the elements are deformed more severely than with $F_2(A)$, and stretch beyond the height of the original square. Under $F_2(A)$, however, the height of the deformed shapes is the same as that of the original. The features of the $F_2(A)$ case are consistent with optical micrographs taken of route A samples, e.g., Segal (1995).

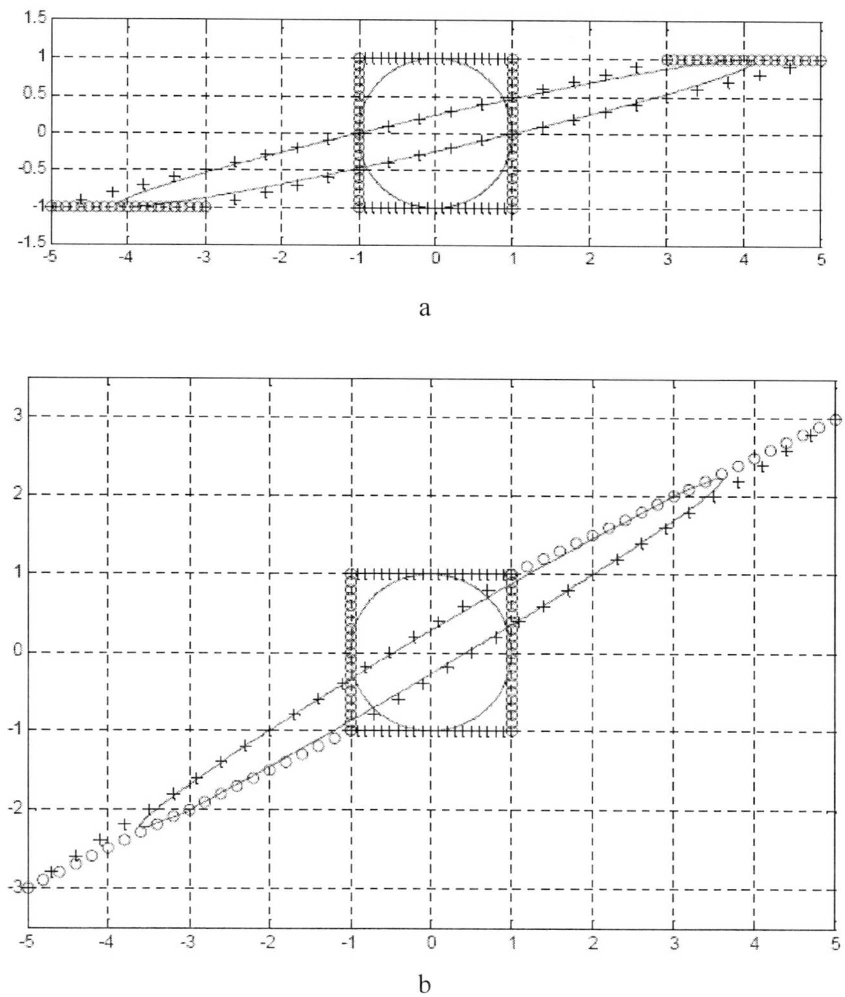

Figure 8.6. Deformation of an original circle inscribed in a square under the ideal case of simple shear along the intersection plane ($\beta_m = 0$), (a) after two pass deformation of route and accounting for the rigid body rotations between passes, and (b) hypothetical case of double shear deformation, not accounting for rigid body rotations between passes. (Beyerlein and Tomé, 2004)

The second way to model the rotations in multi-pass ECAE is to rotate the sample while keeping the die fixed, as is done in a laboratory experiment. This is our preferred method and is the one used in the VPSC code. In this method, the die and hence, location of the shear plane does not change during the ECAE process and L is the same for all passes, just as in the laboratory. To simulate rigid rotation of the billet between passes, both the *crystallographic* (lattice orientation distribution) and *morphological* (grain shape orientation distribution) textures are rotated rigidly relative to the coordinate system attached to the die. The same rotations in Eq.(8.13) are applied, but to transform both the crystallographic and ellipsoid axes of each grain and not F and L. Further details regarding this methodology are found in (Beyerlein et al., 2003).

The advantage of this method lies in its flexibility to model several passes and both homogeneous and inhomogeneous flow per pass, and changes in flow characteristics from pass to pass. When the deformation is inhomogeneous, one needs to further account for the change of the positions of material points in the billet during the pass. In routes C, Bc, and Ba, material points in the top and bottom layers of the sample change from pass to pass, while they are maintained in route A. This indeed has made route A easier to model than the rest.

Chapter 9

TEXTURE CHARACTERIZATION

9.1. IDEAL TEXTURE COMPONENTS FOR SIMPLE SHEAR

In Chapter 6 we mentioned that the textures developed during monotonic rolling (plane strain) or simple shear can be characterized by the relative intensities of so called *ideal texture components*. Hereinafter, we focus on simple shear since it is the characteristic mode of deformation in several cases of severe plastic deformation. Ideal simple shear components are usually defined with respect to a shear plane with its normal parallel to the vertical axis and a positive shear direction along the horizontal axis. This is the case for the FCC pole Figure map in Figure 6.6b. Figures 9.1 and 9.2 show these components for an FCC metal in a (111) pole Figure and for a BCC metal in a (110) pole Figure, respectively, with the shear direction, however, in the negative sense (Li et al., 2005b). As will be seen shortly, this minor alteration will facilitate characterization of textures developed in ECAE.

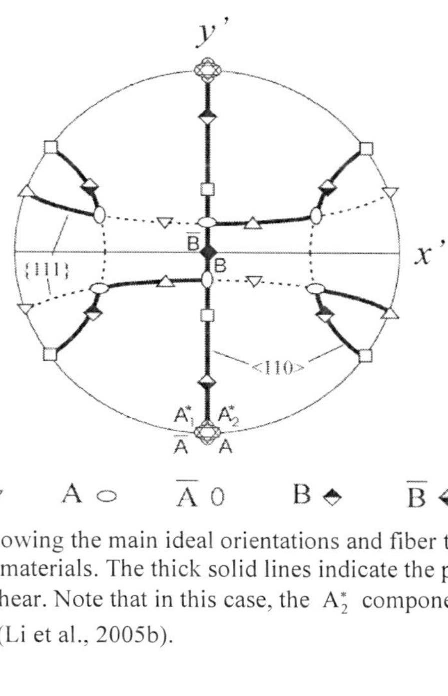

Figure 9.1. (111) pole Figure showing the main ideal orientations and fiber textures associated with simple shear deformation of fcc materials. The thick solid lines indicate the partial fiber textures resulting from *negative* simple shear. Note that in this case, the A_2^* component is much weaker than the other components. Taken from (Li et al., 2005b).

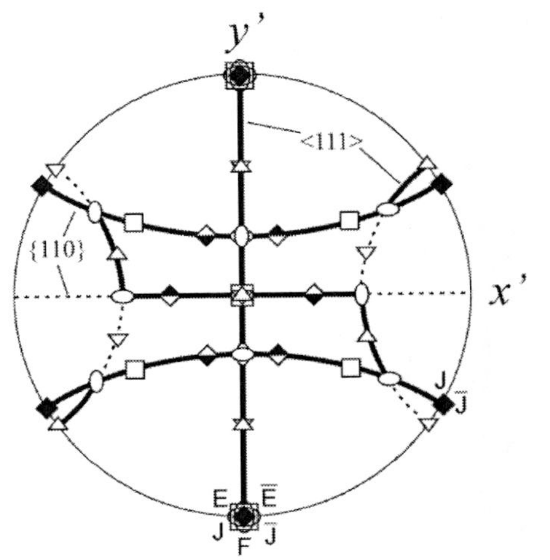

Figure 9.2. (111) pole Figure showing the main ideal orientations and fiber textures associated with simple shear deformation of bcc materials. The thick solid lines indicate the partial fiber textures resulting from *negative* simple shear. Taken from (Li et al., 2005b).

9.2. First-Pass Components

In ECAE, simple shearing occurs in the negative sense on a plane oriented $\Phi/2$ from the billet axis. (For example, when $\Phi = 90°$, the shear plane is oriented 45° from the billet axis, see Figure 38 of Part I.). Accordingly, first-pass ECAE textures can be related to these classic shear components by a straightforward rotation of the shear plane (Li et al., 2005b) about the plane of the die (the 3-axis). Let an ideal orientation $g = (\varphi_1, \phi, \varphi_2)$ be expressed by a set of Euler angles φ_1, ϕ and φ_2, defined with respect to the main axes of the extruded sample, where the 1-axis is parallel to the billet long axis, the 3-axis is normal to the plane of the die, and the 2-axis lies in the plane of the die and is normal to the 1-axis. The ideal shear components for negative shear are rotated accordingly to define what we refer to as 'ideal ECAE components':

$$g_\Phi = \left(\varphi_1 + \Phi/2, \phi, \varphi_2\right) \tag{9.1}$$

The ideal ECAE components for FCC and BCC materials assuming $\Phi = 90°$, for example, are listed in tables 9.1 and 9.2. The subscript Φ signifies the dependency on Φ. In Figure 9.3, these components are also indicated on a PF in three views, the TD (z-plane), ND (y-plane), and ED (x-plane) for both $\Phi = 90°$ and 120° and both an FCC (Figure 9.3a,b) and BCC metal (Figure 9.3c,d) (Li et al., 2005b). For the FCC case, Figure 9.4 shows two sections of the ODF (Beyerlein and Tóth, 2008) for simple shear and an ECAE die with $\Phi = 90°$. These sections contain all ideal components with the exception of the A fiber.

Table 9.1.

Notation	Euler angles[a] (°)			Miller indices[b]			Fibers it belongs to
	φ_1	ϕ	φ_2	ND	ED	TD	
$A^*_{1\Phi}$	80.26/260.26 170.26/350.26	45 90	0 45	$[81\bar{1}]$	$[1\bar{4}4]$	$[011]$	$\{111\}_\Phi$
$A^*_{2\Phi}$	9.74/189.74 99.74/279.74	45 90	0 45	$[1\bar{4}4]$	$[81\bar{1}]$	$[011]$	$\{111\}_\Phi$
A_Φ	45	35.26	45	$[914]$	$[1\ 11\ \bar{5}]$	$[\bar{1}12]$	$\{111\}_\Phi$, $\langle 110\rangle_\Phi$
$\bar{A}_\Phi$	225	35.26	45	$[\bar{1}\ \overline{11}\ 5]$	$[\bar{9}\ \bar{1}\ \bar{4}]$	$[\bar{1}12]$	$\{111\}_\Phi$, $\langle 110\rangle_\Phi$
B_Φ	45/165/285	54.74	45	$[15\ 4\ 11]$	$[7\ 26\ \overline{19}]$	$[\bar{1}11]$	$\langle 110\rangle_\Phi$
$\bar{B}_\Phi$	105/225/345	54.74	45	$[\bar{7}\ \overline{26}\ 19]$	$[\overline{15}\ \bar{4}\ \overline{11}]$	$[\bar{1}11]$	$\langle 110\rangle_\Phi$
C_Φ	135/315 45/225	45 90	0 45	$[334]$	$[22\bar{3}]$	$[\bar{1}10]$	$\langle 110\rangle_\Phi$

[a] Given in the $\varphi_2 = 0°$ and 45° sections only.
[b] approximate values.

Table 9.1. Ideal ECAE orientations and fibers for fcc materials using a $\Phi = 90°$ die (Li et al., 2005a)

Table 9.2.

Notation	Euler angles[a] (°)			Miller indices [b]			Fibers it belongs to
	φ_1	ϕ	φ_2	ND	ED	TD	
$D_{1\Phi}$	99.74/279.74 9.74/189.74	45 90	0 45	$[118]$	$[44\bar{1}]$	$[1\bar{1}0]$	$\langle 111\rangle_\Phi$
$D_{2\Phi}$	170.26/350.26 80.26/260.26	45 90	0 45	$[44\bar{1}]$	$[118]$	$[\bar{1}10]$	$\langle 111\rangle_\Phi$
E_Φ	135	35.26	45	$[914]$	$[\bar{1}\ \overline{11}\ 5]$	$[\bar{1}12]$	$\langle 111\rangle_\Phi$, $\{110\}_\Phi$
$\bar{E}_\Phi$	315	35.26	45	$[\bar{9}\ \bar{1}\ \bar{4}]$	$[1\ 11\ \bar{5}]$	$[\bar{1}12]$	$\langle 111\rangle_\Phi$, $\{110\}_\Phi$
J_Φ	15/135/255	54.74	45	$[15\ 4\ 11]$	$[\bar{7}\ \overline{26}\ 19]$	$[\bar{1}11]$	$\{110\}_\Phi$
$\bar{J}_\Phi$	75/195/315	54.74	45	$[\overline{15}\ \bar{4}\ \overline{11}]$	$[7\ 26\ \overline{19}]$	$[\bar{1}11]$	$\{110\}_\Phi$
F_Φ	45/225 135/315	45 90	0 45	$[334]$	$[\bar{2}\bar{2}3]$	$[\bar{1}10]$	$\{110\}_\Phi$

[a] Given in the $\varphi_2 = 0°$ and 45° sections only.
[b] approximate values.

Table 9.2. Ideal ECAE orientations and fibers for bcc materials using a $\Phi = 90°$ die (Li and Beyerlein, 2005)

Figures 9.3 and 9.4 are meant to provide a standard characterization method for analyzing ECAE textures. In most cases, one-pass ECAE textures can be described well by these ideal ECAE components. An ECAE texture will not necessarily contain all of them, and one distinguishes one texture from another by differences in the relative intensities of these preferred orientations. Moreover texture components in ECAE textures can deviate from these ideal positions. Initial texture effects, deformation twinning, substructure evolution, and deviations from simple shearing are just some of the reasons for such differences.

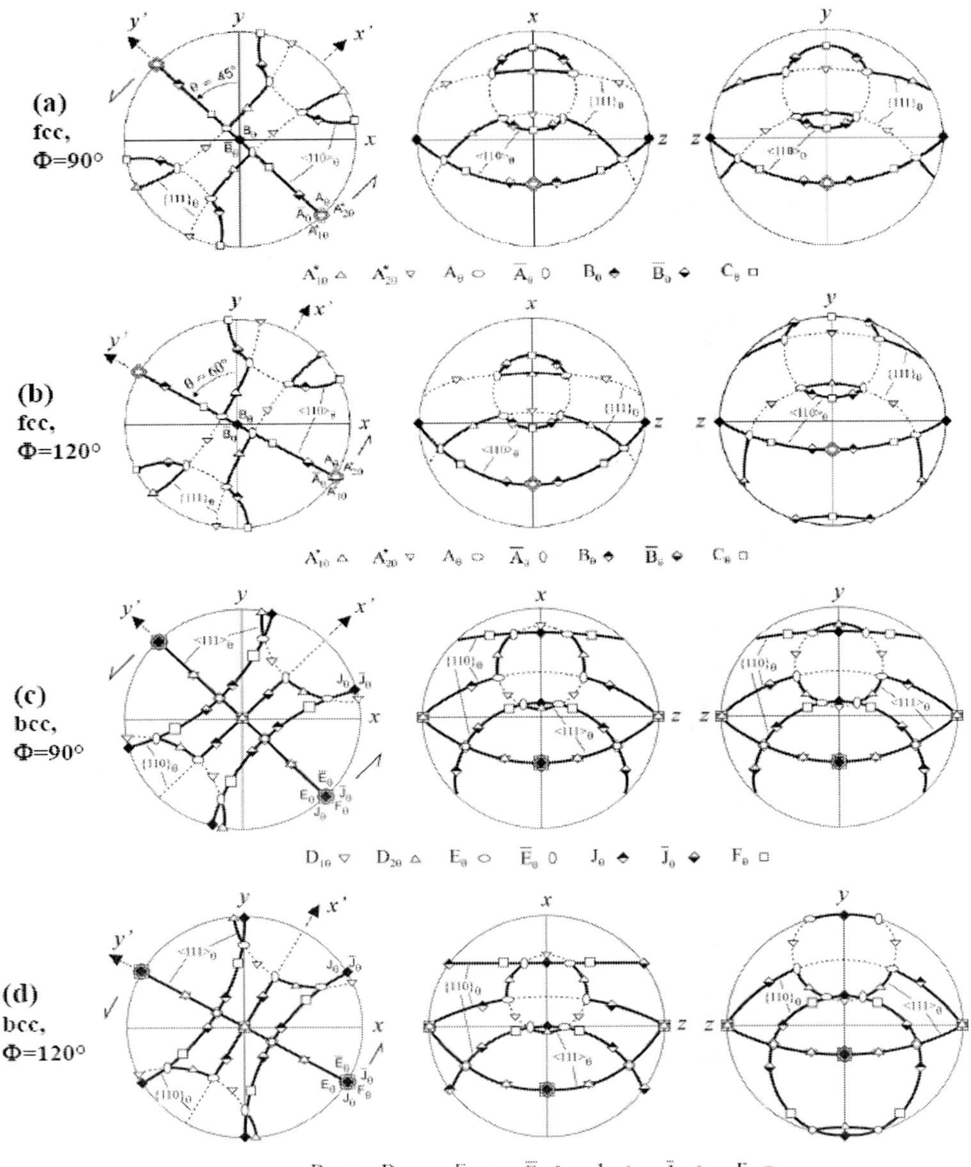

Figure 9.3. (111) pole figures on the TD, ND and ED planes showing the ideal orientations of ECAE textures defined by the ideal simple shear model in (a) fcc and 90° die; (b) fcc and 120° die; (c) bcc and 90° die; (d) bcc and 120° die, Li et al. (Li et al., 2005b).

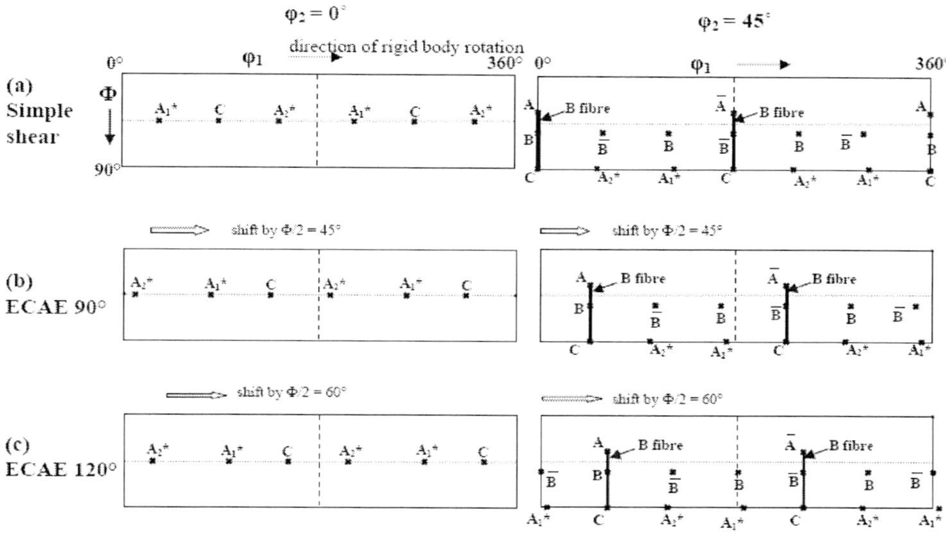

Figure 9.4. Relationship between the ideal positions in simple shear (a) and ECAE deformation for (b) $\Phi = 90°$ and (c) 120° in orientation space using two different sections, $\phi_2 = 0°$ and $\phi = 45°$. Note that any subscript Φ or θ to signify ideal ECAE orientations from ideal simple shear orientations have been removed in this Figure. Recall that for $\Phi = 90°$ the ideal orientations are shifted $\Phi/2 = 45°$ and for $\Phi = 120°$, they are shifted $\Phi/2 = 60°$. Taken from (Beyerlein and Toth, 2008).

9.3. Multiple Passes

Texture evolution after multiple passes is not expected to be described by these preferred orientations due to the strain path changes that take place between passes. For defining components in multi-pass textures it is important to use a fixed laboratory coordinate system, such as in figures 8.1 and 8.2. With respect to the sample, shearing does not necessarily occur on the same plane in multi-pass ECAE, with the exception of route C. In route C, while the shear plane remains the same, the direction of simple shearing is reversed every pass. For route A, in spite of the strain path change, there is, fortunately, a simple relationship between the ideal ECAE orientations and those after subsequent passes. When the sample is re-inserted into the die, the whole ODF shifts in the lab reference system. For route A, the ODF simply shifts according to (Beyerlein and Tóth, 2008):

$$g_{\Phi}(\text{Route A}) = \left(\varphi_1 - \Phi, \phi, \varphi_2\right) \qquad (9.2)$$

The consequences of this shift are different from the one expressed in Eq. (9.1). In the above Eq. (9.2), all grains with stable orientations will be shifted to orientations which are *not* stable when the next pass begins. (One exception is the Φ = 120° die where the B and $\bar{B}$ components are exchanged (Beyerlein and Tóth, 2008).)

For any general route and die angle, another characterization method is to identify new orientation fibers found repeatedly across different routes and from pass to pass. This task has been attempted in a few studies (Ferrasse et al., 2004; Li et al. 2005a; Li and Beyerlein 2005). Table 9.3 lists orientation fibers identified in (Li et al., 2005a) for FCC f_1-f_3 and BCC materials b_1-b_3. In copper, Li et al. (2005a,c) showed that ECAE textures consist of components distributed along three fibers f_1-f_3, and within these fibers, orientations come and go with N and for routes Bc, C, and A. A similar trend was found in IF-steel processed by different Φ and routes A, Bc, Ba, and C (Li et al., 2006c,2007). The components and fibers for different Φ are similar in nature, only differing by a rotation $\theta = \Phi/2$ about the TD. Another convenience in using f_1-f_3 or b_1-b_3 is that deviations in ideal positions (due to deformation changes for instance) translate to shifts in the components along these fibers. Note that f_2 (b_2) is symmetric to f_3 (b_3). Example FCC (Cu Li et al. 2005a,c) and BCC (IF-steel Li et al., 2006c) textures with these fibers indicated are shown in figures 9.5 and 9.6.

Identifying ideal texture components or fibers is useful in characterizing texture, but not in determining the deformation mode that generated them. This is particularly true when the deformation path sequence is complex, such as in multi-pass ECAE, which involves strain path changes. One cannot deduce the ECAE route purely from an analysis of the texture components contained in the final texture after two or more passes. Comparisons between model prediction and measurement are the best way to determine deformation history. Some studies attempt to relate the texture of each route to a different deformation mode, such as rolling, drawing, or shearing on a plane parallel to the exit flow plane or even to the entry flow plane. Such analyses, however, are erroneous due mainly to differences in the rotations and symmetries of the deformation (Beyerlein and Tóth, 2008). Any qualitative similarity between ECAE textures and those of other deformation paths is just a coincidence.

Table 9.3.

Notation of fiber	Orientations in fiber
f_1	$A^*_{1\Phi}$ -A_Φ/ $\overline{A}_\Phi$ - $A^*_{2\Phi}$ $\{111\}_\Phi$ partial fiber
f_2	C_Φ- $\overline{B}_\Phi$ /B_Φ - $\overline{A}_\Phi$ /A_Φ $\langle 110\rangle_\Phi$ partial fiber; $\overline{A}_\Phi$ /A_Φ- $A^*_{1\Phi}$ $\{111\}_\Phi$ partial fiber
f_3	C_Φ -B_Φ/ $\overline{B}_\Phi$ -A_Φ/ $\overline{A}_\Phi$ $\langle 110\rangle_\Phi$ partial fiber; A_Φ/ $\overline{A}_\Phi$ - $A^*_{2\Phi}$ $\{111\}_\Phi$ partial fiber
b_1	$D_{2\Phi}$ -E_Φ -$D_{1\Phi}$ $<111>_\Phi$ partial fiber
b_2	F_Φ -J_Φ - E_Φ $\{110\}_\Phi$ partial fiber; E_Φ -$D_{2\Phi}$ $\langle 111\rangle_\Phi$ partial fiber;
b_3	$D_{1\Phi}$ - $\overline{E}_\Phi$ $<111>_\Phi$ partial fiber; - $\overline{E}_\Phi$ - $\overline{J}_\Phi$ -F_Φ $\{110\}_\Phi$ partial fiber;

Table 9.3. Definition of fibers in fcc (f_1, f_2, f_3) (Li et al., 2005a) and bcc (b_1, b_2, b_3) ECAE textures (Li and Beyerlein, 2005)

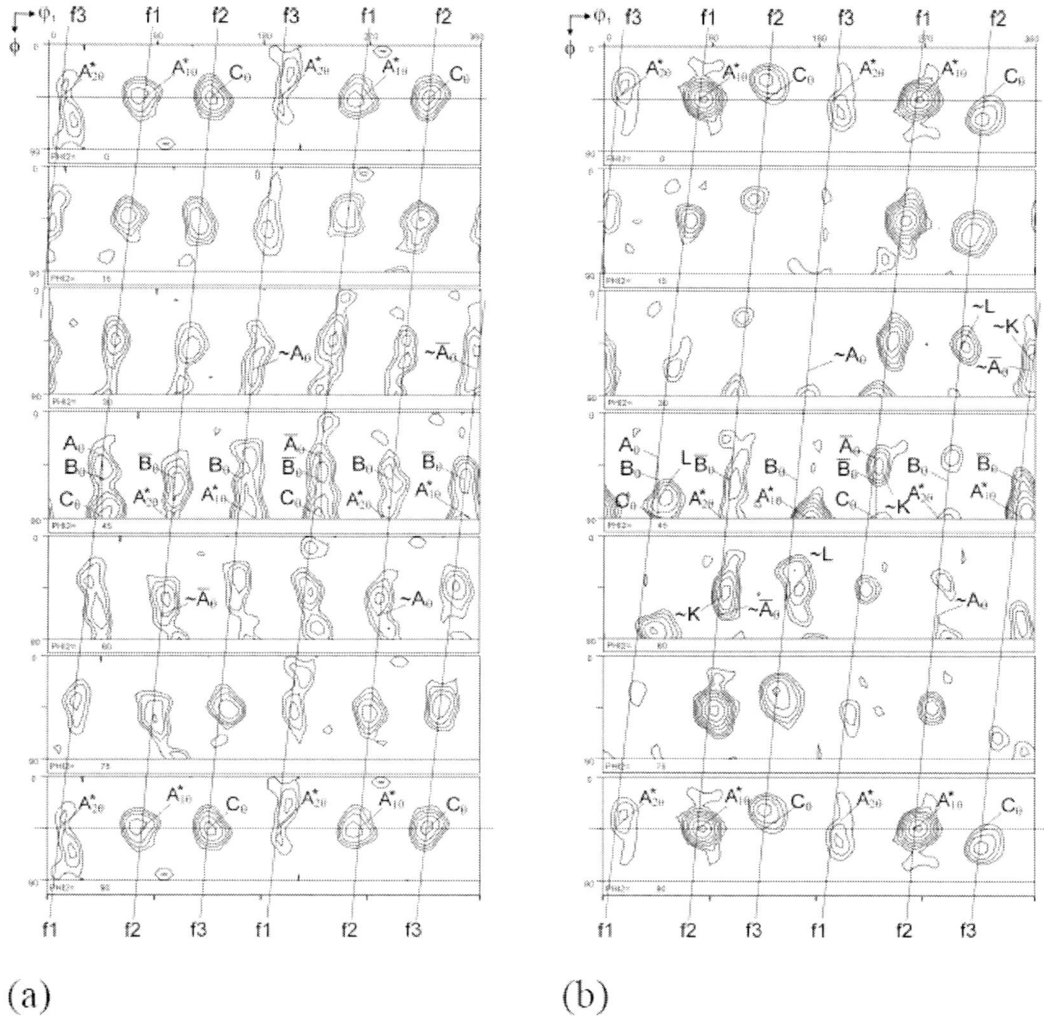

Figure 9.5. φ_2 = constant (0–90° in steps of 15°) ODF sections of (a) 1 pass and (b) 4 pass route Bc textures of Cu showing the ideal orientation fibers f_1-f_3. Contours are 1.4/2/2.8/4/5.6/8/11/16/22/29 mrd. Taken from (Li et al. 2005a,c).

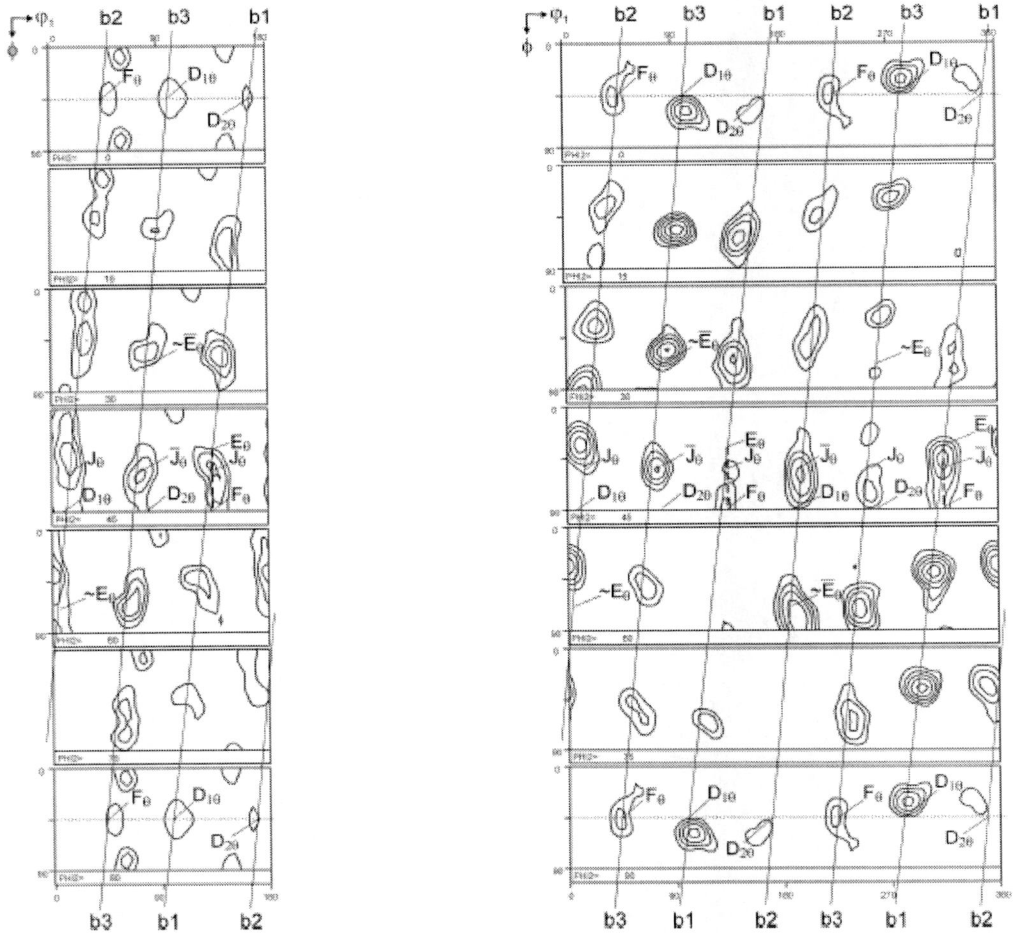

Figure 9.6. φ_2 = constant (0–90° in steps of 15°) ODF sections of (a) 1 pass and (b) 4 pass route Bc textures of IF-steel showing the ideal orientation fibers b_1-b_3. Contours are 1.4/2/2.8/4/5.6/8 mrd. Taken from (Li et al., 2006c)

Chapter 10

TEXTURE EVOLUTION DURING ECAE

10.1. TEXTURE EVOLUTION IN CUBIC METALS

In what concerns texture evolution during ECAE, cubic metals have been the most commonly studied in the literature. For discussion, we return to the previous examples of textures of Cu (FCC) and IF-steel (BCC) after ECAE using a tool angle of $\Phi = 90°$ that were presented in Section 9.3. However as an alternative to the ODF sections in figures 9.5 and 9.6, figures 10.1 and 10.2 show the (111) pole figures of Cu (FCC) and (110) pole figures of IF-steel (BCC) after one, two and four passes of ECAE (Li et al. 2005a; Li et al., 2006c). In both materials, the initial texture was weak. With this starting condition, grain orientations after the first pass are usually found to approach the ideal ECAE positions. As shown, in both Cu and IF-steel the first pass extrusion induces a shear texture in which the ideal components mapped out in Figure 9.3 can be identified. This correspondence occurs because the deformation in one pass is close to monotonically applied simple shear (See Figure 7.4 and Figure 9.3). Deformation in multi-pass ECAE, however, accumulates by shear deformation on different planes and/or different directions with each new pass, and not monotonically in the same plane and in the same sense of direction as assumed in classical simple shear texture characterization. Therefore ECAE textures after two or more passes are not expected to exhibit classical simple shear textures. Figures 10.1 and 10.2 for Cu and IF-steel, respectively, display the textures developed in subsequent passes of routes A and C. As shown, the features of the first-pass texture are maintained. The strengths of the ideal ECAE components, however, change with pass number (Li et al. 2005a; Li et al. 2006c; Beyerlein and Tóth, 2008). Although in subsequent passes of routes Bc and Ba, the first-pass texture is lost, a pattern seems to develop. For route Bc, it has been reported that the ECAE texture after eight and 16 passes are similar to those of four, with only minor differences in the second-order features (Li et al. 2005a). For route Ba, texture evolution followed another pattern: the third-pass texture is approximately the same as the second and fourth-pass textures, but rotated by 180° about the 3-axis (Li et al., 2006c).

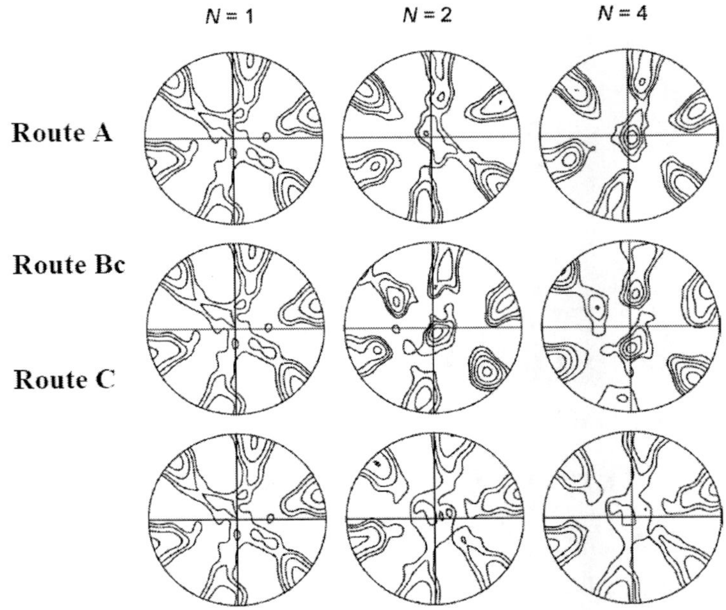

Figure 10.1. (111) pole figures of Cu textures in routes A (Li et al., 2006a), Bc, and C (Li et al. 2005a) for a 90° die with a rounding of $\Psi = 37°$ (see Fig. 8.1), after $N = 1{,}2$ and 4 passes. Pole figures are shown in the transverse (TD) view, in which the extrusion exit 1-axis is horizontal and the entry 2-axis is vertical (See Fig. 8.2). Contour levels are 1/1.4/2/2.8/4 mrd.

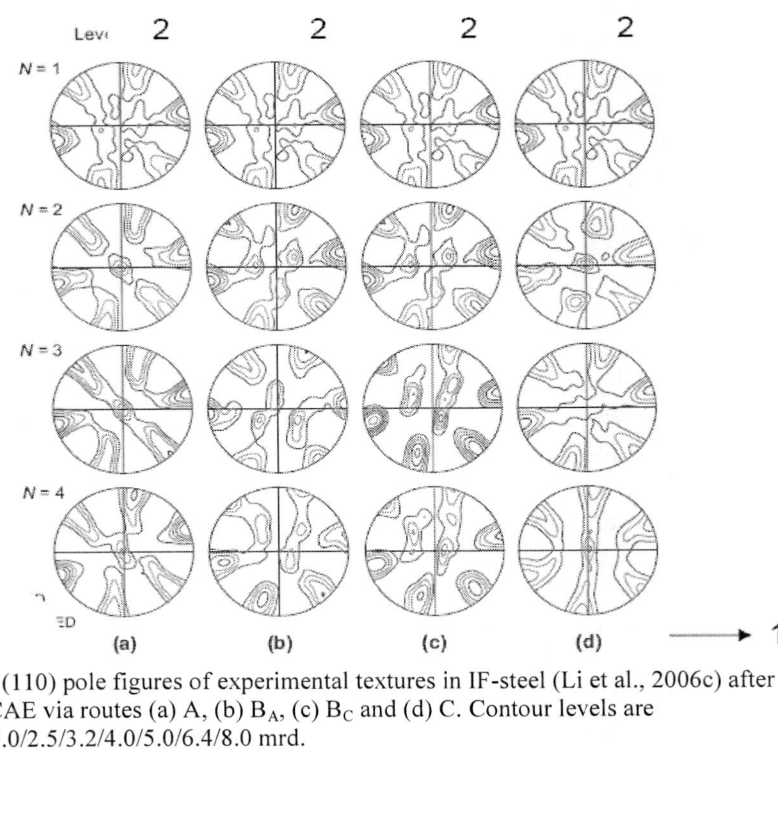

Figure 10.2. (110) pole figures of experimental textures in IF-steel (Li et al., 2006c) after $N = 1$ to 4 passes of ECAE via routes (a) A, (b) B_A, (c) B_C and (d) C. Contour levels are 1.0/1.3/1.6/2.0/2.5/3.2/4.0/5.0/6.4/8.0 mrd.

In the above examples, the samples are processed in a die with $\Phi = 90°$. Changing Φ has several effects but the most dominant one is to rotate the ideal ECAE components as predicted from Eq. (9.1). Consequently a texture developed using an $\Phi_1 = 120°$ die will differ from that developed using a $\Phi_2 = 90°$ die by a rotation of $(\Phi_1 - \Phi_2)/2 = (120° - 90°)/2 = 15°$ about the 3-axis (or TD in Figure 9.3).

10.2. Comparison of FC-Taylor and VPSC Predictions for FCC Metals

In Chapter 7 we review the two polycrystal models that have been used widely in the ECAE literature to model texture evolution: the full-constraints (FC) Taylor model and VPSC. In this section, we briefly compare their predictions for ECAE texture development for a hypothetical FCC metal.

For both models we use the velocity gradient L for simple shearing for $\Phi = 90°$, Eq. (8.7), and the method for multiple-pass modeling discussed in Chapter 8 to predict texture evolution in ECAE up to four passes for the four standard routes: A, Bc, Ba, and C. Calculations are performed for an initially random polycrystalline FCC metal (1000 orientations) with initially spherical grains. The grain splitting scheme is invoked with $R = 5$. A constant plastic CRSS is used for all slip systems, i.e., $\tau_0 = 1$ and τ_1, θ_0, and $\theta_1 = 0$ in Eq. (6.30). For the VPSC calculation, each grain is allowed to deform according to its individual orientation, shape and interaction with the homogeneous, anisotropic effective medium. To first-order, the following predictions will, therefore, apply when the processing conditions are nearly ideal and the material is initially non-textured. (Ideal conditions include rigid-plastic material, sharp inner and outer corners, simple shear and frictionless die surfaces. Backpressure may be applied to promote ideal conditions.) When the ECAE deformation is inhomogeneous, different boundary conditions need to be imposed. The models and predicted texture development for such cases will be discussed in the next chapter.

Figure 10.3 presents the texture results for (a) FC-Taylor and (b) VPSC for $\Phi = 90°$. The two models predict similar textures for the first pass. Differences increase with pass number and depend on route. For large strains (or numbers of passes), certain texture features predicted by the VPSC model are not predicted by the FC-Taylor model. While predictions for route A between the two models are similar, they are noticeably different for route C. The texture intensities are, however, not significantly different for the same strain level.

For a given Φ, ECAE route and pass number make a significant impact on texture evolution. The in-plane, nearly continuous, route A displays a shear texture which intensifies with N. The reversal route C also displays a shear texture in the odd passes and a diffuse texture in the even passes. In the FC-Taylor case, it nearly recovers the initial texture every even pass, whereas in the VPSC case, there is some weak shear components in the even-pass prediction. Routes Bc and Ba show different and mixed-type textures, as expected from the out-of-plane cross-shearing imposed by them. In essence, the texture predictions closely followed what would be expected by the strain path changes associated with each route.

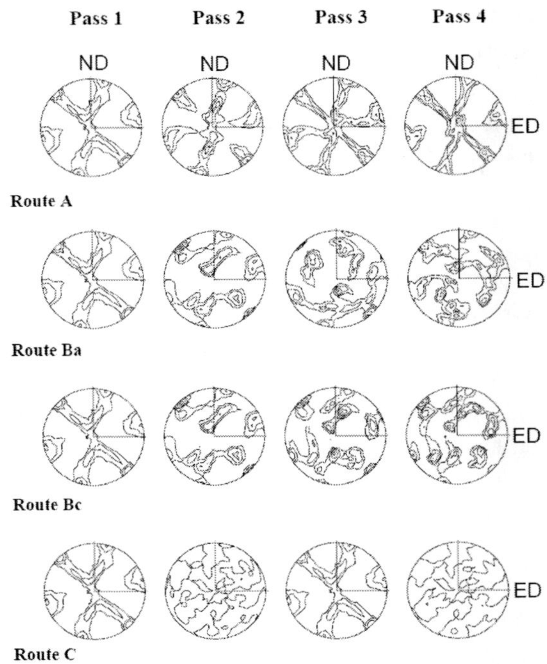

Figure 10.3(a): Texture predictions for routes A, Bc, Ba, and C up to four passes using FC-Taylor. Contour levels are 1/2/4/8/11 mrd.

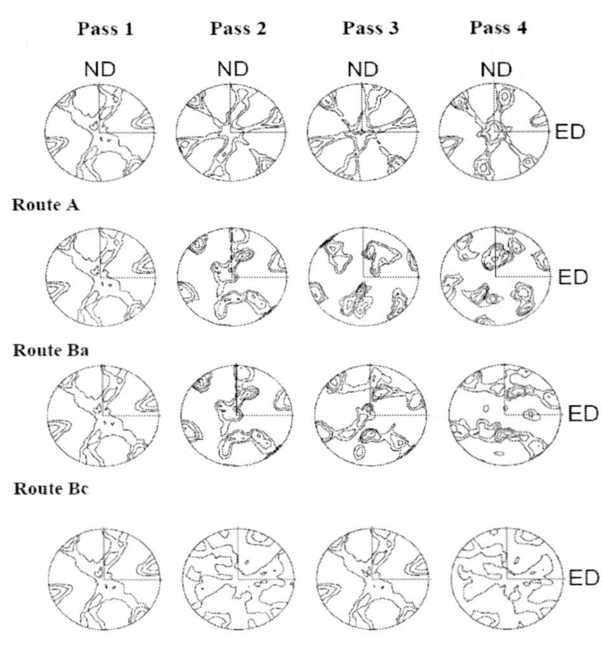

Figure 10.3(b). Texture predictions for routes A, Bc, Ba, and C up to four passes using VPSC. Contour levels are 1/2/4/8/11 mrd.

10.3. Geometric Hardening (and Softening)

In the simulations above, the CRSS values were assumed constant and hence do not evolve with straining. As a consequence, any positive or negative changes in the calculated flow stress will be due to geometric (texture) hardening or softening. Figure 10.4 plots the deformation response corresponding to the VPSC calculations for route Ba in Figure 10.3. The 'flow shear stress' and shear strain γ plotted are the polycrystalline (macroscopic) flow stress and imposed shear strain resolved on the intersection plane of the die. The two features to notice are the jump in flow stress in transition from pass to pass and the softening behavior that immediately follows in each pass. This severe geometric hardening/softening behavior is a direct result of the changes in the input texture due to the rigid rotations between passes associated with route Bc.

To better understand this effect, we show in Figure 10.4 the texture changes associated with both rigid rotations and deformation and superimpose the same resolved shear stress-strain curve calculated in the case of no strain path changes (labeled 'monotonic'). All (111) pole figures in Figure 10.4 are represented in the same channel die reference axes. The top row displays the textures as they enter the die, starting with the initial (non-textured) one. The bottom row shows the textures as they exit the die and correspond to the route Ba textures shown in Figure 10.3. (These calculations are taken from Beyerlein et al. (2003) in which the number of orientations were much less than those used in Figure 10.3).

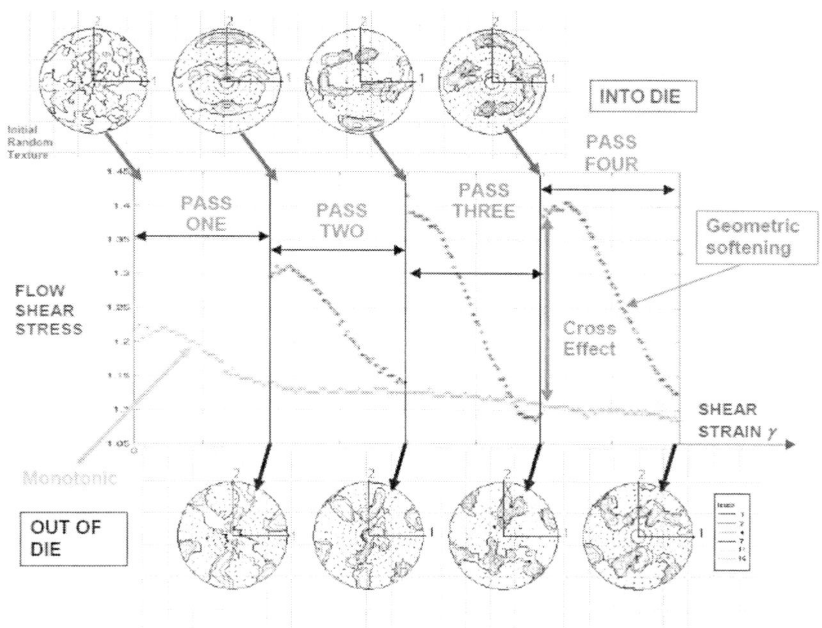

Figure 10.4. VPSC predictions of the flow shear stress versus shear strain resolved in the intersection plane of the die for route Ba. Because work hardening due to defect storage is suppressed in these calculations, the flow behavior is due to geometric (texture) effects alone. The textures as a result of deformation after each pass (bottom) and of the rigid rotations between passes (top) are shown as (111) pole figures with the 3-axis normal to the page. Contour levels are 1/2/4/7/11/16 mrd.

In the first pass, the monotonic and ECAE response coincide. The material is found to geometrically soften as the grains reorient by slip to orientations that are favorable for accommodating the imposed shear deformation. As a result, a shear texture develops. Consequently the material is no longer 'isotropic'. When the sample is rotated prior to the second pass, the entry texture for the second pass has changed from the favorable shear texture that was produced after the first. As a result, most of the grains are in a 'hard' orientation with respect to the shearing imposed by the die and abrupt jump in the flow stress occurs (labeled 'cross effect'). Crystallographic slip induced by the severe ECAE shearing will once again re-orient the grains to an orientation favorable for accommodating the deformation. As they re-orient to easier orientations, the material softens (labeled 'geometric softening'). As shown, this geometric hardening/softening pattern repeats every pass. However, because the entry texture is not the same for every pass, the texture after each pass will not be the same. At least for the route shown (Ba), the geometric softening that develops in each pass is severe, so much that when combined with any work hardening contribution from defect storage, it is likely that the overall material response may still soften. One can expect from this analysis that the geometric contribution to overall material response can be strong in ECAE and will depend on route, Φ, and initial texture.

Other routes were analyzed in (Beyerlein et al., 2003). Because route Bc also involves drastic changes in shearing direction and plane with respect to the sample, it produces similar results. Route C, on the other hand, does not change the shear plane (only reverses the direction) with each pass, and so there will not be an abrupt jump as a result of texture when transitioning from one pass to another.

10.4. Performance Against Experiment

To quantitatively assess the quality of a predicted texture, the simulated Orientation Distribution Function (ODF) $f_{\mathrm{sim}}(g)$ is compared to a measured one $f_{\mathrm{exp}}(g)$, using the normalized texture index $\hat{T}^{\mathrm{d}}$ of the difference ODF (DODF)

$$\hat{T}^{\mathrm{d}} = \frac{T^d}{T} = \frac{\int\left[f_{\mathrm{sim}}(g) - f_{\mathrm{exp}}(g)\right]^2 \mathrm{d}g}{\int\left[f_{\mathrm{exp}}(g)\right]^2 \mathrm{d}g} \tag{10.1}$$

where T^{d} is the texture index of the DODF, i.e., $f_{\mathrm{sim}}(g) - f_{\mathrm{exp}}(g)$. Compared to T^{d}, $\hat{T}^{\mathrm{d}}$ allows direct comparisons between predicted textures with different reference textures, such as those generated from different materials or deformation paths.

The VPSC textures are consistently found to agree better with measured textures after ECAE processing than FC-Taylor (Agnew et al., 1998; Li et al., 2005a, 2005c, 2006c). Using $\hat{T}^{\mathrm{d}}$ as a measure of performance, Li et al. (2005a) found that despite whether finite element or simple shearing was used to model ECAE deformation, VPSC outperformed FC up to 16 passes in Cu, processed by route Bc with $\Phi = 90°$. A similar result was obtained for IF-steel processed also by $\Phi = 90°$ up to four passes (Li et al., 2006c). The VPSC-predicted textures reproduced the positions, as well as the relative intensities of all the main components in the experimental textures.

The Taylor model (FC) is considered to perform best for cubic materials that deform by slip only. When this is the case, there are some ECAE cases under which FC performs adequately. Many earlier works (Ferrasse et al.2004, Baik et al. 2003) observed good agreement with measurement after the first pass only. As far as location of texture components, FC seems to perform better with route A than the other routes (Li et al., 2006a; Ferrasse et al., 2004; Agnew et al., 1998; Baik et al., 2003; Li et al. 2006c,2007). Recent work on IF-steel (Li et al., 2007) for two different die angles suggests that the FC model can perform nearly as well as VPSC if deformation is less severe per pass. For material processed by $\Phi = 120°$, the performance of the two models was similar in routes A, C, and even Bc (Li et al., 2007).

The comparisons made here are not surprising and similar trends between the two have been reported previously with other large strain deformation modes. For FCC metals, early works (Canova et al., 1984; Molinari et al. 1987) show that FC fails to capture effects associated with grain shape at large strain, as do the relaxed constraints Taylor (RC) or VPSC. The latter two models do not require that five independent slip systems be activated, a requirement that is likely not to be the case when strains become large.

The VPSC and FC textures both evolve much more quickly than the experimental textures, yielding higher intensities for the same deformation. Reasons given are typically related to local grain-grain interactions and substructure evolution, leading to inhomogeneities in crystal orientation and in directional hardening within the grains, features which are generally lost in the homogenization assumption of the polycrystal model for large deformation. Also, other deformation mechanisms, including shear band localization, grain boundary sliding, microstructural refinement, and grain rotation can become important during SPD. On the other hand, the processing mechanics are not controlled in most experimental works and hence, the actual stresses and strains are unknown. As texture evolution is sensitive to the applied deformation, comparisons between modeling and experiment should be done for identical loading conditions.

10.5. First-Pass Textures in HCP Metals

Metals with the hexagonal close-packed (HCP) crystal structure deform using a combination of slip and twinning modes. The common HCP slip modes are basal slip $\{0001\}\langle 1\bar{2}10\rangle$, prismatic slip $\{1\bar{1}00\}\langle 11\bar{2}0\rangle$, pyramidal $<a>$ slip $\{10\bar{1}1\}\langle\bar{1}2\bar{1}0\rangle$, first-order pyramidal $<c+a>$ slip $\{10\bar{1}1\}\langle 2\bar{1}\bar{1}\bar{3}\rangle$ and second-order pyramidal $<c+a>$ slip $\{2\bar{1}\bar{1}2\}\langle\bar{2}113\rangle$ (Yoo and Wei, 1966; Partridge, 1967). There are seven twin types, but the frequently observed deformation twinning modes are $\{10\bar{1}2\}<\bar{1}011>$, $\{11\bar{2}2\}<11\bar{2}\bar{3}>$, and $\{10\bar{1}1\}<10\bar{1}2>$ (Yoo, 1969; Partridge, 1967). The slip modes and twin types active in a given HCP metal depend on the metal and can only be determined by experiments. Of those selected, the relative activities of the slip and twinning used in plastic deformation depend on a multitude of factors: strain rate, temperature, grain size, composition, and c/a ratio.

In the past decade most ECAE studies on HCP metals have processed pure Ti, Zr, Be, Mg and their alloys (See Beyerlein and Tóth, 2008 and references therein). The operative mechanisms in Ti, Zr, Mg and Be were generally found to be basal and prismatic in different proportions together with less or negligible activity of pyramidal <c+a>[1] and deformation twinning, although the variation across studies can be large as a result of different initial textures and processing conditions.

Texture development in ECAE depends on the relative activities and type of slip and twinning systems active throughout the extrusion. In most cases, texture in HCP metals evolves in a similar manner during the first pass. The texture has a centro-symmetric pole Figure in which the symmetry plane is misaligned from the shear plane normal. This basic feature of first-pass HCP textures has been reported for Mg alloys (Kim et al. 2003; Agnew et al. 2005a,b), Ti (Shin et al., 2003; Perlovich et al., 2006a; Yu et al., 2004; Him et al., 2002), Be (Li et al., 2008; Beyerlein et al.2008; Field et al. 2002) and Zr (Perlovich et al.; 2006b; Yapici et al., 2008; Yu et al., 2002).

Figure 10.5 shows the (0001) pole figures of pure Ti (transition HCP metal c/a = 1.586) (Alexander et al., 2004) and pure Be (simple HCP metal c/a = 1.581) (Beyerlein et al., 2008; Li et al., 2008) after one pass. The initial texture of the Be was nearly random. The initial texture of the Ti consisted of basal poles that were uniformly distributed normal to the billet axis with a maximum strength of 4 mrd. After a single pass, the textures possess the centro-symmetric appearance typical of most HCP metals. The maxima lie at roughly 35°-45° CCW from the 2-axis (normal to the extrusion axis--see Figure 8.2) for the Be processed by a Φ = 120° die and 22° for the Be process by a Φ = 90° die. This difference is a result of mechanical loading rather than microscale differences. Due to the change in orientation of the shear plane, the first pass texture of samples processed by two different die angles Φ_1 and Φ_2 are expected to differ by a rotation $\Delta\theta = \sim(\Phi_1 - \Phi_2)/2 = (120° - 90°)/2 = 15°$ about axis-3. In the first-pass Ti texture (Φ = 90°), the maxima appear to be split, lying at ~11° and ~33° from the ND axis.

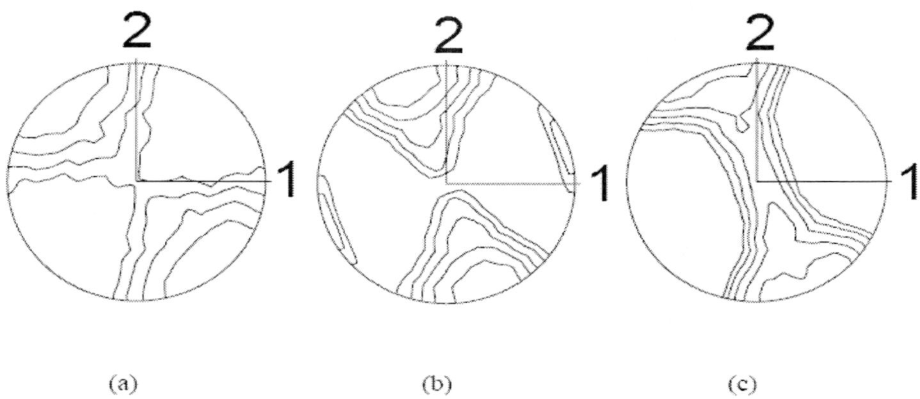

Figure 10.5. Measured (1000) pole figures of first-pass textures of hcp metals processed in our laboratory: (a) pure Be processed at 425 °C with Φ = 120°, (a) pure Be processed at 425 °C with Φ = 90°, and (c) pure Ti processed at 200 °C with Φ = 90°. Data courtesy of D. J. Alexander et al. 2004. Contour levels are 0.7/1.0/1.4/2.0/2.8 mrd.

[1] Ti and Zr exhibit first-order pyramidal <c+a>, whereas Mg and Be exhibit second-order pyramidal <c+a>.

The first-pass texture of HCP metals are similar because in every case the deformation involved pronounced activity of basal slip, with a lesser contribution from prismatic slip and pyramidal slip. For some HCP metals, such as Zr and Ti, the idea of substantial basal slip is in conflict with several experimental results and observations, suggesting that basal slip is not active. However, these tests usually involve single crystals or polycrystals strained from low to moderate strain levels (Kaschner et al., 2006; Akhtar, 1971; Tomé et al. 2001; McCabe et al., 2006; Partridge, 1967). The unexpected prevalence of basal slip in Zr could be a consequence of the severe plastic deformation and deserves more experimental evaluation. This issue will be pursued further in Chapter 13 in which the deformation texture and behavior of ECAE processed Zr are modeled.

10.6. Effect of Slip Activity on HCP Texture Evolution

In simulations, the relative activities of basal, prismatic, and pyramidal <c+a> slip during ECAE can be altered by changing the relative values of their slip resistances, i.e., τ_0 in Eq. (6.30). Figure 10.6 presents the VPSC texture predictions in (0001) pole figures after the first and second pass of route B, Φ = 90°, as a result of different amounts of non-basal slip activity: i) primarily basal slip, ii) basal slip plus moderate pyramidal <c+a> slip activity (first-order) and iii) basal slip plus moderate prismatic slip activity. The calculations are repeated for two starting textures, random and axisymmetric (~4.0 m.r.d.) (see Figure 10.6a,b). In all cases, the first-pass textures possess the centro-symmetry observed in measurement. In the second pass, the symmetry is slightly broken, in some cases more so than others. The symmetry can be attributed to basal slip (case i). The effect of a moderate amount (10-20%) of pyramidal slip (case ii) is to split the maxima in all cases. The effect of a moderate amount (20-30%) prismatic slip (case iii) is more subtle: it tends to align the basal poles perpendicular to the entry and exit directions (center of the pole figures).

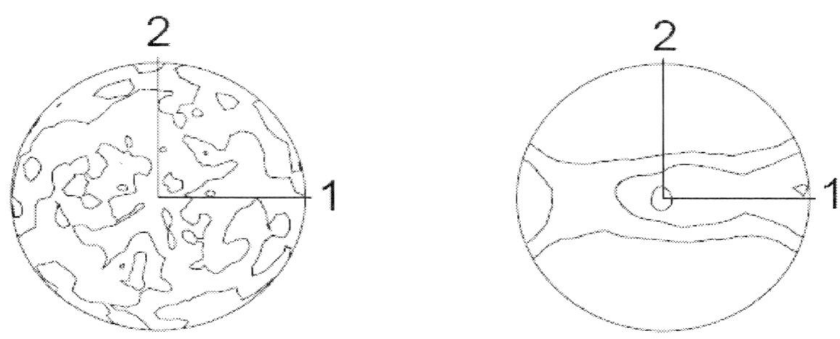

(a) Initial Random Distribution (b) Initial Rolling Texture

Figure 10.6. Continued on next page.

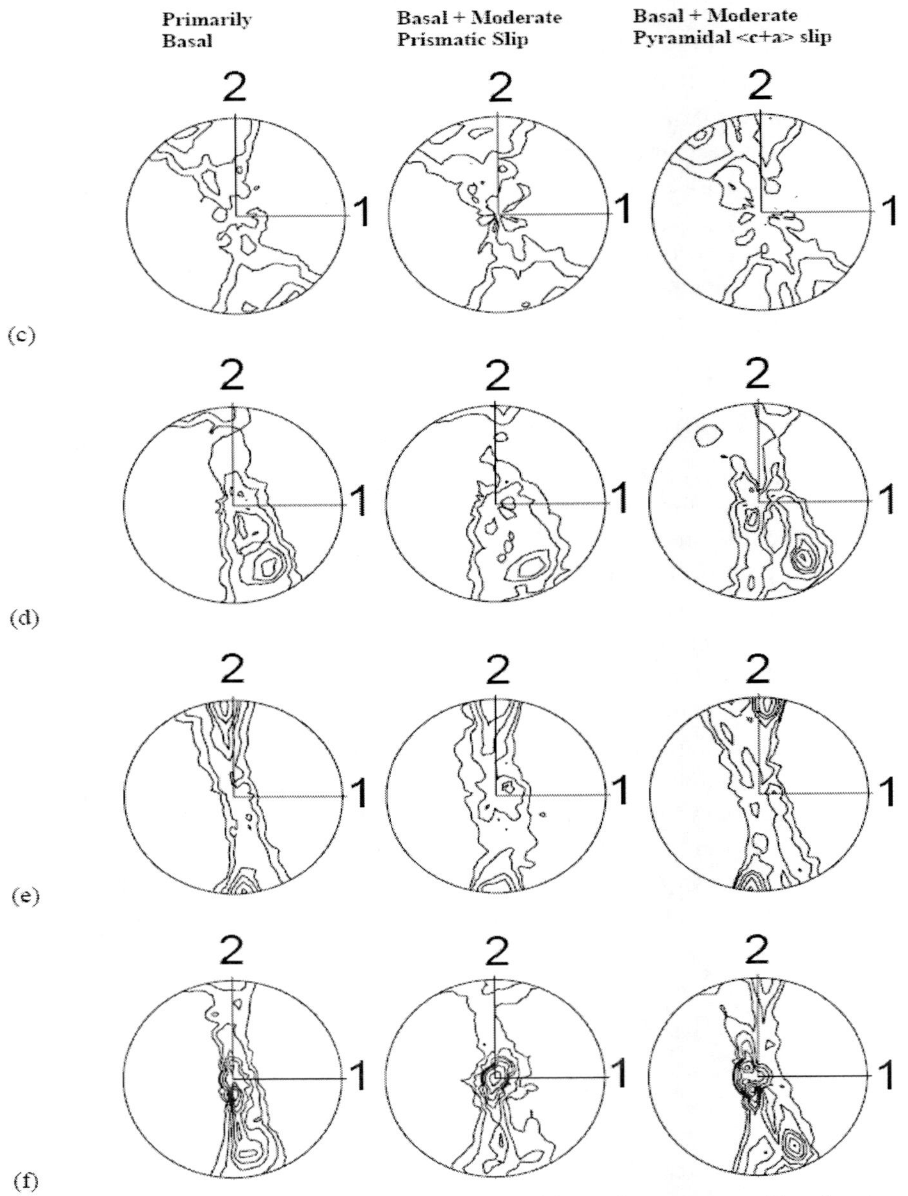

Figure 10.6. Predicted (1000) pole figures of hcp metals with varying amounts of basal and non-basal slip activity: basal, basal + moderate prismatic slip, and basal + moderate pyramidal <c+a> slip. Two starting textures were used: a) initial random distribution and b) initial rolling texture;. c-d) first and second pass route B with initial random texture; e-f) first and second pass route B with initial rolling texture. Contour levels are 1.0/2.0/4.0/5.6/8.0/11.3/16.0 mrd.

Chapter 11

DEVIATIONS FROM SIMPLE SHEARING

11.1. IDEAL FAN MODEL

The simple shear model is attractive in its simplicity and ease of use. This model represents ECAE deformation under ideal, optimal processing conditions, (see Part I). Unfortunately, in many situations, ECAE deformation cannot be generally described according to this model, which assumes concentrated simple shearing along the intersection plane of the two channels. Finite element models (FEM), both 2-D and 3-D (Park and Suh, 2001; Kim et al., 2000, Li et al., 2004b; Bowen et al., 2000; Kim, 2002; Kim et al., 2001; Budilov et al., 2004; Beyerlein et al., 2004, 2005a), deformations of scribed grids (Wu and Baker, 1997), and slip line theory (Segal, 2003; Stoica and Liaw, 2003) reveal that in most cases, plastic deformation takes place in a broad plastic deformation zone (PDZ) of the sample. The size, shape, and deformation field within this PDZ depends mainly on contact friction and die design and less on material flow response, backpressure and pressing rate. The PDZ shrinks and reduces to simple shearing along the intersection plane when the inner and outer corners are sharp, frictionless surface conditions exist, backpressure is applied, and the material is rigid plastic (Park and Suh 2001, Segal 2003). Deviations from these conditions widen the PDZ. Broad PDZs leave the sample in an inhomogeneous deformed state. Inhomogeneous deformation can result in heterogeneity in the final microstructure, texture, and mechanical properties across the sample length, width, and thickness (Li et al., 2004b, Beyerlein et al., 2005a).

For texture and constitutive behavior prediction, it is crucial that the macroscopic deformation characteristics are modeled accurately. Particularly for SPD processes, a multiscale modeling approach that directly couples, for instance, FE, VPSC, subgrain microstructural model and dislocation dynamics, would undeniably be a promising approach. To date, the computational requirements are still too great for fully-coupled multiscale models to be a feasible method for exploratory investigations. One should also keep in mind that any multiscale model is only as good as the constituent models that it links.

An alternative to full-scale computational mechanics simulations of macroscale deformation is analytical mechanics descriptions of the essential characteristics of the deformation history. The first analytical model of ECAE deformation that deviates from simple shearing was derived by Segal (1999) and is described in full in Part I. In this formulation the slip-line field analysis is used, which allows one to forecast the shape and size of the deformation zones associated with different processing parameters and to determine the conditions leading to inhomogeneous versus homogeneous deformation. This analysis assumes a rigid plastic isotropic material. The model predicts that for sharp corner dies ($\Psi = 0$ dies) and non-zero friction conditions a dead metal zone (non-moving material) forms at the corner. Consequently, the PDZ assumes a 'fan' shape with internal angle β_m symmetric about the intersection plane (See Figure 11.1). The angle β_m was related to the die surface friction coefficient. The deformation within the fan could be described as simple shearing on continuously rotating shear planes. Favorable comparisons were made between deformations calculated from slip-line theory, FEM calculations, and measurements of scribed grids.

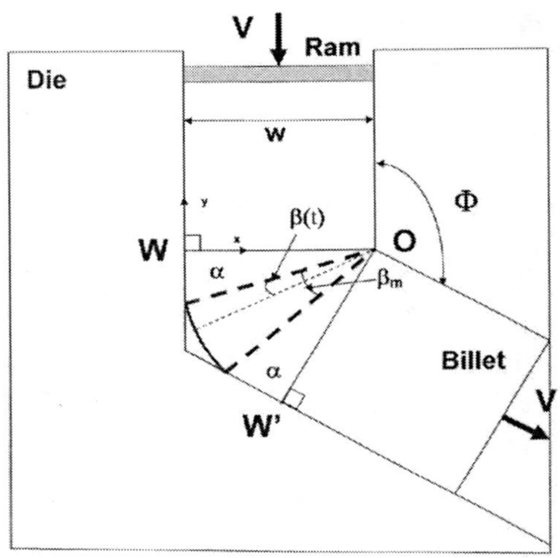

Figure 11.1. Schematic of an ideal fan with angle β_m for the plastic deformation zone formed in a sample extruded in ECAE tooling with die angle Φ.

Polycrystalline and microstructural evolution models require that the imposed deformation is expressed by its deformation gradient F or velocity gradient L (Beyerlein et al., 2003). Beyerlein and Tomé (2004) later derived the corresponding F and L for the fan-shaped PDZ predicted by Segal's slip-line analysis, which assumes that the material is rigid plastic and friction is uniform in both channels. This derivation is found in the next section. It is clear that such a description of the PDZ is still idealized. Because of its symmetrical shape, it produces homogeneous deformation at the end of each pass. For this reason, we refer to this fan-shaped PDZ as an *ideal fan*.

11.2. DEFORMATION AND VELOCITY GRADIENT TENSORS FOR AN IDEAL FAN

The kinematic tensors and their relations are described in Section 6.2 and 8.1 of this book. In the following, expressions for the plastic deformation gradient F and velocity gradient L associated with an ideal fan of internal angle β_m are derived. In previous work by Beyerlein and Tomé (2004), Φ was fixed to be 90°. The model is extended here for any ECAE die angle Φ. A schematic of this fan representing the PDZ for a general ECAE die is illustrated in Figure 11.1. As shown in Figure 11.1, OW and OW' are the through-thickness directions, normal to the channel entry axis and exit axis respectively. A fan-shaped zone is equidistant by an angle α from both sides OW and OW'. This simple representation allows the PDZ to be completely determined by only one parameter α. The inner angle of the fan, β_m, is related to α by

$$\beta_m = \pi - \Phi - 2\alpha \tag{11.1}$$

which ranges from 0 to $\pi - \Phi$ as α decreases from $(\pi - \Phi)/2$ to 0. The fan extends from the inner to the outer corner. In an ideal fan, β_m is independent of initial position of the material point X_0. The fan angle β_m is recognized here to have a complex relationship between material response, die friction, and tool design.

The location of the inner corner of the die is assigned $(O_x, O_y) = (w,w)$. Any given flow line begins at $t = 0$ and at $X_0 = (x_0, y_0, z_0)$ inside the die, i.e. $0 < x_0 < w$, $0 < y_0 < w$. Note that generally when $\alpha < (\pi - \Phi)/2$, the outermost flow line, starting at $x_0 = 0$, does not fill the outer corner. The flow history for any material point traveling through this die is divided into three regions.

In **region 1**, the flow line $\bar{x}(t)$ of a material point traveling at velocity V follows the axes of the channel (vertical in this case) until it intersects the central fan at $t = t_1$ at the first boundary. Time t_1 depends on X_0, α, and V by

$$t_1 = \frac{(y_0 - w) + (w - x_0)\tan\alpha}{V} \tag{11.2}$$

Hence, in region 1, $\bar{x}(t)$ depends on X_0 and t via

$$\bar{x}(t) = [x, y, z] = [x_0, y_0 - Vt, z_0], \qquad 0 \le t \le t_1. \tag{11.3}$$

The deformation gradient F associated with Eq. (11.3) is identity, meaning the material experiences no plastic deformation prior to reaching the central fan at $t = t_1$. Accordingly, the velocity gradient L and the displacement gradient are zero for $t < t_1$.

At $t = t_1$, the material point enters the ideal fan and plastic deformation begins. Within the ideal fan, **region 2**, the flow line $\bar{x}(t)$ is assumed to follow a circular trajectory (see section 3.5, Part I). The tangential velocity along the flow line is reduced to $V^* = RV$, where R is the ratio of the flow velocity in the central fan to the material entering (and exiting) velocity. For incompressible flow $R = \cos\alpha$ (Segal, 2003; Cui et al., 1998) that The material point leaves region 2 at time t_2, when it intersects the second fan boundary, which is

$$t_2 = \frac{1}{V}\left[(y_0 - w) + (w - x_0)\tan\alpha + \frac{\beta_m(w - x_0)}{\cos^2\alpha}\right] \tag{11.4}$$

As time marches forward within region 2, the angular location $\beta(t)$ of the material point from the first boundary (See Figure 11.1) changes according to

$$\beta(t) = \frac{V(t - t_1)\cos^2\alpha}{(w - x_0)} \qquad t_1 \le t \le t_2 \tag{11.5}$$

As the material point travels from OA to OB, $\beta(t)$ increases from $\beta(t_1) = 0$ to $\beta(t_2) = \beta_m$. As plastic deformation does not begin until $t = t_1$, it is convenient to define a shifted time, t',

$$t' = t - t_1 \tag{11.6}$$

Introducing the following time normalization

$$t_n = \frac{(w - x_0)}{V} \tag{11.7}$$

to time t, t_1 and t_2 gives, respectively,

$$T = \frac{t}{t_n}, \quad T' = \frac{t'}{t_n} = \frac{t - t_1}{t_n}, \quad T_2 = \frac{t_2 - t_1}{t_n} = \frac{\beta_m}{\cos^2\alpha} \tag{11.8}$$

so that $\beta(t')$ can be rewritten as

$$\overline{\beta}(T') = \beta(t) = T'\cos^2\alpha \tag{11.9}$$

As a function of $\overline{\beta}(T')$, the flow line for region 2 is

$$\overline{x}(t') = [x, y, z] = [w - (w - x_0)\cos(\alpha + \overline{\beta}(T'))/\cos\alpha,$$
$$w - (w - x_0)\sin(\alpha + \overline{\beta}(T'))/\cos\alpha, z_0], \tag{11.10}$$

From Eq. (11.10) we can calculate $F = \mathrm{d}\overline{x}/\mathrm{d}X_0$ (see Eq. 8.1). Applying normalizations Eqs. (11.8) and (11.9) yields the *normalized* deformation gradient:

$$\overline{F}_{11} = \frac{\cos(\alpha + \overline{\beta})}{\cos\alpha} + \sin(\alpha + \overline{\beta})(\overline{\beta}\sec\alpha + \sin\alpha)$$
$$\overline{F}_{12} = -\cos\alpha\sin(\alpha + \overline{\beta})$$
$$\overline{F}_{21} = \frac{\sin(\alpha + \overline{\beta})}{\cos\alpha} - \cos(\alpha + \overline{\beta})(\overline{\beta}\sec\alpha + \sin\alpha)$$
$$\overline{F}_{22} = \cos\alpha\cos(\alpha + \overline{\beta})$$
$$\overline{F}_{i3} = \overline{F}_{3i} = \delta_{i3} \tag{11.11}$$

where δ_{i3} equals unity if $i = 3$ and 0 otherwise. The deformation gradient $\overline{F}$ is found to only vary with $\overline{\beta}(T')$. One can show that the deformation is incompressible, i.e., $\det\overline{F}(t)=1$, in the interval $0 \leq t' \leq t_2 - t_1$ and for all α since $R = \cos\alpha$. Returning the deformation gradient to real dimensions $F(t')$ (by simply expanding $\overline{\beta}(T')$) using Eqs.(11.2, 11.7-9) finds that $F(t')$ is a function of time t', V, w, x_0, $\beta(t')$, and α.

The region 2 velocity gradient $L = \dot{F}F^{-1}$ and its normalized form $\overline{L}$ are related by

$$\overline{L} = t_n L = \cos^2\alpha \begin{bmatrix} \dfrac{\sin 2(\alpha+\overline{\beta})}{2} & -\cos^2(\alpha+\overline{\beta}) & 0 \\ \sin^2(\alpha+\overline{\beta}) & -\dfrac{\sin 2(\alpha+\overline{\beta})}{2} & 0 \\ 0 & 0 & 0 \end{bmatrix} \tag{11.12}$$

The normalized strain rate tensor $\overline{D}$ and rotation rate tensor or 'spin', $\overline{W}$ can be determined directly from $\overline{L}$ (Eq. 11.12)

$$\overline{L} = t_n L = t_n(D+W) = t_n D + t_n W = \overline{D} + \overline{W} \tag{11.13}$$

$$\overline{D} = \frac{1}{2}\left(\overline{L} + \overline{L}^T\right) = \frac{\cos^2\alpha}{2}\begin{bmatrix} \sin 2(\alpha+\overline{\beta}) & -\cos 2(\alpha+\overline{\beta}) & 0 \\ -\cos 2(\alpha+\overline{\beta}) & -\sin 2(\alpha+\overline{\beta}) & 0 \\ 0 & 0 & 0 \end{bmatrix} \tag{11.14}$$

$$\overline{W} = \frac{1}{2}\left(\overline{L} - \overline{L}^T\right) = \frac{\cos^2\alpha}{2}\begin{bmatrix} 0 & -1 & 0 \\ 1 & 0 & 0 \\ 0 & 0 & 0 \end{bmatrix} \tag{11.15}$$

$\overline{W}$ represents a CCW spin rate which increases as α decreases.

The purpose of defining the deformation and velocity gradients in terms of dimensionless parameters is to remove any unnecessary input parameters from their calculation. The gradient tensors $\overline{F}$ and $\overline{L}$ depend only on β_m, Φ, and vary continuously with $\overline{\beta}(T')$, and only these need to be specified in their calculation. As long as the fan is an ideal one, where β_m is constant and independent of X_0, the final accumulated strain and texture will depend only on β_m and Φ, and likewise, be independent of X_0. Texture evolution is only affected by the relative magnitudes of the components of $\overline{L}$, which only depend on β_m, Φ, and $\overline{\beta}(T')$ and not X_0. Hence the above non-dimensional formulation of fan deformation shows that one need not specify w, $X_0 = (x_0, y_0, z_0)$, and V when calculating texture evolution. The polycrystal model only needs $\overline{L}$ as a function of β_m, Φ, and $\overline{\beta}(T')$.

Determining the actual F and L for a given set of w, $X_0 = (x_0, y_0, z_0)$, and V involves applying Eqs. (11.8) and (11.9) to $\overline{F}$ and $\overline{L}$ in Eqs. (11.11) and (11.12). Having done so, one finds that for fixed β_m and Φ, F and L changes with w, $X_0 = (x_0, y_0, z_0)$, and V, specifically increasing in value from the outer to the inner corner.

When written in the laboratory coordinate system (Figure 8.2), D or $\overline{D}$ Eq (11.14) indicates that there is a compression component in the 2-direction D_{22} which decays and a tensile component in the 1-direction D_{11} which increases as the material travels through the fan in $\overline{\beta}(T')$. There are also non-zero shear components; D_{12} and D_{21}. This type of evolution has been observed in FE calculations (Beyerlein et al., 2005a).

The type of deformation engendered by an ideal fan can still be categorized as simple shearing. When $\pi - \Phi \geq \beta_m > 0$, ideal fan deformation occurs by simple shearing taking place on a series of rotating planes rather than being concentrated on a single plane. The outer boundaries of the fan at $\overline{\beta}(T') = 0$ and $\overline{\beta}(T') = \beta_m$ represent the first and last of these shear planes in the sequence. The velocity gradient $\overline{L}(\overline{\beta})$ in Eq. (11.12) corresponding to the shearing along these boundaries where the flow lines have discontinuities are $\overline{L}(0)$ and $\overline{L}(\beta_m)$ respectively.

The close relation between ideal fan deformation and simple shearing is perhaps expected. First-pass textures of a variety of cubic materials under general processing conditions can usually be related to classical shear components (unless the initial texture is unusually strong and without monoclinic symmetry). When $\beta_m = 0$, the simple shear model is recovered and $\overline{F}$ and $\overline{L}$ reduce to Eqs. (11.6) and (11.7).

For completeness we conclude this section with a description of region 3. As a material point leaves the ideal fan at $t = t_2$, it enters **region 3**, where the flow line follows the axis of the exit channel, the flow is steady, and the velocity returns to V. In region 3 the flow line follows

$$\overline{x}(t) = [w - (w - x_0)\tan\alpha + V(t - t_2),\; w - (w - x_0),\; z_0] \qquad t > t_2, \qquad (11.16)$$

Plastic flow *is* steady in region 3 and the corresponding velocity gradient, denoted as F_{ss}, is solely a function of α and independent of time.

$$F_{ss} = \begin{bmatrix} \Lambda & -1 & 0 \\ 1 & 0 & 0 \\ 0 & 0 & 1 \end{bmatrix} \qquad t \geq t_2 \text{ or } t' \geq t_2 - t_1 \qquad (11.17)$$

where

$$\Lambda = 2\tan\alpha + \frac{\pi - \Phi - 2\alpha}{\cos^2\alpha} \qquad (11.18)$$

Note that in region 3, F_{ss} is incompressible, i.e. $\det F_{ss} = 1$, $L = 0$. Because it is independent of X_0, the deformation is homogeneous across the billet.

11.3. Effect on Texture Evolution

The primary effect of ideal fan deformation ($\beta_m > 0°$) on texture evolution is a deviation in the texture components from the positions expected under ideal simple shearing. In the TD viewpoint (with the TD or 3-axis in Figure 8.2 directed normal to the page), the deviation appears as a rotation or 'tilt' about the TD. For demonstration, Figure 11.2 presents (111) pole figures predicted for an FCC metal after one ECAE pass with various $\Phi = 90°$, 120°, and 135°, and three fan sizes $\beta_m = 18°$ (π/10), 30° (π/6), and 45° (π/4), as well as ideal simple shearing $\beta_m = 0°$.

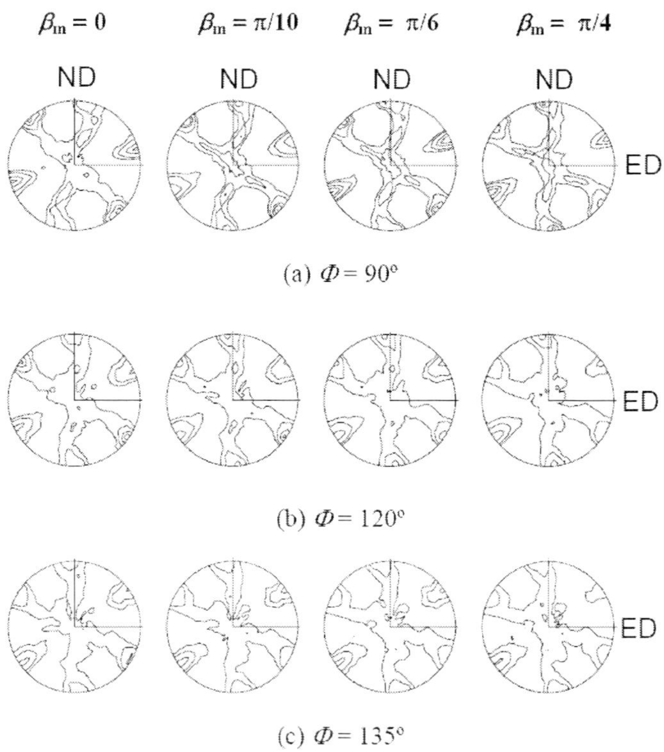

Figure 11.2. (111) pole figures of one-pass textures for increasing fan angles, $\beta_m = 0$, p/10, π/6, and $\pi/4$ for (a) $\Phi = 90°$, (b) 120°, and (c) 135°. Contour levels are 1.0./2.0/4.0/5.6/8.0 mrd.

For all Φ, under $\beta_m = 0°$ (first column in Figure 11.2), a simple shear texture with six maxima develops. Because of the change in shear plane with Φ, textures generated by two die angles Φ_1 and Φ_2 are expected to differ by a rotation of $(\Phi_1 - \Phi_2)/2$ about the 3-axis (Beyerlein and Tóth, 2008). As shown in the first column of Figure 11.2, the $\Phi = 90°$ texture is tilted 15° from the $\Phi = 120°$ texture and 22.5° from the $\Phi = 135°$ texture. Also from this set of ideal ECAE textures (i.e., $\beta_m = 0°$), we can see that the intensity of the textures decreases due to the reduction in total strain with increasing Φ.

As β_m increases above 0°, the texture maintains these features and dependence on Φ. The main effect of β_m is to rotate the entire texture about the 3-axis from the ideal texture of β_m = 0° by an amount that is proportional to β_m. This effect of β_m, however, weakens as Φ increases. Even though β_m may be the same, the strain levels reached within the fan are lower in larger angle Φ dies. This result suggests that deviations from ideal simple shearing deformation are less important as Φ increases, or alternatively, characteristics of the ECAE deformation in larger Φ dies cannot be estimated well from texture measurements due to the lack of sensitivity to β_m. The intensities of the six maxima are also altered with β_m for the same Φ. Any noticeable reduction in texture evolution takes place as β_m increases from $\pi/6$ to $\pi/4$ for all Φ.

As shown in Part I, the central angle β_m depends mainly on contact friction and the tool angle Φ. It can be calculated via Eqs. (3.12), (3.12a), where φ is β_m and η is α in the present notation, or it can be determined experimentally. As shown in Figure 11.2, the deviation is sufficiently sensitive to β_m that β_m can be estimated quite well by comparisons with experimental textures. Under conditions where the deformation characteristics do not change appreciably from pass to pass, then only one texture measurement, e.g., the first-pass texture, is needed to characterize β_m. This estimate for β_m can, in principle, be used in multi-pass simulation without need for more measurements.

Particularly for strain-hardening or rounded corner dies, $\Psi > 0$, it is expected that $\beta_m > 0°$. In such cases, it will be important to use an ideal fan deformation with an appropriate value of β_m rather than ideal simple shearing. For FCC metals, such as Cu and Al, processed by a rounded corner die ($\Psi \sim 37°$), it was demonstrated in (Beyerlein et al., 2005a; Li et al., 2005c) that ideal fan velocity gradient (Eq.(11.7)) with $\beta_m = 18°$ ($\pi/10$) produced similar textures as an FE-derived velocity gradient, both of which were in good agreement with the measurement up to four passes and in better agreement than the ideal simple shear model (Eq.(11.12)). Under non-ideal conditions (such as strain hardening, friction, $\Psi > 0°$), the ideal fan model is a better representation of ECAE deformation than concentrated simple shearing, and just as effective, but more computationally efficient than an FE simulation.

11.4. Effect on Element Distortion

Using the deformation gradient F_{ss} of region 3 (Eq. (11.17)), one can determine the final distortion of any initial element, such as a cube or a sphere, after passing through the die. As described earlier in Chapter 9, although the applied deformation is the same, cubes and spheres do not deform in the same way.

Consider an initial grid element (or cube). In Part I, the evolution of the distorted grid element *abcd* as it passes through different points of the fan was described in Section 3.2.1 and in Figure 3.10. After completely passing through the die, the angle of the distorted parallelogram, θ_r, measured from the axis of the exit channel, depends on β_m and is

$$\theta_r = \tan^{-1}\left(\frac{1}{\Lambda}\right) \tag{11.19a}$$

where

$$\Lambda = 2\tan\alpha + \frac{\beta_m}{\cos^2\alpha} \tag{11.19b}$$

Under the same deformation F_{ss}, a sphere, on the other hand, distorts into an ellipsoid, whose long axis N and short axis S have lengths

$$N^2 = \frac{(\Lambda^2+2)+\Lambda\sqrt{\Lambda^2+4}}{2}$$

$$S^2 = \frac{(\Lambda^2+2)-\Lambda\sqrt{\Lambda^2+4}}{2} \tag{11.20}$$

As a direct result of incompressibility and in-plane deformation, $NS = 1$, leaving the third, medium length axis M, unity for all Λ. Because Λ is a function of β_m only, the aspect ratio, N/S is also a function of β_m only. For instance, when $\Phi = 90°$, N/S ranges from 5.83 for $\beta_m = 0$ to 4.23 for $\beta_m = \pi/2$. A calculation of the eigenvectors of F^TF shows that the long axis of this ellipsoid is inclined at an angle θ_s^* from the axis of the exit channel:

$$\theta_s^* = \tan^{-1}\left(\frac{N^2-(\Lambda^2+1)}{\Lambda}\right)+\Phi-\frac{\pi}{2} \tag{11.21}$$

As shown in Figure 11.3 (Beyerlein and Tomé, 2004) for $\Phi = 90°$, (a) the orientation of the grains $\theta_s = \theta_s^*$. Eq. (11.21), and (b) N/S, Eq. (11.20), are not sensitive to β_m. The same trend applies to all Φ. Therefore, unlike texture components, the characteristics of grain shape are not good indicators for estimating β_m.

11.5. Inhomogeneous Fan

Although a simplification, an ideal fan can reasonably represent the deformation experienced by a large central portion of the billet (Semiatin et al., 2000; Kim, 2002; Kim et al., 2001; Li et al., 2004a). Near the outer surface layers, however, the shape and deformation characteristics of the PDZ deviate from those assumed in the ideal fan model. Common observations are that the fan angle is not constant across the thickness of the sample (Beyerlein et al., 2005a), the fan may be skewed about the intersection plane and closer

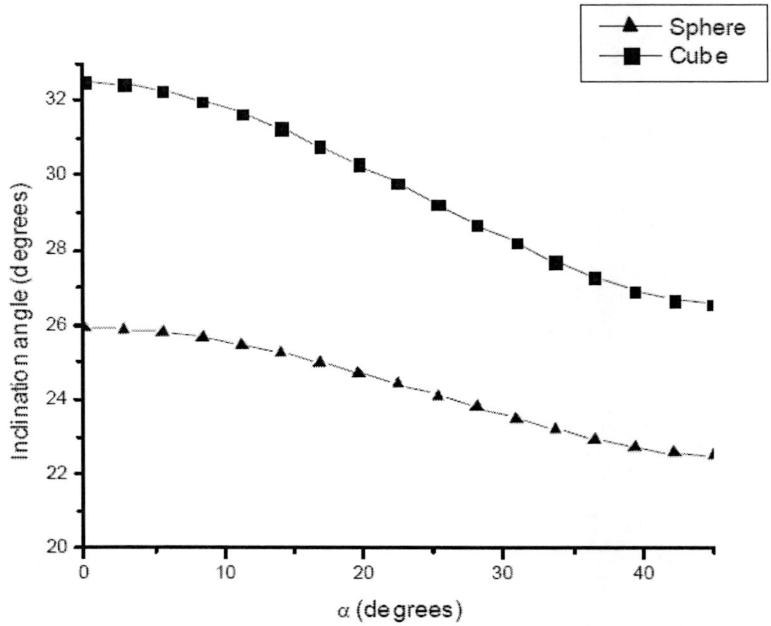

Figure 11.3(a). Changes in inclination angle of a distorted ellipsoid (originally a sphere) θ_s and distorted parallelogram (originally a cube) θ_r versus α (as defined in Fig. 11.1)

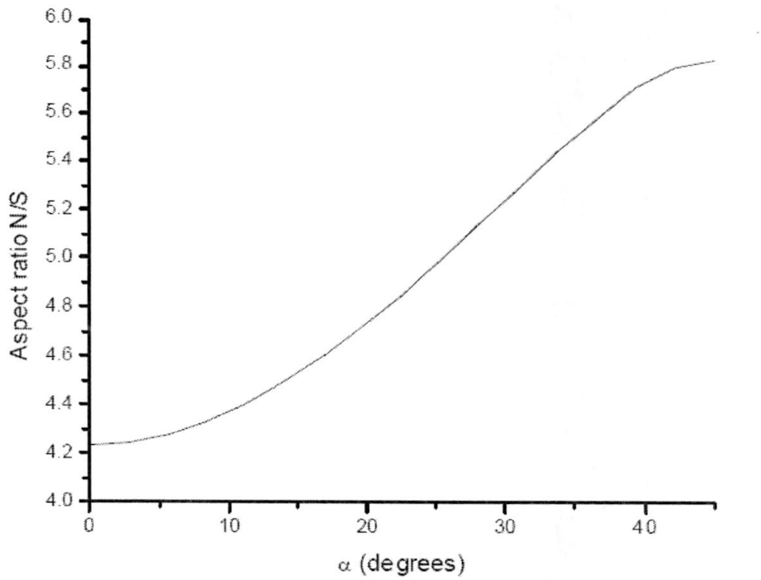

Figure 11.3(b). Changes in aspect ratio *N/S* of the distorted ellipsoid, which was originally a sphere prior to deformation, versus α. Taken from Beyerlein and Tomé, 2004)

to the exit channel (Mahesh et al., 2005; Li et al., 2007), or the fan shape may break down near the outer corner or very close to the inner corner. Many of these features are due to either one or a combination of the following factors: non-uniform friction, high backpressure, certain material behaviors (e.g., strain hardening), corner gap formation and rounded corner dies. Figure 11.4 displays an FE calculation of a strain rate field during ECAE of a strain hardening material, which reveals that the PDZ is skewed (Li et al., 2007).

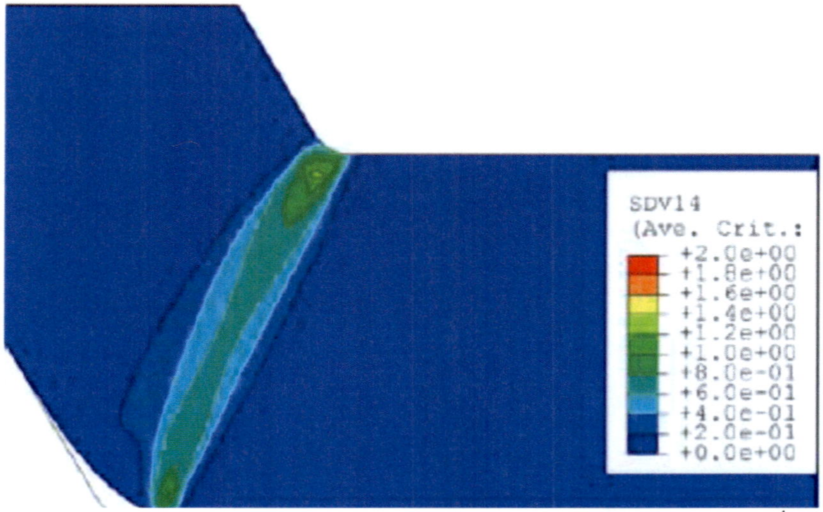

Figure 11.4. Finite element calculated distribution of equivalent plastic strain-rate (in s^{-1}) in a strain hardening material in an ECAE extrusion of the IF-steel using the $\Phi = 120°$ die. The region of high plastic strain rates (plastic deformation zone) is skewed towards the exit channel. (Li et al., 2007)

In this section, we consider the deformation characteristics of an inhomogeneous fan that commonly develops in rounded corner dies or when a corner gap forms between the die and the sample. In these cases, the ideal fan model or, in fact, shear-like deformation, breaks down near the outer corner. Slip-line analyses (Segal, 2003) and FE simulations for rounded corner dies (Kim et al., 2000; Kim 2001; Li et al., 2004b; Dumoulin et al., 2005; Bowen et al., 2000; Kim et al., 2002) show that the PDZ contains two main parts: one, a fan-shaped part covering the central and upper portion of the sample, and two, a broader part of little deformation shearing and mostly rigid body rotation. Deformation in the lower part primarily takes place in two small and localized zones at its entry and exit points (Li et al., 2004a,b). We refer to this model as an inhomogeneous fan because the final texture and accumulated strain depend on the location of the material point.

For the purposes of modeling texture evolution in such cases, we abstract this inhomogeneous fan in two domains and determine the velocity gradients L of each part. As illustrated in Figure 11.5, the upper region is the fan AOB and the lower region is contained in ABCD. The flow line beginning at x_L marks the boundary between the AOB fan region and lower ABCD region. For material points flowing through the fan and along the boundary, $x \geq x_L$, F and L have already been derived and only a value for β_m needs to be provided. For material points flowing through the lower ABCD region, $x < x_L$, the material point traveling from AC to BD experiences localized shear deformation at slip lines AC and BD and rigid body rotation about O in-between. The angle Ψ_{CD} is the angular separation between C and D (Figure 11.5).

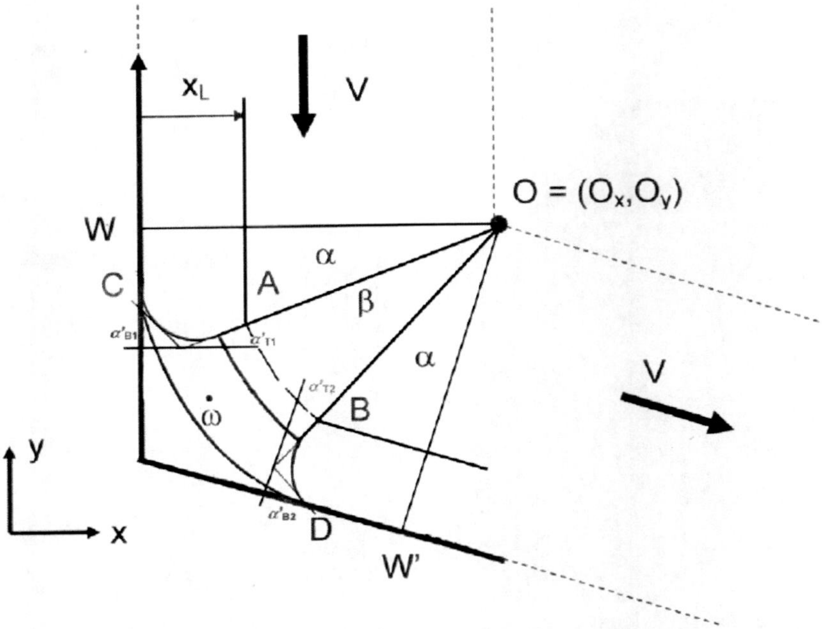

Figure 11.5. Schematic of a two-part plastic deformation zone (PDZ) within a central fan in the upper region and rotating lower region with angular velocity $\dot{\omega}$. The model representations of shear boundary (b) AC and (c) BD defining a' angles are also shown.

The shapes of the curved slip lines AC and BD (Segal, 2003) depends on Φ and Ψ and need to be determined. As a simplification, we describe each slip line by two piecewise segments, one connected to the fan and the other connected to the outer corner (See Figure 11.5). Material points flowing through the former segments of AC and BD are said to be in the transition region and those through the latter, the bottom region. For AC, the transition segment (connecting to the fan) is oriented at α_{T1}' from the horizontal and the bottom segment (connecting to the outer surface) is oriented at ($\pi - \alpha_{B1}$') from the plane inclined parallel to OW_1. Similarly, at BD the transition segment is oriented at α_{T2}' from the vertical and the bottom one is oriented at ($\pi - \alpha_{B2}$') from the vertical. Note that the transition region begins where α'_{T1} and α'_{T2} deviate from α.

A simple shear γ ' and a shear strain rate $\dot{\gamma}$ ' act along each segment. The total shear strain accumulated at each α' boundary is, therefore,

$$\gamma'(\alpha') = \tan(\alpha + \alpha') - \tan\alpha' \tag{11.22}$$

In simulation, one only needs to assign $\dot{\gamma}$ ' and using Eq. (11.22) determine Δt such that in n time steps of Δt, $|\gamma'| = \dot{\gamma}'(n\Delta t)$.

In entering AC, the velocity gradient (in the 1-2-3 coordinate system in Figure 11.2) for material points entering the bottom and transition regions are

$$L_{B1} = \frac{\dot{\gamma}'(\alpha_{B1}')}{2}\begin{bmatrix} -\sin 2\alpha_{B1}' & -2\cos^2\alpha_{B1}' & 0 \\ 2\sin^2\alpha_{B1}' & \sin 2\alpha_{B1}' & 0 \\ 0 & 0 & 0 \end{bmatrix} \quad \text{(entering the bottom region)} \qquad (11.23)$$

and

$$L_{T1} = \frac{\dot{\gamma}'(\alpha_{T1}')}{2}\begin{bmatrix} \sin 2\alpha_{T1}' & -2\cos^2\alpha_{T1}' & 0 \\ 2\sin^2\alpha_{T1}' & -\sin 2\alpha_{T1}' & 0 \\ 0 & 0 & 0 \end{bmatrix} \quad \text{(entering the transition region)} \qquad (11.24)$$

In traveling from AC to BD, the velocity gradient associated with the rigid body rotation $\dot{\omega}$ is

$$L_{RR} = \dot{\omega}\begin{bmatrix} 0 & -1 & 0 \\ 1 & 0 & 0 \\ 0 & 0 & 0 \end{bmatrix} \quad \text{(within the outer corner)} \qquad (11.25)$$

The material experiences no deformation over a time period of $\Psi_{CD}/\dot{\omega}$. Afterwards, it experiences localized shear at BD when it finally exits the lower part. The velocity gradient associated with the bottom and transition regions are

$$L_{B2} = \frac{\dot{\gamma}'(\alpha_{B2}')}{2}\begin{bmatrix} -\sin 2\alpha_{B2}' & -2\sin^2\alpha_{B2}' & 0 \\ 2\cos^2\alpha_{B2}' & \sin 2\alpha_{B2}' & 0 \\ 0 & 0 & 0 \end{bmatrix} \quad \text{(exiting the bottom region)} \qquad (11.26)$$

and

$$L_{T2} = \frac{\dot{\gamma}'(\alpha_{T2}')}{2}\begin{bmatrix} \sin 2\alpha_{T2}' & -2\sin^2\alpha_{T2}' & 0 \\ 2\cos^2\alpha_{T2}' & -\sin 2\alpha_{T2}' & 0 \\ 0 & 0 & 0 \end{bmatrix} \quad \text{(exiting the transition region)} \qquad (11.27)$$

The corresponding accumulated strains for the bottom and transition regions can be determined from integrating the symmetric components of L.

This conceptualization of the deformation of the lower part of an inhomogeneous fan has been tested and proven against the deformation predicted by FE simulations (Bowen et al. 2004; Li et al. 2004b; Beyerlein et al., 2005a). The velocity gradient was also applied as input into VPSC to calculate textures within the transition and bottom regions of the lower part (Beyerlein et al., 2005b; Li et al. 2004a). Figure 11.6 compares the VPSC texture predictions using Eqs. (11.23) and (11.26) with OIM texture measurements taken 400 μm (3% of the sample width) from the center and bottom of a first-pass sample of Cu (Beyerlein et al., 2005a), which was processed using a rounded corner die with $\Psi = 37°$. We observe from the measurement that the bottom and center deformation textures are significantly different suggesting that they experienced different deformation histories. The prediction in Figure 11.6(b) uses $\dot{\omega}$ ~1.6, $\alpha'_{B1} = \alpha'_{B2} = 6^o$ and Ψ_{CD} ~50° and assumes that $\dot{\gamma}$ ' are the same for both slip lines. A low effective strain of 0.19 was calculated for the bottom region. The calculation for texture development in the center used the ideal fan model with $\beta_m = 18^o$ (or $\alpha = 36^o$).

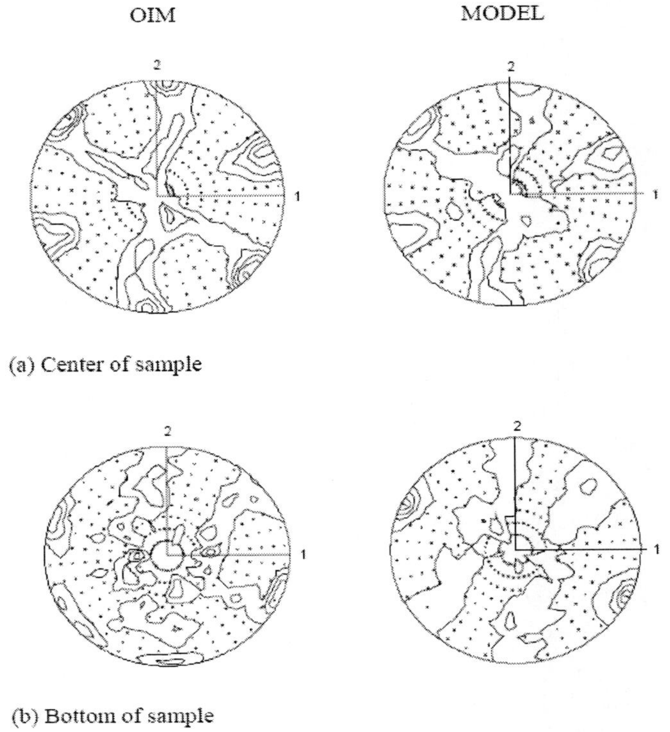

Figure 11.6. (111) pole figures of OIM measured textures after one pass in the central and bottom regions and corresponding VPSC texture predictions (see text for details). The maximum intensities for the central and bottom measured textures are 6.5 and 4.1 mrd. The maximum intensities for the central and bottom predicted texture are 6.9 and 4.8 mrd. Contour lines are 1/2/3/4/6/ 7 mrd.

The agreement supports the idea that material flow in the lower part involves rigid body rotation and a small amount of low shear deformation. Carrying out more calculations with different parameters finds that texture evolution depends mostly on Ψ_{CD} and $\dot{\omega}$ and less so on the details of the deformation at AC and BD. To this inhomogeneous fan model, one may add further refinements, such as varying the simple shear γ ' and shear strain rate $\dot{\gamma}$ ' between AC and BD or between the transition and bottom regions, or increasing the number of segments used to model curved slip lines AC and BD. It is only important to retain the main deformation characteristics respected here.

11.6. Measures of Total Strain

Analytical functions for F and L provide unique and complete descriptions of deformation. Every component of F and L is needed to determine both stretches and rigid body rotations and texture evolution.

For the ideal fan or inhomogeneous fan, F and L evolve as the material point travels through the PDZ. In these cases ($\beta_m > 0$), the deformation involves large strains and continuous rotations of the plane of shearing. Calculating accumulated strain in these cases calls for an integral approach. A time integration over L Eq. (11.12) for an ideal fan leads to the displacement gradient after a single pass:

$$\frac{\partial u}{\partial x}_{pass} = \begin{bmatrix} \frac{1+\sin 2\alpha}{2} & -\left(\frac{\beta_m}{2}+\frac{\sin 2\alpha}{2}+\sin^2\alpha\right) & 0 \\ (1+\beta_m)\tan\alpha+\frac{\beta_m}{2}-\frac{\sin 2\alpha}{2}+\sin^2\alpha & -\left(\frac{1+\sin 2\alpha}{2}\right) & 0 \\ 0 & 0 & 0 \end{bmatrix} \qquad (11.28)$$

Taking the symmetric part of Eq. (11.28) yields the strain tensor.

$$\varepsilon = \begin{bmatrix} \frac{1+\sin 2\alpha}{2} & \frac{(1+\beta_m)\tan\alpha-\sin 2\alpha}{2} & 0 \\ \frac{(1+\beta_m)\tan\alpha-\sin 2\alpha}{2} & -\left(\frac{1+\sin 2\alpha}{2}\right) & 0 \\ 0 & 0 & 0 \end{bmatrix} \qquad (11.29)$$

The skew part of the displacement gradient Eq. (11.28) is the rigid rotation tensor.

$$\omega = \frac{1}{2}\begin{bmatrix} 0 & -\left(\beta_m+2\sin^2\alpha+(1+\beta_m)\tan\alpha\right) & 0 \\ \beta_m+2\sin^2\alpha+(1+\beta_m)\tan\alpha & 0 & 0 \\ 0 & 0 & 0 \end{bmatrix} \qquad (11.30)$$

For $\beta_m = 0$ ($\alpha = (\pi - \Phi)/2$) ε and ω become (Beyerlein and Tomé, 2004)

$$\varepsilon = \cot\left(\frac{\Phi}{2}\right)\begin{bmatrix} \sin\Phi & \cos\Phi & 0 \\ \cos\Phi & -\sin\Phi & 0 \\ 0 & 0 & 0 \end{bmatrix} \qquad \text{(ideal case)} \qquad (11.31)$$

and

$$\omega = \cot\left(\frac{\Phi}{2}\right)\begin{bmatrix} 0 & -1 & 0 \\ 1 & 0 & 0 \\ 0 & 0 & 0 \end{bmatrix} \qquad \text{(ideal case)} \qquad (11.32)$$

For $\Phi = 90°$, the rigid body rotations are counter-clockwise (CCW) 57.4° when $\beta_m = 0°$ and 45° when $\beta_m = 90°$.

From Eq. (11.29) the von Mises equivalent strain for an ideal fan ε_{vM} is found to be:

$$\varepsilon_{vM} = \sqrt{\frac{1}{3}}\sqrt{1 + 2a(1 - b + a) + b^2} \qquad (11.33)$$

where $a = \sin 2\alpha$ and $b = (1 + \beta_m)\tan\alpha$. For $\Phi = 90°$, ε_{vM} increases from 0.5774 from $\beta_m = 90°$ to 1.1547 for $\beta_m = 0°$. The scalar measure Eq. (11.33) is different than the one proposed by Iwahashi et al. (1996):

$$\gamma(\Psi) = 2\cot(\Phi/2 + \Psi/2) + \Psi\csc(\Phi/2 + \Psi/2) \qquad (11.34)$$

where Φ and Ψ are defined in Figure 8.1. Eq. (11.34) is the tangent of the final inclination of a grid element in the exit die. The proof is straightforward: simply replace β_m with Ψ in the expression for the inclination θ_r of a distorted grid element Eq.(11.19). It must be emphasized, however, that generally β_m and Ψ are not equal.

Chapter 12

TEXTURE EVOLUTION IN ROUTE C

12.1. EXPERIMENTAL OBSERVATIONS OF SHEAR TEXTURES

Deformation in cyclic route C is similar to a repetitive large strain shear reversal. In route C the billet is twisted and rotated between passes in such a way that the deformation of the first pass is for the most part reversed in the second pass, that of the third reversed in the fourth pass, and so forth. As would be expected, for metals in which the starting grain shape was equiaxed, the reversal in the second pass (or even-numbered ones) more or less returns the grain shape to its original equiaxed state. However, recovery of the initial texture after the second or even-numbered passes of route C is rarely observed. In most cases, this second-pass texture is not weak at all, but is a well-developed shear texture, much like the first pass. This result has been found in a variety of materials, Cu (Figure 10.1), Al and Al alloys, Fe, IF-steel (Figure 10.2), Ti, Be, and Zr (Isaenkova et al., 2005; Perlovich et al., 2006a,b; Ferrasse et al., 2004, Gibbs, 1998; Tidu et al., 2000; McNelley et al., 2004; Baik et al., 2003; Li et al., 2005a; Li et al., 2006b; Li et al., 2006c,2007). These texture observations in route C samples are not unlike the unaltered shear textures measured after reverse torsion tests in Fe, brass, and Cu, see (Backofen, 1950,and references therein).

Predicting texture evolution in route C has proven to be quite challenging. A straightforward application of most polycrystalline models for the reverse path change during two consecutive passes of route C would not predict this widespread observation in texture evolution. Taylor predicts perfect recovery of the initial texture after even-numbered passes. VPSC, with or without isotropic hardening, predicts near return of the initial texture (See Figure 10.3) (Beyerlein et al., 2003).

12.2. POSSIBLE EXPLANATIONS FOR PERSISTENT SHEAR TEXTURES

There are many possible reasons why the grain reorientation would not follow the reverse path back to their original position: grain-grain interactions, texture heterogeneities, substructure evolution, dislocation storage, and an imperfect reversal at the macroscale. These behaviors cover many lengthscales, from the micron-scale to the macro-scale, indicating that this irreversibility may be the result of a combination of factors. In this section, modeling efforts, which also involve treatments at these lengthscales, are reviewed.

The first explanation listed above is grain-grain interactions. Interactions between neighboring grains can prevent individual grains from exactly retracing their deformation path in reverse shearing. However, this is a weak constraint, especially against the severe strain deformations imposed in each ECAE pass. Polycrystal models that make some account for grain-grain interactions cannot reproduce the second-pass measurements (Beyerlein et al., 2003; Li et al., 2005a). For these reasons, it cannot be responsible for the shear texture components evident in second-pass (and even-numbered) route C textures.

Another reason is concerned with texture measurements made over large areas of the ECAE sample that encompass substantial texture gradients. Li et al. (2005a) demonstrated through modeling how texture inhomogeneities across the billet can alone contribute to shear-like textures after two passes of route C. In their study, the outer corner of the die was rounded, which is known to lead to heterogeneous deformation across the thickness of the billet. The bottom 10-20% of the sample experiences little shear straining, unlike the rest of the same. In the second pass of route C, the bottom and top layers switch positions; the low-deformation bottom region left after the first pass, experiences severe shearing in the second pass and conversely, the previous top layer of the first pass experiences little deformation in the second. The central portion of the sample would more or less experience a reverse deformation in going from the first to the second pass. As a result, after the second pass, the sample is left with a shear texture within the top and bottom layers, and a presumably recovered initial texture in the central portion of the sample. If this speculation is correct, then a bulk texture measurement, such as the neutron diffraction technique used in their work, would produce a shear texture as it would include the shear areas in the outer layers. To test this idea, the heterogeneities in deformation and texture were simulated and the resulting textures averaged over the entire thickness of the model sample. The average two-pass texture was indeed a shear-like texture. Without a doubt heterogeneities in texture development across the sample would bias a bulk texture measurement.

Later, OIM measurements were made on a relatively small area (in which the scan height was 1/20 of the thickness) in the center of the same two-pass Cu sample and they displayed a relatively weak shear texture (Li et al., 2006b). Therefore, even the material in the dead center of the billet, which is more likely to experience homogeneous deformation and a reversal over the two-pass sequence, did not recover the initial texture and formed a shear texture instead. These highly resolved measurements proved that in no part of the sample was the initial texture recovered.

This last result motivates examination of lower length-scale phenomena, such as substructure evolution, single crystal (anisotropic) hardening, and grain refinement, to explain the formation of well-developed shear textures after a reversal.

The role of substructure evolution in shear texture development following a reversal was, in fact, addressed in early reverse torsion studies (Backofen, 1950; Rollett et al., 1988). On the one hand, the argument seems plausible. In the absence of dynamic recrystallization, after large strain deformation, most metals will form organized dislocation substructures. Larger strain levels tend to lead to a rather well-defined substructure, which fragment the grains into smaller misoriented regions (lamellae, bands, subgrains), and to larger misorientations across these regions. The larger the misorientation, the more likely these regions will use different slip systems in reverse deformation than the original grain used in forward deformation.

Consistent with this argument, it was found that the higher the strain level used in the forward and reverse strokes, the more well-developed the shear texture after the reversal. On the other hand, it was argued by Backofen (1950) that the same slip systems would develop the same textures, whether in forward or reverse. Therefore the shear textures after a reversal could not have been produced solely by different slip systems operating in new subgrain domains.

For another reason, it is possible that substructure evolution did not contribute to the shear textures in even-pass route C. In these early reverse torsion studies, the strain levels (per path) at which unaltered shear textures were reported were at least 3.0, much higher than a true strain of 1, typically achieved per ECAE pass. While substructure evolution does take place in a single ECAE pass, substructures will not be as well-developed as those generated in these torsion samples after 3-6 times more strain. No shear textures were found in the torsion samples for lower strain levels, which are comparable to those achieved in a single ECAE pass. In ECAE, the increase in misorientation across substructure boundaries observed in Cu was small, from ~3° after the first pass to 6-8° after the second (Li et al., 2006b) and fourth passes (McNelley et al., 2004). Higher misorientation angles have been reported, but because the variation in misorientation angle across these subgrain boundaries are found to be wide (Li et al., 2006b; Sun et al., 2004), some higher angle boundaries certainly will be present as well. Nonetheless, the average misorientation is low. While these amounts of misorientations may be sufficient to diffuse the texture about the main components and reduce texture intensities, they cannot be fully responsible for well-developed shear texture components after a reversal.

Some recent modeling developments support the hypothesis that misorientations developed after one or a few passes cannot be fully responsible for well-developed shear texture components after two and four passes of route C. Models for substructure evolution that can adequately test this notion involve development of regions (lamellae, bands, subgrains) misoriented from the parent grain, as was done in (Mahesh et al., 2005; Mahesh and Tomé, 2004; Enikeev et al. 2007; Nazarov et al., 2006; Sivakumar and Ortiz, 2004). Implementation of these subgrain models into either a Taylor (Mahesh et al., 2005) or VPSC (Enikeev et al., 2007) formulation were successful in predicting second-pass route C textures that were different than the starting texture and that contained some shear components. However their intensity was still much weaker than the measurement. The main role of substructure in these cases was to diffuse and spread the texture. It is likely that the strains induced per pass in the first two passes of route C are not large enough to increase the substructure misorientations to a degree that would significantly alter bulk texture evolution.

Another phenomenon that needs to be considered is defect accumulation and material hardening that accompanies plastic straining. Stored dislocations, dislocation debris, and organized dislocation patterns become more stable with deformation, such that they cannot be easily untangled or removed by reverse loading, at least not at the same rate as they were formed. Such changes in hardening behavior could prevent the grain from rotating back to its original state when the applied strain is reversed. To test this idea, some VPSC simulations of the first two passes of route C were carried out where the material hardening of each FCC grain was made to follow an extended Voce law Eq. (6.30), described previously in Chapter 6. The resulting textures are shown in Figure 12.1(b-c). In applying this law to an FCC metal, the critical resolved shear stress τ_V for each of the {111}<110> slip systems is made to evolve with the total accumulated strain in the grain Γ, Eq. (6.30). The associated parameters are the same for each slip system, τ_0, τ_1, θ_0, and θ_1 = 20, 80, 300, and 10, respectively. The hardening is isotropic, meaning that all slip systems harden at the same rate, whether they are active or not. In addition to material hardening, geometric hardening is also present, as is true with any polycrystal model. The total hardening (material plus geometric) will evolve differently for each grain according to its crystallographic orientation. We have also included a scheme for grain-grain interactions (Tomé et al., 2002) in which randomly paired grains were made to have the same rotation. Ideal simple shearing (Φ = 90°) is used for both ECAE extrusions as illustrated in Figure 12.1(a). Figure 12.1(b-c) shows the textures resulting from this calculation. After the second pass (Figure 12.1(c)), the initially random texture (1000 orientations) is nearly recovered, in spite of irreversible hardening at the single crystal level, differences in hardening (material plus geometric) among grains of different orientation, and grain-grain interactions.

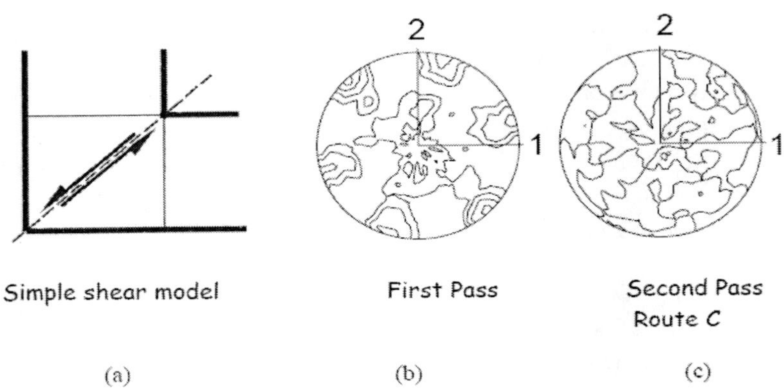

Figure 12.1. VPSC simulated texture evolution in an fcc metal assuming ideal uniform deformation (far left) in both passes of route C for a 90° die. Contour levels are 1/2/3/4 mrd.

For many reasons, grain hardening behavior can be anisotropic; that is, each slip system in the grain hardens at a different rate depending on its relative activity. Metals that deform by several slip modes or by slip and twinning (Beyerlein et al., 2007b; Yapici et al., 2008), are plastically anisotropic. Strain-induced dislocation patterning or substructure evolution can also cause grain hardening to change from isotropic to anisotropic (Beyerlein and Tomé, 2007). For Cu, which deforms by only the {111}<110> slip mode, Li et al. (2006b) tested the effect of anisotropic hardening by affecting the $h^{\alpha\alpha'}$ hardening parameters coupling the four independent slip planes (see ahead Eq. (14.3) and Chapter 14 for discussion). When the active planes were made to harden faster (Beyerlein and Tomé, 2006) or slower than the inactive (latent) ones, shear texture development in the second pass was promoted, unlike the case in which isotropic hardening was used (Figure 12.1(c)). However, it was not as strong as the measurement. Although anisotropic hardening can inhibit perfectly reversed deformation, the shear textures it produces are not as well-developed as those measured in the second (and fourth) pass of route C.

12.3. EFFECT OF IMPERFECT REVERSALS

Up to this point, many micro-scale and meso-scale explanations for shear textures in route C have been reviewed. Although they should not be disregarded, the theoretical textures resulting from these effects alone do not contain as many shear texture components and with the same intensity as the measurements. In support, analogous reverse torsion studies suggest that with a true strain per pass of one or less, shear textures should not develop after the reversal.

Thus far, in these models, single crystal and polycrystalline hardening behavior have been decoupled from the macroscopic calculation of deformation fields imposed by the ECAE die. In reality, microstructural evolution will impart a change in flow stress and work hardening, which, in turn, changes the nature of plastic flow throughout the extrusion. ECAE deformation characteristics are particularly sensitive to strain hardening behavior of the metal. As will be detailed below, this coupling between ECAE deformation and material strain hardening leads us to the final explanation for reverse shear textures: irreversibility in the macroscale deformation characteristics.

In polycrystal modeling, the macroscale deformation is represented by velocity gradients. Typically these velocity gradients are assumed not to change from pass to pass. This assumption was applied in the previous calculations in Figure 12.1. Instead, if macroscale deformation was to reflect changes in plastic behavior as the material is severely deformed in the ECAE process, then it is likely it would not be the same from pass to pass. With this in mind, consider the following, highly probable situation: in the first pass, the metal strain-hardens through stages II to IV, whereas throughout the second pass, its strain-hardening rate is mild and constant (stage IV) or has saturated. Therefore, first-pass deformation characteristics will be different than those of the second pass. From this perspective although the second-pass deformation of route C will still reverse the deformation of the first pass, the two-pass sequence will not be a pure reversal. As a result of this 'imperfect reversal', deformation shear textures could develop after the second pass.

In ECAE, the effect of material work-hardening on the deformation characteristics cannot be separated from the die geometry, such as the outer corner angle □. For sharp corner dies (□ = 0), when the material is strain hardening, a corner gap forms and a fan-like plastic deformation zone (PDZ) is generated as illustrated in Figure 12.2(a). When the material is perfectly-plastic, however, it fills the die, and the deformation is close to the ideal simple shearing depicted in Figure 12.2(b). For a rounded corner die (□ > 0), it is less likely that a corner gap forms and a broad PDZ develops regardless of material behavior. However when □ > 0, the deformation field within the PDZ for the strain hardening material is skewed slightly CCW (Figure 12.3(a)), whereas for the perfectly-plastic material the PDZ field is symmetric (Figure 12.3(b)). This skew effect has been predicted by finite element simulations as shown in Figure 11.4 (Li et al., 2006b; Li et al., 2007) for a strain hardening material.

To investigate these effects on texture evolution, VPSC calculations are performed using the analytical ideal fan model of Eq. (11.12) for the velocity gradients of a PDZ with size $□_m$ = □ – □□□ – 2□, that has developed in a □ = □□□ die. We first consider the deformation sequence for a sharp corner die as illustrated in Figure 12.2 (a-b) (□ = 0). In the first pass, the deformation is a symmetric PDZ ($□_m$ = 18° in Eq. (11.12)) and in the second pass, it is ideal simple shearing. (The effect of the route C 180° rotation between passes is modeled but not shown.) For the same FCC material as in Figure 12.1, the textures are calculated for the first two passes of route C and shown in Figure 12.2 (c-d). Development of a second-pass shear texture under this imperfect reversal is evident in Figure 12.2(d), which is much stronger than the texture developed in Figure 12.1(c) under a perfect reversal. The shear texture in Figure 12.2(d) is the result of the deformation imposed during the second pass. It is not a retained-shear texture from the first pass.

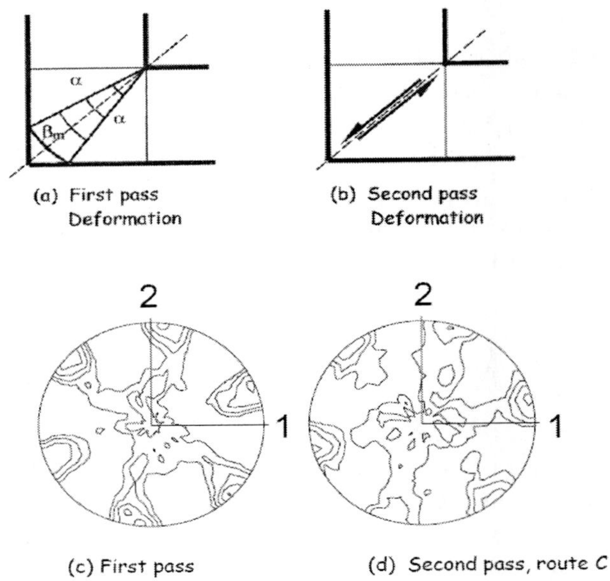

Figure 12.2. Hypothetical sequence of deformation in two passes of ECAE for a 90° die with a sharp outer corner angle. (a) In the first pass, deformation spreads out over a fan-shaped zone due to strain hardening and formation of a corner gap. (b) Second-pass deformation is close to ideal simple shearing because work hardening has saturated. (c) The corresponding VPSC simulated texture evolution in an fcc metal after the first pass and (d) second pass of route C. Contour levels are 1/2/3/4 mrd.

Beyerlein et al. (2008) applied the same deformation sequence in Figure 12.2(a-b) to explain the well-developed shear textures in beryllium (Be) after two passes of route C, processed using a sharp-corner $\Phi = 90°$ die. At the processing temperature of 425° C, Be strain hardens within the first 25% of straining and hence Figure 12.2(a-b) was a likely deformation scenario. Assuming ideal simple shearing in both passes, however, produced a nearly recovered texture after the second pass. Near return to the initially random texture occurred in spite of anisotropic hardening and multiple slip mode deformation (basal, prismatic, and pyramidal slip) used for Be.

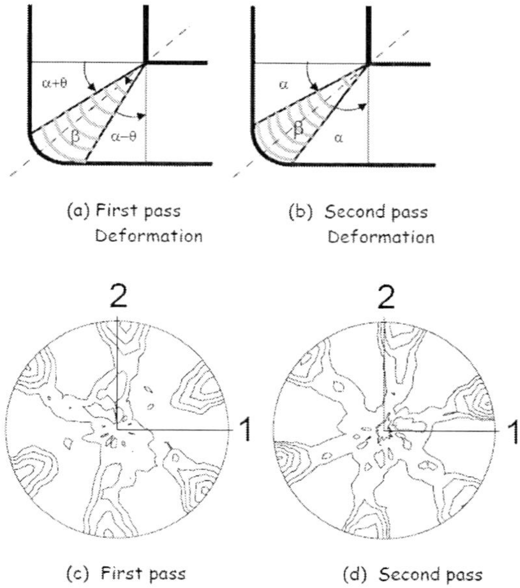

Figure 12.3. Hypothetical sequence of deformation in two passes of ECAE for a 90° die with a rounded outer corner angle. (a) In the first pass, the deformation spreads out over a fan-shaped zone which is skewed counter-clockwise by an angle θ due to strain hardening. (b) Second-pass deformation is close to a symmetric fan because work hardening in the material has saturated and the material fills the rounded outer corner. Subfigures (c)-(d) show the corresponding texture evolution after the (c) first pass and (d) second pass. Contour levels are 1/2/3/4 mrd.

For the model FCC metal, Figure 12.3 repeats the calculation of Figure 12.2 but for a rounded corner die ($\Psi > 0$). In this case, the first-pass deformation is an asymmetric PDZ skewed $\theta = \pi/30$ ($\beta_m = 18°$) due to strain hardening (Figure 12.3(a)) and the second-pass deformation is a symmetric PDZ due to perfect-plastic behavior (Figure 12.3(b)). As in the case of the sharp corner die (Figure 12.2(d)), a deformation shear-texture is generated after the second pass. Mahesh et al. (2005) demonstrated that with this skew effect (□ = □/50), VPSC predictions agree well with the corresponding OIM measurements made in the center of a two-pass route C Cu sample (Li et al., 2006b) processed using a rounded corner die (□ = 37°).

We have reviewed many possible causes for observations of shear textures in even-numbered passes of route C and some of the modeling efforts in the literature. All explanations are valid and can affect texture evolution to varying degrees. Notably, considering simply the differences in macroscale deformation between the first and second pass due to changes in strain hardening behavior is sufficient for capturing the completeness and intensity of shear textures observed in the measurement. In other words, while route C, for the most part, imposes a reverse deformation sequence, it is not a perfect reversal, unless the deformation characteristics from pass to pass are the same. Even the slightest imperfection can lead to a shear texture after the reversal sequence. The imperfect reversals reflect changes in the microstructure, substructure, and single crystal hardening behavior from one to two passes.

The largest difference in deformation characteristics will likely occur between the first and second passes because differences between the second and subsequent passes are likely to be negligible as material hardening nearly saturates at large strains. The deformation between even-number passes greater than two will, therefore, be nearly perfect reversals. When the second-pass texture contains shear components, the fourth-pass texture will recover the second-pass shear texture under a perfectly reversible deformation and so on for the higher *N*. Consequently, the textures after more passes will still be shear textures.

Chapter 13

DEFORMATION TWINNING AND SPD

13.1. INTRODUCTION

Deformation twinning is a plastic deformation mechanism which takes place on a crystallographic plane K and provides shear deformation S in the twin direction η. Twin domains are usually bounded within a crystal but have been observed on occasion to cross grain boundaries. Several features differentiate twinning from slip: twinning is directional (can only shear in one sense), it involves a crystallographic transformation assisted by twinning dislocations, and within the twin domain the crystal lattice is reoriented 180° about twin direction η. For FCC metals, K, η = {111}<112> and for BCC, K, η = {112}<111>. A {111}<112> twin, for instance, reorients the lattice 60° with respect to the original matrix. Compared to cubic metals, twinning is much more prevalent in HCP metals. For an HCP structure, there are seven twin types, the commonly observed ones are $\{10\bar{1}2\}<\bar{1}011>$, $\{11\bar{2}2\}<11\bar{2}\bar{3}>$, and $\{101\bar{1}\}<10\bar{1}\bar{2}>$ (Partridge, 1967; Yoo, 1969; Mendelson, 1969,1970). Which twin mode or modes are active vary substantially from one HCP material to another (Partridge, 1967; Christian and Majahan, 1995). Twin-reorientation depends on twin type and the c/a ratio of the HCP metal. In Mg (c/a = 1.63), for instance, the common $\{10\bar{1}2\}<\bar{1}011>$ twin is activated by a tensile stress along the crystal c-axis and reorients the lattice by 86.6° around $\{10\bar{1}2\}$.

Because twinning shears are based on a crystallographic transformation, there is a direct relation between the twin shear S and the fraction of transformed volume in the crystal:

$$V = \frac{\Delta v^{tw}}{v^{grain}} = \frac{\Delta\gamma^{tw}}{S} \tag{13.1}$$

where V will be used hereinafter to denote the ratio of twin volume fraction Δv^{tw} to the grain volume v^{grain}. The quantity $\Delta\gamma^{tw}$ is the shear strain contributed by the twin. For the {111}<112> FCC twin, S = 0.707, while for the $\{10\bar{1}2\}<\bar{1}011>$ tensile twin in Mg S = 0.146. Clearly, a much larger twin fraction of an HCP grain is required to accommodate the same amount of shear than that in an FCC grain.

In most cases, twinning does not operate in isolation, but in combination with dislocation slip to accommodate the applied deformation. The processes involved in the nucleation and growth of twins and how slip is related to them are not well understood. Most of what is known is based on experimental observations. Some of the conditions found to promote twinning are low temperatures, high strain rates, crystallographic orientation, low stacking fault energies (for FCC metals), large grain sizes (in coarse grained metals), low symmetry crystal structures (such as HCP metals), and high purity. This list is by no means comprehensive and there are many exceptions.

Deformation twins are morphologically quite complex. They can appear as straight-edged lamellae or lenticular, oval domains. They can remain thin or thicken with strain or even overtake the entire grain. Secondary twins can form within primary twins or two intersecting sets of twins can form within the same grain. The misorientation imparted by a twin can be undone by the activation of a secondary twin that reorients the domain back to that of the matrix, i.e., detwinning or reverse twinning (Proust et al., 2008a,b; Brown et al., 2007). For comprehensive reviews on deformation twinning, see (Partridge, 1967; Christian and Majahan, 1995; Yoo, 1969).

Deformation twinning plays an important role in the plastic behavior, crystallographic texture, grain refinement, and fracture of metals. In addition to providing another means of shear deformation, twinning can also alter the choice of slip systems in a crystal in two ways: 1) twin boundaries may act as barriers to slip promoting slip systems parallel and hindering systems non-parallel to the twin boundary and 2) the reoriented twin domains may be favorable for activation of slip systems that were not favorable in the parent.

Most SPD studies focus on cubic metals that deform primarily by slip and less so on metals that deform by multiple slip modes or by both slip and twinning. In FCC SPD metals, twinning has been observed at a small scale, usually in nano-sized grains (Liao et al., 2004; Huang et al., 2006) or close to regions of concentrated strain, such as shear bands (Huang et al., 2006; Manping et al., 2007), or low stacking fault metals (Beyerlein et al., 2007a; Yapici et al., 2004; Huang et al., 2007; Beyerlein et al., 2009b). Observations of twinning in HCP metals in SPD have been reported in commercially pure Ti and Ti-6Al-4V (Shin et al., 2003; Him et al., 2002; Yapici et al., 2006) processed at high temperature (350 °C and above), Zr (Yapici et al., 2008) processed at room temperature, and Mg ZK60 (Lapovok et al., 2005) processed at intermediate (warm ECAE) temperatures (200-300 °C).

This chapter begins with a description of a model for twin reorientation, which is later incorporated in the VPSC polycrystal model. Next, two examples will be considered in which twinning was an important mechanism in determining plastic behavior and texture evolution during ECAE: silver (Ag) which deforms by $\{111\}<110>$ slip and $\{111\}<112>$ twinning, and zirconium (Zr), in which basal slip $\{0001\}\langle 1\bar{2}10\rangle$, prismatic slip $\{1\bar{1}00\}\langle 11\bar{2}0\rangle$, first-order pyramidal $<c+a>$ slip $\{10\bar{1}1\}\langle 2\bar{1}\bar{1}3\rangle$, and $\{10\bar{1}2\}<\bar{1}011>$ twinning will be made available.

13.2. Predominant Twin Reorientation (PTR) Model

Incorporating twinning in polycrystal models requires one to account for the crystallographic reorientation associated with the twin domains. The predominant twin reorientation (PTR) scheme for representing the reorientation by twinning (VanHoutte, 1978; Tomé et al., 1991) fully reorients the grain according to its predominant twin system (PTS) when the volume fraction associated with the activity of this twin exceeds a threshold value V^{thres}. The threshold volume is defined as:

$$V^{thres} = A_0 + A_1 \frac{V^{reor}}{V^{shear}} \tag{13.2}$$

where V^{shear} is the volume fraction of twins in the aggregate based on the contribution of shearing on all twin systems to the accumulated strain, V^{reor} is the volume fraction associated with those grains that are fully reoriented, and A_0 is an 'incubation' volume that the PTS must accumulate in the grain before the grain is considered for reorientation. Initially $V^{reor} = 0$, and $V^{thres} = A_0$. As twinning accumulates V^{reor} approaches V^{shear} and V^{thres} approaches $A_0 + A_1$, a 'saturation' value for V^{thres}.

Within this model, 're-twinning' is allowed, which occurs if the PTS in the grain changes, either because it has been fully reoriented by twinning or because of a strain path change. In re-twinning, the modeling scheme described above is repeated. If the new PTS reaches the threshold value, then the grain is reoriented to the orientation of the new PTS.

In the model, grains either retain their original matrix orientation or fully adopt the orientation of its predominant twin, but in either case they remain homogeneous. Thus, over the polycrystal the PTR model homogenizes the actual twin-matrix microstructure as in a rule-of-mixtures approach and local twin morphologies and grain-grain interaction effects are not included.

It may be necessary to also account for the barrier effect that twin boundaries pose on dislocation or other twin propagation, a feature that cannot be modeled with the PTR scheme. For this purpose, we mention that a more sophisticated 'composite grain' approach has been recently developed, in which a grain is micromechanically treated as a laminated grain with alternating matrix and twin domains, respecting continuity of stress and strain rate components across the matrix-twin boundary (Tomé and Kaschner, 2005; Proust et al., 2007; Proust et al., 2008a; Beyerlein et al. 2009a). The model of a laminated grain, however, is unlikely to hold at large strains, where the grains tend to fragment. As a consequence, the more empirical PTR scheme is likely to provide a better representation of twin reorientation when simulating the ECAE process. The composite grain approach is not used in the simulations presented in this Chapter.

13.3. Example: Silver

Ag has a very low SFE (19 $\pm$ 3 erg/cm^2 or mJ/m^2) (Tadmor and Bernstein, 2004) and twins readily. Studies have shown that twinning occurs in Ag under most room temperature deformation conditions (e.g. rolling, wire-drawing, torsion-testing, indention testing, uniaxial straining) and for most specimen geometries (such as single crystals, thin films, and polycrystalline aggregates regardless of grain size) (Winter, 1974; Matthews 1970; Ma and Clarke, 1995).

Deformation twinning was found to have a major impact on texture development of Ag processed by ECAE (Suwas et al., 2003; Beyerlein et al., 2007a). To characterize ECAE textures, we make use of the so-called ideal ECAE components (Li et al., 2005b) discussed in Chapter 10. These are simply the classical ideal shear components (Canova et al., 1984) shifted in the following way to account for the orientation of the shear plane in the ECAE die.

$$g_{\Phi} = (\varphi_1 + \Phi/2, \phi, \varphi_2)\,. \qquad (13.3)$$

In the above, an ideal orientation $g = (\varphi_1, \phi, \varphi_2)$ defined by the Euler angles is expressed in the die reference system x-y-z, see Figure 9.1. As shown, the transformation to the corresponding ideal ECAE orientation g_{Φ} simply involves an additional $\Phi/2$ rotation to φ_1. As examples, tables 10.1 and 10.2 list these ideal ECAE orientations for $\Phi = 90°$ in FCC and BCC metals, respectively. For more discussion on single and multi-pass ECAE texture characterization, see Chapter 10.

The measured ODFs ($\phi_2 = 0°$ sections only) and (111) pole figures of Ag (Beyerlein et al., 2007a) after the first three passes of route A are shown in figures 13.1 and Figure 13.2(a). It can easily be seen in the ODFs that 1) in the first pass, the $B_{\Phi}/\overline{B}_{\Phi}$ components are the strongest and $A_{1\Phi}$, the weakest, and 2) in the next two passes, the $B_{\Phi}/\overline{B}_{\Phi}$ component strengthens, the $A_{2\Phi}$ component maintains its strength, and the $A_{1\Phi}$ and C_{Φ} components weaken.

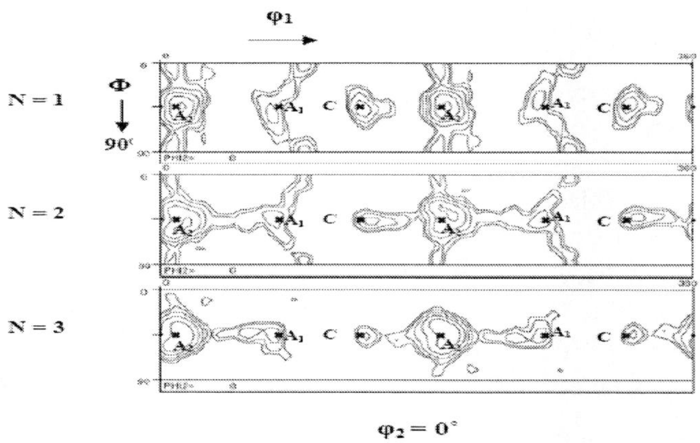

Figure 13.1. Measured $\varphi_1 = 0°$ section of the ODF of silver after one to three passes of route A. Contour levels are 0.8/1/1.4/2/2.8/4/5.6/8 mrd.

In contrast, ECAE textures for processed FCC metals with higher stacking fault energies (e.g., Al, Cu, Ni) that deform by slip only are different from those in figures 13.1 and 13.2. The first-pass textures of such metals contain nearly all of the ideal components (Li et al., 2005b; Tóth et al., 2004) with the C_{Φ} component being the strongest and $A_{2\Phi}$ component being the weakest. In subsequent passes of route A (Li et al., 2006a), C_Φ weakens, $B_\Phi/\overline{B}_\Phi$ strengthens, and $A_{1\Phi}$ and $A_{2\Phi}$ do not vary appreciably.

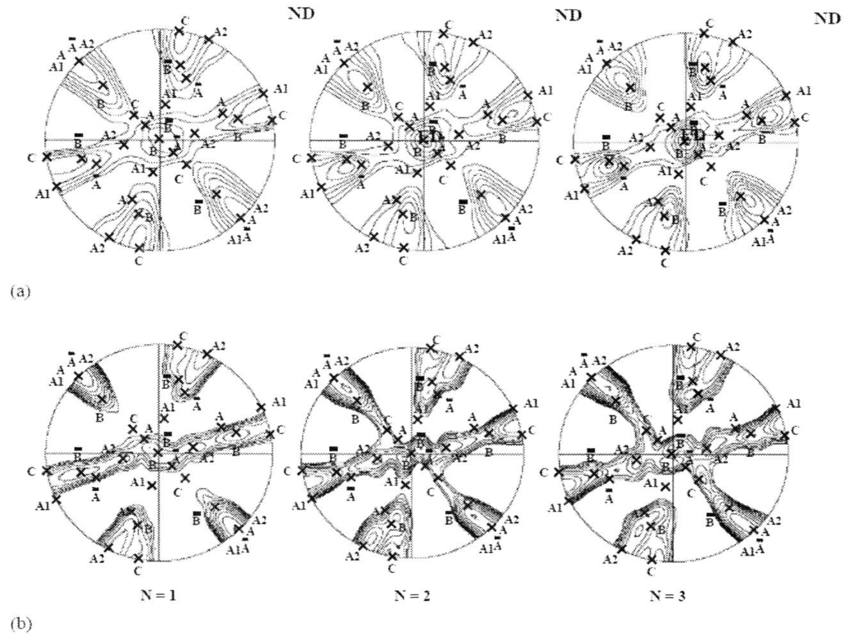

Figure 13.2. (a) Measured and (b) simulated (111) pole figures of silver after one to three passes of route A. Contour levels are 0.8/1/1.3/1.6/2/2.5/3.2/4/5/6.4 mrd.

The ECAE Ag texture results are not unlike the shear textures that develop in other low stacking fault FCC metals under large strain torsion. In Ag, brass, and a Ni-60% Co alloy (Stout et al., 1988; Hughes and Wenk, 1988), for instance, the C component weakens with straining and can even disappear, whereas in Al and Cu, the C component weakens more slowly with straining. Twinning was observed in the Ni-60% Co alloy, but was either unknown or not mentioned in the other studies.

In the ECAE processed Ag samples, twinning was clearly evident. EBSD measurements (Beyerlein et al., 2007a) find that twin fractions and the number of twin boundaries increased with pass number, suggesting that twinning occurred in each pass. The twin boundaries maintained their 60° misorientation in spite of the severe deformation. After three passes, the grain size reduced from 30-40 μm to an average of 600 nm (as low as 200 nm) and throughout, the average grain aspect ratio remained below 2. Lenticular twins were found in both EBSD and TEM images. In TEM, some fine straight ones were also found, as well as suggestions of slip activity within the twins. Twins were found in grains with an A_1 orientation.

As another way to determine which grain orientations twin as well as to ascertain if and why twinning occurs in each pass, VPSC simulations using the PTR model (section 13.2) were performed (Beyerlein et al., 2007a). In ECAE, these simulations assumed simple shearing along the intersection plane of the $\Phi = 90°$ die and the multi-pass ECAE route A process was modeled as described in Chapter 9. To represent the starting Ag metal, the initial polycrystal consisted of spherical grains and 1000 random orientations. Because the aspect ratio of the grains remained at 2 or below, the grain splitting scheme (Chapter 7) was invoked. During the deformation history re-twinning was allowed as many times as required by the reorientation condition. The shear rates of both slip and twinning were calculated using Eq. (6.10), where the critical resistance was governed by the extended Voce law Eq. (6.30). For twinning, this critical resistance represents the resistance to twin propagation and is not intended to model twin nucleation. Nucleation events are believed to be governed by local processes and stress concentrations, and the current model assumes that the required twin nuclei already exist in the material when the threshold for twin propagation is achieved. Table 13.1 lists the Voce parameters associated with slip and twinning. As shown, the critical resistances of these two modes evolve with straining, but in a different manner. They are assigned such that the resistance to twin propagation is always higher than that for slip.

Table 13.1.

Mode	τ_0	τ_1	θ_0	θ_1	$h^{ss'}$	$h^{ss'}$
					slip-slip	slip-twin
Slip	33.0	38.3	65.0	2.0	1.0	0.9
					twin-slip	twin-twin
Twinning	45.0	35.0	50.0	1.25	0.9	1.0

Table 13.1. Voce and latent hardening parameters used to evolve the critical resolved shear stresses of the slip and twinning systems according to Eqs. (6.31) and (6.32).

The parameters of the PTR model were taken as $A_0 = 0.25$ and $A_1 = 0.1$ (see Eq. (13.2)). Although twinning activity may commence prior to $V^{thres} < A_0$, the crystal is not twin reoriented until $V^{thres} = A_0$. It is assumed that when a grain twins—that is, it is replaced with its PTS variant—hardening continues with the same CRSS values that were attained before the reorientation of the original grain.

Figure 13.2(b) show the calculated (111) pole figures, taken from Beyerlein et al. (2007a), after the first three passes of route A. The texture after the first pass exhibits the correct order of intensities among the texture components, with the A_1 component being the weakest. As in the measurement, the predicted textures do not change substantially from pass to pass.

According to the VPSC model and in agreement with the EBSD analysis (Beyerlein et al., 2007a), under shear the grains near the A_1 orientation twin readily and the A_2 and the $B/\overline{B}$ components do not. This orientation dependence of twinning explains why after the first pass the A_1 component is weak, while the A_2 and $B/\overline{B}$ components are the strongest. The C and A_1 components are induced primarily via slip. The model also explains why texture development is similar in each pass. Grains near A_1 orientations twin into orientations near A_2 after the first pass. Upon reinsertion of the sample in the die according to route A, orientation A_2 becomes nearly A_1. Then during the second pass, these A_1-oriented grains twin again. This mechanism is repeated in each pass and as a result, twinning persists in each pass.

In conclusion, the calculations show that the differences between ECAE textures of Ag and higher SFE metals, like Al, are attributed to deformation twinning. The twin fractions grew with pass number, just as observed experimentally. The latent hardening parameter controlled this aspect. Table 13.1 lists the latent hardening parameters between slip and twinning, which were found to be slightly less than unity. (Latent hardening parameters between systems in the same family were set to unity.) This implies that the more active mode hardens slightly faster than the less active one. This type of 'latent' hardening will be discussed further in Chapter 14. As slip was the more active mode, it hardened faster than twinning, which made twinning relatively more favorable as straining proceeded. This increased propensity for twinning was necessary to successfully predict the texture evolution and is well supported by the EBSD observation that the twin fraction increased with pass number.

As discussed in Chapter 7, the Taylor polycrystal model is considered to be applicable to cubic metals and, as shown in Chapter 9, performs comparable to VPSC in predicting the first-pass textures, especially for true strains less than 1 (large Φ). These examples considered slip only. It was found in (Beyerlein et al., 2007a) that when both {111}<110> slip and {111}<112> twinning were made available for Ag, the full-constraints Taylor model predicted larger twin fractions than VPSC.

13.4. EXAMPLE: ZIRCONIUM

The following example illustrates how for non-cubic metals SPD can lead to highly anisotropic plastic behavior. It also demonstrates how combining comprehensive experimental testing with modeling can be used to gain a basic understanding of the material and its deformation mechanisms. Here, experiments and modeling have been applied to investigate the plastic anisotropy of high purity HCP zirconium (Zr) following a single pass of equal channel angular extrusion (ECAE) at room temperature (Yapici et al., 2009). The initial material was a clock-rolled Zr plate[1], which had the basal poles strongly aligned with the through-thickness direction of the plate (see Figure 13.3). Prior to ECAE processing, the samples were annealed at 550°C for 1 hr resulting in an average grain size of 25 μm. As illustrated in Figure 13.3, Zr samples were then ECAE-processed with the basal poles either aligned with (Case I) or perpendicular to (Case II) the extrusion direction (ED). Following ECAE, the processed samples were compressed up to 30% along the three main orthogonal directions (ED, LD, and TD) of the billets. ECAE was conducted using a 90° corner angle ECAE die at room temperature and at an extrusion rate of 0.03 mm/sec corresponding to an approximate strain rate of 10^{-3} s^{-1}. This strain rate was selected to match the strain rate employed in the post-ECAE room temperature compression experiments. Textures were measured after ECAE and after 30% post-ECAE compression testing using the high-pressure preferred orientation (HIPPO) neutron diffractometer at the Los Alamos Neutron Science Center (LANSCE).

Figure 13.4 shows the measured textures for Cases I and II. Irrespective of the initial texture, the extrusion textures always exhibit a basal component rotated ~25° away from the die inlet channel axis (LD). Otherwise the two textures have many dissimilar features. These differences are the result of the slip and twinning modes activated during the ECAE pass. The

[1] Clock rolling consists of performing successive rolling reductions, each one at a different angle from the previous, in order to develop an axi-symmetric final texture. For the Zr studied in Yapici et al. (2009), the clocking rolling processing schedule was classified.

starting Case I texture favors shear accommodation by means of either basal or pyramidal slip. The starting Case II texture favors accommodation via prismatic slip.

As was discussed in Sections 10.5 and 10.6, first-pass textures similar to Figure 13.4 (Case I) have been reported for other HCP materials such as Mg, Be, Ti, and Ti-6Al-4V, (Beyerlein and Toth, 2008) ; and have been attributed to the operation of basal slip. In the case of pure Zr with larger than micron grain size deforming at room temperature, experimental observations indicate that basal slip is not a common deformation mechanism (Akhtar, 1953; Rapperport, 1959). The consistency suggests that the severe plastic deformation processing promotes basal slip in Zr, which is not observed at small strains.

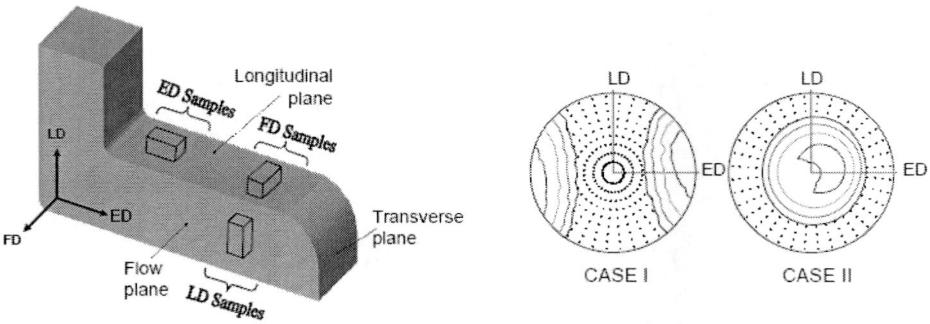

Figure 13.3. Billet representation for ECAE processing and orientation of compression samples on the as-processed billet. ED, LD and FD denote the extrusion, longitudinal and flow directions, respectively. Also shown is the initial billet texture for the two insertion positions. Case I: basal poles are parallel to ED; Case II: basal poles are parallel to FD. Contour levels: 0.5/1.0/ 2.0 / 3.5 / 5.5.

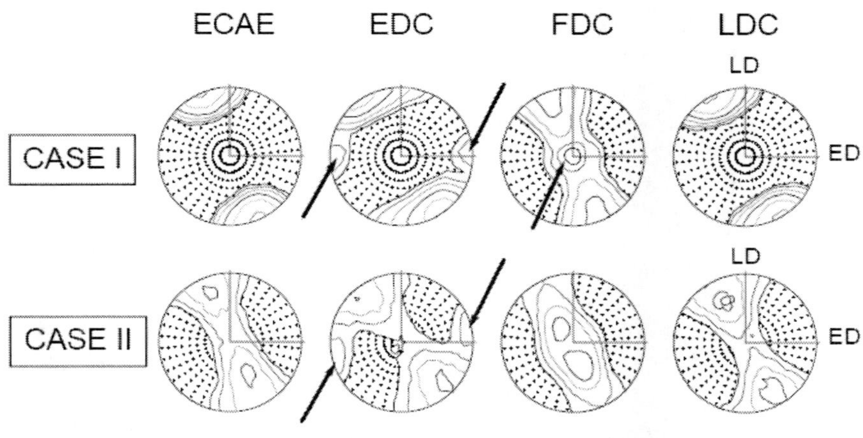

Figure 13.4. Experimental (0002) pole figures for Case I and Case II. Textures resulting from single pass ECAE and from post-ECAE compression to 30% strain: along the Extrusion Direction (EDC), along the Flow Direction (FDC) and along the Longitudinal Direction (LDC). Contour levels: 0.5/1.0/ 2.0 / 3.5 / 5.5/8.0. The arrows show texture components associated with compression-induced twin reorientation.

Figure 13.5 shows the post-ECAE compressive response along the three main sample directions in Cases I and II. In both we observe significant plastic anisotropy in flow stress and hardening and, surprisingly, no evidence of saturation that usually occurs after large plastic straining. Compared to the initial material, the post ECAE flow stresses are generally two times higher and exhibit lower hardening rates. Although the character of the flow responses is not the same for Cases I and II, there are some common features. Compression along the ED results in the lowest flow stress and a low hardening rate. Compression along the FD results in an increasing hardening rate with strain, a common sign of twin activity. Compression along the LD (LDC) results in hardening up to 10% strain and a noticeable softening under further straining. The latter is a consequence of the inhomogeneous deformation of the sample illustrated in Figure 13.6. Compressing the sample texture after ECAE in LDC favors basal slip, which induces a variable shear component that superposes to the compressive strain.

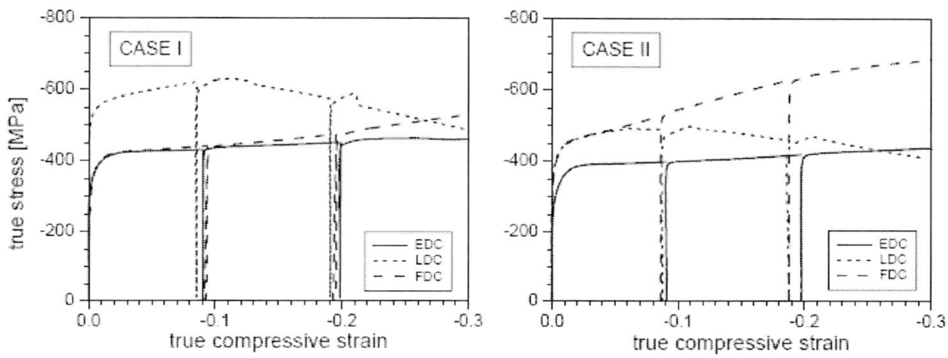

Figure 13.5. Uniaxial compressive response of the ECAE processed Zr along the three orthogonal directions (Extrusion, Longitudinal, Flow) up to 30% plastic strain for Case I and Case II. The apparent softening during LD compression is a consequence of shear superimposed to compression and non-homogeneous deformation of the sample.

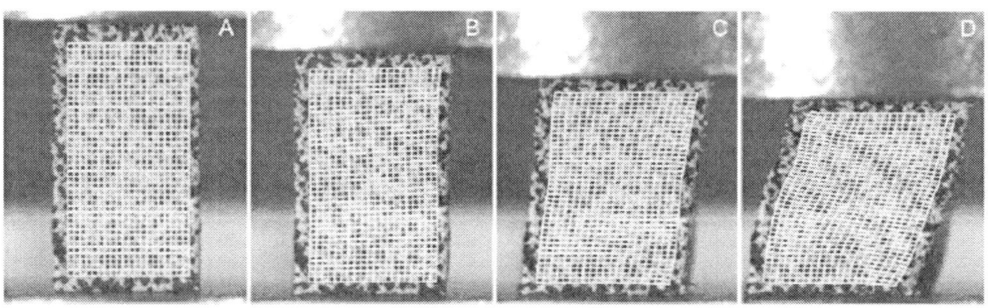

Figure 13.6. Evolution of strain in the sample during LD compression following ECAE for Case I, illustrated by the evolution of a grid superimposed by means of the Digital Image Correlation (DIC) technique (Sutton et al. 2000). A, B, C and D correspond to 0%, 10%, 20% and 30% compression, respectively.

Texture evolution during post-ECAE compression is dictated by the ECAE texture and the loading direction (see Figure 13.4). In both cases, compression along the ED (EDC) results in a spread of the basal poles about the 25° from the LD and development of a weak component along ED. The latter results from twinning, which reorients the c-axis along the compression direction. LDC induces inhomogeneous deformation in the sample (see Figure 13.6), which prevents a straightforward interpretation of the corresponding texture and stress-strain evolution. For samples loaded in FDC compression (FDC), the c-axes of the grains, which are preferentially aligned along the 25° after ECAE, are now placed in tension. As a result, $\{10\bar{1}2\}$ tensile twinning is activated in FDC, which aligns the basal poles along the FD compression axis. Such an effect is particularly evident in Case I.

13.4.1. Modeling ECAE of Zirconium

The VPSC model, combined with extended Voce hardening and the Predominant Twin Reorientation (PTR) scheme, were used to simulate ECAE processing and the subsequent compression tests. The model predicts texture evolution, plastic flow stress versus strain under different post-processing loading conditions, and the relative activities of the different slip and twinning modes for a prescribed deformation path. The activities of each slip and twin mode can be calculated from the model output:

$$ACT(\alpha) = \frac{\sum_{s=1}^{n(\alpha)} \left|\dot{\gamma}^s\right|}{\sum_{\alpha} \sum_{s=1}^{n(\alpha)} \left|\dot{\gamma}^s\right|} \tag{13.4}$$

where α is the mode which contains a set of slip or twin systems s , $\dot{\gamma}^s$ is the slip rate of slip system s, and $n(\alpha)$ is the number of systems in α. The Voce and latent hardening parameters used for evolving the critical resolved shear stresses (CRSS) with plastic straining are summarized in Table 13.2 for each slip and twin mode made available to the simulation: prismatic slip, first-order pyramidal <c+a> slip, basal slip, and tensile twinning. The measured crystallographic texture for the starting sample is represented in the model by 1944 grains. The initial grain shape was taken as spherical and during deformation, the shape of each grain was allowed to evolve separately. A maximum aspect ratio of $R = 4$ was enforced, at which point the grains were split in the manner described in Section 7.4.3.

Table 13.2.

Deformation System	τ_0 (MPa)	τ_1 (MPa)	Θ_0 (MPa)	Θ_1 (MPa)	$h^{s,pri}$	$h^{s,bas}$	$h^{s,pyr}$	$h^{s,twn}$
Prismatic Slip	20	90	400	15	1	1	1	1
Basal Slip	300	0	10	10	1	1	1	1
Pyramidal <c+a> Slip	145	90	400	70	1	1	1	1
Tensile Twinning	100	90	100	35	1	1	1	1

Table 13.2. Voce and latent hardening parameters used in the VPSC simulations for predicting the texture evolution and mechanical response of the ECAE processed Zr (see Eqs. (6.31) and (6.32) for meaning).

In simulating the ECAE process, the deformation state is described by two steps. First, there is an initial compression upset of 10% when the billet is first pushed into the entry channel. Second, the upset is followed by negative simple shear (described by Eq. 8.7) along the intersection plane at 45°, which is appropriate in the present case of a sharp-corner die with sliding walls. In what concerns the subsequent axial compressive tests along one of the three orthogonal directions, we apply strain control in the direction of straining and assume traction-free surfaces in the other two directions, letting the sample expand laterally as dictated by its plastic anisotropy. More details concerning the loading conditions can be found in Yapici et al. (2009).

Figure 13.7 shows the predicted textures for Case I and II at three stages: after the axial compressive upset of the sample in the jacket, after ECAE, and after subsequent compression in the ED and FD. (LDC textures are not calculated because deformation in these samples was non-homogeneous as shown in Figure 13.6) The calculated texture evolution is a direct consequence of the slip and twinning mechanisms activated during deformation. Figure 13.8 shows the calculated relative activities of slip and twinning as defined by Eq. 13.4. During the upset, activities are the same in both cases and twinning reorients about 30% of the volume fraction. For Case I the model predicts substantial pyramidal slip and prismatic slip, and an increasing contribution of basal slip after 20% accumulated strain. Twinning makes a smaller but non-negligible contribution compared to these slip modes bringing the total final accumulated twin volume fraction to approximately 70% (see Figure 13.8). In Case II, prismatic slip is the main deformation mode throughout and, as in Case I, basal slip becomes a relevant deformation mechanism after 20% accumulated strain (see Figure 13.8), while twinning activity is minor.

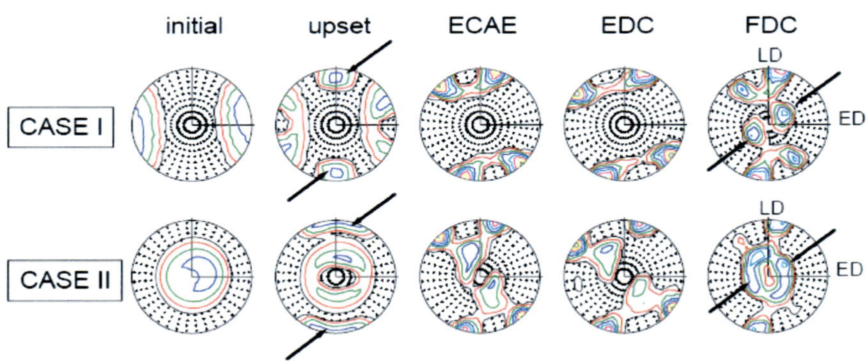

Figure 13.7. Predicted (0002) pole figures for Case I and Case II. Textures shown are: initial, after 10% upset compression in the die, upset plus ECAE, and post-ECAE compression to 30% strain along the Extrusion Direction (EDC) and along the Flow Direction (FDC). Pole figures corresponding to compression along the Longitudinal Direction (LDC) are not shown because the inhomogeneous deformation (see Fig. 4) prevents comparison with experimental texture. Contour levels: 0.5/1.0/2.0/3.5/5.5/8.0/11/15. The arrows show texture components associated with compression-induced twin reorientation.

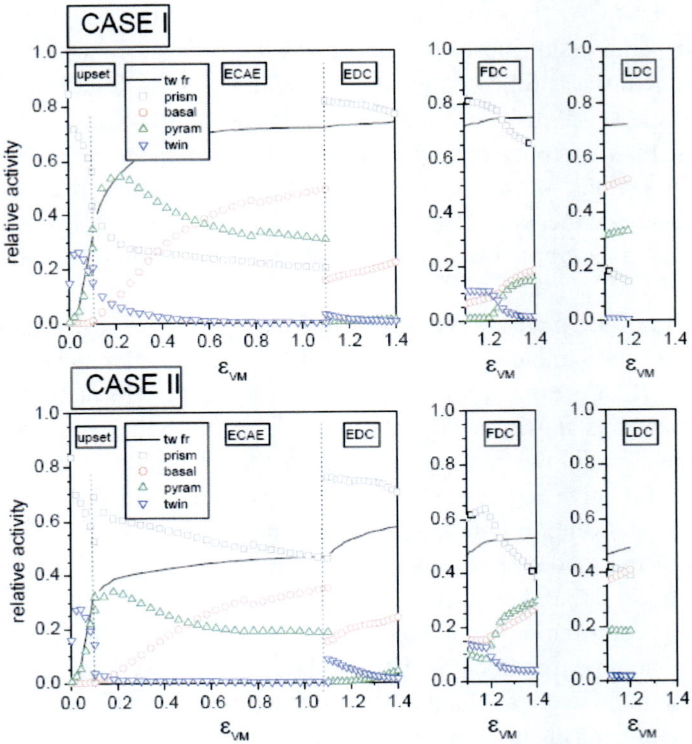

Figure 13.8. Predicted relative activity of the operative deformation mechanisms during compression upset, ECAE, and post-ECAE axial compression, for Case I and Case II, as a function of accumulated Von Mises strain. The evolution of the twin volume fraction is indicated by a solid line. For Longitudinal Direction Compression (LDC) the simulation was carried up to 10% strain, the point at which the experimental sample starts deforming non-homogeneously (see Fig. 13.6).

13.4.2. Post-ECAE Compression Modeling

When modeling post-ECAE compression the challenge is not only to predict texture evolution, but to predict the subsequent stress-strain evolution in each of the three orthogonal testing directions for Cases I and II. Figure 13.9 compares the predicted compressive flow response with the measurement for each individual curve in Cases I and II. The corresponding textures and activities can be found in figures 13.7 and 13.8. For Case I, the model effectively predicts the observed plastic anisotropy and hardening rates. In Case II, the qualitative features of the deformation response are correctly simulated, although the quantitative agreement is not very good.

For Case I, the model indicates that basal, prismatic, and pyramidal slip, along with tensile twinning, are activated. Notably basal slip makes a significant contribution in all three loading directions. The relative activity of each deformation mode strongly depends on the compression direction. As in the experiment, the predicted EDC and FDC curves start at about the same flow stress, but with further straining, the FDC response exceeds the EDC response due to a higher hardening rate. Initially, both load tests activate primarily prismatic slip and some basal slip. FDC, however, also activates pyramidal slip and tensile twinning (Figure 13.8) since it applies compression along the c-axes of those grains oriented parallel to

FD (activating pyramidal) and tension along the c-axes of those parallel to the LD (activating tensile twinning) simultaneously. EDC, however, does not activate either of these harder modes. As a result, the FDC response has a higher hardening rate than that in EDC. LDC directly compresses most of the grains along the c-axis at the beginning of post-ECAE compression, which activates the harder pyramidal slip mode and leads to the highest flow stress of the three directions.

In Case II, basal, prismatic, and pyramidal slip and tensile twinning are activated, although prismatic slip plays a larger role compared to that in Case I, especially for LDC. In agreement with the measurement, the calculated Case II EDC response has the lowest yield stress and hardening rate (Figure 13.9). The modeling results indicate that this response arises because EDC is mostly accommodated by the softer prismatic slip. Also consistent with the measurement, FDC has the highest hardening rate and flow stress. Although prismatic slip dominates, the texture also requires pyramidal slip and twinning to be activated. These relatively harder modes of deformation effectively raise the hardening rate. LDC directly compresses some grains along the c-axis, which activates pyramidal slip, and in turn, produces the highest flow stress, as in Case I. However, the LDC flow stress of Case II is much lower than that of Case I since fewer grains are oriented in this hard orientation after ECAE in Case II than in Case I.

Agreement between the measured and predicted post-deformation textures, even if not perfect, further confirms the operative deformation mechanisms suggested by the model. For 30% strain imposed in compression, only the basal pole reorientation induced by twinning is expected to change the texture noticeably. The activation of twinning in the FDC cases creates a texture component aligned with compression axes (FD), which is present in both the prediction (Figure 13.7) and the experiments. In contrast, no change in the basal pole orientation is expected when prismatic slip is active, and only small changes in texture evolution are expected from basal and pyramidal slip. Therefore, the observed texture evolution is only possible with twinning.

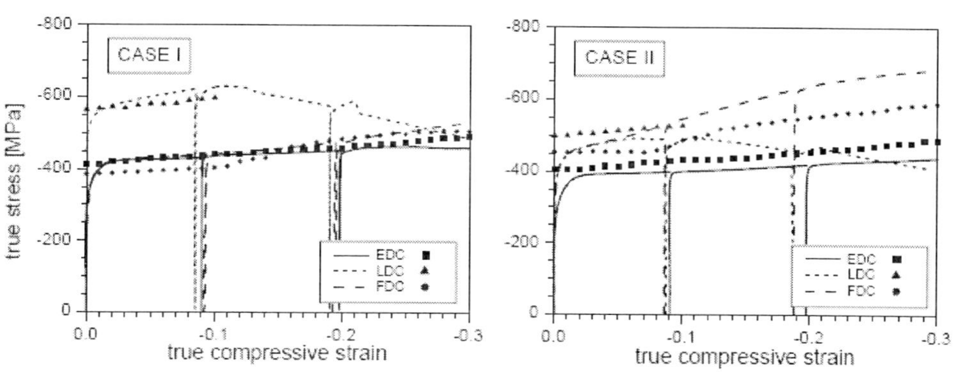

Figure 13.9. Predicted stress-strain response during post-ECAE compression along ED, FD, and LD directions of the billet, for Case I and Case II. Also shown are the corresponding experimental results. Predicted LDC was carried up to 10%, when experimental deformation becomes non-homogeneous. The lines represent the experiments and symbols indicate the predictions.

The results discussed above lead to two interesting conclusions that go against the conventional wisdom for Zr: 1) basal slip semms to be ~~is~~ active during SPD, despite no evidence of it at small and medium strain levels and 2) tensile twinning remains active after

SPD, even though the grain size is reduced below one micron. The improved predictions when basal slip is included in the model supports the first conclusion. Moreover the fact that prismatic slip is still favored over basal slip in Zr at medium strains is also consistent with previous characterization of Zr up to 20-30% strain (Kaschner et al. 2006). In addition, the measured Zr single-pass ECAE texture is similar to that measured in other hcp metals (Mg, Be, Ti, Zr, and Ti-6Al-4V; see Beyerlein and Toth, 2009), where it has been unequivocally related to basal slip activity. Yapici et al. (2009) speculate that the unexpected prevalence of basal slip in pure Zr could be a consequence of the large strains and/or hydrostatic pressure imposed by ECAE. As for the activation of $\{10\bar{1}2\}$ tensile twinning at large strain, its presence is clear from the observed texture and hardening rate evolution, whenever a tensile stress component was applied along the basal poles (e.g., FD compression). Again, the rationale for this observation is based on the generation of large internal stresses during ECAE that activate twinning regardless of grain size.

Chapter 14

MECHANICAL RESPONSE OF SPD'ED METALS

14.1. SINGLE CRYSTAL HARDENING

When governed solely by dislocation slip, the plastic stress-strain (σ-ε) response of a metal is conventionally described by sequential stages of work hardening $\partial\sigma/\partial\varepsilon$ with strain. Depending on its initial orientation, the work-hardening response of a single crystal can begin either with a low work hardening stage, stage I, also called the easy glide regime associated with single or planar slip, or with a relatively higher work hardening stage II, associated with multi-slip. The work hardening response of a polycrystal, on the other hand, usually begins with stage II. In single and polycrystals, the stage II response softens with further straining into stage III. Stage III can be followed by stage IV, characterized by a constant, relatively low value of $\partial\sigma/\partial\varepsilon$. A rigorous development and discussion on the stages of work hardening and links to dislocation processes can be found in (Kocks et al., 1975; Kocks and Mecking 2003; Kuhlmann-Wilsdorf, 1962).

The constitutive shear stress τ – shear strain Γ response for a single slip system ($\tau-\Gamma$) is more challenging to measure and it is customary to assume that it has the same form as that of the polycrystal; that is, it begins with stage II and follows through to stages III and IV. In the model described here, we use the extended Voce law (Tomé et al., 1984), which was introduced in Chapter 6, for the constitutive response of each available slip system. This law provides a general mathematical form for the hardening response ($\tau-\Gamma$) for each slip system with accumulated shear strain Γ, and leads to a polycrystal response that spans from stage II to stage IV:

$$\tau_V^s(\Gamma)=\tau_0^s+\left(\tau_1^s+\theta_1^s\Gamma\right)\left[1-\exp\left(-\theta_0^s\Gamma/\tau_1^s\right)\right] \qquad (14.1)$$

The superscript s on the four Voce parameters τ_0, τ_1, θ_0, and θ_1 indicates that they may depend on slip system s. It is common to set these equal for systems belonging to the same crystallographic family, but to vary them from one slip mode or twin mode to another (e.g., between prismatic and basal slip in an HCP crystal or between {110} slip and {112} slip in a BCC crystal).

The schematic in Figure 6.5 illustrates a Voce curve according to Eq. (14.1) and the effects of each of its four parameters. As shown, τ_0 is the initial resistance, θ_0 is the stage II hardening rate, $\tau_0 + \tau_1$ is roughly the stage III 'saturation stress', and θ_1 is the stage IV hardening rate. Conventional thermal activation models would relate τ_0 to friction, alloying, and grain size, θ_0 to a predominance of dislocation multiplication; τ_1 to predominance of dynamic recovery; and θ_1 possibly to dislocation patterning.

A common measure of accumulated shear in a grain is Γ, calculated by the sum of the accumulated shears on all slip systems via:

$$\Gamma = \int_0^t \sum_s \left|\dot{\gamma}^s\right| dt . \tag{14.2}$$

where γ^s is the shear rate on s, a scalar quantity, which can be positive or negative. (It can be calculated, for instance, using Eq. (6.10).) Accumulated shear Γ is also a scalar quantity and does not reflect the mode of straining or strain paths used to accumulate it. It is also independent of slip activity, defined as the number and distribution of active slip systems in a grain. Note also that Γ accounts for all slip in the grain, not just that of a particular system s. Because in the above model $\tau_V^s(\Gamma)$ is a function of Γ and not γ^s, some coupling between the hardening of s and entire slip activity in the grain via Γ is, therefore, introduced. Hardening models that relate τ_V^s to a total dislocation density ρ and not to a total strain Γ accumulated in the grain are no different in this respect. In a similar fashion, a single population ρ is calculated either for each grain (crystal plasticity) or for the entire material (continuum formulation). In this case, ρ, like Γ, has no direct relation to details of slip activity and changes in slip activity. Hardening in all slip systems are coupled indirectly through ρ.

Because of these assumptions, both strain-based and ρ-based models will apply best when the strain path does not change in the course of deformation. Relating the hardness of each slip system to a total Γ or ρ is consistent with the idea that hardening evolution is due to an increase in dislocation storage with straining. For deformations in which the loading conditions are fixed (e.g., monotonic), stored dislocations accumulate monotonically with straining and at a non-increasing rate. The forest of dislocations is assumed to be spatially homogeneous and to be mostly dependent on the level of strain (in the grain) and less so on the details of slip activity.

Recall from Chapter 6 that in multi-slip conditions, the hardening rate of system s is related to the activities of the other systems s' via:

$$\dot{\tau}_h^s = \frac{\partial \tau_V^s}{\partial \Gamma} h^{ss'} \dot{\gamma}^{s'} \tag{14.3}$$

where $h^{ss'}$ are latent hardening parameters that couple the hardening on s with slip activity in s'. It is common practice to set self-hardening to unity, $h^{ss'} = 1$ when $s = s'$. For the off diagonal parameters, $s \neq s'$, $h^{ss'}$ can vary from 0 to infinity. For $s \neq s'$, $h^{ss'} = 1$ represents isotropic hardening and $h^{ss'} \neq 1$ anisotropic hardening. In isotropic hardening all systems harden equally, whether they are active or latent. Consequently, grain hardening will be weakly affected by a change in its slip activity that accompanies a macroscale change in strain path. For the latter case, this is not true; in anisotropic hardening ($h^{ss'} \neq 1$), a strain path change would affect the calculated plastic response. The effect would entirely depend on the choice of $h^{ss'}$ (Beyerlein et al., 2005b). In anisotropic hardening, there are two possibilities: 1) $h^{ss'} < 1$ which we term 'active' hardening because in this case, the more active systems harden faster than the latent ones and 2) $h^{ss'} > 1$, which is commonly referred to as 'latent hardening', in which the latent systems harden faster.

14.2. SPD-Induced Plastic Anisotropy

The deformation behavior of metals processed by SPD can exhibit what appear to be transients in flow stress, in which hardening rates can be positive, negative, or zero over appreciable amounts of strain. The usual work hardening characteristics, in which work hardening decreases or remains constant with strain, are found to resume after such transients have faded. In some cases, usually after large pre-strains, the straining periods over which such transients persist are so long that resumption of normal work hardening is not observed.

These transients in work hardening are found to depend on the direction and sense of loading of the post-test, e.g., (Yapici et al., 2007, 2009; Beyerlein et al., 2005b; Alexander and Beyerlein, 2005; Beyerlein and Tomé, 2007). Tension testing along the ED produces a plastic response with linear hardening and no transients, whereas compression testing along the same direction produces one with softening or negligible hardening (Yapici et al., 2007). Axial testing of the sample normal to the die plane (transverse) leads to an immediate spike in the flow stress followed by worksoftening (Alexander and Beyerlein, 2005; Yapici et al., 2007; Beyerlein and Tomé, 2007). Because of the directional dependence of these transients, the SPD metal is plastically anisotropic.

Transients are a reflection of rearrangements in the microstructure from small-scale changes in dislocation processes to meso-scale changes in texture evolution. The amount and type of rearrangement required depends on the orientation relationship between the first strain path and second one and the dislocation substructure developed during pre-straining. If pre-straining is small, the transients are less severe. If the material is annealed after processing such that the deformation-induced dislocation microstructure is removed, the transients disappear. The transients cannot be attributed solely to anisotropic hardening during pre-straining. Changes in latent hardening parameters alone (Beyerlein et al., 2005b), for instance, will not be sufficient to explain their characteristics. The hardening model, whether strain-based or dislocation-based, must also account for the dynamic re-arrangement of dislocations during reloading.

At the individual grain length scale, two extremes in reloading can be considered: one in which straining remains on the same slip plane, but is reversed, and another in which straining changes from occurring predominantly on one slip plane to predominantly on another, non-parallel to the previous one.

In a reversal reload, the Bauschinger effect is the first transient and is associated with a drop in yield stress. Also included in the Bauschinger effect is the rapid work-hardening that immediately follows 'reverse yield'. This work-hardening rate is characteristically much higher than that achieved during pre-straining. After the Bauschinger effect has already been relieved with some amount of reverse straining, secondary transients may appear. These are not usually considered part of the Bauschinger effect and are the result of different mechanisms. They occur after large prestrains and the severity and duration of these secondary transients increase with pre-straining. They vary more widely in character than the Bauschinger effect; in some cases, the secondary transients are extended strain periods of work-softening or of reduced hardening, while in others, the reload response after the Bauschinger effect overshoots the extrapolated forward one (Stout and Rollett, 1990). The former is characteristic of medium to high SFE pure metals and the latter of metals with inclusions or second phases.

Consider as an example, the response of ECAE-processed copper when compressed along the ED. The measured responses after one to 16 passes of route Bc are shown in Figure 14.1. Macroscopically, the deformation sequence from the last ECAE extrusion to the ED compression test can be viewed as a pseudo-reversal. During an ECAE extrusion, there is a normal tensile strain component along the ED (see Chapter 12, Eq. (12.31)). The subsequent ED compression test reverses the ECAE straining action. Simulations of this deformation sequence (Beyerlein and Tomé, 2007) also find that in a majority of the grains, slip activity was reversed. The ED compression responses after each pass in Figure 14.1 are characteristic of a reversal response. A secondary transient of an extended plateau region of little to no work hardening is apparent. This transient is typical of pure FCC metals (high to medium SFE) which have been pre-strained to large strain levels. Without an analogous continuous straining curve to serve as a reference, a Bauschinger effect is, however, not readily apparent in these measurements. The single crystal constitutive model described in the following sections does suggest that there is a Bauschinger effect and it is responsible for a lower yield stress in ED compression than the yield stress in other directions. The Bauschinger effect will be modeled in Section 14.7.

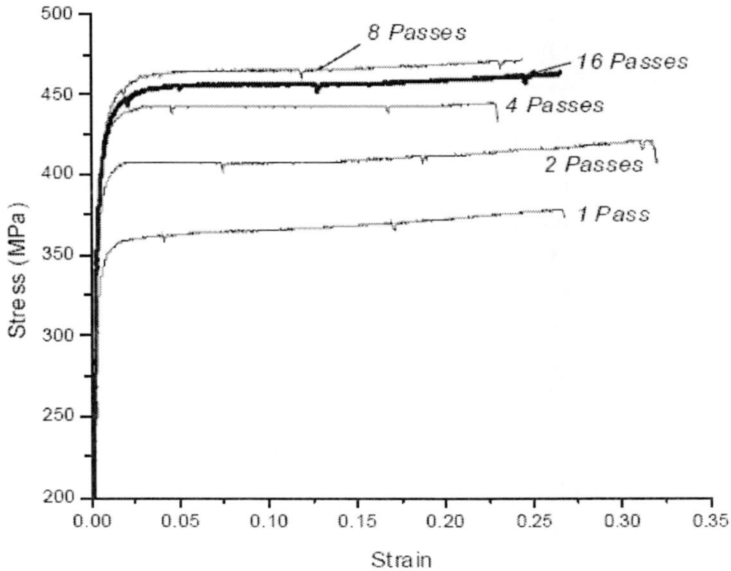

Figure 14.1. Measured responses of ECAE copper samples in compression along the extrusion direction (ED) after 1 to 16 passes of route Bc (Alexander and Beyerlein, 2005; Beyerlein et al., 2007). The responses exhibit extended plateau regions characteristic of a reversal re-load.

In the event that the strain path change is orthogonal to the previous one, slip activity within a grain can take place on different planes than those active before. If previous slip activity occurred predominantly on one or two slip planes and the pre-strain was large, it is likely that substructure develops, in the form of extended planar dislocation walls more or less aligned along the most (or more) active plane(s) (Mahesh et al., 2004; Xue et al., 2007). The new slip activity after the strain path change will be impeded by the previously generated substructure and may be forced to locally 'cut-through' it in order to proceed. In their attempt to channel through, they would encounter resistance, proportional to the 'strength' of the sub-boundaries. This strength is proportional to the dislocation density collected within the boundary ρ_{sub}, which will decay as the penetrating dislocations react, untangle, annihilate, or dissolve ρ_{sub}. Concomitantly, the resistance to dislocation glide also reduces. This two-step process of encountering and overcoming the existing substructure is thought to lead, macroscopically, to an immediate increase in flow stress followed by a softening response. This 'cut-through mechanism' will be modeled in Section 14.8.

14.3. Slip Plane Vector Analysis

To provide measures of slip activity and direction of slip, a slip-plane-vector analysis was developed by Beyerlein and Tomé (2007). Below it is described for an FCC metal. Its application to other crystal structures should be readily apparent.

In the slip-plane vector analysis, the three slip systems on each of the four independent {111} planes will be reduced to a single slip action. These four slip planes are designated as $\alpha, \alpha' = 1, 2, 3, 4$. A shear rate vector $\dot{\bar{v}}^{\alpha}$ and shear vector $\bar{v}^{\alpha}$ on slip plane α are defined as

$$\dot{\bar{v}}^{\alpha} = \sum_{s \in \alpha} \bar{b}^{s} \dot{\gamma}^{s} \qquad \text{and} \qquad \bar{v}^{\alpha} = \int_{0}^{t} \dot{\bar{v}}^{\alpha} dt \tag{14.4}$$

where $\bar{b}^{s}$ is the normalized Burgers vector of slip system s expressed in crystal axes, such that the vectors are independent of crystallographic rotations. The bar above the symbol indicates that the quantity is a vector. The shear vectors in Eq. (14.4) are contained in plane α and weigh the relative contributions of the three slip systems on the plane α. For instance if slip system s dominates then $\bar{v}^{\alpha}$ will be nearly parallel to $\bar{b}^{s}$, whereas if two slip systems s and s' dominate, then $\bar{v}^{\alpha}$ will lie in-between $\bar{b}^{s}$ and $\bar{b}^{s'}$. The associated scalar shear rate and shear are defined by the norms of $\dot{\bar{v}}^{\alpha}$ and $\bar{v}^{\alpha}$:

$$\dot{v}^{\alpha} = \left\| \dot{\bar{v}}^{\alpha} \right\| \qquad \text{and} \qquad v^{\alpha} = \left\| \bar{v}^{\alpha} \right\| \tag{14.5}$$

Total scalar shear rates and shear are defined as sums over the plane specific ones:

$$\dot{v} = \sum_{\alpha=1}^{4} \dot{v}^{\alpha} \qquad \text{and} \qquad v = \sum_{\alpha=1}^{4} v^{\alpha} \tag{14.6}$$

14.4. Strain Path Change Indicators

Using the vector analysis of Beyerlein and Tomé (2007), local strain path changes can be characterized by defining certain quantities that couple activity in different planes. For example, reversal on slip plane α occurs at the instant when

$$P^{\alpha\alpha} = \frac{\overline{v}^{\alpha}}{v} \cdot \frac{\dot{\overline{v}}^{\alpha}}{\dot{v}} < 0 \qquad -1 \leq P^{\alpha\alpha} \leq 1 \tag{14.7}$$

is satisfied. Let $P_0^{\alpha\alpha}$ be the magnitude of $|P^{\alpha\alpha}|$ at the first instant Eq. (14.7) is satisfied.

A significant change in activity from plane α' to plane α when going from strain increment n to n+1 can be detected by first calculating:

$$X^{\alpha\alpha'} = \left\| \frac{\dot{\overline{v}}^{\alpha,n+1}}{\dot{v}} \times \frac{\overline{v}^{\alpha'}}{v} \right\| - \left\| \frac{\dot{\overline{v}}^{\alpha,n}}{\dot{v}} \times \frac{\overline{v}^{\alpha'}}{v} \right\| \tag{14.8}$$

For the cut-through mechanism (Section 14.2), a necessary condition is that $X^{\alpha\alpha'} > 0$ and $\alpha \neq \alpha'$. When $\alpha \neq \alpha'$, the larger $X^{\alpha\alpha'}$, the greater the shift in slip activity from being most active on plane α' during the previous strain path to being most active on plane α in the new strain path. Accordingly, significant changes are associated with larger values of $X^{\alpha\alpha'}$. These are the conditions under which the cut-through mechanism is expected to operate and hence small values of $X^{\alpha\alpha'}$ are not of interest. For later modeling of cut-through, it will be convenient to define $X_0^{\alpha\alpha'}$, where $X_0^{\alpha\alpha'} = X^{\alpha\alpha'}$ when $X^{\alpha\alpha'} > X_c$ and zero otherwise. Geometric considerations motivate X_c to be around 0.15 (Beyerlein and Tomé, 2007).

14.5. Anisotropic Slip Resistance

A microstructurally-based single crystal hardening law applicable to strain path changes in FCC materials was developed in (Beyerlein and Tomé, 2007). The model includes dislocation storage via the Voce law and adds two strain-path-change mechanisms: a reversal-effect, which lowers the density of stored dislocations, and a cut-through mechanism, in which newly gliding dislocations are resisted by and must 'cut-through and dissolve' previously generated substructure. In (Yapici et al., 2007) the law was modified to couple these two mechanisms.

In keeping with the "slip-plane viewpoint" (Section 14.3), the three slip systems contained on slip plane α, $s \in \alpha$, have the same slip resistance, denoted as τ_c^{α}.

$$\tau_c^s = \tau_c^\alpha, \qquad s \in \alpha \tag{14.9}$$

where τ_c^α contains several components:

$$\tau_c^\alpha = \tau_h^\alpha + \Delta\tau_{cut}^\alpha + \Delta\tau_{rev}^\alpha \tag{14.10}$$

The first term on the right hand side, τ_h^α, depends on overall dislocation evolution. This term was described earlier in Section 14.1. The second and third terms are the changes in hardness as a result of the two strain-path-change mechanisms mentioned in Section 14.2. When there is a change in slip activity from one plane α' to another α (Eq. (14.8)) new slip on plane α must overcome the substructure left behind by prior slip activity on the other planes $\alpha \neq \alpha'$, leading to an additional resistance to slip $\Delta\tau_{cut}^\alpha$ (> 0). A reversal on slip plane α (Eq. (14.7)) involves deformation mechanisms in which small populations of dislocations that are immobile in forward glide become mobile in reverse glide. They lead to a lower slip resistance $\Delta\tau_{rev}^\alpha$ (≤ 0) compared to that at the end of forward glide.

14.6. Latent Hardening

Earlier the hardening rate in system s was related to hardening on slip system s' via Eq. (14.3). This relationship is modified in the slip vector analysis. The resistance τ_h^α on slip system s contained on plane α ($s \in \alpha$) evolves with the slip rate on its plane and that of the other slip planes α' by

$$\dot{\tau}_h^\alpha = \frac{\partial \tau_V^s}{\partial \Gamma} \hbar^{\alpha\alpha'} \dot{\nu}^{\alpha'} \qquad s \in \alpha \tag{14.11}$$

For the parameters $\hbar^{\alpha\alpha'}$ between independent slip planes α and α', consider the following function:

$$\hbar^{\alpha\alpha'} = (1 - h_{sat})e^{-\delta^{\alpha\alpha'}} + h_{sat} \tag{14.12}$$

where $\hbar^{\alpha\alpha'}$ evolves with deformation through

$$\delta^{\alpha\alpha'} = \frac{\left|\nu^\alpha - \nu^{\alpha'}\right|}{\nu_0} \tag{14.13}$$

Both h_{sat} and ν_0 are material parameters. The parameter ν_0 characterizes the strain at which $\hbar^{\alpha\alpha'}$ changes from 1, isotropic hardening, to a saturation value h_{sat}, anisotropic hardening when $h_{sat} \neq 1$ means anisotropic hardening. Equation (14.12) is intended to model the influence of microstructural evolution on slip interactions in the following way:

a) when $\alpha = \alpha'$, $\delta^{\alpha\alpha'} = 0$ and $\hbar^{\alpha\alpha'} = 1$;

b) when $\alpha \neq \alpha'$ and when these two planes are equally active, i.e., $\nu^{\alpha} \approx \nu^{\alpha'}$, then $\hbar^{\alpha\alpha'} \sim 1$, meaning that non-coplanar slip systems with nearly equivalent activities harden equally.

c) when $\alpha \neq \alpha'$ and for small strains $| \nu^{\alpha} - \nu^{\alpha'} | << \nu_0$, $\hbar^{\alpha\alpha'} \sim 1$. In this case the grain substructure is relatively weak, consisting mainly of a three-dimensional network of tangles and cells formed by statistical trapping and with low misorientations (< 1°). This structure will lead to isotropic hardening as all systems are likely to encounter resistance from them;

d) when $\alpha \neq \alpha'$ and after large strains in either plane α or plane α', but not both, then $| \nu^{\alpha} - \nu^{\beta} | > \nu_0$, and hardening via Eqs. (14.11-13) turns from isotropic to anisotropic. In pure FCC metals with medium to high SFE, planar dislocation boundaries tend to form along the most active slip planes (Winther et al., 2004; Xue et al., 2007; Huang and Liu, 1998; Mahesh et al., 2004; Miyamoto et al., 2004) after large strains. Assuming that this leads to an anisotropic hardening condition, in which the active planes harden faster than the latent ones, h_{sat} is set to less than one.

14.7. Change in Resistance due to a Reversal

The Bauschinger effect is related to the backstress generated in pre-straining. More specifically in forward loading, elastic internal stresses - called backstresses - are generated, which oppose dislocation motion. At any given instant, there is a small fraction of dislocations ρ_B for which the backstress negates the applied stress, preventing them from gliding forward. When the applied stress on the glide plane is reversed, it becomes aligned with the backstress. Consequently, with the backstress aiding reverse glide, many of these previously immobile ρ_B dislocations can glide in reverse at a lower applied stress. If sufficient fractions of grains experience a strain reversal and activate glide of ρ_B dislocations, then macroscopically plastic flow would begin at a lower stress than at the end of forward loading, leading to a Bauschinger effect.

X-ray diffraction measurements by Wilson et al. (1990) show that the internal residual strains generated in forward straining decay exponentially in reverse straining. Accordingly, the residual 'backstrain' on slip plane α would be:

$$\nu^{\alpha}_{back} = \nu_{peak} \exp\left(-\frac{\nu^{\alpha}_{new}}{\nu_B}\right) \quad , \tag{14.14}$$

where ν^{α}_{new} is the strain accumulated since reversal, ν_B is the reversal strain required to negate the back-strain (around 0.5%-2% for pure metals), and ν_{peak} is the peak value. Being an elastic internal stress, the backstress is approximately μv^{α}_{back}, where μ is the shear modulus. The amount of ρ_B previously held up by the back stress will be proportional to v^{α}_{back} and to the local re-orientation, with a pure reversal remobilizing the most ρ_B. The reversal indicator $P^{\alpha\alpha}_0$ defined by Eq. (14.7) serves as a excellent measure of the degree of reversal. As such, it can be applied to determine, for instance, the decrement $\Delta\tau^{\alpha}_B$ in the slip resistance due to the glide of ρ_B in reverse in the following way:

$$\Delta\tau^{\alpha}_B = -P^{\alpha\alpha}_0 \mu v^{\alpha}_{back}, \qquad \text{(no sum)} \qquad (14.15)$$

There is a second population of glissile dislocations ρ_R that glide with reverse straining. While in forward loading, they become trapped or piled up in the substructure periphery (Kocks et al., 1980) or other barriers, in reverse loading they are released from entrapment, become mobile again, and contribute to plastic flow. Because these reverse gliding ρ_R dislocations were already available in the microstructure, they are called *available dislocations* (Vincze et al., 2005). The reverse glide of available dislocations can result in worksoftening or plateau regions of no hardening. In a reversal the glide of surplus ρ_R would remove the need for the generation of new dislocations and hence dislocation multiplication would slow down. Dynamic recovery processes, such as dislocation annihilation, would, on the other hand, not slow down but continue in reverse as in forward deformation. With the dominance of dynamic recovery, the dislocation storage rate would stop decreasing or drop. It is also possible that the weaker substructures (such as cells) would dissolve as well, but this apparently does not happen in all reversals. Some studies have seen, via TEM, dissolution of microstructure (Hasegawa et al., 1975), whereas others have not (Vincze et al., 2005), yet in both cases transient reductions in hardening were observed.

The fraction of ρ^{α}_R released with respect to ρ^{α}_{tot} (the total density on slip plane α) with reverse straining on slip plane α is:

$$f\left(v^{\alpha}_{new}\right) = \int_0^{v_{new}} f'(v)dv = \frac{\int_0^{v_{new}} \partial\rho_R / \partial v \, dv}{\rho^{\alpha}_{tot}} \qquad \text{where } f' \geq 0 \qquad (14.16)$$

where $\partial\rho_R / \partial v$ is the release rate. As shown in Figure 14.2, $\partial\rho_R / \partial v$ initially rises as reverse straining releases ρ^{α}_R and then $\partial\rho_R / \partial v$ must eventually fall as further straining exhausts the supply of ρ^{α}_R. In any function of this shape, there are two main variables, the peak value $\partial\rho^*_R / \partial v$ and the decay rate. An increase in $\partial\rho^*_R / \partial v$ and a decrease in the decay rate both serve to increase the area under the curve corresponding to the total ρ^{α}_R available in the microstructure at the start of the reversal. The density ρ^{α}_R, therefore, would depend on material and pre-strain. For instance, one can expect more ρ^{α}_R to be collected near obstacles and hence ρ^{α}_R would increase with obstacle content (e.g., inclusions, solutes, participates, alloying elements). For the same reason, ρ^{α}_R may also increase with pre-straining as pre-straining increases the stored dislocation and substructure content in the grains. Prior studies on Cu and Al (Yapici et al., 2007; Beyerlein et al., 2007b; Beyerlein and Tomé, 2007) suggests that pre-strain increases ρ_R^α , specifically by extending the decay length of the $\partial\rho_R / \partial v$ vs. v_{new} curve (Figure 14.2) but not affecting its peak $\partial\rho_R^* / \partial v$ value.

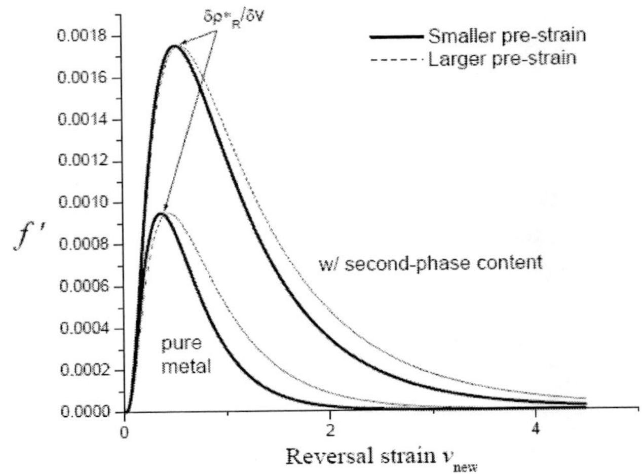

Figure 14.2. Illustration of the release function f' and how it changes with v_{new}, which is roughly proportional to v_{R}.

In a reversal, the release of ρ_R^α will alter the slip resistance by an amount $\Delta\tau_R^\alpha$. This difference $\Delta\tau_R^\alpha$ is made proportional to the normalized release rate f' and to the degree of reversal $P_0^{\alpha\alpha}$:

$$\Delta\tau_R^\alpha = \mu P_0^{\alpha\alpha} f'\left(v_{new}^\alpha\right) \tag{14.17}$$

The general form of $\partial\rho_R / \partial v$ and the influences of microstructure and pre-strain on $\partial\rho_R / \partial v$ are key assumptions in the model which, thus far, are found to work well. To operate the model, an empirical mathematical function for f' was proposed (Beyerlein and Tomé, 2007), which follows the form in Figure 14.2 and increases the decay length with pre-strain. This success, however, should not hinder development of a more physically based model.

Finally, the $\Delta\tau_{rev}^\alpha$ in Eq. (14.10) is the sum of $\Delta\tau_B^\alpha$ and $\Delta\tau_R^\alpha$, Eqs. (14.15) and (14.17):

$$\tau_{rev}^\alpha = \Delta\tau_B^\alpha + \Delta\tau_R^\alpha, \tag{14.18}$$

14.8. Change in Resistance due to the Cut-Through Mechanism

The conditions for the cut-through mechanism are a significant shift in slip activity from one or two slip planes to another ($X^{\alpha\alpha'} > X_c$ in Eq. (14.8)) and large pre-strains sufficient for developing a well-defined substructure. In particular when substructure evolves in the form of distinct dislocation walls or boundaries, it is important to account for their directionality. Here it is assumed that they form along the most active planes and become well developed after a sufficient amount of strain. Let ρ_{sub} be the dislocation density in the sub-boundaries and further let it be associated with a slip plane α, ρ^{α}_{sub}. At low strain levels, the density ρ^{α}_{sub} will be negligible until a certain strain is reached, $\nu^{\alpha} = \nu_1$, after which it will grow with ν^{α} and eventually saturate at $\nu^{\alpha} = \nu_2$, as recovery processes within the substructure dominant. The resistance posed by this sub-boundary is calculated as

$$\tau^{\alpha}_{d} = k\mu \frac{b}{d} \ln\left(\frac{d}{b}\right) \quad \propto \quad k\mu b\sqrt{\rho^{\alpha}_{sub}} \ln\left(\frac{1}{b\sqrt{\rho^{\alpha}_{sub}}}\right) \tag{14.19}$$

where b/d is the ratio of the Burgers vector to the dislocation spacing in the boundary, which is proportional to $b\sqrt{\rho^{\alpha}_{sub}}$ (approximate). Eq. (14.19) is an extended Taylor law believed to be applicable to large dislocation densities (Basinski and Basinski, 1979), such as those that would constitute deformation substructures. Its contribution to hardening by substructures was demonstrated in dislocation dynamics simulations (Madec et al., 2003). The constant $k = 0.086$ and is defined such that Eq. (14.19) approaches the classic Taylor law $\mu b\sqrt{\rho}$ for small ρ.

In a dislocation density formulation, ρ_{sub} can be calculated, such as in (Capolungo et al., 2009a,b). Here, for simplicity, the following non-linear empirical relationship is used:

$$b\sqrt{\rho^{\alpha}_{sub}} \approx \left(\frac{b}{d}\right)^{(\alpha)} = \frac{k_c}{1+\exp(-2(2\nu^{\alpha} - \nu_2 - \nu_1)/(\nu_2 - \nu_1))} \tag{14.20}$$

As illustrated in Figure 14.3, this form respects the onset ($\nu^{\alpha} = \nu_1$), development ($\nu_1 < \nu^{\alpha} < \nu_2$), and recovery ($\nu^{\alpha} > \nu_2$) involved in substructure development.

In the event of a strain path change, (when $X^{\alpha\alpha'} > X_c$) the initial resistance $\Delta\tau^{\alpha}_{cut,0}$ experienced by α-dislocations posed by all boundaries α' non-coplanar to α is.

$$\Delta\tau^{\alpha}_{cut,0} = \sum_{\alpha' \neq \alpha} X_0^{\alpha\alpha'} \tau^{\alpha'}_{d} \tag{14.21}$$

where we recall that $X_0^{\alpha\alpha'} = X^{\alpha\alpha'}$ when $X^{\alpha\alpha'} > X_c$ and zero otherwise (Eq. (14.8)). With further shear straining on the α plane, the α-dislocation population can either push through boundaries α' or react with the dislocations in boundaries α', lowering the resistance for subsequent slip. Accordingly, $\Delta\tau^{\alpha}_{cut}$ will decay at a rate proportional to $\dot{\nu}^{\alpha}$:

$$\Delta\dot{\tau}^{\alpha}_{cut} = -\omega\Delta\tau^{\alpha}_{cut}\dot{\nu}^{\alpha} \tag{14.22}$$

where ω characterizes the rate at which newly activated glide by α locally dissolves the substructure.

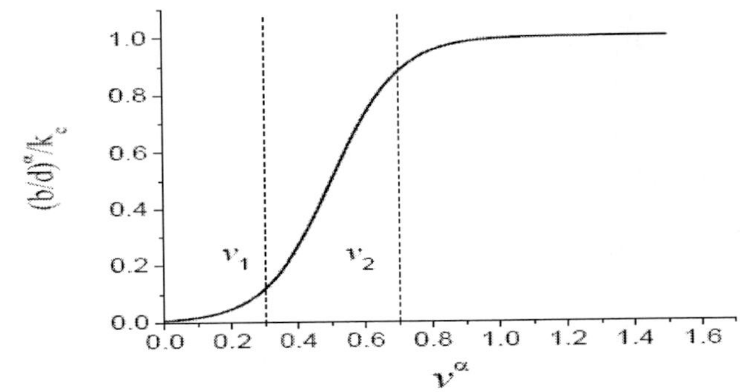

Figure 14.3. Relationship between the boundary quantity *b*/*d* (related to dislocation density) along plane α and ν^{α}, the accumulated shear on plane α. The meaning of ν_1 and ν_2 are also shown. See Eq. (14.20).

Note that $\Delta\tau^{\alpha}_{B}$, $\Delta\tau^{\alpha}_{R}$, and $\Delta\tau^{\alpha}_{cut}$ have a preceeding '*Δ*' to indicate that these are perturbations from the usual work hardening term τ^{α}_{h}. If an individual slip plane does not undergo a reversal ($P^{\alpha\alpha} \geq 0$) or a grain does not experience a significant shift in slip activity ($X^{\alpha\alpha'} \sim 0$), the constitutive law does not change from that of the Voce law. Otherwise, the corresponding *Δ*-stress contribution becomes nonzero. These *Δ*-stresses include one of the strain path indicators, either $P^{\alpha\alpha}_{0}$ or $X^{\alpha\alpha'}_{0}$. In doing so, they are zero when there is no strain path change and they become non-zero and increase with the severity of the strain path change. When and if the indicator is satisfied depends on the slip activity within a grain, its orientation, and individual deformation history, and therefore, may not necessarily coincide with the macroscopic deformation.

In some grains it is possible that both the cut-through and reversal mechanisms will be active. One way of identifying a 'mixed' grain is when one of its four crystallographic planes meets the criteria of $P^{\beta\beta}_{0} < -0.15$ and $X^{\alpha\alpha'} > 0.15$ (Yapici et al., 2007), where ± 0.15 is set arbitrarily. Satisfying such conditions indicates new slip activity on α is cutting through another plane α', of different orientation, undergoing a reversal. Planes undergoing a reversal have their stored dislocation densities reduced and it is anticipated that such planes would pose much less resistance to slip on non-coplanar planes. In other words, the cut-through effect would be diminished. As a conservative approach, in mixed grains $\tau^{\alpha}_{cut,0}$ is set to zero.

When the single crystal model described here is implemented into VPSC, the resulting constitutive model is able to calculate the anisotropic plastic response of SPD-processed metals in subsequent mechanical testing. In addition to texture evolution and dislocation storage, three distinct, strain-path-change mechanisms contribute to the predicted anisotropy: (1) reverse glide of dislocations previously held by backstresses, (2) reverse glide of dislocations previously held by barriers, and (3) the barrier effect and dissolution of substructure by new dislocation slip activity.

14.9. APPLICATION: PLASTIC ANISOTROPY OF COPPER

In this section, the model described above is applied to a series of uniaxial tests that were carried out by (Yapici et al., 2007) on pure polycrystalline Cu processed by one pass ECAE. After one pass, the processed samples were tested in three orthogonal directions (ED, LD, and FD), both in tension and compression, as illustrated in Figure 14.4. The billet cross-section (~2.5cm×2.5cm) was sufficiently large so that tension testing in directions normal to the billet axis (LD and FD) was possible. The starting material is practically isotropic and does not exhibit tension-compression asymmetry prior to SPD. The simulation begins using the initial measured texture represented by 2369 orientations, Voce parameters fit to axial tests on the starting material (see Table 14.1), $\hbar^{\alpha\alpha'}$ parameters for anisotropic 'active' hardening (Eqs. (14.11,14.12)), and spherical grains. ECAE and the subsequent mechanical test are then simulated sequentially and continuously. Deformation provided by a single ECAE pass is modeled using the velocity gradient of the ideal fan model with β_m = 10° Eq. (12.12). Estimations of the cut-through (Table 14.2) and reversal parameters (Table 14.3) are made such that a single set of parameters could effectively model the flow response in all six tests.

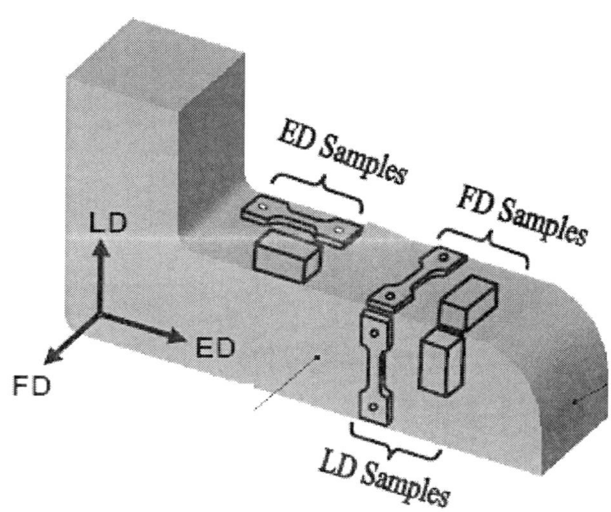

Figure 14.4. Schematic of the copper billet during an ECAE extrusion. After the extrusion, mechanical test samples were cut from the center of the billet in the orientations indicated. For illustration only, these are enlarged and depicted on the sample surface. ED, LD and FD denote the extrusion, longitudinal and flow directions, respectively.

Table 14.1.

Shear Modulus μ [GPa]	Grain size [μm]	τ_0 [MPa]	τ_1 [MPa]	θ_0 [MPa]	θ_1 [MPa]	h_{sat}	ν_0
42	25	156	0	0	9	0.6	0.01

Table 14.1. Material parameters used for forest (Voce) and latent hardening of Cu.

Table 14.2.

Symbol	Estimatio	Description
k_c	6.5 x 10^{-2}	Dislocation density coefficient
□	12	Rate of cut-through
ν_1	0.3	Strain at which wall formation begins
ν_2	0.7	Strain at which wall formation is completed

Table 14.2. Material parameters for the cut-through mechanism in Cu.

Table 14.3.

Symbol	Estimate	Description
μ_R	□□□□	Lognormal scale parameter
σ_R	0.85	Lognormal shape parameter
ϑ_B	15	Rate of saturation for reverse shear
$\nu_{B,sat}$	2.8 x 10^{-3}	Maximum shear available for Bauschinger effect
f_{peak}	0.95x10^{-3}	Peak value in release rate function
ν_B	0.01	Strain for decay of internal stress
ν_R	0.4	Measure of pre-strain in release rate function

Table 14.3. Material parameters used for the reversal mechanism in Cu.

Figure 14.5 compares the calculated plastic behavior with the measured ones in the (a) ED, (b) LD, and (c) FD directions. It is observed that the plastic response depends on the direction and sense of loading. In both tension and compression, the flow stresses in the FD are the highest, followed by LD and then ED. There is a tension-compression asymmetry in all directions being the most pronounced in FD. The asymmetry is directional; in the FD, tension stresses are lower than the compression stresses, while in the LD and ED the tension stresses are higher than the compression stresses.

As shown in Figure 14.5, the agreement is very good and, based on the effectiveness of the model, several interpretations can be made. Table 14.4 lists the calculated weight percentage of grains in the polycrystal that experience only reversal, only cut-through, or both mechanisms for each mechanical test. The grains experiencing no strain-path-change mechanisms are not included. ED tension experienced virtually no strain path change and contained the least amount of grains with active strain path change mechanisms. Its response is that of linear hardening, much like one that would be expected if it were possible to continuously strain the material in the same direction as the ECAE pass. ED compression, on the other hand, had the highest number of grains undergoing a reversal. LD tension had the second highest number of reversal grains. Consequently, these two responses exhibit a lower yield stress than that of ED tension and characteristics suggesting a Bauschinger effect. For instance the ED compression response has an extended plateau region of suppressed hardening, a characteristic response of metals during large strain reversals. Compression and tension in the FD both invoked the cut-through mechanism in a large number of grains. As a result, their flow responses exhibit an initial spike in stress followed by appreciable work-softening.

Table 14.4.

	ED Tens	LD Tens	FD Tens	ED Comp	LD Comp	FD Comp
Reversal Only	0.0%	39.8%	23.9%	51.6%	0.0%	0.5%
Cut-Through Only	17.1%	25.9%	51.2%	17.2%	27.8%	60.1%
Mixed Only	0.0%	2.1%	9.3%	0.7%	0.0%	0.1%

Table 14.4. Weight percentage of grains experiencing different strain-path-change mechanisms during the post-straining of ECAE processed Cu.

The tension-compression asymmetry in the ED and FD can be explained by strain path change effects. In ED and FD, the tension response is lower than the compression response because of the higher fraction of grains undergoing reversal during tension. In FD, the tension response also had the highest fraction of 'mixed' grains, which contributed its lower flow stress compared to FD-compression. Recall that in mixed grains, the cut-through effect is removed due to the reduction in stored dislocation density by simultaneous reversals. The agreement suggests that reversal mechanisms weaken previously developed dislocation walls and locally diminish the strength of these obstacles against new gliding dislocations. If this interaction between the cut-through and reversal mechanism within mixed grains were removed and hence the two mechanisms made to operate independently, the tension-compression asymmetry in FD observed in measurement would not be predicted (Yapici et al., 2007). (The other calculations were unaffected because the percentage of mixed grains in these other tests was small.)

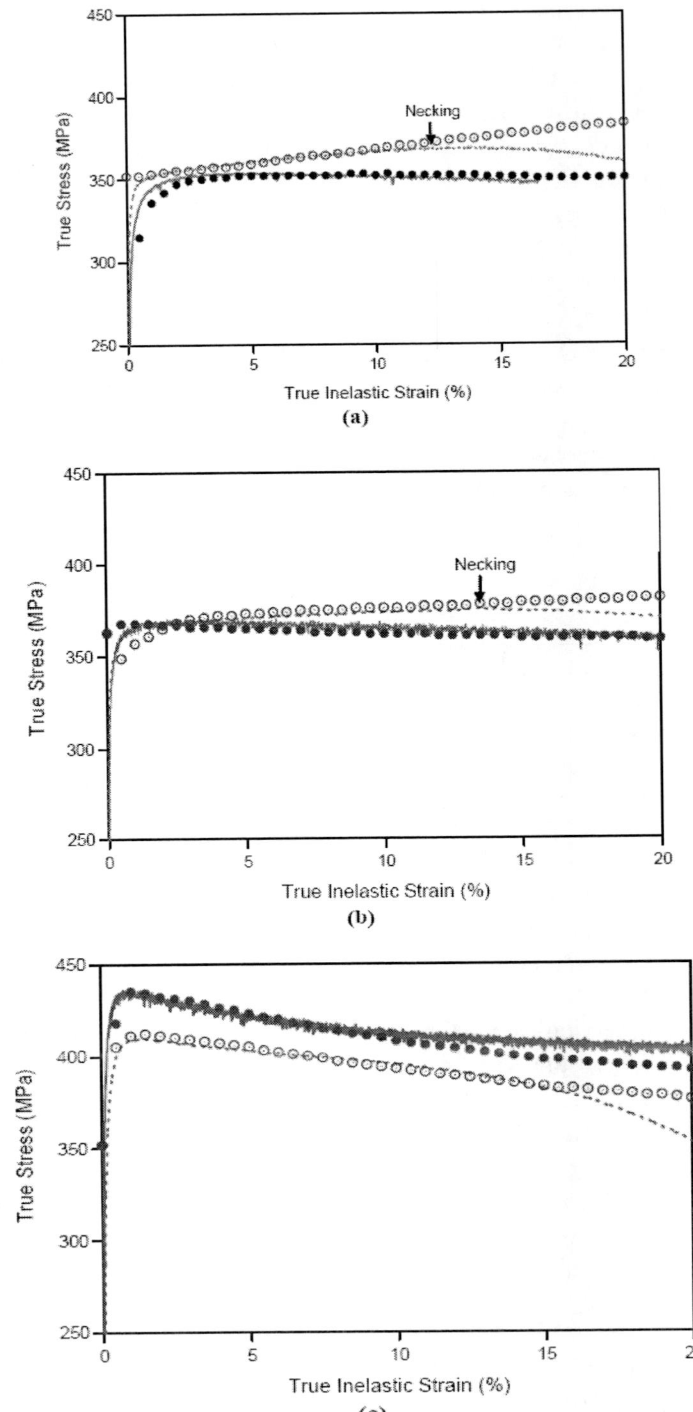

Figure 14.5. Experimental and simulated tension-compression behavior of ECAE processed copper during axial monotonic straining along (a) ED, (b) LD and (c) FD, where ED, LD and FD denote the extrusion, longitudinal and flow directions as indicated on Fig. 14.4.

Texture evolution is included in all these simulations and played a particularly relevant role in the LD tests. The tension-compression asymmetry in the LD can be attributed to a competition between texture evolution and strain path change effects. Due to texture evolution, the hardening rate in LD tension would exceed that in LD compression. Due to the higher percentage of reversal grains in LD tension than LD compression (Table 14.4), the hardening rate in LD compression would exceed that in LD tension. As shown in Figure 14.5(b), texture (geometric) hardening prevails and after 5% straining, the LD tension response is higher than the LD compression response.

In this application of the model, we have shown that the changes in flow stress, work hardening, and plastic anisotropy observed in processed samples can be attributed to the texture and microstructural changes developed during ECAE and to strain path change effects induced when going from the extrusion to the axial mechanical test. These factors will certainly play a role in the plastic behavior of other SPD processed metals and non-negligible plastic anisotropy in flow stress and work-hardening can be expected.

The single crystal hardening law featured in this section builds upon a phenomenological Voce law in which slip resistance is a function of accumulated strain, not dislocation density. Using the same philosophy the changes in slip resistance due to changes in slip activity and concomitant dislocation rearrangement are also modeled as a function of strain and strain rate measures. As a consequence, while the processes being modeled are based on dislocation motion and storage, they are not a direct function of them. Certainly the model presented in the previous sections can be made more physical if these same resistances were functions of dislocation density and substructure evolution. Although such an endeavor is highly recommended, models for dislocation density and substructure evolution are, however, not without their own uncertainties and multitude of parameters. If such a pursuit is undertaken, the important features of the current anisotropic single crystal law that should be retained are 1) the way in which hardening evolution is related to the directionality and magnitude of shear in each slip plane of each grain, 2) the coupling between the previous shear activity with new shear activity induced by changes in strain path, and 3) the introduction of strain-path-change matrix indicators, which characterize and quantify strain path changes (e.g., reversal, continuous, cross) in individual grains.

REFERENCES

Agnew, S.R.; Kocks, U.F.; Hartwig, K.T.; Weertman, J.R. Texture evolution during equal channel angular forging and dislocation boundary misorientations after severe plastic deformation In *Proceedings of the 19th Risø International Symposium on Materials Science*; Carstensen, J. V.; Leffers, T.; Lorentzen, T.; Pedersen, O. B.; Sørensen, B. F.; Winther, G.; Risø National Laboratory: Roskilde, Denmark, 1998; Vol. 19, pp 201-207.

Agnew, S.R.; Mehrotra, P.; Lillo, T.M.; Stoica, G.M.; Liaw, P.K. Texture evolution of five wrought magnesium alloys during route A equal channel angular extrusion: Experiments and simulations *Acta Mater.* 2005a, *53*, 3135-46.

Agnew, S.R.; Mehrotra, P.; Lillo, T.M.; Stoica, G.M.; Liaw, P.K. Crystallographic texture evolution of three wrought magnesium alloys during equal channel angular extrusion *Mater. Sci. Engng.* A 2005b, *408*, 72-78.

Akhtar A. Basal slip in zirconium Acta Metall 1973;21:1.

Akhtar, A.; Teghtsoonian, A. Plastic deformation of zirconium single crystals *Acta Metall.* 1971, *19*, 655-63.

Alexander, D. J.; Vogel, S.; Li, S. 2004a, unpublished data.

Alexander, D. J.; Beyerlein, I. J. Anisotropy in mechanical properties of high-purity copper processed by equal channel angular extrusion *Mater. Sci. Engng.* A 2005, *410-411*, 480-484.

Alexander, D. J.; Beyerlein, I. J. Mechanical properties of high-purity copper processed by equal channel angular extrusion In *Ultra Fine-Grain Symposium III*; Zhu, Y. T.; Langdon, T. G.; Valiev, R. Z.; Semiatin, S. L.; Shin, D. H.; Lowe, T. C.; The Minerals, Metals & Materials Society: Warrendale, PA, 2004b; Vol. 3, pp 517-522.

Asaro R.J.; Needleman, A. Texture development and strain hardening in rate dependent polycrystals *Acta Metall.* 1985, *33*, 923-953.

Backofen, W.A. The torsion texture of copper *Trans. AIME* 1950, *188*, 1454-9.

Baik, S. C.; Estrin, Y.; Hellmig, R. J.; Jeong H-T.; Brokmeier, H-G.; Kim, H.S. Plastic anisotropy in aluminum and copper pre-strained by equal channel angular extrusion. *Z. Metallkd.* 2003, *94*, 1189-98.

Basinski, S. J.; Basinski,, Z. S. *Dislocations in Solids*; North-Holland, Amsterdam, 1979; Vol. 4, pp 261-362.

Beaudoin, A.J.; Engler, O. Deformation Processing: Texture Evolution, In *Encyclopedia of Materials: Science and Technology;* Elsevier Science Ltd.: Amsterdam, The Netherlands, 2001; pp 2014-22.

Beausir, B.; Tóth, L.S.; Neale, K.W. Ideal orientations and persistence characteristics of hexagonal close packed crystals in simple shear *Acta Mater.* 2007, *55*, 2695-2705.

Beyerlein, I.J.; Tomé, C.N. Modeling transients in the mechanical response of copper due to strain path changes *Int. J. Plast.* 2007, *23*, 640-64.

Beyerlein, I.J.; Tóth, L.S.; Tomé, C.N.; Suwas, S. Role of twinning on texture evolution of silver during equal channel angular extrusion *Phil. Mag*. 2007a, *87*, 885-906.

Beyerlein, I.J.; Field, R.D.; Hartwig, K.T.; Necker, C.T., Texture development in two-pass ECAE-processed beryllium *J. Mater. Sci.* 2008, *43*, 7465-7473.

Beyerlein, I. J.; Li, S.; Alexander, D. J.; Necker, C. T.; Tomé, C. N.; Bourke, M. A. M.; Heterogeneity in texture development in single pass equal channel angular extrusion. In *Ultra Fine-Grain Symposium III*; Zhu, Y. T.; Langdon, T. G.; Valiev, R. Z.; Semiatin, S. L.; Shin, D. H.; Lowe, T. C.; The Minerals, Metals & Materials Society: Warrendale, PA, 2004; Vol. 3, pp 185-192.

Beyerlein, I. J.; Tomé, C. N. Modeling directional anisotropy in grains subjected to large strains and strain path changes, In *Ultrafine Grained Materials IV*; Zhu, Y. T.; Langdon, T. G.; Horita, Z.; Zehetbauer, M. J.; Semiatin, S. L.; Lowe, T.C.; The Minerals, Metals & Materials Society: Warrendale, PA, 2006; Vol. 4, pp. 63-71.

Beyerlein, I. J.; Toth, L. S. Texture evolution in equal channel angular extrusion *Prog. Mater. Sci.*, 2008, *54*, 427-510.

Beyerlein, I.J.; Lebensohn, R.A.; Tomé, C.N. Modeling texture and microstructural evolution in the equal channel angular extrusion process *Mater. Sci. Engng. A* 2003, *345*, 122-38.

Beyerlein, I.J.; Li, S.; Necker, C.T.; Alexander, D.J.; Tomé, C.N. Non-uniform microstructure and texture evolution during equal channel angular extrusion *Phil. Mag.* 2005a, *85*, 1359-94

Beyerlein, I.J.; Tomé, C.N. Analytical modeling of material flow in equal channel angular extrusion (ECAE) *Mater. Sci. Engng. A* 2004, *380*, 171-90.

Beyerlein, I. J.; Tomé, C.N. A dislocation-based constitutive law for pure Zr including temperature effects I*nt. J. Plast.* 2008, *24*, 867-95.

Beyerlein, I.J.; Tomé, C.N. Modeling compression reloads in copper pre-strained by equal channel angular extrusion (ECAE) *Mater. Sci. Engng. Tech.* 2005, *36*, 541-545.

Beyerlein, I. J.; Li, S.; Alexander, D. J. Modeling the plastic anisotropy in pure copper after one pass ECAE *Mater. Sci. Engng.* A 2005b, *410-411*, 201-206.

Beyerlein, I. J.; Alexander, D. J.; Tomé, C. N. Plastic anisotropy in aluminum and copper pre-strained by equal channel angular extrusion (ECAE) *J. Mater. Sci.* 2007b, *42*, 1733-1750.

Beyerlein, I. J.; L. Capolungo, L; Yapici, G. G.; Tomé, C. N.; Karaman, I. Multi-scale modeling of texture evolution in beryllium and zirconium during equal channel angular extrusion, In: *Ductility of Bulk Nanostructured Materials*, Y. Zhao and X. Liao, eds, In press, 2009a.

Beyerlein, I. J.; Mara, N. A.; Bhattacharyya, D.; Alexander, D. J.; Necker, C. T. Texture evolution via combined slip and deformation twinning in rolled silver-copper eutectic nanocomposite. Int. J. Plast. 2009b, submitted.

Bishop, J.F.W.; Hill, R. A theory of the plastic distortion of a polycrystalline aggregate under combined stresses *Phil. Mag.* 1951a, *42*, 414-427.

Bishop, J.F.W.; Hill, R. A theoretical derivation of the plastic properties of a polycrystalline face-centred metal *Phil. Mag.* 1951b, *42*, 1298-1307.

Bonet, J.; Wood, R. D. *Nonlinear continuum mechanics for finite element analysis*; Cambridge University Press: Cambridge, UK, 1997.

Bowen, J.R.; Gholinia, A.; Roberts, S.M.; Prangnell, P.B. Analysis of the billet deformation behaviour in equal channel angular extrusion *Mater. Sci. Engng. A* 2000, *287*, 87-99.

Brown, D.W.; Jain, A.; Agnew, S.R.; Clausen, B. Twinning and detwinning during cyclic deformation of Mg alloy AZ31B *Mater. Sci. Forum* 2007, *539-543*, 3407-3413.

Budilov, I. N.; Alexandrov, I. V.; Lukaschuk, Y. V.; Beyerlein, I. J.; Zhernakov, V. S. Three dimensional modeling of plastic deformation flow during ECAP In *Ultra Fine-Grain Symposium III*; Zhu, Y. T.; Langdon, T. G.; Valiev, R. Z.; Semiatin, S. L.; Shin, D. H.; Lowe, T. C.; The Minerals, Metals & Materials Society: Warrendale, PA, 2004; Vol. 3, pp 193-198.

Bunge, H.J. Texture Analysis in Materials Science – Mathematical Methods; Butterworths: London, 1982.

Canova, G.R.; Kocks, U.F.; Jonas, J.J. Theory of torsion texture development *Acta Metall.* 1984, *32*, 211-26.

Capolungo, L.; Beyerlein, I. J.; Tomé, C. N. Slip-assisted twin growth in hexagonal close packed metals *Scripta Mater.* 2009a, *60,* 32-35.

Capolungo, L.; Beyerlein, I.J.; Kaschner, G.C.; Tomé, C.N. On the interaction between slip dislocations and twins in HCP Zr *Mater. Sci. Engng.* 2009b, *A 513-514,* 42-51.

Christian, J.W.; Mahajan, S. Deformation twinning *Prog. Mater. Sci.* 1995, *39*, 1-157.

Cui, H. J.; Goforth, R. E.; Hartwig. K. T. The three dimensional simulation of flow pattern in equal-channel angular extrusion *J. Mater.-E*, Aug. 1998, *50*, 1-4.

Dawson, P.R.; MacEwen, S.R.; Wu, P.D. Advances in sheet metal forming analyses: dealing with mechanical anisotropy from crystallographic texture *Int. Mater. Rev.* 2003, *48*, 86-122.

Dumoulin, S.; Roven, H.J.; Werenskiold, J.C.; Valberg, H.S. Finite element modeling of equal channel angular pressing: Effect of material properties, friction and die geometry *Mater. Sci. Engng. A* 2005, *410-411*, 248-51.

Engler, O.; Huh, M.Y.; Tomé, C.N. A study of through-thickness texture gradients in rolled sheets *Metall. Mater. Trans. A* 2000, *31*, 2299-2315.

Enikeev, N.A.; Abdullin, M.F.; Nazarov, A.A.; Beyerlein, I.J. Modelling grain refinement in fcc metals during equal-channel angular pressing by route "C" *Int. J. Mater. Res.* 2007, *98*, 167-71.

Eshelby, J.D. The determination of the elastic field of an ellipsoidal inclusion and related problems *Proc. Roy. Soc. London A* 1957, *241*, 376-396.

Fang, D.R.; Zhang, Z.F.; Wu, S.D.; Huang, C.X.; Zhang, H.; Li, J.J.; Huang, C.X.; Zhang, H.; Zhaob N.Q.; Li J.J. Effect of equal channel angular pressing on tensile properties and fracture modes of casting Al-Cu alloys *Mater. Sci. Engng. A* 2006, *426*, 305-13.

Ferrasse, S.; Segal, V.M.; Kalidindi, S.R.; Alford, F. Texture evolution during equal channel angular extrusion Part I. Effect of route, number of passes and initial texture *Mater. Sci. Engng. A* 2004, *368*, 28-40

Field, R.D.; Hartwig, K.T.; Necker, C.T.; Bingert, J.F.; Agnew, SR. Equal-channel angular extrusion of beryllium *Metall. Mater. Trans. A* 2002, *33*, 965-72.

Gholinia, A.; Bate., P.; Prangnell, P.B. Modelling texture development during equal channel angular extrusion of aluminium *Acta Mater.* 2002, *50*, 2121-36.

Gibbs, M.A.; Hartwig, K.T.; Cornwell, L.R.; Goforth, R.E.; Payzant, E.A. Texture formation in bulk iron processed by simple shear *Scripta Mater.* 1998, *39*, 1699-704.

Gurtin, M. E. *An Introduction to Continuum Mechanics*; Academic Press: San Diego, CA, 1981.

Han, W.Z.; Zhang, Z.F.; Wu, S.D.; Li, S.X.; Wang, Y.D. Anisotropic compressive properties of iron subjected to single-pass equal-channel angular pressing *Phil. Mag. Lett.* 2006, *86*, 435-41.

Han, W.Z.; Zhang, Z.F.; Wu, S.D.; Li, S.X. Nature of shear flow lines in equal-channel angular-pressed metals and alloys *Phil. Mag. Lett.* 2007, *87*, 735-41.

Han, B.Q.; Lavernia, E.J.; Mohamed, F.A. Mechanical properties of iron processed by severe plastic deformation *Metall. Mater. Trans.* A 2003, *34*, 71-83.

Hasegawa, T.; Yakou, T.; Karashima, S. Deformation behavior and dislocation structures upon stress reversals in polycrystalline aluminum. *Mater. Sci. Engng.* 1975, *20*, 267-276.

Hill, R. Generalized constitutive relations for incremental deformation of metal crystals by multislip *J. Mech. Phys. Solids* 1966, *14*, 95-102.

Him, Y.S.; Kim, J.; Kim, I.; Shin, D.H. In *Ultra Fine-Grain Symposium II* ; Zhu, Y.T. ; Langdon, T. G. ; Mishra, R. S. ; Semiatin, S. L. ; Saran, M. J. ; Lowe, T. C. ; The Minerals, Metals & Materials Society: Warrendale, PA, 2002; Vol. 2, pp 99-104.

Honneff, H.; Mecking, H. Analysis of the deformation texture at different rolling conditions, In *The Sixth International Conference on Texture of Materials;* Nagashima, S.; The Iron and Steel Institute of Japan: Tokyo, Japan, 1981; Vol. 6, pp 347-352.

Hosford, W.F. *The Mechanics of Crystals and Textured Polycrystals*; Oxford Engineering Science Series; Oxford Univ. Press: New York, NY, 1993.

Huang, X.; Liu, Q. Determination of crystallographic and macroscopic orientation of planar structures in TEM *Ultramicroscopy* 1998, *74*, 123-130.

Huang, C.X. ; Yang, G. ; Deng, B. ; Wu, S.D. ; Li , S.X. ; Zhang, Z.F. Formation mechanism of nanostructures in austenitic stainless steel during equal channel angular pressing *Phil. Mag.* 2007, *87*, 4949-71.

Huang, C.X.; Wang, K.; Wu, S.D.; Zhang, Z.F.; Li, G.Y.; Li , S.X. Deformation twinning in polycrystalline copper at room temperature and low strain rate *Acta Mater.* 2006, *54*, 655-65.

Hughes, D.A.; Lebensohn, R.A.; Wenk, H.R.; Kumar, A. Stacking fault energy and microstructure effects on torsion texture evolution *Proc. Roy. Soc. London A* 1996, *456*, 921-53.

Hughes, D. A.; Wenk, H. R. In *The Eighth International Conference on Textures of Materials*; Kallend, J.S.; Gottstein, G.; The Minerals, Metals, & Materials Society: Warrendale, PA, 1988; Vol. 8, pp 455-460.

Hutchinson, J.W. Bounds and self-consistent estimates for creep of polycrystalline materials *Proc. Roy. Soc. London A* 1976, *348*, 101-127.

Isaenkova, M.; Perlovich, Y.; Fesenko, V.; Grekhov, M.; Alexandrov, I. V.; Beyerlein, I.J. Formation of inhomogeneous texture and structure in metal materials under equal-channel angular pressing *Mater. Sci. Forum* (ICOTOM-14) 2005, *495-497*, 827-32.

Iwahashi, Y.; Wang, J.T.; Horita, Z.; Nemoto, M.; Langdon, T.G. Principle of equal channel angular pressing for the processing of ultra-fine grained materials *Scripta Mater.* 1996, *35*, 143-6.

Kaschner, G.C.; Tomé, C.N.; Beyerlein, I.J.; Vogel, S.C.; Brown, D.W.; McCabe, R.J. Role of twinning in the hardening response of Zr during temperature reloads *Acta Mater.* 2006, *54*, 2887-2896.

Kamachi, M.; Furukawa, M.; Horita, Z.; Langdon, T.G. A model investigation of the shearing characteristics in equal-channel angular pressing *Mater. Sci. Engng. A* 2003, *347*, 223-30.

Karaman, I.; Sehitoglu, H.; Beaudoin, A.J.; Chumlyakov, Y.I.; Maier H.J.; Tomé, C.N. Modeling the deformation behavior of Hadfield steel single and polycrystals due to twinning and slip *Acta Mater.* 2000, *48*, 2031-2047.

Kim, W.J.; Hong, S.I.; Kim, Y.S.; Min, S.H.; Jeong, H.T.; Lee, J.D. Texture development and its effect on mechanical properties of an AZ61 Mg alloy fabricated by equal channel angular pressing *Acta Mater.* 2003, *51*, 3293-307.

Kim, H. S.; Seo, H. H.; Hong, S. I. On the die corner gap formation in equal channel angular pressing *Mat. Sci. Engng. A* 2000, *291*, 86-90.

Kim, H. S. Evaluation of strain rate during equal channel angular pressing. *J. Mater. Res.* 2002, 17, 172-179.

Kim, H. S. Hong, S. I. Seo, M. H. Effects of strain hardenability and strain-rate sensitivity on the plastic flow and deformation homogeneity during equal channel angular pressing. *J. Mater. Res.* 2001, *16*, 856-864

Kim, H.S. Finite element analysis of equal channel angular pressing using a round corner die *Mater. Sci. Engng. A* 2001, *315*, 122-8.

Kim, H.S.; Seo, M.H.; Hong, S.I. Finite element analysis of equal channel angular pressing of strain rate sensitive metals *J. Mater. Proc. Tech.* 2002, *130-131*, 497-503.

Kocks, U. F.; Argon, A. S.; Ashby, M. F. Thermodynamics and kinetics of slip *Prog. Mater. Sci.* 1975, *19*, 1 -288.

Kocks, U.F.; Canova, G.R.; Jonas, J.J. Yield vectors in FCC crystals *Acta Metall.* 1983, *31*, 1243-1252.

Kocks, U. F.; Hasegawa, T.; Scattergood, R. O. Origin of cell walls of lattice misorientations during deformation. *Scripta Metall.* 1980, *14*, 449-454.

Kocks, U.F.; Mecking, H. Physics and phenomenology of strain hardening: the FCC case *Prog. Mater. Sci.* 2003, *48*, 171-273.

Kocks, U.F.; Tomé C.N.; Wenk, H.-R. Texture and Anisotropy - Preferred Orientations in Polycrystals and their Effect on Material Properties; Cambridge University Press (1998)

Kok, S.; Beaudoin A.J.; Tortorelli, D.A. A polycrystal plasticity model based on the mechanical threshold *Int. J. Plast.* 2002, *18*, 715-741.

Kuhlmann-Wilsdorf, D. A new theory of work hardening *Trans. Metall. Soc. AIME* 1962, *224*, 1047-1061.

Lapovok, R.; Thomson, P.F.; Cottam, R.; Estrin, Y. The effect of grain refinement by warm equal channel angular extrusion on room temperature twinning in magnesium alloy ZK60 *J. Mater. Sci.* 2005, *40*, 1699-708.

Lebensohn, R.A.; Tomé, C.N. Ponte Castañeda, P. Self-consistent modeling of the mechanical behavior of viscoplastic polycrystals incorporating intragranular field fluctuations *Phil. Mag.* 2007, *87*, 4287-4322

Lebensohn, R.A.; Tomé, C.N. A self-consistent anisotropic approach for the simulation of plastic deformation and texture development of polycrystals: Application to zirconium alloys *Acta Metall. Mater.* 1993, *41*, 2611-2624.

Li, S. *Acta Mater*. Orientation stability in equal channel angular extrusion. Part II: hexagonal close-packed materials. 2008, *56*, 1031-1043.

Li, S.; Alexander, D.J.; Beyerlein, I.J.; Brown, D.W. Texture evolution during multi-pass equal channel angular extrusion of beryllium. In *The Fifthteenth International Conference on Textures of Materials*; Rollett, A. D.; Carnegie Mellon University: Pittsburgh, PA, 2008; Vol. 15, CD-ROM.

Li, S.; Beyerlein, I.J. Modelling texture evolution in equal channel angular extrusion of bcc materials: effects of processing route and initial texture *Model. Simul. Mater. Sci. Engng.* 2005, *13*, 509-30.

Li, S.; Beyerlein, I.J.; Alexander, D.J. Characterization of deformation textures in pure copper processed by equal channel angular extrusion via route A *Mater. Sci. Engng. A* 2006a, *431*, 339-45.

Li, S.; Beyerlein, I.J.; Alexander, D.J.; Vogel, S.C. Texture evolution during multi-pass equal channel angular extrusion of copper: Neutron diffraction characterization and polycrystal modeling *Acta Mater.* 2005a, *53*, 2111-25.

Li, S.; Beyerlein, I.J.; Alexander, D.J.; Vogel, S.C. Texture evolution during equal channel angular extrusion: Effect of initial texture from experiment and simulation *Scripta Mater.* 2005c, *52*, 1099-104.

Li, S.; Beyerlein, I.J.; Bourke, M.A.M. Texture formation during equal channel angular extrusion of fcc and bcc materials: comparison with simple shear *Mater. Sci. Engng. A* 2005b, *394*, 66-77.

Li, S.; Beyerlein, I.J.; Necker, C.T. On the development of microstructure and texture heterogeneity in ECAE via route C *Acta Mater.* 2006b, *54*, 1397-408.

Li, S.; Beyerlein, I.J.; Necker, C.T.; Alexander, D.J.; Bourke, M.A.M. Heterogeneity of deformation texture in equal channel angular extrusion of copper *Acta Mater.* 2004a, *52*, 4859-75.

Li, S.; Bourke, M. A. M.; Beyerlein, I. J.; Alexander, D. J.; Clausen, B. Finite element analysis of plastic deformation and deformation zones in equal channel angular extrusion *Mat. Sci. Engng. A* 2004b, 382, 217-236.

Li, S., Gazder, A.A.; Beyerlein, I.J.; Davies, C.H.J.; Pereloma, EV. Microstructure and texture evolution during equal channel angular extrusion of interstitial-free steel: Effects of die angle and processing route *Acta Mater.* 2007, *55*, 1017-32.

Li, S.; Gazder, A.A.; Beyerlein, I.J.; Pereloma, E.V.; Davies, C.H.J. Effect of processing route on microstructure and texture development in equal channel angular extrusion of interstitial-free steel *Acta Mater.* 2006c, *54*, 1087-10

Liao, X.Z.; Zhao, Y.H.; Srinivasan, S.G.; Zhu, Y.T.; Valiev, R.Z.; Gunderov, D.V. Deformation twinning in nanocrystalline copper at room temperature and low strain rate *Appl. Phys. Lett.* 2004, *84*, 592-4.

Liu, Y.S.; Delannay, L.; Van Houtte, P. Application of the Lamel model for simulating cold rolling texture in molybdenum sheet *Acta Mater.* 2002, *50*, 1849-56.

Ma, Q.; Clarke, D. R. Size-dependent hardness of silver single crystals. *J. Mater. Res.* 1995, *10*, 853-863.

Madec, R.; Devincre, B.; Kubin, L.; Hoc, T.; Rodney, D. The role of collinear interaction in dislocation-induced hardening *Science* 2003, *301*, 1879-1882.

Mahesh, S.; Beyerlein, I.J.; Tomé, C.N. Loading- and substructure-induced irreversibility in texture during route C equal channel angular extrusion *Scripta Mater.* 2005, *53*, 965-9.

Mahesh, S.; Tomé, C.N. Deformation banding under arbitrary monotonic loading in cubic metals *Phil. Mag.* 2004, *84*, 3517-46.

Mahesh, S.; Tomé, C. N.; McCabe, R. J.; Kaschner, G. C.; Beyerlein, I. J.; Misra, A. Application of a substructure-based hardening model to copper under loading path changes *Metall. Mater. Trans. A* 2004, *35*, 3763-3774.

Manping, L.; Roven, H.J.; Yingda, Y. Deformation twins in ultrafine grained commercial aluminum *Int. J. Mater. Res.* 2007, *98*, 184-90.

Matthews, J. W. Role of deformation twins in fracture of single-crystal films. *Acta Metall.* 1970, *18*, 175.

McCabe, R.J.; Cerreta, E.K.; Misra, A.; Kaschner, G.C.; Tomé, C.N. Effects of texture, temperature and strain on the deformation modes of zirconium *Phil. Mag.* 2006, *86*, 3595-611.

McNelley, T.R.; Swisher, D.L. Deformation banding and grain refinement during ECAP of aluminum In *Ultra Fine-Grain Symposium III*; Zhu, Y. T.; Langdon, T. G.; Valiev, R. Z.; Semiatin, S. L.; Shin, D. H.; Lowe, T. C.; The Minerals, Metals & Materials Society: Warrendale, PA, 2004; Vol. 3, pp 89-94.

Mecking, H.; Kocks, U. F. Kinetics of flow and strain-hardening *Acta Metall.* 1981, 29, 1865-1875.

Mendelson , S. Dislocation dissociations in hcp metals *J. Appl. Phys.* 1970, *41*, 1893-1910.

Mendelson , S. Zonal dislocation and twin lamellae in hcp metals *Mater. Sci. Engng. A* 1969, *4*, 231-242.

Miyamoto, H.; Erb, U.; Koyama, T.; Mimaki, T.; Vinogradov, A.; Hashimoto, S. Microstructure and texture development of copper single crystals deformed by equal-channel angular pressing *Phil. Mag. Lett.* 2004, *84*, 235-243.

Molinari, A.; Canova, G.R.; Ahzi, S. A self-consistent approach of the large deformation polycrystal viscoplasticity *Acta Metall.* 1987, *35*, 2983-2994.

Montheillet, F.; Cohen, M.; Jonas, J.J. Axial stresses and texture development during the torsion testing of Al, Cu and alpha-Fe *Acta Metall.* 1984, *32*, 2077-89.

Nazarov, A.A.; Enikeev, N.A.; Romanov, A.E.; Orlova, T.S.; Alexandrov, I.V.; Beyerlein, I.J.; Valiev, R.Z. Analysis of substructure evolution during simple shear of polycrystals by means of a combined viscoplastic self-consistent and disclination modeling approach *Acta Mater.* 2006, *54*, 985-95.

Park, J.-W.; Suh, J.-Y. Effect of die shape on the deformation behavior in equal-channel angular pressing *Metall. Mater. Trans. A* 2001, *32*, 3007-3014.

Partridge, P.G. The crystallography and deformation modes of hcp metals *Metall. Rev.* 1967, *12*, 169-94.

Perlovich, Y.; Isaenkova, M.; Fesenko, V.; Grekhov, M.; Yu, S.H.; Hwang, S.K.; Shin, D.H. Features of texture and structure development in zirconium under equal channel angular pressing *Mater. Sci. Forum* 2006b, *503-504*, 859-64.

Perlovich, Y.; Isaenkova, M.; Fesenko, V.; Grekhov, M.; Alexandrov, I.; Beyerlein, I.J. Formation of texture and structure in rods of copper and titanium under equal-channel angular pressing *Mater. Sci. Forum* 2006a, *503-504*, 853-858.

Proust, G.; Tomé, C. N.; Kaschner, G. C. Modeling texture, twinning and hardening evolution during deformation of hexagonal materials *Acta Mater.* 2007, *55*, 2137-2148.

Proust, G.; Tomé, C.N.; Jain, A. ; Agnew, S.R. Modeling the effect of twinning and detwinning during strain path changes of Mg alloy AZ31 *Int. J. Plast.* 2009, *25*, 861-880.

Proust, G.; Kaschner G. C.; Beyerlein, I. J.; Clausen, B.; Brown, D. W.; McCabe, R.J.; Tomé, C. N. Detwinning of high-purity Zirconium: *in-situ* neutron diffraction experiments *Exper. Mech.* 2008 DOI 10.1007/s11340-008-9213-6

Rapperport EJ. Room temperature deformation processes in zirconium Acta Metallurgica 1959;7:254.

Rollett, A.D.; Lowe, T.C.; Kocks, U.F.; Stout, M.G. The microstructure and texture of torsion-reverse torsion experiments. In *The Eighth International Conference on Textures of Materials*; Kallend, J.S.; Gottstein, G.; The Minerals, Metals, & Materials Society: Warrendale, PA, 1988; Vol. 8, pp 473-478.

Segal, V.M. Materials processing by simple shear *Mater. Sci. Engng. A* 1995, *197*, 157-64.

Segal, V.M. Equal channel angular extrusion: from macromechanics to structure formation *Mater. Sci. Engng. A* 1999, *271*, 322-33.

Segal, V. M. Slip line solutions, deformation mode and loading history during equal channel angular extrusion *Mater. Sci. Engng A* 2003, *345*, 36-46.

Semiatin, S. L.; Delo, D. P.; Shell, E. B. The effect of material properties and tooling design on deformation and fracture during equal channel angular extrusion. *Acta Mater.* 2000, *48*, 1841-1851

Shan, A.D.; Moon, I.G.; Ko, H.S.; Park, J.W. Direct observation of shear deformation during equal channel angular pressing of pure aluminium *Scripta Mater.* 1999, *41*, 353-357.

Shin, D.H.; Kim, I.; Kim, J.; Kim, Y.S.; Semiatin, S.L. Microstructure development during equal-channel angular pressing of titanium *Acta Mater.* 2003, *51*, 983-96.

Sivakumar, S.M.; Ortiz, M. Microstructure evolution in the equal channel angular extrusion process *Comp. Meth. App. Mech. Engng.* 2004, *193*, 5177-94.

Stoica, G. M.; Liaw, P. K. Evaluating the induced strain during equal channel angular processing *Mater. Life. Sci. Engng., TMS* (Liaw et al. eds) 2003, 119-133.

Stout, M. G.; Kallend, J. S.; Kocks, U. F.; Przystupa, M. A.; Rollett, A. D. Material dependence of deformation texture development in various deformation modes. In *The Eighth International Conference on Textures of Materials*; Kallend, J.S.; Gottstein, G.; The Minerals, Metals, & Materials Society: Warrendale, PA, 1988; Vol. 8, pp 479-485.

Stout, M. G.; Rollett, A. D. Large strain Bauschinger effects in FCC metals and alloys *Metall. Trans. A* 1990, *21*, 3201-3213.

Sun, P-L.; Kao, P-W.; Chang, C.P. Effect of deformation route on microstructural development in aluminum processed by equal channel angular extrusion *Metall. Mater. Trans. A* 2004, *35*, 1359-68.

Sutton MA; McNeill SR; Helm JD; Chao YJ. Advances in two-dimensional and three-dimensional computer vision. Heidelberger Platz 3, D-14197 Berlin, Germany: Springer-Verlag Berlin, 2000.

Suwas, S.; Tóth, L. S.; Fundenberger, J. J.; Eberhardt, A.; Skrotzki, W. Evolution of crystallographic texture during equal channel angular extrusion of silver *Scripta Mater.* 2003, *49*, 1203-1208.

Suzuki, T.; Vinogradov, A.; Hashimoto, S. Strengthening enhancement and deformation behavior of gold after ECAP *Mater. Trans.* 2004, *45*, 2200-8.

Tadmor, E. B.; Bernstein, N. A first-principles measure for the twinnability of FCC metals. *J. Mech. Phys. Solids* 2004, *52*, 2507-2519.

Tidu, A.; Wagner, F.; Huang, W.H.; Kao, P.W.; Chang, C.P.; Grosdidier, T. X-ray characterisation of size, strain and texture inhomogeneities in ultra fine grained copper processed by equal channel angular extrusion *J. Phys. IV France* 2000, *10*, 211-7.

Tomé, C.N.; Canova, G.R.; Kocks, U.F.; Christodoulou, N.; Jonas, J.J. The relation between macroscopic and microscopic strain hardening in FCC polycrystals *Acta Metall.* 1984, *32*, 1637-53.

Tomé, C. N.; Kaschner, G. C. Modeling texture, twinning and hardening evolution during deformation of hexagonal materials *Mat. Sci. Forum* 2005, *495-497*, 1001-1006.

Tomé, C. N.; Kocks, U. F. The yield surface of HCP crystals *Acta Metall.* 1985, *33*, 603-621

Tomé, C. N.; Lebensohn, R. A.; Kocks, U. F. A model for texture development dominated by deformation twinning: application to zirconium alloys *Acta Metall. Mater.* 1991, *39*, 2667-2680.

Tomé, C.N.; Lebensohn, R.A.; Necker, C.T. Mechanical anisotropy and grain interaction in recrystallized aluminium *Metall. Mater. Trans. A* 2002, *33*, 2635-48.

Tomé, C.N.; Maudlin, P.J.; Lebensohn, R.A.; Kaschner, G.C. Mechanical response of zirconium—I. Derivation of a polycrystal constitutive law and finite element analysis *Acta Mater.* 2001, *49*, 3085-96.

Tóth, L. S.; Arruffat-Massion, R.; Germain, L.; Baik, S. C.; Suwas, S. Analysis of texture evolution in equal channel angular extrusion of copper using a new flow field. *Acta Mater.* 2004, *52*, 1885-1898.

Van Houtte, P. Simulation of the rolling and shear texture of brass by the Taylor theory adapted for mechanical twinning *Acta Metall.* 1978, *26*, 591-604.

Vincze, G.; Rauch, E. E.; Gracio, J. J.; Barlat, F.; Lopes, A. B. A comparison of the mechanical behavior of an AA1050 and a low carbon steel deformed upon strain reversal *Acta Mater.* 2005, 53, 1005-1013.

Wilson, D. V.; Zandrahimi, M.; Roberts, W. T. Effects of changes in strain path on work-hardening in CP aluminum and an Al-Cu-Mg alloy. *Acta Metall. Mater.* 1990, *38*, 215-226.

Winter, R. E. Determination of twinning stresses in (100) silver films. *Thin Solid Films* 1974, *24*, 341.

Winther, G.; Huang, X.; Godfrey, A.; Hansen, N. Critical comparison of dislocation boundary alignment studied by TEM and EBSD: Technical issues and theoretical consequences *Acta Mater.* 2004, *52*, 4437-4446.

Wu, Y.; Baker, I. An experimental study of equal channel angular extrusion *Scripta Mater.* 1997, *37*, 437-42

Xue, Q.; Beyerlein, I. J.; Alexander, D. J. Mechanisms for initial grain refinement in OFHC copper in equal channel angular extrusion *Acta Mater.* 2007, *55*, 655-668.

Yapici, G.G.; Karaman, I.; Luo, Z-P. Mechanical twinning and texture evolution in severely deformed Ti-6Al-4V at high temperatures *Acta Mater.* 2006, *54*, 3755-71.

Yapici, G.G.; Beyerlein, I.J.; Karaman, I.; Tomé, C.N.; Vogel, S.; Liu, C. Plastic flow anisotropy of pure Zr after severe plastic deformation at room temperature *Acta Mater.* 2009, *57*,, 4855-65.

Yapici, G. G.; Beyerlein, I. J.; Karaman, I.; Tomé, C. N. Tension-compression asymmetry in severely deformed pure copper *Acta Mater.* 2007, *55*, 4603-4613.

Yapici, G.G.; Karaman, I.; Luo, Z.P.; Maier, H.J.; Chumlyakov, Y.I. Microstructural refinement and deformation twinning during severe plastic deformation of 316L stainless steel at high temperatures *J. Mater. Res.* 2004, *19*, 2268-78.

Yoo, M. H. Interactions of slip dislocations with twins in hcp metals *Trans. AIME* 1969, *245*, 2051-2060.

Yoo, M.H.; Wei, C.T. Slip modes of hexagonal-close-packed metals *Phil. Mag.* 1966, *14*, 573-87.

Yu, S.H.; Shin, D.H.; Park, N.J.; Huh, M.Y.; Hwang, S.K. Evolution of texture in Zr and Ti grain-refined by ECAP *Mater. Sci. Forum* 2002, *408-412*, 661-6.

Yu, S.H.; Shin, D.H.; Hwang, S.K. Evolution of grain boundary character distribution in ECA-pressed pure titanium In *Ultra Fine-Grain Symposium III*; Zhu, Y. T.; Langdon, T. G.; Valiev, R. Z.; Semiatin, S. L.; Shin, D. H.; Lowe, T. C.; The Minerals, Metals & Materials Society: Warrendale, PA, 2004; Vol. 3, pp 227-34.

Yu, S.H.; Chun, Y.B.; Hwang, S.K.; Shin, D.H. Texture development and Monte-Carlo simulation of microstructure evolution in pure Zr grain-refined by equal channel angular pressing *Phil. Mag.* 2005

PART III. STRUCTURE AND PROPERTIES OF MATERIALS PRODUCED BY SEVERE PLASTIC DEFORMATION

V.N. Chuvil'deev[1] *and V.I. Kopylov*[2]

[1] Physical-Technical Research Institute of N.I.Lobachevsky State University of Nizhniy Novgorod, Prospekt Gagarina 23/3; Nizhniy Novgorod, 603950, Russia.

[2] Physical-Technical Institute of National Academy of Sciences of Belarus, Ulitsa Kuprevicha 10, Minsk, 220141, Belarus.

PREFACE

Part III describes the structure, physical and mechanical properties of nano- and microcrystalline (NMC) materials[20] produced by severe plastic deformation (SPD) and, in particular, by equal-channed angular extrusion (ECAE). The number of publications covering this problem has extensively grown over the last few years. Rather than overviewing the published results, it seems more important to offer our readers a general approach to the description of the structure and properties of NMC materials.

The text below is a brief description of key results obtained by our team over the recent several years. It presents theoretical research and new models describing NMC materials. In order to substantiate the most important findings, we we6 used our own experimental data. In cases when they were insufficient, we based ourselves on the results of our colleagues.

It is supposed that the readers are familiar with the problems of SPD. Parts I and II describe mechanics and mathematical modelling of SPD. The emphasis of Part III is placed on the description of the fundamentals of plastic deformation and structure evolution theory, as well as on the examination of specific properties of NMC ECAE materials.

It is known that the text should either be clear or rigorous – it is impossible to combine these two requirements. Choosing our style, we preferred clarity over rigor (however, the clarity is always questionable). This affected the form of presentation of the material. As a rule, for clarity of presentation, each chapter includes three blocks: (i) the problem description with a brief analysis of available experimental data and theoretical results, (ii) formulation of a model including the description of approach and key assumptions and derivation of basic relations, and (iii) comparison of the modeling results with experimental data.

Part III contains eleven chapters which are subdivided into three sections. The first Section (Chapters 15 to 17) covers the theoretical basis and the analysis offor describing the physical and mechanical properties of NMC materials, and includes the description of their structural defects: grain boundaries, dislocations, and vacancies. Special attention is given to the theory of nonequilibrium grain boundaries (TNGB), which constitutes an important key for the understanding of specific features of the NMC materials behavior.

The second Section (Chapters 18 to 21) covers structure evolution during deformation, heating, and annealing of NMC materials. The following processes are discussed: recovery (Chapter 18), recrystallization in pure metals (Chapter 20), and recrystallization in alloys with

[20] The definition of nano- and microcrystalline materials (NMC) used in this Part is identical to ultra-fine grained (UFG) materials used in Parts I, II.

a low content of alloying elements (Chapter 21). In Chapter 19, a model of the grain refinement limit at ECAE is described.

In the third Section (Chapters 22 to 25), the analysis of mechanical properties of NMC materials is performed, including peculiarities of their superplastic behavior (Chapters 22, 23), the problems of strength (Chapter 24) and plasticity (Chapter 25) at room temperature.

This text was written with a hearty support and contribution of our colleagues at the Physical-Technical Institute of the National Academy of Science of Belarus (PTI NASB), Physical-Technical Research Institute (PTRI) and Physical Department (PD) of the N. I. Lobachevsky State University of Nizhni Novgorod (UNN), Russia, and the Nizhni Novgorod branch of the A. A. Blagonravov Machine Science Institute of the Russian Academy of Science.

Many chapters are based on the works written in cooperation with our colleagues from the Metal Physics Laboratory of PTRI UNN and PD UNN. It is the authors' pleasant duty to express their gratitude to Prof. Vladimir N. Perevezentsev (Chapters 16, 20), Dr. Olga E. Pirozhnikova (Chapters 15, 18, 19, 22, 23), Dr. Mikhail Yu. Gryaznov (Chapters 22, 25), Dr. Elena S. Smirnova (Chapters 15, 20, 21), Dr. Alexey V. Nokhrin (Chapters 18-20, 24-25), Yury G. Lopatin (Chapters 18-20), Anna V. Schavleva (Chapters 21, 22), and Anatoly N. Sysoev (Chapters 22, 25). Special acknowledgment goes to Dr. Nokhrin for his help and titanic work in preparing and discussing numerous variants of this text. Many thanks are due to Ekaterina A. Pugacheva for translating this Part into English and to Boris B. Khina (PTI NASB) for valuable work on editing the scientific English and presentation style.

PROLEGOMENA

NMC metals produced by SPD constitute a new class of materials. In Materials Science, a new research object should be examined in four key aspects: *Processing Technology*, *Structure* characterization, evaluation of *Properties*, and *Modeling* of the Processing-Structure-Properties relations (Cahn, 2001).

We shall start our description by examining the issue of *structure* and its connection with the *processing*. Being a versatile technology of Materials Processing, ECAE offers a wide range of possibilities to control deformation parameters and deformation conditions – and, respectively, material structure. Granular and dislocation structures, as well as morphological and crystallographic texture of material can be controlled by changing the degree of deformation per pass, deformation mode, strain rate and temperature, a number of passes, and type of processing routes (see Parts I, II). Since it is impossible to cover the vast range of possible materials/structures produced by ECAE, for this Part, we have chosen to study one type, namely NMC materials and structures.

The typical size of structural elements of NMC materials ranges from 50 to 1,000 nm (see Figure 1). Adjacent areas on the size scale of structural elements are occupied by microcrystalline (MC) materials with a typical grain size exceeding of approximately 10 μm (right) and by nanocrystalline (NC) materials with a typical grain size of less than 100 nm (left). However, the submicron size of a structural element alone is insufficient for the definition of NMC materials. Such materials should be distinguished from microfragmented structures produced by common deformation techniques, which are actually composed of *fragments* misoriented by small angles (Rybin, 1986). In contrast, NMC structures have substantial misorientation of their elements, which can only be induced by high levels of homogeneous shear. In its turn, this deformation mode is attained only by multi-pass ECAE under optimum temperature and strain rate conditions. It should be noted that non-optimal conditions of ECAE may also lead to the formation of microfragmented (not microgranular) structure with the identical size of structural elements. Therefore, the structure of materials produced by ECAE should be characterized by a grain boundary misorientation spectrum. At present, there is a large number of papers devoted to studying the misorientation spectrum of NMC materials, which appeared after the development of the EBSD technique. It has been shown that the fraction of high-angle boundaries (and their misorientation angle) in aluminum alloys (Berta et al. 2005; Bowen et al. 2002; Cabibbo et al. 2005; Chowdhury et al. 2008; El-Danaf, 2008; Kaibyshev et al. 2005; Sun et al. 2004), magnesium alloys (Eddahbi et al. 2005; Jin et al. 2006; Kim et al. 2002; Suwas et al. 2007), copper (Huang et al. 2004; Mishin et al.

2003; Mihra et al. 2007; Waryoba e3t al, 2008; Molodova et al. 2007), nickel (Kolobov et al. 2001c; Valiev et al. 2006), vanadium (Jiang et al. 2008) and other metals monotonically increases with deformation and, under optimum ECAE conditions, can reach 80-85% of the total number of boundaries.

Figure 1. Microstructure of materials after ECAE: a) Al-4.5%Mg-0.22%Sc-0.15%Zr, 6 passes, T=473 K, Route D (atomic-force microscopy); b) nickel, 8 passes, 293 K Route D (transmission electron microscopy); c) titanium, 8 passes, T=673 K, Route D (transmission electron microscopy); d) copper, 12 passes, T=293 K, Route D (transmission electron microscopy).

This work is focused on the nano- and microcrystalline materials having a predominantly high–angle misorientation spectrum of grain boundaries (HAGB) which are hereinafter refered to as NMC materials. The quantitative criterion for a submicron grained material being an NMC material is the proximity of the misorientation spectrum of its grain boundaries to the spectrum of randomly misoriented ensemble. In the case of their agreement, it is possible to talk about a real polycrystalline material with NMC structure produced by ECAE. The fundamental importance of HAGB domination in the grain boundary misorientation spectrum is connected with the outstanding role, which they play in the formation of the unique properties of NMC materials. As shown below, a special property of HAGB (as opposed to low-angle grain boundaries of fragments) is their ability to transit to a nonequilibrium statenonequilibricity during ECAE and maintain such a state for a certain time after deformation. This brings about specific physical and mechanical properties of NMC materials.

When describing the NMC material *structure*, attention should also be given to a special condition of lattice defects, *viz.* vacancies and dislocations. Although the role of such defects, as opposed to the role of nonequilibrium grain boundaries, is not crucial, there are several important features in their behavior. As for point defects in NMC materials, unlike the ones in coarse-grained materials, there is an extremely high concentration of vacancies after ECAE, which sometimes reaches $C_{vac}=10^{-4}$ (Setman et al. 2008; Cao et al. 2008). Since the nonequilibrium vacancies are long-term recovered at room or elevated temperature, their contribution to the properties of NMC materials proves toshould be insignificant.

Other interesting features are observed in the dislocation subsystems of NMC materials. First, it is a high dislocation density, up to ~10^{15} m^{-2}, which is by two orders of magnitude above usual (Noskova et al. 2003; Valiev et al. 2006). Second, due to a fine grain size in NMC materials, the functioning of regular intragranular dislocations sources is complicatedhindered. The key dislocation emission mechanism in NMC materials is their emission from grain boundaries. This can lead to the appearance of special mesostructural zones: the so-called zero dislocation charge areas whose sizes exceed the grain size. That is opposed to coarse-grained materials, where the area of zero dislocation charge is close to the grain size. Third, notwithstanding a high dislocation density, due to a fine grain size, there are very few dislocations in grain body, and they do not form complex ensembles and pile-ups in the lattice and can move freely inside grains. Finally, most important is that both dislocation emission and the kinetics of their movement and annihilation are determined by the interaction of dislocations with grain boundaries (Chuvil'deev, 2004). This is the key to understanding of deformation behavior of NMC materials.

Basically, examination of the lattice defect subsystems is impossible without accounting for their interaction processes. When describing this interaction, it is important to single out a main subsystem of defects, which plays a critical role in the material behavior. An example of such approach is Ashby's deformation mechanism maps (Frost et al. 1982). In case of NMC materials, all types of defects contribute to the strain behavior and structural evolution; however, nonequilibrium grain boundaries play a crucial role. Studies of nonequilibrium grain boundaries and their interaction with lattice dislocations (Chuvil'deev, 2004) can give a key to understanding of the structure and properties of NMC materials.

Thus, from the *structural* point of view, the *subject* of our *research* is *homogeneous* nano- and microcrystalline structure with predominantly high-angle and nonequilibrium grain boundaries.

The given definition has one more structural parameter – structural homogeneity. Homogeneity is important for practical applications of materials produced by ECAE. As shown in Part I, a simple shear deformation mode is a necessary condition to provide uniform deformation and uniform structure during ECAE. However, in a number of papers, substantial inhomogeneity of the material structure after ECAE was reported (e.g., (Xia et al. 2005)). In our view, this is connected with the fact that in these works ECAE is used in name, but not in substance. Owing to the great efforts of ECAE neophytes, an opinion has spread that the essence of ECAE is a mere extrusion of a billet through two intersecting channels of the same cross-section (Valiev et al. 2000). The materials produced in this way are may be heterogeneous and fragmented rather than granulated, which results in lower properties. Such ECAE definition is, of course, insufficient. This definition lacks keywords, which reflect the essence of the process – "extrusion of a blank through … provided that simple-shear homogeneous deformation is guaranteed". A need for ensuring a simple-shear deformation mode makes ECAE a nontrivial process and necessitates the development of special engineering techniques for implementing ECAE.

Thus, the *subject* of our *investigation* from the point of view of *technology* of its production can be determined as follows – materials subjected to homogeneous severe deformation in simple-shear conditions.

Another important issue is the value of simple shear deformation, which should be reached to create NMC structure with required *properties*. For example, in order to ensure high superplastic properties, it is necessary to create a material with a highest possible fraction of HAGB, because grain boundary sliding (GBS) can occur only in such grain boundaries. Usually, it is necessary to perform several ECAE passes depending on the nature of material, processing temperature, strain rate and deformation per pass. A high fraction of HAGB is also necessary to provide a simultaneous increase in the strength and plasticity of NMC materials (see Chapter 25). On the other hand, for materials with high creep resistance, the structure with a large fraction of low-angle boundaries should be created to prevent Coble diffusion creep. In more complex cases, ECAE is able to ensure a more delicate grain boundary design to manage such properties of NMC materials like fatigue and some others. Therefore, for each type of problems, it is necessary to determine a specific mode of ECAE processing and subsequent heat treatment.

In analyzing the *properties* of NMC materials, of particular interest are nontrivial experimental results of fundamental nature, which are usually introduced as new effects.

Here, an "effect" means a nonlinear behavior (change) of a certain parameter characterizing a physical property with a linear change of its structural parameters or parameters of external conditions.

Physical effects that may appear in NMC materials can be classified as follows:

1 *Threshold effects* are observed only when a small size of structural elements comparable with a typical value of the scale of a transfer process has been reached (Gleiter, 2000; Gusev, 2007; Suzdalev, 2006). Among them are quantum effects due to quantum confinement in semiconductor structures observed after a certain (nanometer) size of quantum dots has been attained. In NMC materials produced by ECAE, threshold effects were not observed.

2 *Offset effects*, such as changes in structurally sensitive characteristics of magnetic, electrical, thermal and elastic properties after ECAE. In NMC materials, these phenomena are a consequence of structure refinement, modification of texture and an increase of the lattice dislocation density and vacancy concentration. Sometimes, it is difficult to separate the exact role of each type of defects, and hence a comparison with the data known for severely strained metals produced by traditional processing methods may be helpful. Let us consider certain offset effects relating to the physical properties of materials after severe plastic deformation.

Magnetic Properties

A change of the Curie temperature by 36°, i.e. by ≈6%, and a shift of the saturation magnetization in NMC nickel and iron by more than 30% were claimed in (Valiev et al. 1992; Valiev et al. 2000). However, these results were not confirmed by subsequent publications. We had also conducted a series of special measurements on similar Ni and Fe samples and did not find any changes of these parameters. In our opinion (see also Deryagin et al. 2002; Elsukov et al. 1999), the "effects" discovered in the works (Valiev et al. 1992; Valiev et al. 2000) result from insufficient experimental accuracy (see also Deryagin et al. 2002; Elsukov et al. 1999). Indeed, it is difficult to expect any noticeable change in the Curie temperature since the second-kind phase transition is associated with a rearrangement of the electronic subsystem. Such rearrangements are quite possible in dislocation cores and grain boundaries; however, in case of NMC structures, the fraction of material in such defect cores is insignificant and cannot noticeably influence the relevant magnetic properties. Moreover, heating to the Curie temperature (630-670 K for nickel) recovers the defect structure, dislocation density and grain size to their regular levels (Valiev et al. 2000; Noskova et al. 2003; Chuvil'deev et al. 2003b). Thus, there are no physical reasons for changes of the Curie temperature.

A number of works (Noskova et al. 2003; Valiev et al. 2000; Mulyokov et al. 1990) reported another offset effect consisting in a 2-4 times increase of coercitive force after severe plastic deformation. Our measurements of coercitive force for NMC iron (N=8, T=293 K, Route D) and nickel (N=8, T=293 K, Route D) also show a twofold to fourfold increase. A change of the coercitive force and its recovery are typical of severely strained structures obtained by different processing techniques (Bida et al. 2002) and are not a specific effect connected with the NMC structure.

Electrical Resistance

It is known that the length of the electron free path in metals does not exceed several nanometers (Kittel, 1962; Shmatko et al. 1987). Since this size is significantly less than the typical structural elements of NMC materials (more than 100 nm), ECAE cannot affect fundamental electrical properties of materials. However, a noticeable increase (offset) of electrical resistance, e.g. up to 10% in copper at a room temperature, is often observed in experiments (Makarov, 1999). This is due to a high density of lattice defects in NMC materials, which affects electrical conductivity. The changes of electrical conductivity can be calculated using the known values of resistivity of lattice defects. For example, for copper the resistivity of defects is, $\Delta\rho=2\cdot10^{-8}$ Ohm·m for 1% of vacancies, $\Delta\rho=2.8\cdot10^{-19}$ Ohm·cm^3 for dislocations and $\Delta\rho\sim4\cdot10^{-12}$ Ohm·cm^2 for grain boundaries (Shmatko et al. 1987). For the limiting values of defect concentration in NMC copper (the grain size is d~200 nm, dislocation density is $\rho_V\sim10^{11}$ cm^{-2} and the concentration of vacancies is $C_{vac}=10^{-4}$), the increase of electrical resistance at a room temperature is about $2.5\cdot10^{-9}$ Ohm·cm, which is close to 10% of the standard value. The corresponding contributions of grain boundaries (~$2\cdot10^{-9}$ Ohm·m), dislocations ($2.3\cdot10^{-10}$ Ohm·m) and vacancies ($2\cdot10^{-10}$ Ohm·m) show that grain boundaries play a key role in the experimentally measured change of electrical resistance. These estimates demonstrate that a shift of electrical resistance after ECAE can be trivially described within traditional concepts of the influence of defects on electrical conductivity of metals, and it is not a specific feature of NMC materials.

Heat Capacity

The heat capacity of NMC materials is usually studied by differential scanning calorimetry (DSC) (Valiev et al. 2006; Noskova et al. 2003; Valiev et al. 2000; Amirkhanov et al. 1998). Stored energy in NMC materials with a grain size of ~0.1÷0.2 μm is around 1 J/gr (Goto et al. 2008; Cao et al. 2008; Setman et al. 2008). Such a high value also demonstrates a certain offset effect. In accordance with classical concepts (Gorelik et al. 2005), the energy stored during heating is connected with the recovery of defect structure. The energy per unit length of a lattice dislocation is $\sim Gb^2/4\pi(1-\nu))\ln(R_x/Sr_d)$ (Hirth et al. 1969), whereas the energy per unit area of a grain boundary is $\gamma_b \sim Gb/24$ (Orlov et al. 1980). Using these values together with experimental data of the grain size and dislocation density, it is easy to determine the contribution of each type of defects. Additional source of energy is the vacancy concentration. The energy of vacancy formation is ~10 kT_m (Cottrell, 1959; Orlov et al. 1983) and their concentration in severely strained metals is $C_{vac} \sim 10^{-4}$ (Cao et al. 2008; Setman et al. 2008). Estimations show that the observed offset of stored energy is a trivial effect relating to a high density of defects in NMC materials. This is also supported by the fact that DSC curves of stored energy (and the scale of the effect) after ECAE are similar to ordinary DSC curves for deformed materials (Gorelik et al. 2005; Benchabane et al. 2008).

Thus, the materials subjected to ECAE indeed display an offset of parameters characterizing physical properties as compared to severely deformed materials. However, they are trivial and conform to the expected – and easily calculated within classical concepts – changes of structurally sensitive parameters with the accumulation of defects in the material. The scale of these changes is close to that in materials deformed by other methods. The observed offsets are connected not only with fine grain size of NMC materials, but also with high density of lattice dislocations and vacancies. It should be noted that examination of such phenomena can be helpful in estimating the density of certain defects, especially when using a combination of different methods. Since these effects do not require a special explanation, they will not be further studied in detail.

3 *"Resonance" effects* appear in the case when the time of external influence coincides with the characteristic time of a transfer process, which provides effective energy relaxation. An example of such effects in coarse-grained materials is the occurrence of internal friction peaks (Kê, 1947; Eds. Blanter, 1991) at certain frequencies (for a given temperature) or at a certain temperature (for a constant frequency).

Let us estimate the possibility of resonance effects in NMC materials. A typical time of external influence (deformation) is inversely proportional to the strain rate $\dot{\varepsilon}$, i.e. $\tau_{def} \sim 1/\dot{\varepsilon}$. For ECAE with $\dot{\varepsilon}=10^{-1}\div1\ s^{-1}$, this time is $\tau_{def}\sim$1-10 sec. The characteristic time for cold working and for high-strain-rate superplastic deformation of NMC materials can be estimated as $10^2\div10^4\ s^{-1}$ and $1\div10^2$ s, respectively. In order to perform plastic deformation in solid state, accommodation processes should work effectively. Such accommodation in NMC materials is provided mainly by grain boundary diffusion. The resonance effects may appear if the diffusion time t_{diff} is comparable with the deformation time τ_{def}. For grain boundary diffusion

$$t_{diff} \approx d^3 / \delta D_b^* \quad (1)$$

where δ=2b is the grain boundary width, $D_b^* = D_{b0}^* \exp(-Q_b^* / kT)$ is the effective diffusion coefficient in nonequilibrium grain boundaries, D_{b0}^* is the preexponential factor, and Q_b^* is the activation energy for diffusion in nonequilibrium grain boundaries. Using Eq. (1), the specific temperature T, at which resonance effects become possible, can be determined for any material and grain size.

Resonance effects will be accounted for when considering the grain refinement limit (Chapter 19), optimum modes of superplastic deformation of NMC materials (Chapter 22) and in some other cases (Chapter 25).

4 *Dimensional additivity effects* are observed in NMC materials when a few process parameters depend on grain size d. In the case when a change of d leads to changes in differently directed parameters, it is possible that a certain d* would provide maximum or minimum dependence of process parameters on grain size.

An example of such effects is the superplastic behavior of NMC materials shown in the Figure 2. Here, the maximum plasticity is observed not in metals with smaller grain size – as it ensues from traditional representations of superplasticity theory predicting a continious increase of plasticity with reducing grain size (Novikov et al. 1981; Kaibyshev, 1987; Edington, 1976; Nieh et al. 1997), but in materials with an "intermediate" grain size d of ~ 1 μm (Islamgaliev et al. 2003; Sergueva et al. 2005; Islamgaliev et al. 2006; Kang et al. 2008) This result can be explained (Chapter 22) using a model of nonequilibrium grain boundaries in NMC materials.

Another example of the dimensional additivity effect is an extraordinary ability of NMC materials to exhibit simultaneously a high strength and good ductility during tensile testing at room temperature (Segal et al. 1981; Valiev et al. 2001; Yamashita et al. 2001; Wang et al. 2002; Kim et al. 2004; Chuvil'deev et al. 2008b). In coarse-grained materials, an increase of strength is usually accompanied by reduction of plasticity. In order to illustrate this effect, a chart of parameter $A_\Sigma = \sigma_S \varepsilon_{max}$ versus grain size d is shown in Figure 3, where σ_s is the ultimate tensile stress and ε_{max} is the maximum elongation to failure. A significant increase of parameter A_Σ is observed for some NMC materials within the grain size range of 0.5-1 μm. This effect will be analyzed in detail in Chapter 25.

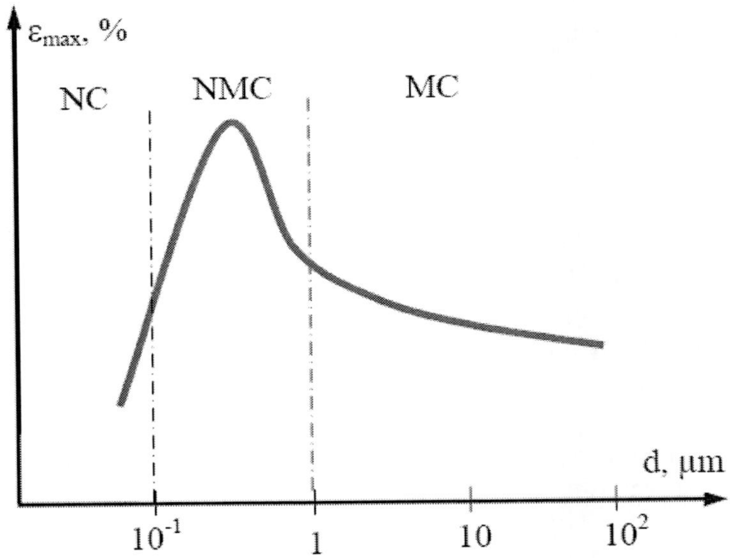

Figure 2. Typical maximum elongation-to-failure dependence (ε_{max}) on grain size (d) for superplastic materials.

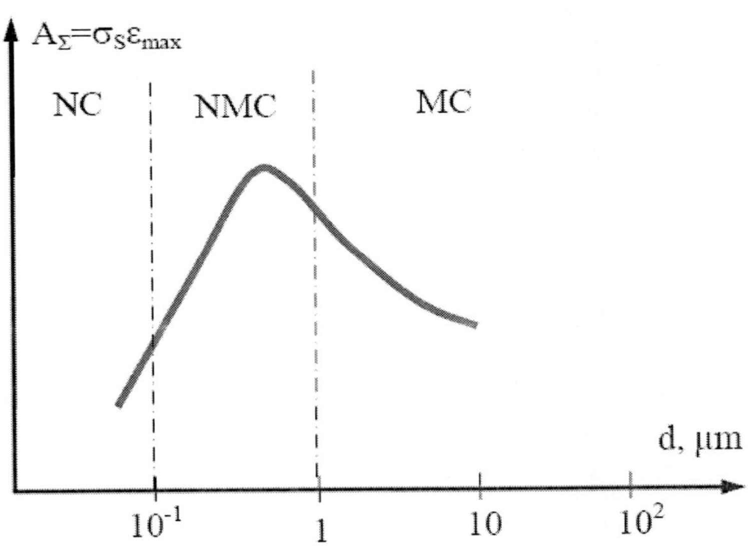

Figure 3. Simultaneous improvement of strength and ductility in NMC materials at room temperature.

The third example widely discussed in the literature is the dependence of yield stress on grain size for NMC materials. Most works describe this relation by the classical Hall-Petch formula $\sigma_y=\sigma_0+K/d^{-1/2}$ (Gutierrez-Urrutia et al. 2007; Furukawa et al. 1997; Zhilyaev et al. 2008; Molodova et al. 2007; del Valle et al. 2006). However, in some cases this law is not valid for NMC materials: some works (Degtyarev et al. 2007; Kim et al. 2003; Shafler et al.

2004; Wei et al. 2006) note that σ_y insignificantly depends on d while other works (Meyers et al. 2006; Conrad et al. 2005; Wei et al. 2006) state that the smaller d is, the smaller σ_y becomes. Detailed analysis of this problem is given in Chapter 24.

The above examples clearly demostrate that NMC materials possess distinctive properties and exhibit a wide range of specific effects. Some of these effects are evident and can be estimated in terms of the theory of defects. Some other effects, mainly those connected with mechanical properties of NMC materials, are unexpected. Below it will be shown that both "resonance" effects and dimensional additivity effects are connected with nonequilibrium nature of grain boundaries in NMC materials. Thus, defining the *subject* of our investigation – NMC materials from the point of view of their *properties*, it can be suggested that their specific properties are due to grain boundary properties, to be more exact – as will be shown later – by nonequilibrium grain boundary properties.

Let us discuss this factor in more detail. As noted earlier, NMC materials lie between the microcrystalline (MC) and nanocrystalline (NC) materials. The properties of MC and coarse-grained materials only insignificantly depend on the structure of grain boundaries because the key role is played by intragranular dislocations and dislocation ensembles (except for such phenomena as intergranular fracture at elevated temperatures and some others). Specific properties of nanocrystalline (NC) materials manifest themselves when the threshold effects mentioned above are initiated and electron and/or lattice structures have been changed (Gleiter et al. 2000; Gusev, 2007). For NC materials, the lattice properties rather than the properties of grain boundaries are the key factor. Although nonequilibrium nonequilibricity of grain boundaries can easily be created in nano-grained materials, it is difficult to maintain them thatit for a long time due to a high rate of recovery processes.

Thus, it is only in NMC materials that nonequilibrium grain boundaries play the key role in determining their properties. This allows us to refine the definition of NMC materials prepared by SPD from the point of view of their *properties* – they are the materials whose properties are principally determined by the properties of nonequilibrium grain boundaries.

Having defined the specifics character of the *subject*, we shall now briefly describe it from the point of view of its description (modeling) *methods*.

The models used in this Part to describe processes of deformation, evolution of structure and properties of NMC materials are models written in terms of defects theory. The key processes regulating behavior of NMC materials do not take place in the lattice (grains), as it happens with coarse-grain materials, but at grain boundaries. In this case, main types of defects determining the nature of these processes are not dislocations and vacancies (as in coarse-grain materials), but internal interface (boundaries). The key peculiarities of processes at grain boundaries are due to their interaction with embedded lattice dislocations and point defects. However, within the classic theory of defects, the methods describing such interactions are insufficiently developed. In order to model these processes, we shall use original approaches developed in the theory of nonequilibrium grain boundaries (Chuvil'deev, 2004) and briefly described in Chapter 15.

SECTION I. STRUCTURAL DEFECTS IN NMC MATERIALS

Chapter 15

NONEQUILIBRIUM GRAIN BOUNDARIES

15.1. CONCEPT OF NONEQUILIBRIUM GRAIN BOUNDARIES

We would like to start this part of the book with an extensive quotation from a remarkable book by Robert Cahn «The Coming of Materials Science» (Cahn, 2001, section 5.1.2.1): «... As MSE (Materials Science and Engineering) became ever more quantitative and less handwaiving in its approach, one feature became steadily more central – the power of surprise. Scientists learned when something they had observed was mystifying... in a word, surprising... or, what often came to the same thing, when an observation was wildly at variance with the relevant theory. The importance of this *surprise factor* goes back to Pasteur, who defined the origin of scientific creativity as being "savoir s'étonner à propos" (to know when to be astonished with a purpose in view)... *Because* the feature that occasions surprise is so unexpected the scientist who has drawn the unavoidable conclusion often has a sustained fight on his hands».

Thus, in order to achieve maximum productivity, we should accentuate surprising, nontrivial results. In particular, the theory of nonequilibrium grain boundaries described below suggests that a slight action on grain boundaries, which insignificantly changes their energy, can lead to substantial (enormous) changes in their diffusion and diffusion-controlled properties such as migration mobility, GBS intensity, segregation ability, etc. Apparently, this contradicts traditional linear models where substantial changes of properties usually require a large energy input. This Chapter covers the theory of nonequilibrium grain boundaries in NMC materials which leads to new, surprising results.

The conception of nonequilibrium grain boundaries was first described by Grabski and Korski (Grabski et al. 1970) who showed that interaction of boundaries with lattice dislocations changed their energy and resulted in specific diffusion properties. The term 'nonequilibrium grain boundaries' (NGB) was introduced by Pumphrey and Gleiter (Pumphrey et al. 1975). When studying nonequilibrium grain structures that are formed during recrystallization, grain boundary sliding, and adsorption of vacancies and lattice dislocations by boundaries, they found that (i) atomic mobility in NGB is higher than in low-energy boundaries; (ii) the difference in their structure is connected with forces that appear in nonequilibrium boundaries; (iii) the structure of a nonequilibrium grain boundary is not amorphous. In general, this definition of nonequilibrium grain boundaries has not changed so far. It is widely used with regard to grain boundaries, which possess excess energy and/or display anomalies in kinetic (diffusion) properties, for describing such phenomena as (i) grain-boundary interaction with lattice dislocations (Gleiter, 1979; Kwiecinski et al. 1991; Lojkowski et al. 1981, 1988; Lojkowsky, 1991; Styczynska et al. 1985; Varin, 1979, 1980, 1982; Watanabe, 1997; Valiev et al. 1980a,b; Valiev et al. 1983a,b; Orlov et al. 1980; Perevezentsev et al. 1992), (ii) migration (Valiev et al 1990a; Hillert et al. 1978; Smidoda et al. 1978), (iii) emission of dislocations and point defects (Gleiter et al. 1980; Gleiter, 1981; Gottschalk et al. 1980)), (iv) superplastic deformation (Perevezentsev et al. 1992), (v) diffusion properties of nano materials (Siegel, 1993; Gleiter, 2000), and (vi) properties of materials obtained by severe plastic deformation (Kolobov et al. 2001c; Chuvil'deev et al 2002; Valiev et al. 2006).

Despite a wide use, the term "nonequilibrium grain boundaries" has not been unambiguously determined. It is worth noting that any grain boundary, being a lattice defect and possessing excess free energy, is a nonequilibrium object. Therefore, one should specify a measure of nonequilibrium and its initial, or reference, level. When describing the nonequilibrium statenonequilibricity of boundaries with embedded defects (according to the definition of (Grabski et al. 1970) and (Pumphrey et al. 1975)), it is necessary to define a reference level and indicate the value of "additional" nonequilibrium.

In a number of works, the boundary energy is used to characterize the level of nonequilibrium statenonequilibricity. In these works, it is assumed that an equilibrium boundary is a boundary with a minimal free energy (at given external conditions). Similarly, a boundary with excess energy is considered as a nonequilibrium one (Pumphrey et al. 1975). This definition is based on the Borisov's model, which links the energy of a boundary to the activation energy of grain boundary diffusion (Borisov et al. 1964). In Borisov's model, it is implied that the diffusion micromechanisms in the crystal lattice and in the grain boundary are similar. The difference in their diffusion properties is explained by a reduction of the activation energy for atomic migration in a boundary due to the presence of excess free energy γ_b. The relation between grain-boundary diffusion parameters, *viz.* the activation energy Q_b and pre-exponent D_{b0}, and the corresponding values for bulk diffusion, Q_v and D_{v0}, is determined as (Borisov et al. 1964):

$$Q_b = Q_V + kT\ln(D_{v0}/D_{b0}) - 2\gamma_b b^2 \qquad (15.1)$$

Then, a change in the diffusion properties of a boundary with embedded lattice defects is due to the increase of its free energy (from γ_b to γ_b^*) by the energy of accumulated dislocations $\Delta\gamma_b$ (Lojkowski, 1991):

$$\gamma_b^* = \gamma_b + \Delta\gamma \; \gamma_b^* = \gamma_b + \Delta\gamma \,. \qquad (15.2)$$

Inserting Eq. (15.2) into Eq. (15.1), we obtain an expression for activation energy Q_b^* and diffusion coefficient D_b^* in a nonequilibrium boundary:

$$Q_b^* = Q_b - \Delta\gamma b^2;\ D_b^* = D_b \exp(\Delta\gamma\ b^2 / kT) \qquad (15.3)$$

In Eq. (15.3), it is assumed that the preexponential factors in the diffusion coefficient for the equilibrium (D_{b0}) and nonequilibrium (D_{b0}^*) boundary are equal. It is evident that a noticeable rise of the grain-boundary self-diffusion coefficient under deformation requires a significant increase in the boundary energy per unit area, $\Delta\gamma_b$. For example, in order to change D_b by two orders of magnitude at $T=0.5T_m$, $\Delta\gamma^* b^2$ should reach 2 kT_m, where T_m is the melting temperature and k is the Boltzmann constant. Value. 2 kT_m is comparable with the free energy of an equilibrium boundary at this temperature: $\gamma_b b^2 = (2.5\div3)kT_m$ (see tables 15.1 and 15.4).

Let us estimate the density of grain boundary dislocations ρ_b^*, which could provide such a high level of excess energy $\Delta\gamma^*$. The value of $\Delta\gamma$ is determined by the energy of lattice dislocation cores embedded into the boundary and the energy of their short-range elastic fields:

$$\Delta\gamma = \rho_b G b^2 / 2\pi + \rho_b G b^2 \ln(1/\rho_b S_d)/2\pi, \qquad (15.4)$$

where Sr_d is the dislocation-core radius. Using a typical value of the parameter $Gb^3/kT_m \cong 50$ (Frost et al. 1982), it is easy to see that the required level of excess energy ($\Delta\gamma^* b^2 \cong 2kT_m$) in grain boundaries corresponds to the dislocation density $\rho_b^* \cong 3 \cdot 10^8$ m^{-1}. Apparently, such a high density of dislocations is not reachable at high-temperature deformation. In fact, the experimentally evaluated (Varin, 1982, 1986; Lojkowski, 1991) dislocation density in grain boundaries under similar conditions does not exceed 10^7 m^{-1}, which is by an order of magnitude less than necessary. This analysis shows that the Borisov's model (Borisov et al. 1964) encounters serious difficulties both in defining excess energy as a measure of nonequilibrium statenonequilibricity and in explaning a substantial acceleration of grain boundary diffusion.

In a number of works (Valiev et al. 1983b, Valiev et al. 1989, Valiev et al. 1990a), nonequilibrium boundaries are defined as the boundaries with long-range stress fields. However, it is easily seen that the presence of a long-range elastic field is neither necessary nor sufficient for the nonequilibrium statenonequilibricity. First, a situation is observed where grain boundaries interact with intensive fluxes of point defects, similarly to what occurs in alloys under irradiation. It is known that diffusion properties of such boundaries under certain conditions may substantially differ from regular boundaries, and a notion of nonequilibrium is used to describe these phenomena. At the same time, the stress fields associated with such boundaries will not be the long-range fields, because they are similar in nature to the short-range elastic fields around point defects and pores.

Now, wWe shall consider the interaction of grain boundaries with dislocations, which form long-range internal stress fields. We suppose that the dislocations are embedded in a grain boundary, but their cores do not delocalize in the boundary. Such a situation is possible at low temperatures, at high "contamination" of grain boundaries by impurities (Lartigue et al. 1983), or at a specific special misorientation (Varin, 1979). In these cases, the boundary, more specifically – dislocations distributed in grain boundaries, is a source of long-range fields, but this does not change their diffusion properties. The absence of the grain-boundary nonequilibrium statenonequilibricity is still more evident when the long-range stress fields along near the boundaries are generated by triple junction dislocations or disclinations (Perevezentsev et al. 1983; Rybin et al. 1985; Perevezentsev et al. 1992).

Thus, neither the value of excess energy nor the level of elastic field can be considered as a measure of nonequilibrium statenonequilibricity of grain boundaries. In our opinion, it is boundary free volume that can become such measure. As mentioned above, grain boundaries have a lower atomic order and a certain excess volume in comparison with an ideal lattice. The grain-boundary free volume at zero temperature is a geometrically necessary free volume that determines the initial (zero) level of nonequilibrium. Vacancies and dislocations can also contain free volume (Friedel, 1964). The absorption of such defects by a boundary leads to an increase of its the boundary's free volume. A boundary whose free volume exceeds the geometrically necessary value (i.e., a boundary with excess free volume) can be called a nonequilibrium boundary. In that case, the level of nonequilibrium is characterized by the value of such excess free volume (Chuvil'deev, 2004).

As will be shown later, the boundary energy demonstrates a weak linear dependence on the free volume, while the grain-boundary diffusivity exhibits a strong, exponential dependence. The account of this difference in the influence of free volume on γ_b and D_b allows to explain why embedding a relatively small number of lattice dislocations into the boundary can lead to substantial changes in its diffusion properties, whereas the free energy associated with the boundary does not noticeably increase.

Thus, the use of free volume to describe a nonequilibrium statenonequilibricity of grain boundaries seems to be a constructive approach. The former can serve as a generic parameter, which characterizes the structure and properties of equilibrium and nonequilibrium boundaries and, in particular, determines their thermodynamic and diffusion properties. In order to formulate a theory of NGB within this approach, it is necessary, first of all, to develop a model of equilibrium boundaries where the free volume is a main structural parameter.

15.2. Model of Equilibrium Grain Boundaries

15.2.1. Preliminary Remarks

In this section, random or nonspecial grain boundaries are considered. Although such boundaries constitute the majority of boundaries in real polycrystalline materials, the knowledge about their structure is limited. Over the past decades, substantial progress has been achieved in studying grain boundary structure and properties (Orlov et al. 1980; Kosevich et al. 1980; Bokstein et al. 1986, 1987; Ed. Bokstein, 1988; Kaibyshev, 1987; Kopetskiy et al. 1987; Kolobov, 1998; Eds. Flewitt, 2001). However, the most part of experimental and theoretical works on grain boundaries is devoted to the description of special grain boundaries, which have regular ordered structure. A number of models was proposed for random boundaries: the Bernal's polyhedron model, Lee disclination model, composite-structure model (an attempt to interpret the structure of a boundary with arbitrary misorientation angle as a combination of atomic groups typical of special grain boundaries), base unit model, etc. (Orlov et al. 1980; Shtremel et al. 1999; Fionova, 1992; Ziman, 1979; Ashby et al. 1978; Li, 1972; Chalmers et al. 1971; Sutton et al. 1983). Note that the above models describe the boundary structure only at T=0 K and do not admit generalization to arbitrary temperature. To date, the known attempts to use them as the basis for calculating thermodynamic parameters of a boundary at high temperatures and grain boundary diffusion processes were unsuccessful.

Regarding the problem of boundary description in a wide context, it should be marked that the condition of grain boundary physics reflects the general condition of the physics of disordered systems, which presently has no general or formal description methods. (According to Ziman, the physics of disordered systems «does not have the nucleus of postulates, which could form a logical connection with any established fact… here, a single mathematical principle, just as powerful as Bloch theorem (for crystalline systems), has not yet been discovered.» (Ziman, 1979)).

Thereby, grain boundary physics currently develops within the phenomenological approach. The requirements to phenomenological models are widely known. A model should be as general as possible, quite simple and obvious, and should ensure quantitative comparison with experimental data. Unfortunately, there have not been any phenomenological models of high-angle grain boundaries, which satisfy these requirements. At the same time, within the phenomenological approach, different authors at different times expressed several profound ideas, which served as the basis for the grain boundary theory described below. First and foremost, these are boundary island structure representations, as well as an assumption of a possibility of describing boundary properties using the notion of free volume.

The idea of an island structure of grain boundaries was first expressed by Mott (Mott et al. 1948) and then developed by Kê and Gifkins (Kê, 1949; Gifkins, 1967). It is based on the analogy between a disordered boundary structure and a melt (liquid metal). Since the boundary properties differ greatly from those of a supercooled melt (for the former, the diffusivity is substantially lower while the viscosity is higher), it can be assumed that the "liquid phase" occupies only a part of the grain boundary area, and either the solid phase (according to Mott) or the liquid phase (according to Kê) forms special islands in the boundary. So far, the idea of island structure of a grain boundary has not been disproved.

The idea of free volume was first used to describe grain boundaries by Seeger and Schottky (Seeger et al. 1959) and further developed by many authors (Aaron et al. 1972; Frost et al. 1980; Wolf, 1989a; Knizhnik, 1982; Chuvil'deev, 2004). From the geometrical viewpoint, the grain boundary area, due to lesser atomic ordering, has a certain amount of excess volume as compared to an ideal crystal lattice. Thus, the free volume is determined as a difference between the volume confined within a closed surface, encompassing a grain-boundary segment and a certain amount of adjacent bulk material, and the volume of an ideal crystal containing the same number of atoms (Knizhnik, 1982).

It seems reasonable to use the notion of free volume for building a general theory of grain boundaries, because, on the one hand, it is a microscopic parameter reflecting the grain boundary structure and, on the other hand, it is a phenomenological parameter that can be linked to thermodynamic constants and determined using experimental methods.

15.2.2. Grain Boundary Structure, Energy, and Entropy

Experimental Data on Free Energy and Entropy of Grain Boundaries

Experimental values of specific surface energy ($\gamma_b(T)$) of nonspecial grain boundaries in pure metals at temperature T are listed in Table 15.1. For fcc metals, a typical value of nondimensional parameter $\gamma_b b^2/kT_m$ at $T=T_m$, which corresponds to the surface energy for the area equal to "atomic area" $\pi r_l^2 \sim b^2$ where r_l is the ion radius, is close to 2. For bcc metals, this value is slightly lower.

The temperature dependence of grain boundary energy and the associated boundary entropy S_b is insufficienly investigated. At present, there are data only for some fcc metals. Experimental and calculated values of S_b for Al, Ag and Cu are given in Table 15.2. It is seen that nondimensional parameter $S_b b^2/k$ for the listed materials lies in the range of 2 to 3.

Grain Boundary Structure

The model of a random boundary is based on the combination of two conceptions: (i) the island structure of a grain boundary and (ii) the free volume associated with a boundary.

The structure of a boundary is presented as a mixture of islands of "solid" (S) and "liquid" (L) phases (Figure 15.1). In such models, in order to describe the boundary structure within such approach, it is necessary to determine the atomic structure of these phases, the size and shape of S- and L-islands, their volume fractions and the nature of boundariesing between the islands.

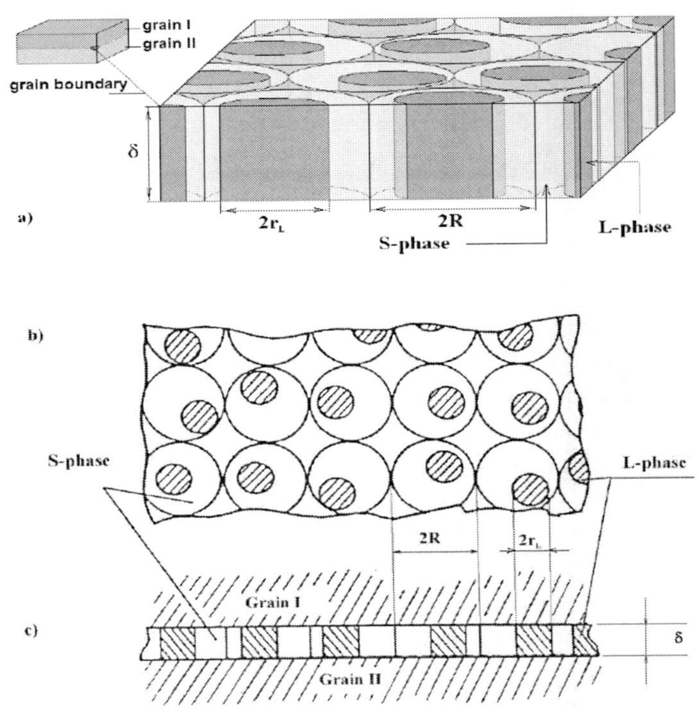

Figure 15.1. Continued on next page.

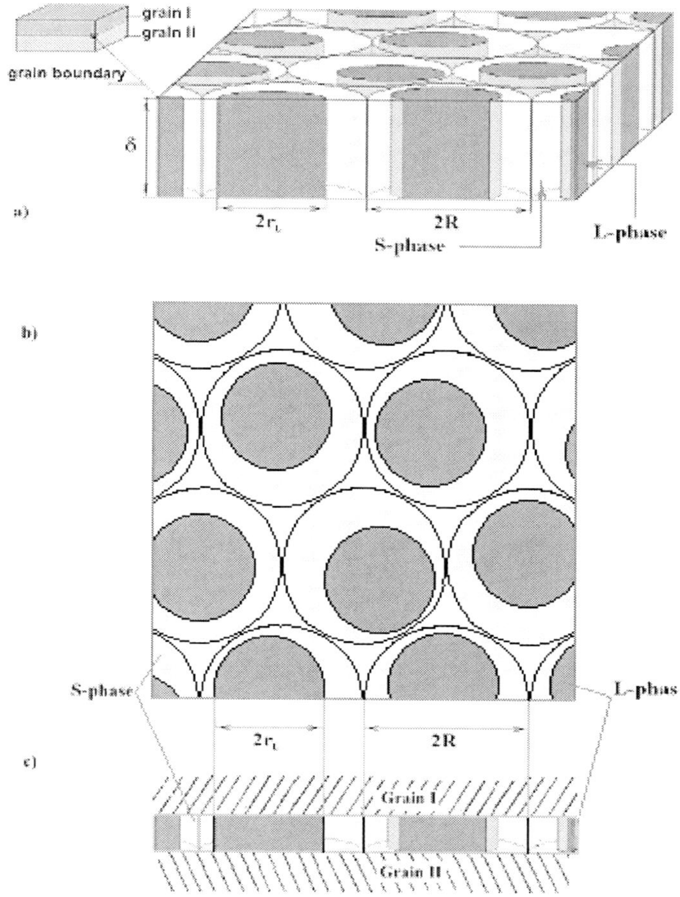

Figure 15.1. The island model of grain boundary structure(R is the radius of S-cell of the boundary; r_L a_L is the radius of L-phase island of the boundary): stereographic projection (a), plane view (b), and side view (c).

Table 15.1. Experimental values of specific grain boundary energy in metals

Metal	$\gamma_b(T_b^0)$, erg/cm^2		$\Delta T_b/T_m$	$\gamma_L(T_m)$	γ_b^L	$\gamma_b b^2/kT_m$
Ag	360 (1,173 K)	(Missol, 1978)	0.049	916	366.4	1.77
	400 (1,173 K)	(McLean, 1957)				
	450 (1,203 K)	(Gleiter et al. 1972)				
Au	360 (1,123 K)	(Gleiter et al. 1972)	0.021	1128	451.2	1.83
	406 (1,308 K)	(Missol, 1978)				
Cu	550 (1,303 K)	(Missol, 1978)	0.039	1325	530	1.93
	595 (1,273 K)	(McLean, 1957)				
	600 (1,073 K)	(McLean, 1957)				

Table 15.1. (Continued)

Metal	$\gamma_b(T_b^0)$, erg/cm^2		$\Delta T_b/T_m$	$\gamma_L(T_m)$	γ_b^L	$\gamma_b b^2/kT_m$
	640 (1,223 K)	(McLean, 1957)				
	860 (1,173 K)	(Gleiter et al. 1972)				
Al	405 (623 K)	(Missol, 1978)	0.035	865	346	2.125
	600 (0 K)	(Hirth et al. 1969)				
	625 (0 K)	(Hirth, et al 1969)				
	380	(Orlov et al. 1980)				
	335 (900 K)	(Orlov et al. 1980)				
Pb	210 (610 K)	(Missol, 1978)	0.047	451	180	1.90
	130 (573 K)	(Gleiter et al. 1972)				
Ni	728 (1,573 K)	(Missol et al. 1978)	0.087	1770	708	1.90
	840	(Orlov et al. 1980)				
	690	(Hirth et al. 1969)				
Pt	780	(Martin et al. 1976)		1865	746	2.14
	1000	(Orlov et al. 1980)				
V	-		-	1950	780	1.58
Nb	756 (2,523 K)	(Chuvil'deev, 1996c)	0.08	1900	760	1.64
Zn	255 (433 K)	(Missol, 1978)	0.366	785	314	2.38

Comments to Table 15.1. For many materials, the experimental values of $\gamma_b(T)$ given by different authors vary greatly . In order to choose a reasonable $\gamma_b(T)$ value, we used a known empirical relation for the grain boundary energy at $T=T_m$:

$$\gamma_b^L(T \approx T_m) = 0.4\gamma_L(T_m), \tag{15.5}$$

where $\gamma_L(T_m)$ is the surface energy of a melt at $T=T_m$. The values of γ_b^L calculated by Eq. (15.5), are listed in column 4. Comparison of γ_b^L and $\gamma_b(T=T_m)$ shows that Eq. (15.5) gives a reasonable estimate of $\gamma_b(T=T_m)$ for most of the fcc metals (with the exception of Pb, for which the experimental value $\gamma_b(T=573\text{ K}) = 130$ erg/cm^2 (Gleiter, 1972) is used) and for Nb (a bcc metal). Thus, Eq. (15.5) can be used to evaluate $\gamma_b(T_m)$ for other bcc metals, in particular, V, for which there are no reliable data. Eq. (15.5) was also used to estimate $\gamma_b(T=T_m)$ for hcp metals. It should be outlined that the value of $\gamma_b(T=433\text{ K}) = 255$ erg/cm^2 for Zn given in (Missol, 1978) is close to the stacking fault energy, $\gamma_b = 250$ erg/cm^2 in this material (Smagorinskiy, 1992) and probably corresponds to a special grain boundary.

Table 15.2. Grain boundary entropy S_b

Metal	S_b, erg/cm^2 K	$S_b b^2/k$
Al	0.2÷0.3 (Hasson et al. 1971)	1.1÷1.63
	0.3 (Chadwick et al. 1976)	1.63
	0.289; 0.479; 382; 0.525 (Prowan et al. 1977)	1.57; 2.6;2.076; 2.85
Au	0.38 (Ewing et al. 1972)	2.262
	0.494 (Prowan et al. 1977)	2.94
Cu	0.433; 0.387; 0.404; 0.409; 0.371 (Prowan et al. 1977)	2.15; 1.92; 2; 2.03; 1.94

We shall assume that the L-phase has a disordered structure similar to a melt. In order to describe its structure, the notion of free volume is used. We will also assume that the free volume of L-phase, V_L^b, is equal to that of a melt ΔV_m, i.e.and equal to the relative volume change at crystal melting:

$$V_L^b = \Delta V_m \tag{15.6}$$

On the other hand, the S-phase has an ordered structure, and its free volume V_S^b V_S^b, similarly to that of an ideal crystal lattice, is close to zero:

$$V_S^b \; V_S^b \cong 0 \tag{15.7}$$

If the boundary free volume V_b is known, the relative volume α_S and α_L occupied by islands of S- and L-phases, respectively, is determined as

$$\alpha_L = V_b / V_L^b ; \; \alpha_S = 1 - \alpha_L \tag{15.8}$$

At V_S^b V_S=0, the relative boundary free volume $\alpha = V_b/\Delta V_m$ appears to be equal to α_L:

$$\alpha = \alpha_L = V_b / V_L^b = V_b/\Delta V_m \tag{15.9}$$

Since $\Delta V_m \cong 5\%$ for most of metallic materials, and the boundary free volume is typically $V_b \cong 1$-2%, the volume fraction occupied by L-islands appears to be below 50%: for $V_L \cong 5\%$ and $V_b \cong 2\%$, $\alpha = 0.4$. In this case, the L-phase forms "lakes" within the boundary, while the most part of the latter is occupied by the S-phase[21].

For simplicity, L-phase islands will be presented by cylinders having height δ=2b and radius $r_L a_L$, and randomly distributed in the boundary (Figure 15.1). In order to describe this structure, the volume of the boundary is subdivided into cylindrical domains of radius R and height δ. The centers of L islands can be located at any point of these domains with an equal

[21] From this point of view, it would be more correct to talk about the "lake" model rather than the "island" model, but here, we adhere to the traditional terminology introduced by Mott (Mott, 1948).

probability. Then, the relative area that corresponds to the relative volume occupied by L-phase islands is determined by Eq. (15.8), and R can be expressed through V_b and ΔV_m or, using Eq. (15.9), through α:

$$R = r_L(\Delta V_m/V_b)^{1/2} = r_L/\alpha^{1/2} \tag{15.10}$$

Free Energy of a Grain Boundary

A general expression for the grain-boundary free energy γ_b is given as

$$\gamma_b = \gamma_b^0 - TS_b \tag{15.11}$$

Here γ_b, γ_b^0, and S_b are respectively the free energy, enthalpy, and entropy of the boundary per unit area. In order to calculate the energy γ_b of a boundary having the above-described island structure, let us estimate the free energy associated with the islands of S and L phases and their interfaces. Consider an elementary grain-boundary volume as a cylindrical domain of radius R and height δ=2b. This domain contains a cylindrical L-phase island whith volume $V_1 = \delta\pi r_L^2$ and lateral surface area $s_1 = 2\pi r_L\delta$, which separates the S- and L-phases within the boundary. The upper and lower surfaces of total area $s_2 = 2\pi r_L^2$ separate the L-phase from the solid phase of adjacent grains. Moreover, this domain contains a boundary S-phase segment with the surface area $s_3 = \pi(R^2 - r_L^2)$. Then, the free energy corresponding to the grain-boundary segment with surface area $s_0 = \pi R^2$ is

$$\pi R^2\gamma_b = s_1\gamma_{S/L}^{(1)} + s_2\gamma_{S/L}^{(2)} + s_3\gamma_0 + V_1U_1 - \pi R^2 S_b^c T \tag{15.12}$$

The first term in the right-hand side of Eq. (15.12) is associated with the interface between L- and S-phases (lateral surface of the L-phase cylinder). The second term refers to the "L-phase/crystalline grain" interface (upper and lower surfaces of the L-phase cylinder), where $\gamma_{S/L}^{(1)}$ and $\gamma_{S/L}^{(2)}$ are the corresponding interface energies per unit surface area. The third term describes the surface energy associated with the boundary S-segment, where γ_0 is the corresponding surface energy per unit area. The fourth term describes the excess energy of L-phase, where U_1 is the excess free energy per unit volume. In the fifth (entropic) term, S_b^c is the configurational entropy of the grain boundary per unit area that depends on the number of positions which the L-phase island can occupy within the boundary:

$$S_b^c b^2/k = -(1-\alpha)\ln(1-\alpha) - \alpha\ln\alpha \tag{15.13}$$

Substituting V_1, s_1, s_2 and s_3 into Eq. (15.13), it gives

$$\pi R^2\gamma_b = \delta\pi r_1^2 U_1 + 2\pi r_L^2\gamma_{S/L}^{(1)} + 2\delta\pi r_L\gamma_{S/L}^{(2)} + \pi(R^2 - r_L^2)\gamma_0 - \pi R^2 TS_b^c \tag{15.14}$$

Dividing both parts of Eq. (15.14) by πR^2 and using Eqs. (15.9) and (15.13), we obtain the relationship between the boundary energy per unit area γ_b and the relative boundary free volume α:

$$\gamma_b = \alpha(\delta U_l + 2\gamma^{(1)}_{S/L} + 2\gamma^{(2)}_{S/L}(\delta / r_L) - \gamma_0) + \gamma_0 - TS^c_b \tag{15.15}$$

In order to proceed to the quantitative analysis of relation between γ_b and α, it is necessary to introduce additional assumptions with regard to parameters U_l, $\gamma^{(1)}_{S/L}$, $\gamma^{(2)}_{S/L}$ and γ_0 which enter Eq. (15.15). For evaluating $\gamma^{(1)}_{S/L}$ and $\gamma^{(2)}_{S/L}$, the solid/liquid interfacial energy $\gamma_{S/L}$ is approximately:

$$\gamma^{(1)}_{S/L} \approx \gamma^{(2)}_{S/L} \approx \gamma_{S/L} \tag{15.16}$$

It is further assumed that the specific excess energy of the L-phase island, U_l, is close to the work of melt-island generation per unit volume. Neglecting the difference between the heat capacity of a solid and a liquid phase (i.e., ΔC_p=0), U_l can be defined as:

$$U_l = \lambda\rho \tag{15.17}$$

where λ is the latent heat of fusion (J/kg), ρ is the density of metal. From Eqs. (15.15)–(15.17) , one can find the expression for the specific surface energy of a grain boundary

$$\gamma_b = \alpha\left[\delta\lambda\rho + 2\gamma_{S/L}\left(1 + \frac{\delta}{r_L}\right) - \gamma_0\right] + \gamma_0 - TS^c_b \tag{15.18}$$

Temperature Dependence of Grain Boundary Energy. Boundary Enthalpy and Entropy

As seen from Eq. (15.18), the temperature dependence of grain boundary energy is due to that of thermodynamic parameters ρ, γ_0, $\gamma_{S/L}$ and structural factor α (relative free volume of the boundary). The specific energy of a solid/liquid interface is (Skripov, 1984)

$$\gamma_{S/L} = \gamma^0_{S/L} - TS_{S/L}, \tag{15.19}$$

where the value of $S_{S/L}$ is high, as so that $T_m S_{S/L}$ constitutes about 80% of $\gamma^0_{S/L}$ (Chuvil'deev, 2004). Having a substantial effect of temperature on $\gamma_{S/L}$, temperature dependences of $\rho(T)$ and $\gamma_0(T)$ can be neglected at a first approximation.

Let us examine the effect of temperature on the free volume of a grain boundary. In order to describe the temperature dependence of α ($\alpha(T)=V_b(T)/\Delta V_m$), we will induce structural V^0_b and thermal V^T_b components of the free volume (and α^0 and α^T, respectively)

$$V_b(T) = V^0_b + V^T_b \tag{15.20}$$

The structural or geometrically necessary component V_b^0 depends on atomic arrangement of the boundary area and, apparently, on crystallographic parameters of the adjacent crystallites (grains), as well as their alignment and orientation with respect to the boundary plane. For random (nonspecial) boundaries this value is not equal to zero at T=0. The thermal component V_b^T appears at T>0 depending on a difference of the material density in the boundary and in the grain volume. Currently, there are no reliable data on V_b^0 and V_b^T, nor on the thermal expansion coefficients for the grain-boundary area, which are necessary for evaluating V_b^T. Therefore, in order to roughly estimate the thermal component of free volume, we suppose that the volume coefficient of thermal expansion of the grain-boundary L-phase is equal to that of a melt β_L, while the thermal expansion coefficient for the S-phase corresponds to that for solid phase β_S. Then, V_b^T can be evaluated as

$$V_b^T = \alpha_L T(\beta_L - \beta_S) = \alpha_L T\Delta\beta \qquad (15.21)$$

Eqs. (15.21), (15.20) together with Eqs.(15.8), (15.920) finally give

$$\alpha = \alpha^0/(1 - \Delta\beta T/\Delta V_m) \qquad (15.22)$$

It is known that for most of the metallic materials the difference between coefficients β_L and β_S, i.e. $\Delta\beta$, is small $\Delta\beta \leq (0.02\div0.1)\beta_S$. Typically, $\beta_S T_m/\Delta V_m = 1\div4$, hence $\Delta\beta T/\Delta V_m \ll 1$. Therefore, the rough estimation of α is

$$\alpha = \alpha^0(1 + \Delta\beta T/\Delta V_m)\, \alpha = \alpha^0\left[1 + \frac{\Delta\beta}{\beta_S}\left(\frac{\beta_S T_m}{\Delta V_m}\right)\frac{T}{T_m}\right]$$

(15.23)

The experimental values of $\beta_S(T_m)$, $\beta_L(T_m)$ and parameter ΔV_m are shown in Table 15.3. It can be seen that $\beta_S(T_m) \cong \beta_L(T_m)$. Therefore, in most cases the second term within brackets of Eq. (15.23) is substantially less than a unity and can be neglected. That means that the temperature exerts insignificant influence on the free volume

$$\alpha = \alpha^0 \qquad (15.24)$$

Table 15.3. Thermodynamic, elastic and structural parameters of metals

Metal	$\frac{\gamma^0_{S/L} b^2}{kT_m}$	$\frac{S_{S/L} b^2}{k}$	$\frac{\lambda^* \rho \Omega}{kT_m}$	$\beta_S(T_m) \times 10$	$\beta_L(T_m) \times 10$	ΔV_m	G (T=300K), GPa
	(Chuvil'deev, 1996b)		(Larikov et al. 1985)	(Zinov'ev, 1989)		(Ed.Grigor'ev, 1991)	(Tihonov et al. 1986)
Ag	0.847	0.733	1.1	1.048	1.13	3.8	30.3
Au	0.918	0.747	1.12	0.803	0.94	5	31
Cu	0.9	0.767	1.15	1.045	1.13	4.32	48.5
Al	1.158	0.92	1.38	1	0.92	6.5	24.45
Pb	0.8176	0.653	0.98	0.63	0.66	3.5	8
Ni	0.9713	0.833	1.25	1.434	1.48	4.5	71.6
Pt	1.023	0.853	1.28	1.1	-	5	59
V	0.9376	0.78	1.17	1.24	-	6	47
Nb	0.944	0.78	1.17	1	-	6	22.6
Zn	1.06	0.873	1.31		1.0	4.7	36

Then, γ_b can be found from Eqs. (15.18), (15.19) and (15.24):

$$\gamma_b = \left[\alpha\left(\delta\lambda\rho + 2\gamma^0_{S/L}\left(1+\frac{\delta}{r_L}\right) - \gamma_0\right) + \gamma_0\right] - T\left[2\alpha S_{S/L}\left(1+\frac{\delta}{r_L}\right) + S^c_b\right] \tag{15.25}$$

In Eq.(15.25), the first part corresponds to the grain-boundary enthalpy γ^0_b, while the second part is the GB entropy S_b:

$$\gamma^0_b = \alpha\left[\delta\lambda\rho + 2\gamma^0_{S/L}\left(1+\frac{\delta}{r_L}\right) - \gamma_0\right] + \gamma_0; \tag{15.26}$$

$$S_b = 2\alpha S_{S/L}\left(1+\frac{\delta}{r_L}\right) - \frac{k}{b^2}\left[(1-\alpha)\ln(1-\alpha) + \alpha\ln\alpha\right] \tag{15.27}$$

These formulae for γ_b, γ^0_b and $S_{S/L}$ contain only one non-strictly determined parameter: the size of the L-phase island $r_L a_L$. It is naturally to assume that the radius of the island does not exceed the size of the first coordination sphere $r_L a_L = b$ and its diameter is equal to boundary thickness $2r_L 2a_L = \delta$ where $\delta = 2b$. In that case, the boundary free energy, enthalpy and entropy can be rewritten in simple and general forms:

$$\gamma_b(T) = \alpha(\lambda\rho\delta + 6\gamma^0_{S/L} - \gamma_0) + \gamma_0 - TS_b; \tag{15.28}$$

$$\gamma^0_b = \alpha(\gamma\rho\delta + 6\gamma^0_{S/L} - \gamma_0) + \gamma_0; \tag{15.29}$$

$$S_b = 6\alpha S_{S/L} + S^0_b \tag{15.30}$$

As seen from Eqs. (15.28)- (15.30), the grain-boundary enthalpy and entropy depend on the material thermodynamic constants λ, $\gamma^0_{S/L}$, $S_{S/L}$, and on two structural parameters: relative free volume of the boundary α and the specific surface energy of S-phase, γ_0. It should be noted that the above expressions allow one to evaluate the boundary free volume α using the experimentally measured grain-boundary free energy $\gamma_b(T)$ because Eqs. (15.28) – (15.30) give

$$\alpha = \frac{\gamma_b(T) + TS^c_b - \gamma_0}{\lambda\rho\delta + 6(\gamma^0_{S/L} - TS_{S/L}) - \gamma_0} \tag{15.31}$$

Table 15.4. Experimental and theoretical values of grain boundary parameters

Metal	$\frac{\gamma_b(T_b)b^2}{kT_m}$ (Missol, 1978)	$\Delta T_b/T_m$	$\frac{\gamma_0 b^2}{kT_m}$	α	$\frac{S_b b^2}{k}$	$\frac{\gamma_b^0 b^2}{kT_m}$
Ag	1.77	0.049	1.9	0.355	2.21	4.1
Au	1.83	0.021	1.644	0.375	2.35	4.23
Cu	1.93	0.039	1.95	0.36	2.3	4.32
Al	2.125	0.035	1.5	0.38	2.76	4.98
Pb	1.9	0.047	2.14	0.36	2.14	4.14
Ni	1.9	0.087	1.92	0.37	2.51	4.63
Pt	2.14	0	1.8	0.36	2.55	4.7
V	1.58	0	1.38	0.36	2.4	4
Nb	1.64	0	1.125	0.4	2.543	4.18
Zn	2.38	0	2.22	0.34	2.42	4.8

Comparison with Experimental Data

Direct comparison of theoretical results with experimental data is complicated, since, by now, the values of the free volume associated with grain boundaries in metals are not known. Therefore, we have to limit ourselves to evaluating the relative free volume $\alpha(\gamma_b)$ by Eq. (15.31) using the known values of $\gamma_b(T)$ (see Table 15.1) and thermodynamic parameters (Table 15.3). The results are presented in Table 15.4. These values may be compared with the relative free volume $\alpha(D_b)$ calculated from the known data on the grain boundary diffusivity to demonstrate a good agreement between $\alpha(\gamma_b)$ and $\alpha(D_b)$. Additionaly, it should be noted that Eqs. (15.28) – (15.31) qualitatively agree with the results reported in literature. In particular, the quasilinear dependence of the boundary energy on free volume seen in these equations corresponds to linear dependence $\gamma_b(\Delta V_b)$ revealed by computer simulation of high-angle boundaries in copper, silver and α-iron (Wolf, 1989b; Wolf, 1990). A qualitative agreement is also observed between the grain-boundary entropy S_b calculated by Eq. (15.30) (see Table 15.4) and the experimental and theoretical values of S_b listed in Table 15.2.

15.2.3. Diffusion in Equilibrium Grain Boundaries

Experimental data on the value of activation enerhy Q_b and the value of preexponential factor D_{b0} of grain boundary self-diffusion are given in Table 15.5.

Table 15.5. Experimental data on diffusion in pure metals

Metal	D_{b0}, $sm^2/s \cdot 10^{-1}$	Q_b^e, kJ/mol		$\frac{Q_b^e}{kT_m}$	D_{L0} $sm^2/s\, 10^4$ Chuvil'deev, 1996a	$\frac{Q_L}{kT_m}$ Chuvil'deev, 1996a
Ag	0.84	90	Frost et al. 1982	8.74	5.8	3.13
	0.11	79	Ed. Larikov, 1987	7.67		
	0.02	64	Ed. Larikov, 1987	6.214		
	1.2	90.3	Ed. Smithells, 1976	8.78		
	0.26	77.3	Ed. Smithells, 1976	7.5		
Au		104.4		9.376	2.84	3.14
Cu	0.98	104	Frost et al. 1982	9.2	14.6	3.6
Al	8.7	84	Frost et al. 1982	10.8	14.9	3.8
Pb	13	66	Frost et al. 1982	13.2	9.15	3.71
	0.87	44	Frost et al. 1982	8.8		
Ni	0.7	115	Frost et al. 1982	8.04	9.7	3.3
	8.1	129	Ed. Larikov, 1987	9.02		
	6.2	131	Ed. Larikov, 1987	9.16		
	0.175	118.5	Ed. Smithells, 1976	8.29		
	0.08	107.5	Ed. Smithells, 1976	7.53		
	4.4	171	Ed. Smithells, 1976	12.0		
	5.2	188.5	Ed. Smithells, 1976	13.2		
	200	153	Ed. Smithells, 1976	10.8		
Pt		154.8		9.1	3.14	3.2
V	9.5	209	Frost et al. 1982	11.61	111	5.54
Nb	8.7	263	Frost et al. 1982	11.5	40	5
Zn	2.4	60.5	Frost et al. 1982	10.6	12	4.15

Typically, the micromechanism of grain-boundary diffusion is similar to that of the lattice diffusion and can be described as a sequence of atomic jumps via vacancies (Bokstein et al. 1986; Kaur et al. 1989). Despite the fact that the vacancy model of grain-boundary diffusion is based on the well-developed theory of volume diffusion, it cannot be used to calculate the key parameters of GB self-diffusion process such as the activation energy Q_b and preexponential factor D_{b0}. That results from an incorrect use of parameters of equilibrium vacancies in disordered structures of high-angle boundaries. A computer simulation (Ballufi et al. 1981; Ciccotti et al. 1990) shows that the vacancies embedded in a random grain boundary are unstable at elevated temperature and are quickly delocalized. These delocalized vacancies manifest different properties than the lattice vacancies, and can not hardly be used for evaluation of a diffusion mass transfer in GBs.

Diffusion in the Island Structure of a Boundary

Consider the diffusion mass transfer in grain boundaries having the above-described island structure. It is evident that due to differences in structures, the L and S phases have different diffusion properties. We will assume that the diffusion parameters of the L-phase, which has the structure of a melt, correspond to those of a liquid metal, and the diffusion properties of the quasicrystalline "solid" S-phase are similar to those of the lattice diffusion. Diffusion mechanisms in a boundary comprising a mixture of the L- and S-phases depend substantially on their volume fractions that is on the grain-boundary free volume. If the boundary free volume is small ($\alpha \to 0$), i.e. almost all its area is occupied by S-phase, parameters of the grain-boundary diffusion should be close to those of the lattice diffusion. On the contrary, when the free volume is high ($\alpha \to \alpha^*$), the islands of L-phase merge into a continuous L-phase cluster, and the grain-boundary diffusivity approaches the value typical of a melt. In intermediate cases ($0<\alpha<\alpha^*$), the L-phase presents separate islands whose centers are randomly distributed over the boundary plane (see Figure 15.1). As islands of L-phase are surrounded by areas of S-phase with low diffusivity, diffusion mass transfer becomes complicated and should be considered in more details.

First, note that the size of a random L-phase island can grow spontaneously due to thermal fluctuations, and, at a certain value, $r_* r_*$, it will touch an adjacent L-phase island. The concentration of such areas in the boundary is proportional to exp(-F/kT), where F is the free energy necessary for the generation of an island with critical size r_*

$$F = Q_l - TS_l \tag{15.32}$$

Here Q_l and S_l are the enthalpy and entropy of this process. When L-islands contact each other, the diffusion mass transfer proceeds in the liquid phase and mobility M_l of atoms in the fluctuation area is determined by the expression:

$$M_l \approx D_{L0} \exp(-Q_L/kT) \tag{15.33}$$

Within this approach, the grain boundary diffusion coefficient D_b is proportional to the concentration of critically-sized areas, which is expressed as $\exp(-F_l/kT)$, and the atomic diffusion mobility in these areas M_l

$$D_b \approx \varphi D_{L0} \exp(-Q_L/kT)\exp(-F_l/kT), \tag{15.34}$$

where φ is a correlation factor (we use $\varphi=0.1$ as a model parameter). Combining Eqs. (15.32)-(15.34) and using the equilibrium grain-boundary diffusion coefficient $D_b=D_{b0}\exp(-Q_b/kT)$, the parameters for grain boundary self-diffusion can be expressed through those for diffusion in a melt:

$$D_{b0} = \varphi D_{L0} \exp(S_l/k), \tag{15.35}$$

$$Q_b = Q_L + Q_l \tag{15.36}$$

Free Energy of Grain Boundary Diffusion

In order to determine the free energy fluctuation F_l (and respectively, Q_l and S_l), we assume that

$$F = F_m + F_\gamma, \tag{15.37}$$

where F_m is the energy necessary to melt the S-phase area encompassing the initial island of L-phase, and F_γ is the energy associated with the solid/liquid interface around the new L-phase island having critical size r_* r_*. In order to evaluate F_m, the following expression can be used[22)]

$$F_m = \lambda\rho(1 - T/T_m)(V^* - V^0) - \gamma_0(s_S^* - s_S^0) \quad (15.38)$$

Here V^0 and s_S^0 are the volume and the surface area of an initial (cylindrical) island of L-phase, V^* and s_S^* are the corresponding parameters of the critically-sized island. The second term in Eq. (15.38) reflects the fact that the L-phase area of grain boundary has specific surface energy γ_0.

Similar formula for F_γ is:

$$F_\gamma = \gamma_{S/L}(s_S^* - s_S^0) = (\gamma_{S/L}^0 - TS_{S/L})(s_S^* - s_S^0) \quad (15.39)$$

Here, s_S^* and s_S^0 is the full surface area of the new and initial L-island, correspondingly. The term $(-\gamma_{S/L}s_S^0)$ describes the energy necessary forof disappearance of the solid/liquid interface area around the initial L-island at fluctuation. From Eqs. (15.37) - (15.39) and (15.32), enthalpy and entropy corresponding to the fluctuation increase of the volume of L island from V^0 to $V^*$are

$$Q_1 = \lambda\rho(V^* - V^0) - \gamma_0(s_S^* - s_S^0) + \gamma_{S/L}^0(s_S^* - s_S^0) \quad (15.40)$$

$$S_1 = \lambda\rho(V^* - V^0)/T_m + S_{S/L}(s_S^* - s_S^0) \quad (15.41)$$

Eqs. (15.40), (15.41), (15.35), and (15.36) give the parameters of grain boundary self-diffusion

$$D_{b0} = \varphi D_{L0}\exp[(\lambda\rho(V^* - V^0)/T_m + S_{S/L}(s_S^* - s_S^0))/k], \quad (15.42)$$

$$Q_b = \lambda\rho(V^* - V^0) - \gamma(s^* - s^0) + \gamma_{S/L}^0(s_S^* - s_S^0) + Q_L \quad (15.43)$$

In order to numerically evaluate parameters Q_b, D_{b0} and calculate the grain boundary diffusion coefficient D_b, it is necessary to discuss the shape and sizes of L-phase islands. As above, these islands are considered to be cylinders with height δ and radius $r_L a_L$. It is assumed that the islands originating at fluctuations have a similar shape and radiusi r_* r_*. As $V^* = \pi r_*^2\delta$, $V^0 = \pi r_L^2\delta$, $s^* = \pi r_*^2$, $s^0 = \pi r_L^2$, $s_S^* = 2\pi r_*^2 + 2\pi r_*\delta$, $s_S^0 = 2\pi r_L^2 + 2\pi r_L\delta$, and Eqs. (15.42), (15.43) become

22 In calculating the free energy fluctuation, the contribution of elastic forces should be accounted for. However, as will be shown later, in most cases thye size of fluctuation area r_* r_*<2b is small, and the elastic energy can be neglected in the first approximation.

$$Q_b = \pi(\lambda\rho\delta - \gamma_0 + 2\gamma^0_{S/L})w + 2\pi\gamma^0_{S/L}\delta(r_* - r_L) + Q_L, \qquad (15.44)$$

$$D_{b0} = \varphi D_{L0}\exp\left[(\pi(\lambda\rho\delta/T_m + 2S_{S/L})w + 2\pi S_{S/L}\delta(r_* - r_L))/k\right], \qquad (15.45)$$

where $w = r_*^2 - r_L^2$. To evaluate the critical size of an L-phase island r_*, at which two adjacent L-phase islands come into contact (Figure 15.2), we shall examine the ensemble of L-phase islands randomly distributed over the grain boundary plane. A parameter α^* is introduced to characterize the critical volume fraction of L-phase at which these islands can contact each other with a sufficiently high probability. Then, α^* is interpreted as a percolation threshold and can be determined using the percolation theory (Efros, 1982). For a two-dimensional net of randomly distributed domains of a conductive and nonconductive phase, the percolation threshold is 0.5 (Efros, 1982). That means that the L-phase islands will come into contact at α^*=0.5. The island's critical size r_* is connected with α^* by an obvious relation $\alpha^* = r_*^2/R^2$. In accordance with Eq. (15.10), r_* can be expressed through the relative boundary free volume α

$$r_* = r_L(\alpha^*/\alpha)^{1/2} \qquad (15.46)$$

By combining Eqs. (15.44) – (15.46), the activation energy and preexponential factor of grain-boundary self-diffusion (here δ=2b) are, respectively,

$$Q_b = (r_L/b)^2[W_1(\alpha^*/\alpha - 1) + W_2((\alpha^*/\alpha)^{1/2} - 1)] + Q_L, \qquad (15.47)$$

$$D_{b0} = \varphi D_{L0}\exp\{(r_L/b)^2[Z_1(\alpha^*/\alpha - 1) + Z_2((\alpha^*/\alpha)^{1/2} - 1)]/k\}, \qquad (15.48)$$

where

$$W_1 = 2\pi(\lambda\rho b^2\delta/2 + \gamma^0_{S/L}b^2 - \gamma_0 b^2/2);\; W_2 = 4\pi\gamma^0_{S/L}b^2;$$
$$Z_1 = 2\pi(\lambda\rho b^2\delta/2T_m + S_{S/L}b^2);\; Z_2 = 4\pi S_{S/L}b^2$$

Parameters Q_b and D_{b0} in these equations are expressed through the material thermodynamic constants λ, $\gamma^0_{S/L}$, $S_{S/L}$ and structural characteristics of a grain boundary: specific energy of the S-phase areas γ_0 and relative free volume α. From Eqs. (15.47) and (15.48), the boundary free volume can be expressed through diffusion parameters Q_b and D_{b0}:

$$\alpha = 0.5/[(F_1^2 + F_2)^{1/2} - F_1]^2 \qquad (15.49)$$

If Q_b is known, then

$$F_1 = \gamma^0_{S/L}/W_1;\; F_2 = W_1^{-1}[W_1 + W_2 + (Q_b - Q_L)/(r_L/b)^2].$$

If D_{b0} is known, then

$$F_1 = S_{S/L} / Z_1;\ F_2 = Z_1^{-1}[Z_1 + Z_2 + (\ln(D_{b0} / \varphi D_{L0}))/(r_L / b)^2]$$

Eqs. (15.47) and (15.48) can be also used to connect the activation energy of grain boundary self-diffusion Q_b with the preexponent D_{b0}. As a general expression for such a relation is lengthy, we present an approximate formula

$$Q_b = Q_L + (W_1/Z_1)\ \ln(D_{b0}/\varphi D_{L0}), \tag{15.50}$$

which is correct when $\gamma^0_{S/L} \approx S_{S/L}T_m$; $W_1(\alpha^*/\alpha-1) >> 2\gamma^0_{S/L}\ [(\alpha^*/\alpha)^{1/2}-1]$. Typical values $Q_L \approx 3.5\ kT_m$, $D_{b0} \approx 10^{-2}\ cm^2/s$, $D_{L0} \approx 10^{-4}\ cm^2/s$, $\varphi=0.1$ and $W_1/Z_1 \approx 1$ give for the activation energy $Q_b=10.4\ kT_m$.

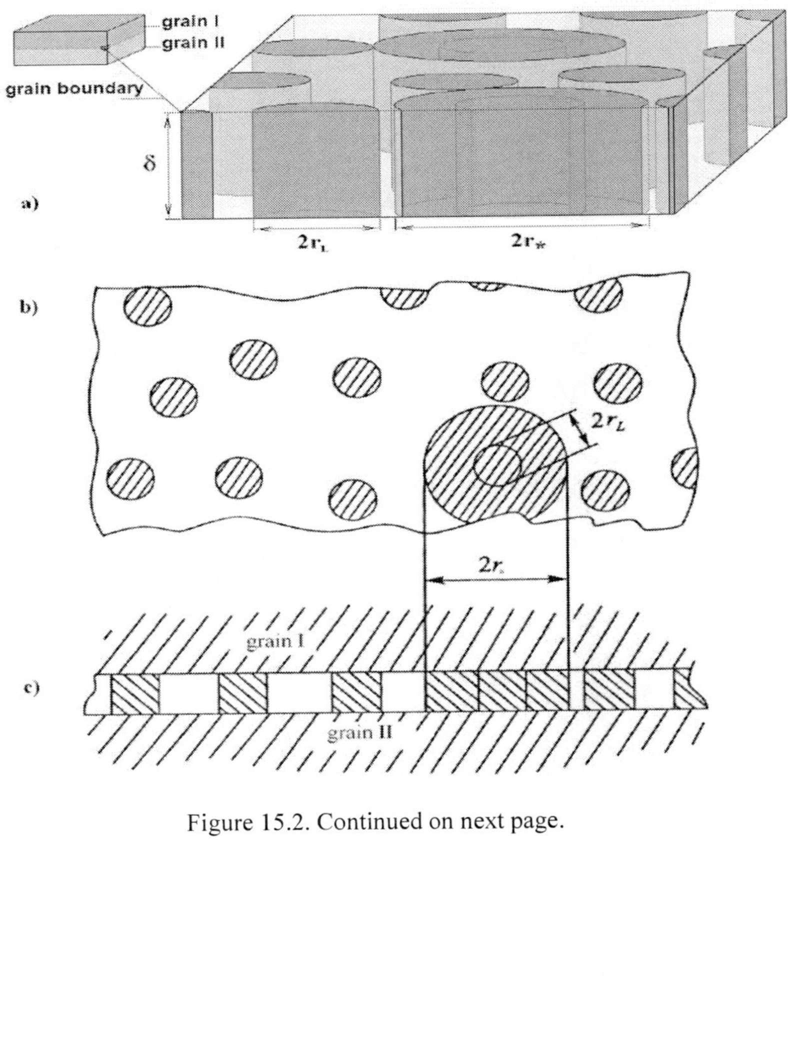

Figure 15.2. Continued on next page.

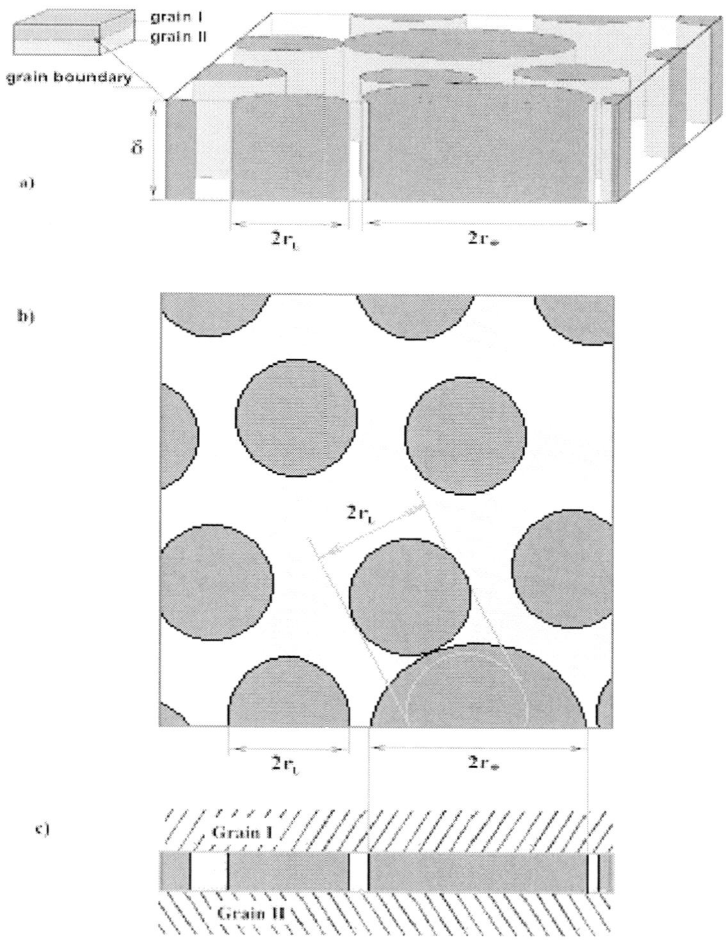

Figure 15.2. Fluctuation increase of the size of an L- phase island before contact with an adjacent L island (r_L a_L is the initial radius of L island; r_* is the critical radius of L island): stereographic projection (a), plane view (b), and side view (c).

An important comment should be added. As seen from Eqs. (15.47), (15.48), and (15.50), when the free volume approximates to the critical value α^*, the activation energy for grain boundary diffusion becomes identical to that for diffusion in a melt (Q_L) because $Q_b^*=Q_L$ at $\alpha=\alpha^*$. This means that at a high free volume ($\alpha\to\alpha^*$) the grain boundary diffusion is high and similar to liquid-metal diffusion at the given temperature (see Figure 15.3). However, it is not an instantaneous grain-boundary "melting" or a jump of the viscosity. As seen from Eq. (15.47), with the increase of α, the activation energy gradually changes from Q_b to Q_L. The corresponding transition from $Q_b\approx9\ kT_m$ to $Q_L\approx3\ kT_m$ at $T=0.5T_m$ implies a change in the diffusion coefficient by 2-3 orders of magnitude, which greatly influences all properties related to the grain boundary diffusion.

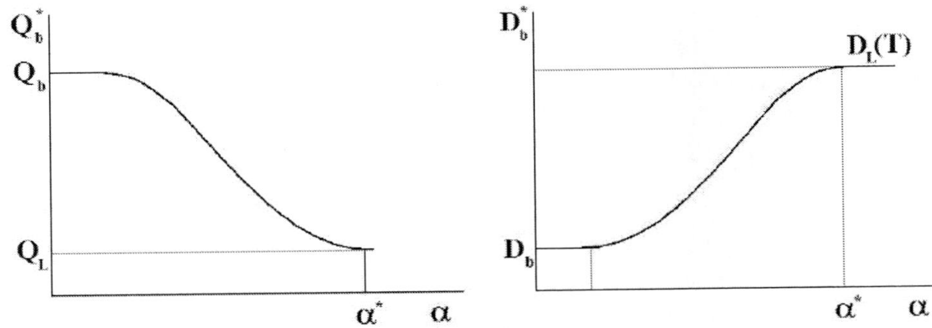

Figure 15.3. Influence of free volume on diffusion properties of grain boundaries.

15.2.4. Comparison with Experimental Data

I. Temperature Effects of the Grain-Boundary Diffusion and Activation Energy

Comparing theoretical and experimental parameters for grain boundary self-diffusion is rather difficult, because currently there are no reliable experimental data on the boundary free volume α. Besides, there is considerable scattering of experimental values of Q_b and D_{b0} (see Table 15.6). To reduce the uncertainty in determination of some parameters, we use the following comparison procedure:

1 The theoretical value of Q_b^t were calculated from Eq. (15.47) with the previously found value of free volume $\alpha(\gamma_b)$ (see Table 15.6, column 1);

2 The obtained theoretical results were compared with the experimental value of Q_b^e by Frost and Ashby (Frost et al. 1982). As seen from Table 15.6 (columns 2 and 3), almost for all materials Q_b^e differ from Q_b^t no more than 20%;

Table 15.6. Theoretical Q_b^t and experimental Q_b^e (Frost et al. 1982) values of the activation energy for grain boundary self-diffusion and the free volume of grain boundary

Metal	$\alpha(\gamma_b)$	$\frac{Q_b^t}{kT_m}$	$\frac{Q_b^e}{kT_m}$	Q_b^e, kJ/mol	D_{b0}^e	$\alpha(D_{b0})$	$\alpha(Q_b)$
Ag	0.355	8.7	8.74	90	$8.4\cdot10^{-2}$	0.357	0.354
Au	0.375	8.3	9.34	104.4	-	-	0.355
Cu	0.36	9.3	9.2	104	$9.8\cdot10^{-2}$	0.374	0.362
Al	0.38	10.4	10.8	84	$8.7\cdot10^{-1}$	0.362	0.375
Pb	0.36	8	8.8	44	$8.7\cdot10^{-2}$	0.35	0.344
Ni	0.37	9	9	129	$8.1\cdot10^{-1}$	0.348	0.37
Pt	0.36	10.1	9.1	154.8	-	-	0.376
V	0.36	11.9	11.6	209	$9.5\cdot10^{-1}$	0.368	0.365
Nb	0.4	9.4	11.5	263	$8.7\cdot10^{-1}$	0.356	0.365
Zn	0.34	11.9	10.6	60.5	$2.4\cdot10^{-1}$	0.368	0.36

3 Parameter D_{b0} for the related value of Q_b^e was found from Table 15.6 (column 6), and the value of $\alpha(D_{b0})$ (Table 15.6, column 7) was calculated using Eq. (15.49);

4 This value of $\alpha(D_{b0})$ was compared with the previously found $\alpha(\gamma_b)$. As one can see from Table 15.6 (columns 2 and 7), the values of $\alpha(D_{b0})$ and $\alpha(\gamma_b)$ agree within a 20-percent accuracy for most of the examined materials.

The observed good agreement between $\alpha(\gamma_b)$, $\alpha(D_{b0})$ and $\alpha(Q_b)$ (Table 15.6, column 8) confirms Eqs. (15.47), (15.48) for activation energy Q_b and preexponent D_{b0} of grain boundary self-diffusion in metals.

II. Effect of Pressure on Diffusion.
Activation Volume of Grain Boundary Self-Diffusion

When describing the experimental dependence of pressure P on diffusion $D_b(P)$, it is usually assumed that hydrostatic pressure changes the activation energy of grain boundary diffusion (or migration) in accordance with equation

$$Q_b(P) = Q_b + PV_p \tag{15.51}$$

Here Q_b is the activation energy for self-diffusion at P=0 and V_p is the activation volume. Within this approach, the activation volume V_p is a key factor which is determined as

$$V_p = -kT(d\ln D_b/dP) \tag{15.52}$$

In order to evaluate V_p in free volume models, let us examine the structure and properties of a high-angle grain boundary in a crystal subjected to hydrostatic compression. Generally, hydrostatic compression changes the thermodynamic parameters of a material (T_m, λ, $\gamma_{S/L}$, Q_L, etc.), which enter expressions for γ_b and D_b and hence affect the boundary free volume α. We limit ourselves to a case when the thermodynamic constants do not change under pressure, and the latter influences only free volume V_b. Apparently, this approximation is correct for relatively small pressures satisfying the condition $P/K_p \leq 10^{-1}$, where K_p is the compression modulus. Under external hydrostatic pressure P, the boundary free volume V_b is reduced by $\Delta V(P)$

$$V_b^P = V_b - \Delta V(P) \text{ or } \alpha^P = \alpha - \Delta\alpha(P), \tag{15.53}$$

where

$$\alpha^P = V_b^P / \Delta V_m, \ \Delta\alpha(P) = \Delta V(P)/\Delta V_m$$

Correspondingly, the L-phaseL-phase size r_L^P will differ from the equilibrium size r_L (see Eq. 15.10) r_L:

$$r_L^p = r_L(1 - \Delta\alpha(P)/\alpha)^{1/2}$$
$$r_L^p = r_L(1 - \Delta\alpha(P)/\alpha)^{1/2}$$

(15.54)

Due to free volume reduction, the enthalpy and entropy of such boundaries will decrease. To calculate them, α^p and r_L^p should be inserted in Eqs. (15.26) and (15.27) instead of equilibrium values α and $r_L a_L$:

$$\gamma_L^{0p} = \alpha^p(\lambda\delta\rho + 2\gamma_{S/L}^0(1+\delta/r_L^p) - \gamma_0) + \gamma_0 = \alpha^p A_p + \gamma_0$$
$$\gamma_L^{0p} = \alpha^p(\lambda\delta\rho + 2\gamma_{S/L}^0(1+\delta/r_L^p) - \gamma_0) + \gamma_0 = \alpha^p A_p + \gamma_0, \quad (15.55)$$

$$S_b^p = \alpha^p(1+\delta/r_L^p)S_{S/L} + S_b^c(\alpha^p) \quad (15.56)$$

In case of a small change in free volume, $\Delta\alpha(P)/\alpha \ll 1$, expressions for γ_b^{0p} and S_b^p are simplified:

$$\gamma_b^{0p} = (\alpha - \Delta\alpha(P))A_p + \gamma_0 = \gamma_b^0 - A_p\Delta\alpha(P), \quad (15.57)$$

$$S_b^p = (\alpha - \Delta\alpha(P))(1+\delta/r_L)S_{S/L} + S_b^c(\alpha) = S_b + \alpha^p(1+\delta/r_L)S_{S/L} \quad (15.58)$$

At $\Delta\alpha(P)/\alpha \ll 1$, $r_L^p \approx r_L$; $S_b^c(\alpha^p) \approx S_b^c(\alpha)$. In this approximation, the free energy of a boundary $\gamma_b^p = \gamma_b^0 - TS_b^p$ may be expressed through equilibrium free energy γ_b:

$$\gamma_b^p = \gamma_b - \Delta\alpha(P)(A_p - 2TS_{S/L}(1+\delta/r_L)) \quad (15.59)$$

The reduction of the boundary energy $\Delta\gamma_p = \gamma_b - \gamma_b^p$ due to a change in its free volume is proportional to $\Delta\alpha(P)$:

$$\Delta\gamma = \Delta\alpha(P)[A_p - 2TS_{S/L}(1+\delta/r_L)] \quad (15.60)$$

To describe diffusion in a grain boundary having free volume $\Delta\alpha(P)$, the equilibrium parameter α in Eqs. (15.48), (15.49) should be replaced by α^p:

$$D_b^p = D_{b0}^p \exp(-Q_b^p/kT) \quad \text{at } \alpha^p < \alpha_A \quad (15.61)$$

Here, preexponential factor D_{b0}^p is:

$$D_{b0}^p = \varphi D_{L0} \exp\{[Z_1(\alpha^*/[\alpha - \Delta\alpha(P)] - 1) + Z_2([\alpha^*/[\alpha - \Delta\alpha(P)]]^{1/2} - 1)]/k\} \quad (15.62)$$

The activation energy of self-diffusion

$$Q_b^p = W_1(\alpha^*/(\alpha-\Delta\alpha(P))-1)+W_2([\alpha^*/(\alpha-\Delta\alpha(P))]^{1/2}-1)+Q_L \qquad (15.63)$$

Parameter α_A in Eq. (15.61) is equal to $\alpha_A=\alpha^*-\alpha$. In the case of small free-volume changes $\Delta\alpha<<\alpha$, $\Delta\alpha(P)<\alpha_A$), Eq. (15.61) gives

$$D_b^p = D_b \exp(\Delta\alpha(P)/\alpha_B); \qquad \Delta\alpha(P)<\alpha_A;\ \Delta\alpha(P)/\alpha<<1. \qquad (15.64)$$

Here, parameter α_B is

$$\alpha_B^{-1} = \frac{\alpha^*}{\alpha^2}\left[\frac{W_1}{kT}-\frac{Z_1}{k}+\frac{1}{2}\left(\frac{W_2}{kT}-\frac{Z_2}{k}\right)\left(\frac{\alpha}{\alpha^*}\right)^{1/2}\right]. \qquad (15.65)$$

For typical values of thermodynamic characteristics listed in tables 15.3 and 15.4, $\alpha=0.35\div0.40$ (see Table 15.6) and at $T=0.5T_m$ one will have $\alpha_B\approx2\cdot10^{-2}$.

To obtain an explicit dependence of the grain boundary energy γ_b and diffusion parameters on P, we will use an empirical equation (Ed. Grigor'ev, 1991) to change the volume when applying pressure

$$\Delta V_b(P) = -a_pP + b_pP^2, \qquad (15.66)$$

where a_p and b_p are the coefficients. At pressures $P<10^9$ Pa ($P/K_p \le 10^{-1}$), this equation gives a satisfactory approximation to experiments. Moreover, in most cases the contribution of the quadratic term in Eq. (15.66) does not exceed 10% and can be neglected:

$$\Delta V_b(P) = P/K_p, \qquad (15.67)$$

$$\Delta\alpha(P) = P/K_p\cdot\Delta V_m. \qquad (15.68)$$

At typical value $2r_L2a_L\approx2b\approx\delta$, Eq.(15.68) together with Eq. (15.60) gives

$$\Delta\gamma/\gamma_b = \frac{P}{K_p}\frac{1}{\Delta V_m}(\lambda\rho\delta+6(\gamma_{S/L}^0-S_{S/L}T)-\gamma_0)/\gamma_b. \qquad (15.69)$$

The influence of pressure on the grain boundary diffusion coefficient can be found from Eqs. (15.68) and (15.64):

$$D_b(P) = D_b\exp\left(-\frac{P}{K_p\Delta V_m\alpha_B}\right). \qquad (15.70)$$

To compare theoretical expression (15.70) with experimental data, the activation volume for GB diffusion V_p is determined from Eqs. (15.52), (15.65):

$$V_p = \left(\frac{kT_m}{K_p\Omega}\right)\frac{1}{\Delta V_m}\left[\frac{\alpha^*}{\alpha^2}\left[\left(\frac{W_1}{kT}-\frac{Z_1}{k}\right)+\frac{1}{2}\left(\frac{W_2}{kT}-\frac{Z_2}{k}\right)\left(\frac{\alpha}{\alpha^*}\right)^{1/2}\right]\left(\frac{T}{T_m}\right)\right]. \qquad (15.71)$$

The volume V_p depends on the thermodynamic parameters ΔV_m, T_m, $\gamma_{S/L}$, elastic constant K_p and the boundary free volume α. Since α is connected with the misorientation angle of a grain boundary and its plane orientation, V_p is also linked to these parameters.

Comparison with Experimental Data

For a quantitative comparison of expressions (15.69), (15.70) and (15.71) with experiment, reliable data on the thermodynamic parameters of materials (including $\gamma^0_{S/L}$, $S_{S/L}$) and the grain-boundary free volume α should be known. However, there are no experimental data on α. Therefore, only a qualitative comparison of the theory with experiment can be currently performed.

First, it should be noted that the model gives an exponential dependence of diffusion coefficient D_b on pressure P (see Eq. (15.70)), which is observed in experiments. Second, it predicts a linear dependence of the boundary energy on P (see Eq. (15.69)). The activation volume V_p can be evaluated using Eq. (15.71). For typical values of $K_p\Omega/kT_m\approx1.5\cdot10^2$, $\Delta V_m=5\cdot10^{-2}$, temperature $T/T_m=0.5$, $\alpha_B^{-1}=10\div50$ (see Eq. (15.65)); and temperature $T/T_m=0.5$, that gives $V_p/\Omega\approx0.67\div3.3$. This result agrees well with experimental data for V_p lying in the range $V_p = (0.5\div5.5)\Omega$ (Kedves et al. 1989; Erdelyi et al. 1987; Friesel et al. 1991; Lojkowski et al. 1993; Vieregge et al. 1991; Straumal et al. 1984; Martin et al. 1967). Third, the temperature dependence of activation volume V_p from Eq. (15.71) can be presented as

$$V_p(T) = A^T(\alpha) - B^T(\alpha)T, \qquad (15.72)$$

where

$$A^T = \frac{kT_m}{K_p\Omega}\frac{1}{\Delta V_m}\left(\frac{\alpha^*}{\alpha^2}\frac{W_1}{k}+\frac{1}{2}\left(\frac{\alpha^*}{\alpha}\right)^{1/2}\frac{W_2}{k}\right),\quad B^T = \frac{kT_m}{K_p\Omega}\frac{1}{\Delta V_m}\left(\frac{\alpha^*}{\alpha^2}\frac{Z_1}{kT_m}+\frac{1}{2}\left(\frac{\alpha^*}{\alpha}\right)^{1/2}\frac{Z_2}{kT_m}\right).$$

As seen from the obtained expression, $V_p(T)$ linearly decreases with temperature. Such behavior was experimentally observed for Cu-7.5% In alloy in (Friesel et al. 1991; Lojkowski et al. 1993).

15.3. Model of Nonequilibrium Grain Boundaries

15.3.1. Energy and Entropy of Nonequilibrium Grain Boundaries

According to the present approach, the distinction in properties between a nonequilibrium and equilibrium grain boundary is mainly due to the difference in their free volume. Consider a grain boundary whose total free volume V^S is formed by the equilibrium (V_b) and excess (V^+) terms:

$$V^S = V_b + V^+; \qquad \alpha^S = \alpha + \Delta\alpha, \tag{15.73}$$

where $\alpha^S = V^S/\Delta V_m$, $\Delta\alpha = V^+/\Delta V_m$. The structure of such a boundary at $\alpha^S < \alpha^*$ can be represented as system of L-phase islands surrounded by S-phase areas. The size of L-phase areas differs from the equilibrium size r_L:

$$r_L^S = r_L(1 + \Delta\alpha/\alpha)^{1/2}. \tag{15.74}$$

The enthalpy γ_b^{0+} and entropy S_b^+ of such boundaries will increase. Correspondingly, replacement of parameters α and r_L in the above-stated relevant expressions by α^S and r_L^S gives:

$$\gamma_b^{0+} = \alpha^S(\lambda\delta\rho + 2\gamma_{S/L}^0(1 + \delta/r_L^S) - \gamma_0) + \gamma_0, \tag{15.75}$$

$$S_b^+ = \alpha^S(1 + \delta/r_L^S)S_{S/L} + S_b^c(\alpha^S). \tag{15.76}$$

For a small change in free volume $\Delta\alpha/\alpha << 1$, $r_L^S \sim r_L$, $S_b^c(\alpha^S) \sim S_b^c(\alpha)$, and γ_b^{0+} and S_b^+ are:

$$\gamma_b^{0+} = (\alpha + \Delta\alpha)A + \gamma_0 = \gamma_b^0 + A\Delta\alpha, \tag{15.77}$$

$$S_b^+ = (\alpha + \Delta\alpha)(1 + \delta/r_L)S_{S/L} + S_b^c(\alpha) = S_b + \Delta\alpha(1 + \delta/r_L)S_{S/L}. \tag{15.78}$$

$$A = \lambda\delta\rho + 2\gamma_{S/L}^0(1 + \delta/r_L) - \gamma_0. \tag{15.79}$$

In this approximation, free energy $\gamma_b^* = \gamma_b^{0+} - TS_b^+$ of a boundary having excess free volume is expressed through equilibrium free energy γ_b:

$$\gamma_b^* = \gamma_b + \Delta\alpha((\lambda\delta\rho + 2(\gamma_{S/L}^0 - TS_{S/L})(1 + \delta/r_L) - \gamma_0)). \tag{15.80}$$

Due to changes in the boundary's free volume, the energy increment $\Delta\gamma = \gamma_b^* - \gamma_b$ becomes proportional to $\Delta\alpha$:

$$\Delta\gamma_b = \Delta\alpha(\lambda\delta\rho + 2(\gamma_{S/L}^0 - TS_{S/L})(1 + \delta/r_L) - \gamma_0). \tag{15.81}$$

From Eq. (15.81), the boundary energy linearly depends on the excess free volume. A relative change in free volume $\Delta\gamma/\gamma_b$ can be calculated using Eq. (15.81). Since typically $2r_L \cong 2b \cong \delta$, Eq. (15.81) can be rewritten as

$$\Delta\gamma/\gamma_b = \Delta\alpha(\lambda\rho\delta + 6(\gamma_{S/L}^0 - S_{S/L}T) - \gamma_0)/\gamma_b. \tag{15.82}$$

15.3.2. Diffusion in Nonequilibrium Grain Boundaries

To describe the diffusion in a grain boundary with excess free volume $\Delta\alpha$, in the above expressions for D_b, Q_b and D_{b0} parameter α should be replaced by α^S:

$$D_b^* = D_L, \qquad Q_b^* = Q_L, \qquad D_{b0}^* = D_{L0} \qquad \text{at } \Delta\alpha \geq \alpha_A, \qquad (15.83)$$

$$D_b^* = D_{b0}^* \exp(-Q_b^* / kT) \qquad \text{at } \Delta\alpha < \alpha_A. \qquad (15.84)$$

Here, preexponential factor D_{b0}^* is determined by Eq. (15.48):

$$D_{b0}^* = \varphi D_{L0} \exp\{[Z_1(\alpha^* /(\alpha + \Delta\alpha) - 1) + Z_2([\alpha^* /(\alpha + \Delta\alpha)]^{1/2} - 1)]/k\}.$$
(15.85)

The activation energy for self-diffusion is calculated using Eq. (15.47):

$$Q_b^* = W_1(\alpha^* /(\alpha + \Delta\alpha) - 1) + W_2([\alpha^* /(\alpha + \Delta\alpha)]^{1/2} - 1) + Q_L. \qquad (15.86)$$

Parameter α_A in Eqs. (15.83) and (15.84) is

$$\alpha_A = \alpha^* - \alpha. \qquad (15.87)$$

For a small free-volume increment $\Delta\alpha << \alpha$, $\Delta\alpha < \alpha_A$, Eq. (15.83) for D_b^* takes a simpler form:

$$D_b^* = D_b \exp(\Delta\alpha/\alpha_B); \; \Delta\alpha < \alpha_A; \; \Delta\alpha/\alpha << 1. \qquad (15.88)$$

Eqs. (15.81), 15.83-15.86, (15.88) can be used to evaluate the effect of excess free volume on the free energy γ_b^* and diffusion coefficient D_b^* of NGB. NFor numerical results evaluation, were obtained at shall take temperature $T = 0.5T_m$ for the typical excess grain-boundary free volumes $\Delta\alpha_1$=0.05 and $\Delta\alpha_2$=0.1. The corresponding increments of the free volume at $\Delta V_m \cong 5 \cdot 10^{-2}$ are $V_{b1}^* \cong 0.25\%$ and $V_{b2}^* \cong 0.5\%$. For the initial free volume $\alpha \cong 0.35$, the chosen values of $\Delta\alpha$ meet the conditions $\Delta\alpha < \alpha$ and $\Delta\alpha + \alpha < \alpha^*$ and approximate Eqs. (15.81) and (15.88) can be used. Then, using values of thermodynamic constants from Table 15.7, Eq. (15.87) gives $\Delta\gamma/\gamma_b(\Delta\alpha_1)=0.125$ for $\Delta\alpha_1$= 0.05 and $\Delta\gamma/\gamma_b(\Delta\alpha_2)=0.25$ for $\Delta\alpha_2$ = 0.1. Thus, the related increase of the boundary energy is 12.5% and 25%.

Similar calculations were performed for the boundary diffusion. In accordance with Eq. (15.88), the diffusion coefficient increases with the free volume $\Delta\alpha$ in proportionally to $\exp(\Delta\alpha/\alpha_B)$. For $\Delta\alpha_1$=0.05, $\Delta\alpha_2$=0.1 and α_B=0.02, diffusions coefficients are $D_b^*(\Delta\alpha_1)=12.2D_b$ and $D_b^*(\Delta\alpha_2)=148D_b$. Therefore, an insignificant change in the boundary energy leads to a substantial increase in diffusivity. This model describes the abnormal grain boundary diffusion due to a change of a boundary free volume. Related analysis will be presented for accumulation of the free volume and diffusion in boundaries induced by lattice dislocations.

Table 15.7. Parameter values used in calculations

Parameters	Notation	Value
Activation energy of self-diffusion in a melt	Q_L	$3.6\ kT_m$
Preexponential factor for diffusion coefficient in a melt	D_{L0}	$8\cdot10^{-4}\ cm^2/s$
Self-diffusion coefficient in a melt ($T=T_m$)	D_L	$1.5\cdot10^{-5}\ cm^2/s$
Activation energy for grain-boundary self-diffusion	Q_b	$9\ kT_m$
Preexponential factor for grain-boundary self-diffusion coefficient	D_{b0}	$2\cdot10^{-2}\ cm^2/s$
Grain-boundary self-diffusion coefficient ($T=0.5T_m$)	D_b	$3\cdot10^{-10}\ cm^2/s$
Enthalpy of the solid/liquid interface	$\gamma^0_{S/L}$ $\gamma^0_{S/L}$	$\gamma^0_{S/L}\ b^2=1\ kT_m$
Entropy of the solid/liquid interface	$S_{S/L}$	$S_{S/L}b^2=0.8\ k$
Parameters	Notation	Value
Grain boundary enthalpy ($\alpha=0.35$)	$\gamma^0_b\ \gamma^0_b$	$\gamma^0_b\ \gamma^0_b\ b^2=4.1\ kT_m$
Grain boundary entropy ($\alpha=0.35$)	S_b	$S_b b^2\cong2.33\ k$
Free energy of a grain boundary ($T=0.5T_m$)	γ_b	$\gamma_b b^2\cong2.9\ kT_m$
Free energy of the grain-boundary S-phase	γ_0	$\gamma_0 b^2\cong1.4kT_m$
Free volume of a grain boundary	V_b	$1.75\cdot10^{-2}$
Relative free volume of a grain boundary	α	0.35
Critical free volume of a grain boundary	α^*	0.5
Excess free volume of a grain boundary	$\Delta\alpha$	0.05; 0.1
Burgers vector	b	$2.5\cdot10^{-8}$ cm
Grain boundary width	δ	$5\cdot10^{-8}$ cm
Atomic volume	Ω	$1.6\cdot10^{-23}\ cm^3$
Typical grain size	d	$d/b\cong10^4$
Volume jump at the melting point	ΔV_m	$5\cdot10^{-2}$
Specific heat of fusion Density	$\lambda,\ \rho$ $\lambda,\ \rho$	$\lambda\rho b^3=1.5\ kT_m$
Melting point Shear modulus	$T_m,\ G$ $T_m,\ G$	$G\Omega/kT_m=50$

15.3.3. Grain Boundary Diffusion in Cores of Delocalized Dislocations

As it is known, a certain excess volume V_0 in bulk metals is associated with the core of lattice dislocations. Its value per unit length of dislocation is (Fridel, 1964)

$$V_0 = \psi b^2, \tag{15.89}$$

where $\psi \cong 1$ is a geometrical factor. In grain boundaries, the cores of lattice dislocations with Burgers vector b can be presented as a continuum set of dislocations with Burgers vector density w^0 satisfying the condition $\int_0^{r_d} w^0 dr = b \int_0^{S_d} w^0 dr = b$ where rS_d is a radius of the dislocation core. Within this formalism, free volume V_0 in cores of a continuum set of dislocations w^0 can be characterized by the free volume density v [cm^2/cm] that satisfies the condition $\int_0^{r_d} v dr = V_0 \int_0^{S_d} v dr = V_0$. For a uniform distribution of the Burgers vector density in the dislocation core [0, Sr_d], $w^0 = b/rS_d$ and $v = V_0/rS_d$. Using Eq. (15.89), an additional relation between w_0 and v is

$$v = w^0 V_0/b = \psi w^0 b. \tag{15.90}$$

When a lattice dislocation is embedded into non-special (random) grain boundary, its core is delocalized. That means that plastic incompatibility, originally localized over distance $r^0S_d \sim 2b$, spreads with time along the boundary over a distance r_d which is substantially larger than $S_d 2b$. This process can be described as a diffusion-controlled spreading of core-constituting continuum dislocations. A typical time for such process t_{diff} can be written in the following form (Perevezentsev et al. 1991; Kuzhidlovsky et al. 1998):

$$t_{diff} \cong A_1 (r_d/b)^3 \, kT / D_b \delta G, \tag{15.91}$$

where $A_1 \cong 10$. Other delocalization models give similar formulae (Kuzhidlovsky et al. 1998). Redistribution of core-forming continuum dislocations in the boundary during delocalization will lead to redistribution of the free volume associated with the dislocation core. It is natural to assume that during delocalization the free volume, initially localized in the area $S_d r_d^0$, is stretched along the boundary with a diffusion rate of core-forming continuum dislocation spreading. This process should be considered in detail.

Lattice dislocations embedded into the boundary form orientational misfit dislocations (OMD) whose Burgers vector Δb can be presented as a sum of normal Δb_n and tangential Δb_t components with respect to the plane of the components' boundary. As a result of core delocalization of a singular OMD, a certain distribution of tangential and normal continual dislocations appears in the boundary. Their Burgers vector densities w_t^0 and w_n^0 should satisfy evident conditions $\int_0^{rd} w_n^0 dr = \Delta b_n$ and $\int w_t^0 dr = \Delta b_t$ $\int_0^{r} w_n^0 dr = \Delta b_n$ and $\int w_n^0 dr = \Delta b_t$.

Using Eqs. (15.89), (15.90), an additional free volume in the boundary connected with these dislocations is generally determined as

$$V^{+} = \psi \frac{b}{\delta}(w_t^0 + w_n^0) = \psi_1 (w_t^0 + w_n^0).$$

(15.92)

At the initial time, the free volume located in the dislocation core of radius Sr_d is

$$V_0^{+} = \psi_1 b / r_d \; V_0^{+} = \psi_1 b / S_d.$$

(15.93)

For typical values ψ=1, δ=2b, ψ_1=ψb/δ≅0.5 and rS_d=2b, we have V_0^{+}=0.25. During spreading, the core radius r(t) grows and at arbitrary time t the free volume of the dislocation core is

$$V_r^{+} = \psi_1 b / r_d(t) \; V_r^{+} = \psi_1 b / r(t).$$

(15.94)

At r(t) = 100b, V_r^{+}=5·10^{-3}. Therefore, in the area of a delocalized dislocation core, the cumulative boundary free volume V^S differs from the initial equilibrium value V_b depending on V_r^{+}. Presently, it is possible to perform the exact calculation of the cumulative free volume for a random boundary in the area of the OMD core. For simplicity, we assume the additivity of free volumes V_b and V_r^{+}:

$$V_r^S = V_b + V_r^{+}. \quad (15.95)$$

Apparently, this approximation is true only for random boundaries having misorientation angles far from specificspecial values. Dividing both parts of Eq. (15.95) by ΔV_m and introducing notations $\alpha_r^S = V_r^S / \Delta V_m$ and $\Delta\alpha_r = V_r^{+} / \Delta V_m$, we induce a more convenient parameter $\alpha = V_b/\Delta V_m$ as a relative free volume of a grain boundary:

$$\alpha_r^S = \alpha + \Delta\alpha_r \,. \quad (15.96)$$

Obtained relations connect the grain boundary self-diffusion with the values of equilibrium α and excess $\Delta\alpha$ free volume. For a small excess free volume $\Delta\alpha_r \ll \alpha$ and $\Delta\alpha_r < \alpha_A$, the grain boundary diffusion coefficient D_b^* follows from Eq. (15.87):

$$D_b^* = D_b \exp(\Delta\alpha_r / \alpha_B); \qquad \Delta\alpha_r/\alpha \ll 1,\ \Delta\alpha_r/\alpha_A < 1. \quad (15.97)$$

In the case when the excess free volume of a grain boundary slightly exceeds the critical level $\Delta\alpha_r \geq \alpha_A$, diffusion corresponds to those of a melt (see Eq. (15.82)):

$$D_b^* = D_L \qquad \Delta\alpha_r/\alpha_A \geq 1. \tag{15.98}$$

Inserting Eq. (15.94) in Eqs. (15.97), (15.98) gives the dependence of the grain boundary diffusion coefficient in a delocalized dislocation core D_{br}^* on the core width r(t):

$$D_{br}^* = D_L, \; r(t) \leq r_A; \tag{15.99}$$

$$D_{br}^* = D_b \exp(r_B / r), \qquad r > r_A. \; D_{br}^* = D_b \exp(r_B / r(t)), \qquad r(t) > r_A \tag{15.100}$$

Here r_A characterizes the size of dislocation core when the cumulative boundary free volume α_r^S exceeds the critical value α^*. Within this area the grain-boundary diffusion corresponds to those of a melt. From Eqs. (15.94), (15.86), r_A at $\alpha+\Delta\alpha_r=\alpha^*$ is

$$r_A = \psi_1 b / V_m(\alpha^* - \alpha) = \psi_1 b / V_m \alpha_A. \tag{15.101}$$

For typical values $\alpha=0.4$, $\alpha^*=0.5$, $\psi_1=0.5$ and $\Delta V_m=5\cdot10^{-2}$, $r_A=100b$.

Parameter r_B in Eq. (15.100) is a radius of area where a dislocation can affect diffusion. In fact, in accordance with this equation, if the radius of the dislocation core is within a range $r_A<r<r_B$, the diffusion coefficient D_{br}^* in the dislocation core substantially – by the factor of $\exp(r_B/r)$ – differs from the diffusion coefficient D_b outside the core. At a large core radius $r>r_B$, this difference becomes less substantial and at $r>>r_B$ $D_{br}^* \cong D_b$. Eqs. (15.94), (15.96) give for the radius r_B

$$r_B = \psi_1 b / \alpha_B V_m. \tag{15.102}$$

At typical values $\psi_1=0.5$, $\alpha_B=0.02$ and $\Delta V_m=5\cdot10^{-2}$, $r_B=500b$.

Now we can evaluate a mean value of the grain boundary diffusion coefficient $\langle D_{br}^* \rangle$ in the area of a delocalized dislocation core spreading from the initial radius $S_d r_d^0$ to a radius $r_1>r_B$. For this purpose, the area $[S_d r_d^0, r_1]$ is conventionally divided into three ranges $[S_d r_d^0, r_A]$, $[r_A, r_B]$ and $[r_B, r_1]$, which are further named ranges *l, d* and *s*, respectively. It is assumed that the mean value of the diffusion coefficient D_{bl} inside range *l* is equal to that in a melt:

$$D_{b\ell}^* = D_L, \qquad r < r_A; \; \alpha > \alpha_A. \tag{15.103}$$

In the area *d*, the mean value of grain boundary diffusion coefficient is determined by the approximate expression followed from Eq. (15.97):

$$D^*_{bd} = D_b \frac{\alpha_B}{\alpha_A - \alpha_B}(e^{\alpha_A/\alpha_B} - e)\cdot \tag{15.104}$$

For the above-stated parameters α_B=0.02 and α_A=0.1, $D^*_{bd} \approx 36D_b$. In the area *s*, the mean diffusion coefficient D^*_{bS} (see Eq. (15.97)) is:

$$D^*_{bS} = D_b \frac{\alpha_B}{\alpha_B - \alpha_1}(e - e^{\alpha_1/\alpha_B}). \tag{15.105}$$

At $\alpha_1 \to 0$, Eq. (15.97) gives $D^*_{bS} \cong 1.7D_b$. The mean diffusion coefficient D^*_{bl} in the area [rS_d, r_B], where diffusion is affected by dislocations, is determined by an obvious approximate relation

$$D^*_{bl} = \frac{D^*_{b\ell} D^*_{bd}}{D^*_{b\ell}(1 - r_A/r_B)^2 + D^*_{bd}(r_A/r_B)^2}. \tag{15.106}$$

At the above-stated parameters and $D^*_{bl} >> D^*_{bd}$, Eq. (15.106) gives $D^*_{bl} \approx 45D_b$. The mean diffusion coefficient $< D^*_{br} >$ throughout the range [$S_d r^0_d$, r_1] can be calculated from formula:

$$< D^*_{br} > = \frac{D^*_{bl} D^*_{bS}}{D^*_{bl}(1 - r_B/r_1)^2 + D^*_{bS}(r_B/r_1)^2}. \tag{15.107}$$

It is interesting to evaluate the characteristic time t_{diff} during which the diffusion properties of a boundary along the core of an embedded dislocation will be noticeably different from the diffusion properties of an equilibrium boundary. Parameter t_{diff} will be called "a time of dislocation effect on grain boundary diffusion". As a first approximation, t_{diff} can be estimated as the time of spreading of the dislocation core to the size r_B (see Eq. (15.102)) since r_B is the radius of the dislocation influence on diffusion. Then, t_{diff} is

$$t_{diff} = A_1 (r_B/b)^3 kT / D^*_{bl}\delta G. \tag{15.108}$$

For typical values A_1=10, r_B=500b, D^*_{bl}=45D_b, $D_b(T=0.5T_m)=3\cdot 10^{-10}$ cm^2/s, Gb^3/kT=100 and δ=2b, Eq. (15.108), gives $t_{diff}=2\cdot 10^{-1}$ s.

Let us compare t_{diff} with the delocalization time t_d. It is known that during t_d the submicroscopic contrast from a dislocation embedded into a grain boundary disappears. That happens when the dislocation core spreads to a typical distance r_d which corresponds to 1 to 3 extinction sizes $r_{dm} \approx 6\cdot 10^{-8}$ m. Using Eqs. (15.91), (15.108), the ratio t_{diff}/t_d can be presented as

$$t_{diff}/t_d = (r_B/r_{dm})^3. \tag{15.109}$$

At typical values $r_B \cong 500$ b and $r_{dm} \cong 240b$, $t_{diff}/t_d \approx 8$, i.e. the time of dislocation effect on diffusion can substantially exceed the time of dislocation delocalization in the grain boundary. Thus, during long time after the disappearance of diffraction contrast of a dislocation embedded into a grain boundary, the boundary can retain high diffusivity in the area of the delocalized-dislocation core.

15.3.4. Diffusion in Grain Boundaries Bombarded by Lattice Dislocations

Kinetics of Defect Accumulation in Grain Boundaries

When lattice dislocations are embedded into the boundary, a system of orientational misfit dislocations (OMD), which are characterized by linear density ρ_b, appears in the boundary. The delocalization process results in the tangential and normal delocalized dislocations, which are described by the local density of Burgers vectors w_n and w_t, respectively. The additional free volume V^+ connected with these defects in general case should beis proportional to the cumulative density of dislocations in the boundary w^Σ:

$$V^+ = \psi_1(\rho_b \Delta b_n + \rho_b \Delta b_t + w_t^0 + w_n^0). \qquad (15.110)$$

As will be shown below, for high-angle boundaries the value of w_n^0 does not contribute to the grain-boundary free volume, and Eq. (15.110) can be rewritten as $V^+ = \psi_1(\rho_b \Delta b + w_t^0)$, where $\Delta b = \Delta b_n + \Delta b_t$. Dividing this expression by ΔV_m, and denoting $w_0 = \Delta V_m/\psi_1$, the relative excess free volume $\Delta\alpha = V^+/\Delta V_m$ is

$$\Delta\alpha = (\rho_b \Delta b + w_t)/w_0 = w^\Sigma / w_0. \qquad (15.111)$$

Thus, the excess free volume $\Delta\alpha$ in a boundary bombarded by lattice dislocations can be found by studying the kinetics of accumulation of OMD and the products of their delocalization in the boundary.

Kinetics of OMD Accumulation

The linear density ρ_b of OMD in a boundary bombarded by lattice dislocation flows during intragranular deformation with the strain rate $\dot{\varepsilon}_V$ can be described by a kinetic equation

$$\dot{\rho}_b = \xi \dot{\varepsilon}_V / b - \rho_b / t_1. \qquad (15.112)$$

Here t_1 is the typical time for the transformation of OMD into normal and tangential delocalized dislocations. It is evident that in the defect structure of the boundary containing a system of orientational misfit dislocations, t_1 is a time of diffusional spreading of a delocalized dislocation core to distance a, which is close to spacing between OMDs that is $a_\rho \sim \rho_b^{-1}$. A typical time of spreading (see Eq. 15.91) is

$$t_1 \cong A_1 a_\rho^3 kT / D_b \delta G \Omega. \tag{15.113}$$

To estimate the time t_1, it is necessary to account for the fact that delocalization process is close to diffusion mass transfer in a boundary containing excess free volume; hence, the diffusion coefficient in Eq. (15.113) should be equal to D_b^*. Inserting $a_\rho = \rho_b^{-1}$ and $D_b = D_b^*$ into Eq. (15.113), the latter is rewritten as:

$$t_1 = A_1 kT / \rho_b^3 D_b^* \delta G \Omega \tag{15.114}$$

Here diffusion coefficient D_b^* is determined by Eqs. (15.97), (15.98), (15.111). In general, D_b^* depends both on the OMD density ρ_b and on the density of the Burgers vector of delocalized dislocations w_t. A related analytical expression for D_b^* is cumbersome, so only several particular cases will be considered below.

First, we examine the situation where the density of Burgers vector of delocalized dislocations is small in comparison with the density of the OMD Burgers vector, i.e. $\rho_b \Delta b >> w_t$. Then, according to Eq. (15.111), formula for $\Delta\alpha$ is

$$\Delta\alpha = \rho_b \Delta b / w_0. \tag{15.115}$$

Eqs. (15.115),(15.97), (15.98) give for the diffusion coefficient D_b^*:

$$D_b^* = D_L, \qquad \rho_b \Delta b > \alpha_A w_0; \tag{15.116}$$

$$D_b^* = D_b \exp(\rho_b \Delta b / \alpha_B w_0), \qquad \rho_b \Delta b << \alpha_A w_0. \tag{15.117}$$

The solution of Eqs. (15.112)– (15.117) defines accumulation in the boundary OMD including their stationary density ρ_b^{st} and the necessary time. For convenience, a situation when the excess free volume attains the critical value α_A, and the grain-boundary diffusion coefficient approximates to that in the L-phase will be called the "L-mode" (see Eq. (15.116)). Similarly, a situation when the diffusion coefficient is determined mainly by the S-phase will be termed as the "S-mode" (see Eq. (15.117)). For the L-mode, the stationary OMD density ρ_b^{st} is obtained by combining Eqs. (15.116), (15.114), (15.112):

$$(\rho_b^{st} \Delta b)_L = (\xi \dot{\varepsilon}_v kT / D_L \delta G)^{1/4}. \tag{15.118}$$

Equating $(\rho_b^{st} \Delta b)_L$ to α_A gives the critical intragranular strain rate $\dot{\varepsilon}_{vA}$ which corresponds to transition to the L-mode:

$$\dot{\varepsilon}_{vA} = (D_L \delta G / kT \xi A_1)(\alpha_A w_0)^4. \tag{15.119}$$

The characteristic time t_{rA} necessary to attain the stationary OMD density $\rho_b^{st}\Delta b$ can be found by inserting $a_\rho=(\rho_b^{st})^{-1}$ and $D_b=D_L$ into Eq. (15.113):

$$t_{rA}=\frac{1}{(\xi\dot{\varepsilon}_v)^{3/4}}\left(\frac{A_1kT}{D_L\delta G}\right)^{1/4}. \tag{15.120}$$

For the parameters of Table 15.7, $\dot{\varepsilon}_{vA}$=3·10^2 s^{-1} and at $\dot{\varepsilon}_v$=6·10^2 s^{-1} > $\dot{\varepsilon}_{vA}$, the stationary density OMD and the characteristic time are $\rho_b^{st}\Delta b\cong1.2\cdot10^{-2}$ and $t_{rA}\cong2.2\cdot10^{-4}$ s.

In the S-mode, when $\rho_b\Delta b<\alpha_A$ and the grain boundary diffusion coefficient D_b^* is determined by Eq. (15.117), the stationary density of OMD ($\rho_b^{st}\Delta b$) is calculated from Eqs. (15.112) and (15.114):

$$\begin{aligned}&4\ln(\rho_b^{st}\Delta b)+(\rho_b^{st}\Delta b)/w_0\alpha_B=\ln(\xi\dot{\varepsilon}_vkTA_1/D_b\delta G),\\&\rho_b^{st}\Delta b<\alpha_Aw_0.\end{aligned} \tag{15.121}$$

When $D_b^*\approx D_b$, Eq. (15.221) can be presented in a general form:

$$\rho_b^{st}\Delta b=(g_1\dot{\varepsilon}_v/D_b^*)^{1/4}, \tag{15.222}$$

where

$$g_1=\xi_1A_1(kT/G\Omega)(b^3/\delta).$$

Similarly, by equating $\rho_b^{st}\Delta b$ to α_A, the rate of intragranular deformation $\dot{\varepsilon}_{vB}$ $\dot{\varepsilon}_{vA}$ satisfying to Eq. (15.121) is

$$\dot{\varepsilon}_{vB}=(D_b\delta G/kT\xi A_1)(\alpha_Aw_0)^4\exp(\alpha_A/\alpha_B). \tag{15.123}$$

Typical time t_{rB} during which the OMD density attains a stationary value is determined from Eqs. (15.114), (15.118), (15.121):

$$t_{rB}=\rho_b^{st}\Delta b/\xi\dot{\varepsilon}_vt. \tag{15.124}$$

For parameters of Table 15.7, $\dot{\varepsilon}_{vB}\cong4.4\cdot10$ s^{-1}; at $\dot{\varepsilon}_v\cong$ 10 s^{-1} < $\dot{\varepsilon}_{vB}$, $\rho_b^{st}\Delta b=6.1\cdot10^{-3}$ and $t_{rB}=1.2\cdot10^{-3}$ s; at $\dot{\varepsilon}_v=10^{-3}$ s^{-1}, $\rho_b^{st}\Delta b=1.85\cdot10^{-3}$ and t_{rB}=3.7 s. It should be noted that the deformation rate $\dot{\varepsilon}_{vB}$ from Eq. (15.123) determines the upper limit of strain rates corresponding to the S-mode. It does not coincide with the strain rate $\dot{\varepsilon}_{vA}$ from Eq. (15.119) that defines the lower limit of strain rates for the the L-mode. That results from the fact that

Eq. (15.117) for the diffusion coefficient is valid if $\alpha<\alpha^*$ and $\rho_b\Delta b<<\alpha_A w_0$. To evaluate the density of grain-boundary defects and the diffusion coefficient in the intermediate range of strain rates $\dot{\varepsilon}_{vB}<\dot{\varepsilon}<\dot{\varepsilon}_{vA}$, it is necessary to find a formal solution of Eqs. (15.111)-(15.117). Obviously, that the coefficient of deformation-stimulated diffusion D_b^* in this rate range has an upper limit equal to D_L (at $\dot{\varepsilon}\cong\dot{\varepsilon}_{vA}$), while the lower limit of D_b^* should be calculated from Eq. (15.117) at $\dot{\varepsilon}=\dot{\varepsilon}_{vB}$.

Accumulation Kinetics of the Products of OMD Delocalization

A kinetic equation for accumulation of tangential components of delocalized dislocationstangential components of Burgers vector of delocalized dislocations w_t in boundary is:

$$\dot{w}_t=\xi\dot{\varepsilon}_v-w_t/t_2. \tag{15.125}$$

Here, parameter t_2 presents a typical time of the withdrawal of tangential components of delocalized dislocationstangential components of Burgers vector of delocalized dislocations from a grain boundary. Several processes that control the rate of this withdrawal can be identified. First and foremost, it is grain boundary sliding (GBS). Tangential components w_t interacts with grain boundary sliding and the rate of their withdrawal is proportional to the grain boundaryintragranular strain rate $\dot{\varepsilon}_b$. In this case, $t_{2A}=1/\dot{\varepsilon}_b$. An expression for the grain boundary strain rate $\dot{\varepsilon}_b$was suggested in (Perevezentsev et al. 1985b):

$$\varepsilon_b=10^2(\sigma/G)^2(G\Omega/kT)D_b^*\delta/d^2b. \tag{15.126}$$

Along with the mechanism of GBS, a diffusion mechanism of «withdrawal» of tangential components of delocalized dislocationstangential components of Burgers vector of delocalized dislocations is possible. In this case, characteristic time t_{2B} can be calculated as the climb time of an "effective" dislocation which has the Burgers vector Δb_t, climbs into an adjacent grain, and annihilates a similar dislocation of the opposite sign from other boundaries (Perevezentsev, 1987):

$$t_{2B}^{-1}=C_1(G\Omega/kT)w_t\delta D_b^*/\Delta b_t d^2, \tag{15.127}$$

where $C_1\cong50$ is a numerical factor. As both these processes occur simultaneously, the characteristic time t_2 can be evaluated as

$$t_2 = t_{2A}\, t_{2B} / (t_{2A} + t_{2B}). \tag{15.128}$$

The values of t_{2A} and t_{2B} are equal at $w_t=w_t^*=10^2(\sigma/G)^2/C_1$. Typically, $C_1=50$, $(\sigma/G) = 10^{-4}\div10^{-3}$ and w_t^* W_t^* lies in the range from $2\cdot10^{-6}$ to $2\cdot10^{-8}$. This value is insignificant;, because therefore, Iin most cases, $w_t >> w_t^*$ W_t^* and $t_2 \cong t_{2B}$. Thuserefore, Eq. (15.127) can be used for rough estimation of t_2.

To determine the characteristic time t_2 and parameter w_t, it is necessary to consider the dependence of grain boundary diffusion coefficient D_b^* on the density of defects in the boundary. Previously, we have examined the situation when the density of defects in the boundary was fully defined by $\rho_b \Delta b$. Now, let us consider another extreme case when the density of delocalized dislocations is much higher than that of OMD, i.e. $w_t >> \rho_b \Delta b$ and $w^{\Sigma} \cong w_t$. As above, diffusion conditions for the L-mode and the S-mode will be examined. Transition to the L-mode is occurs when the density of defects in the boundary satisfies condition $w_t > \alpha_A w_0$. Inserting $D_b^* = D_L$ in Eq. (15.127) and using Eqs. (15.125), (15.128), the stationary value of density of tangential components of delocalized dislocationstangential components of Burgers vector of delocalized dislocations is

$$w_t^{st} = (\xi \dot{\varepsilon}_v kT / C_1 D_L \delta G)^{1/2} (\Delta b_t / b)(d/b),$$
$$w_t^{st} > \alpha_A w_0, \; w_t^{st} > \rho_b \Delta b. \tag{15.129}$$

Using $W_t^{st} = \alpha_A w_0$, we can find the strain rate $\dot{\varepsilon}_{vC}$ during in case of diffusion transition to the L-mode:

$$\dot{\varepsilon}_{vC} = (C_1 D_L \delta G / kT\xi)(b/\Delta b)^2 (b/d)^2 (\alpha_A w_0)^2. \tag{15.130}$$

The typical time t_{rC}, when the density of tangential components of delocalized dislocationstangential components of Burgers vector of delocalized dislocations attains the stationary value W_t^{st}, is determined from Eq. (15.127) at $w_t = W_t^{st}$ and $D_b^* = D_L$:

$$t_{rC}^{-1} = (\xi \dot{\varepsilon}_v kT / GD_L \delta)^{1/2} (b/d). \tag{15.131}$$

For parameters of Table 15.7, $\dot{\varepsilon}_{vC} \cong 6 \cdot 10$ s^{-1}. Then, at $\dot{\varepsilon} \cong 6 \cdot 10^2 > \dot{\varepsilon}_{vC}$, $w_t^{st} \cong 3.2 \cdot 10^{-2}$ and $t_2 \cong 7 \cdot 10^{-5}$ s.

In the S-mode, which takes place at $w_t << \alpha_A w_0$, the grain boundary diffusion coefficient is given by Eq. (15.117). Combining Eqs. (15.117), (15.127), (15.125), (15.128), we obtain a formula for the stationary density of tangential components of delocalized dislocationstangential components of Burgers vector of delocalized dislocations in the boundary:

$$2\ln(w_t^{st} / w_0 \alpha_B) + (w_t^{st} / w_0 \alpha_B) = \ln\left(\frac{\xi \dot{\varepsilon}_v (\Delta b_t / b)^2 (d/b)^2 kT}{C_1 D_b \delta G} \right). \tag{15.132}$$

In the first approximation, $D_b^* \approx D_b$ and Eq. (15.132) is reduced to a simple form

$$w_t^{st} = d(g_2\dot{\varepsilon}_v / D_b^*)^{1/2}, \qquad (15.133)$$

Where $g_2 = (\xi_2 / C_1)(kT / G\Omega)(\delta / b)$ is a numerical factor. From condition $w_t^{st}=\alpha_A w_0$ it is easy to determine the maximum strain rate $\dot{\varepsilon}_{vD}$ at which Eq. (15.132) is valid:

$$\dot{\varepsilon}_{vD} = \frac{C_1 D_b \delta G}{\xi(\Delta b_t / b)^2 (d/b)^2}(\alpha_A w_0)^2 \exp(\alpha_A / \alpha_B)\, \dot{\varepsilon}_{vD} = \frac{C_1 D_b \delta G}{\xi(\Delta b_t / b)^2 (d/b)^2}(\alpha_A w_0)^2 \cdot \qquad (15.134)$$

The typical time t_{rD}, when w_t reaches its stationary value w_t^{st} W_t^{st}, is determined from Eq. (15.127) using the following formula:

$$t_{rD} = w_t^{st} / \xi\dot{\varepsilon}_v (\Delta b_t / b) \cdot \qquad (15.135)$$

For parameters of Table 15.7, $\dot{\varepsilon}_{vD}\cong 10$ s^{-1}. Then, calculations give at $\dot{\varepsilon}=1$ s^{-1}, values $w_t^{st}=7\cdot 10^{-3}$ and $t_{rD}\cong 2.8\cdot 10^{-2}$ s; at $\dot{\varepsilon}=10^{-3}$ s^{-1}, values w_t^{st} $W_t^{st}=10^{-3}$ and $t_{rD}\cong 4$ s.

Normal components of orientational delocalized misfit dislocations, which are characterized by the density of Burgers vector w_n, form an additional misorientation $\omega\sim w_n$ in a boundary. It is assumed that a change in misorientation angle of a random boundary due to this additional misorientation does not lead to significant changes in its structure and free volume. This assumption is supported by a well-known experimental fact that the free energy of random boundaries virtually does not depend on their misorientation angle. In this case, the contribution of normal components of delocalized dislocations w_n to the additional free volume V_n^* induced by defects can be neglected, and

$$V_n^* = \psi_1 w_n \cong 0. \qquad (15.136)$$

Although normal components of delocalized dislocations do not influence the grain boundary diffusion, their role in the boundary's behavior is important, as it will be shown later. Hence, it is necessary to examine the kinetics of their accumulation in grain boundaries during intragranular deformation. The kinetic equation describing this process is

$$\dot{w}_n = \xi_3\dot{\varepsilon}_v - w_n / t_3 . \qquad (15.137)$$

where t_3 is a typical withdrawal time of normal components of delocalized dislocations from grain boundaries. These normal components of delocalized dislocations distributed in the boundaries form an additional misorientation angle (ω) which is proportional to w_n and results in the generation of disclination at triple points of grain boundaries. Sat a single boundary, such disclinations can be described as biaxial disclination dipoles with the power ω and arm d (Figure 15.4).

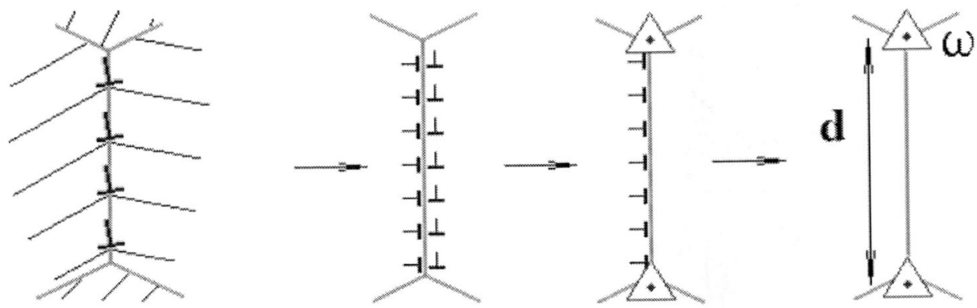

Figure 15.4. Evolution of normal components of delocalized dislocations in a grain boundary: Generation of a biaxial disclination dipole at grain boundary.
1. Generation of orientational misfit dislocations
2. OMD delocalization and formation of continual distribution of normal and tangential components of burgers vector of delocalized dislocations.
3. Withdrawal of tangential components of Burgers vector of delocalized dislocations.
4. Biaxial dislocation dipole at grain boundaries.

Detailed description of tThe kinetics of diffusional accommodation of defects was studied in (Perevezentsev, 1992). In the first approximation, this process can be described as a diffusion climb along grain boundaries of a "superdislocation" with Burgers vector B=ωd on a distance ~*d* until its annihilation with a similar dislocation of the opposite sign under internal stress fieldresulting in a stress field

$$t_3 \approx d / V_c, \tag{15.138}$$

whHere,

$$V_c = A_c(\sigma_i \Omega / kT)(b/d)(D_b^* / B), A_c = 2\pi / \ln(d/b) \sim 1$$

(Hirth, 1969). $A_c = 2\pi / \ln(d/b) \sim 1$ (Hirth, 1969) uUsing $\sigma_i=(GB/d)/2\pi$ and B=ωd in the above expressions, we obtain:

$$t_3 = A_1 \left(\frac{d}{b}\right)^3 \frac{b^3}{\delta D_b^*} \frac{kT}{G\Omega}. \tag{15.139}$$

The value of parameter A_1 will constitute approximately 10 Ffor typical values $d=10^3 b$, $G\Omega/kT_m=50$, $A_1=10$ (see also Kuzhidlovsky et al. 1988) in NMC materials and the equilibrium grain boundary diffusion coefficient $D_{b0}=10^{-6}$ m^2/s ($Q_b=9$ kT_m) at $T=0.5T_m$, formula (15.139) gives $t_3 \cong 30$ s. Comparison of characteristic relaxation times t_1, t_2 and t_3 from Eqs. (15.114), (15.127), (15.139), correspondingly, for reasonable material characteristics showed that time t_3 is the largest and the diffusional withdrawal of triple junction defects is the slowest recovery process that usually takes place during deformation and post deformation annealing.

Diffusion in Grain Boundaries Bombarded by Lattice Dislocations

The analysis of the defect accumulation in a boundary unveils the mechanisms of deformation-stimulated diffusion in grain boundaries. As above, consider L-mode and S-mode grain boundary diffusion for conditions of high and low strain rates, respectively. In the L-mode, the boundary diffusion coefficient is identical to a melt:

$$D_b^* = D_L \quad \text{at } \dot{\varepsilon}_v > \min\{\dot{\varepsilon}_{vA}, \dot{\varepsilon}_{vC}\}. \tag{15.140}$$

In S-mode, the coefficient of deformation-stimulated diffusion D_b^* D_b^* is proportional to the equilibrium diffusion coefficient D_b that exponentially depends on the cumulative density of defects w^{Σ} in the boundary

$$D_b^* = D_b \exp(w^{\Sigma} / w_0 \alpha_B) \quad \text{at } \dot{\varepsilon}_v < \min\{\dot{\varepsilon}_{vB}, \dot{\varepsilon}_{vD}\} \tag{15.141}$$

For small grain sizes $d<d_1$, the approximate relationship $w^{\Sigma} \cong \rho_b \Delta b$ is satisfied, while at $d>d_1$ it is possible to use approximation $w^{\Sigma} \cong w_t$. Parameter d_1 is determined from the condition $\rho_b \Delta b = w_t$ that leads to

$$d_1 = \frac{b}{\Delta b_t} \frac{1}{\rho_b \Delta b} (C_1 A_1)^{1/2}. \tag{15.142}$$

For material characteristics of Table 15.7) and $\rho_b \Delta b \cong 4.5 \cdot 10^{-3}$, one can find $d_1 \sim 10^4 b \cong 2.5$ μm. In materials with the grain size $d>d_1$, $w^{\Sigma} \cong w_t^{st}$ and Eq. (15.141) finally gives

$$D_b^* = D_b \exp(w_t^{st} / w_0 \alpha_B), \tag{15.143}$$

When the grain size $d<d_1$, $w^{\Sigma} \cong \rho_b^{st} \Delta b$ and similar equation for D_b^* D_b^* is

$$D_b^* \; D_b^* = D_b \exp(\rho_b^{st} \Delta b / w_0 \alpha_B); \tag{15.144}$$

15.3.5. Discussion

Now we can briefly discuss one of the most important issues of deformation-stimulated grain boundary diffusion which is fundamentally important for processes of high-temperature deformation and annealing. This problem was formulated in the classic work of Pumphrey and Gleiter (Pumphrey et al. 1975) and has not yet received the ultimate answer. The subject matter is that at elevated temperatures ($T \geq 0.5 T_m$), delocalization of lattice dislocations in boundaries occurs during short time t_l that is substantially less than a typical time of deformation ($\tau_{def} \sim \dot{\varepsilon}^{-1}$) and migration ($t_m = \Delta d / V_m$). For example, for stainless steel at T=1323

K $t_l \cong 10^{-4}$ s^{-1} (Pumphrey et al. 1975). Therefore, the question is: how can such rapid delocalization process affects processes of grain boundary sliding and migration? This problem can be solved within the present model. In fact, typical delocalization time t_l of a single dislocation in a grain boundary at $T>0.5T_m$ is $10^{-3} \div 10^{-4}$ s. A single dislocation can influence boundary diffusion for slightly longer time; however, its delocalization does not have a long-term effect on the grain boundary diffusion.

AHowever, a different situation occurs when a grain boundary is continuously bombarded by dislocation flow. For deformation, the flow is proportional to the itragranular strain rate $\dot{\varepsilon}_v$, i.e. $I_v \sim \dot{\varepsilon}_v /b$; for recrystallization, the flow is proportional to the migration rate V_m and the dislocation density ρ_v, i.e. $I_v \sim V_m \rho_v$. As a result, both the OMD delocalization products and the additional free volume ΔV are accumulated in the boundary. After a period of time t_r *), the density of boundary defects and the additional free volume reach stationary values. That leads to a certain significant change of the boundary structure free volume and to a substantial change of diffusion properties. The relaxation time when diffusion in grain boundaries remains different from ordinary diffusion is substantially longer than the delocalization time of a single dislocation.

Therefore, delocalization at elevated temperatures does occur quicklyover a short period of time, however the defects accumulated during delocalization change boundary free volume and, consequently, boundary diffusion properties. Such state can be kept over a long period of time.

15.4. Summary. Key Equations of Nonequilibrium Grain Boundary Theory

According to the theory of nonequilibrium grain boundaries (NGB), a change of the boundary free volume $(\alpha + \Delta\alpha)$ due to embedded defects (vacancies, foreign atomsalloyed additives, dislocations) is responsible for diffusion properties of boundaries and diffusion-controlled evolution of microcrystalline structures. To describe a behavior of NGB, the diffusion coefficient D_b^* $D_b^*(\alpha+\Delta\alpha)$ should be used instead of the ordinary grain boundary diffusion coefficient $D_b(\alpha)$:

$$D_b(\alpha) \to D_b^* = D_b^*(\alpha + \Delta\alpha)\ D_b(\alpha) \quad \to \quad D_b^* = D_b^*(\alpha + \Delta\alpha). \tag{15.145}$$

A dependence of the grain boundary diffusion coefficient D_b^* D_b^* on external conditions (strain ε, strain rate $\dot{\varepsilon}$, temperature T, pressure P and time t) can be described by models for kinetics accumulation of the excess free volume $\Delta\alpha$ in grain boundaries:

$$\Delta\alpha = f(\dot{\varepsilon}, \varepsilon, T, t)\ \Delta\alpha = f_2(\dot{\varepsilon}, \varepsilon, T, t) \tag{15.146}$$

---

*) As shown above, the characteristic time t_r at $\dot{\varepsilon}_v = 10^{-3}$ s^{-1} is $t_r \cong 1$ s.

Interaction of grain boundaries with lattice dislocations, which at elevated temperature is accompanied by delocalization of cores of lattice dislocations, affects one of the most effective ways of changing the excess free volume of grain boundaries. During interaction, orientational misfit dislocations (OMD) and the products of their delocalization are formed at grain boundaries. These defects have a profound influence on grain boundary properties. Delocalized dislocations contribute their free volume to boundaries and change the GB energy and diffusion properties. The energy of grain boundaries increases insignificantly (by severalmax 10÷20% percent) because of the its linear dependence on the excess free volume whereas the diffusion coefficient increases by the factor of 10^2-10^3thousands of percent since it depends exponentially on w_t ρ_b.excess free volume.

As it will be shown below, the defects distributed in grain boundaries create short-range (w_t and ρ_b) and long-range (w_n) stress fields, which influence intragranular deformation. The defects accumulated at grain boundaries and triple junctions affect the grain-boundary mobility and thus change the nature of their migration.

The lattice dislocations embedded in grain boundaries play an important role in recovery, recrystallization and deformation processes of NMC materials. The kinetics of defect accumulation in grain boundaries is described by the following equations:

(1) Temperature effect of nonequilibrium grain-boundary diffusion coefficient:

$$D_b^* = D_{b0}^* \exp(-Q_b^* / kT) \tag{15.147}$$

(2) Activation energy of nonequilibrium grain-boundary diffusion:

$$Q_b^* = W_1(\alpha^* /(\alpha + \Delta\alpha) - 1) + W_2([\alpha^* /(\alpha + \Delta\alpha)]^{1/2} - 1) + Q_L \tag{15.148}$$

(3) Preexponential factor for the diffusion coefficient in NGB:

$$D_{b0}^* = \varphi D_{L0} \exp\{Z_1(\alpha^* /(\alpha + \Delta\alpha) - 1) + Z_2([\alpha^* /(\alpha + \Delta\alpha)]^{1/2} - 1)\} \tag{15.149}$$

(4) The dependence of excess free volume $\Delta\alpha$ on the density of defects distributed in the boundary (OMD density ρ_b and the Burgers vector density of tangential components of delocalized dislocationstangential components of Burgers vector of delocalized dislocations w_t):

$$\Delta\alpha = (\rho_b \Delta b + w_t) / w_0 . \tag{15.150}$$

(5) The accumulation kinetics of orientational misfit dislocations in grain boundaries in the presence of the flow of lattice dislocations I:

$$\dot{\rho}_b = I - \rho_b / t_1 . \tag{15.151}$$

(6) The accumulation kinetics of Burgers vector for tangential components of delocalized dislocationstangential components of Burgers vector of delocalized dislocations :

$$\dot{w}_t = \rho_b \Delta b_t / t_1 - w_t / t_2. \tag{15.152}$$

(7) The accumulation kinetics of Burgers vector for normal components of delocalized dislocations:

$$\dot{w}_n = \rho_b \Delta b_n / t_1 - w_n / t_3. \tag{15.153}$$

This equation also describes the kinetics of accumulation of disclinations at grain junctions, because junction disclination power $\omega \sim \omega_n$.

(8) The time of diffusion withdrawal of orientational misfit dislocations from a boundary:

$$t_1 = A_1 kT / (\rho_b \Delta b)^3 D_b^* \delta G \tag{15.154}$$

(9) The time of diffusion withdrawal of tangential components of delocalized dislocationstangential components of Burgers vector of delocalized dislocations from a boundary:

$$t_2 = (d/b)^2 kT / C_1 w_t D_b^* \delta G. \tag{15.155}$$

(10) The time of diffusion withdrawal of disclination defects from grain junctions, which is connected with normal components of delocalized dislocations:

$$t_3 = A_1 (d/b)^3 kT / D_b^* \delta G \tag{15.156}$$

Further, these equations will be used to predict behavior of NGB and properties of NMC materials.

Chapter 16

DISLOCATIONS IN NMC METALS

16.1. PROBLEM DESCRIPTION

Average dislocation density in NMC materials is $10^{10}\div10^{11}$ cm^{-2} (Noskova et al. 2003; Valiev et al. 2006). In comparison with a typical dislocation density in coarse-grained materials ($\sim10^9$ cm^{-2}), these values are high. Then the average distance between dislocations is $r_\perp \sim \rho_v^{-1/2} \sim 10^{-5}\div3\cdot10^{-6}$ cm, which is close to a typical grain size ($d^*\sim10^{-5}$ cm) in NMC materials. The proximity of $r_\perp$ to d^* means that a grain in a NMC metal contains only a few dislocations. This significantly differs from the coarse-grained materials where the number of dislocations is so high that they form complex ensembles inside grains.

When describing the dislocation subsystem in NMC materials, it is necessary to examine at least three key issues:

1. Sources and sinks of dislocations in NMC materials. In our opinion, the features of ECAE deformation of NMC materials are strongly connected with the nature of dislocation sources, which will be considered in Section 16.2.
2. Dislocation pile-ups, which will be studied in Chapter 24.
3. Distribution of "dislocation charges" in NMC materials, which has not received much attention in literature. In our view, this factor is responsible for certain peculiarities of strain behavior of NMC materials and for the evolution of their structure at annealing.

As noted above, a grain in a NMC material contains only one or several dislocations. These dislocations are generated from grain boundaries. On the contrary, in coarse-grained materials the number of dislocations in grains is large, and they appear as a result of functioning of intragranular sources such as the Frank-Read source (Fridel, 1964; Hirth et al. 1969). Then, the total dislocation charge in a grain is always equal to zero (Figure 16.1).

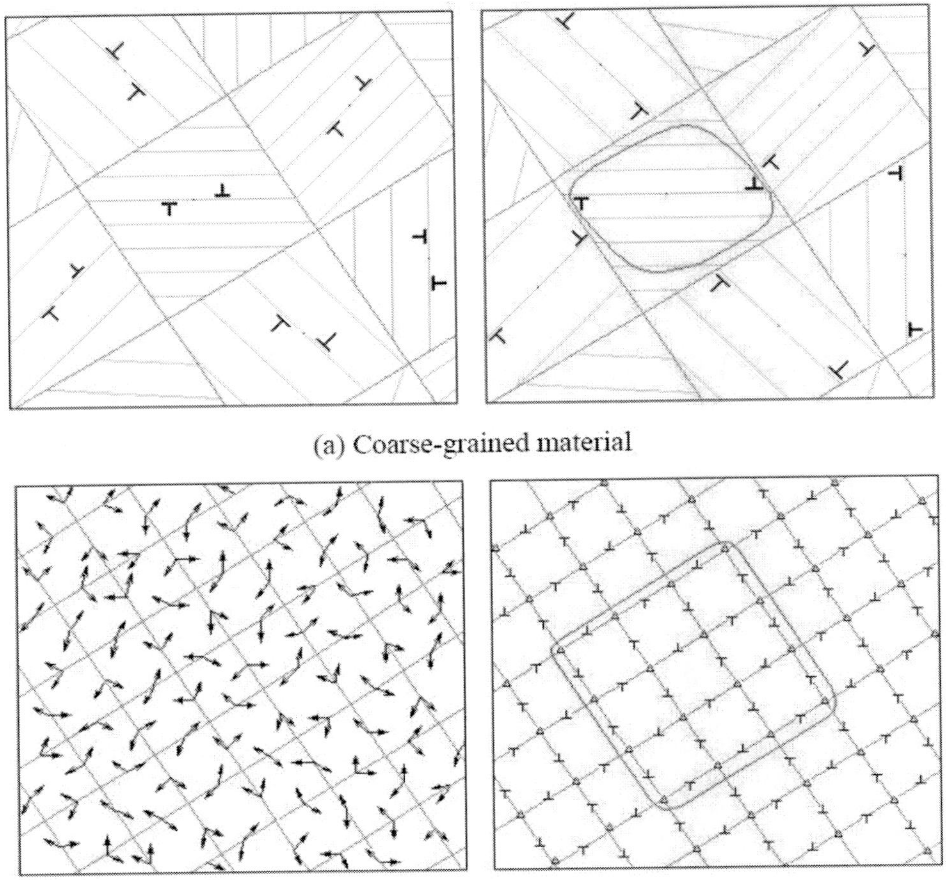

(b) NMC material

Figure 16.1. Formation of dislocation structure in coarse-grained (a) and NMC (b) materials. The thick line outlines the "zero charge zone".

In case of dislocation emission from grain boundaries, the situation with the charge can be different. Each of the grains receives a dislocation with its own Burgers vector (by definition, these vectors, b_1 and b_2, in adjacent grains are not equal). A differentialce vector $\Delta\vec{b} = \vec{b}_1 - \vec{b}_2$ of orientational misfit dislocations (OMD) remains at the grain boundary (Figure 16.1). OMD in grain boundaries will quickly delocalize under ECAE (see Chapter 15). The normal component of delocalized OMD will contribute to the grain boundary misorientation and change the power of triple junction disclinations, while the tangential component of delocalized OMD will pass to triple junctions and annihilate with a similar defect having the opposite-sign.

Since the kinetics of the dislocation charge accumulation in grain boundaries and in grains are not identical, a situation is possible when the dislocation charge inside a grain is nonzero. In this case, the charge is effectively "nulled" in a volume containing several grains, and a certain additional "over-grain" system, which we shall term as a "zero charge zone", appears in the material. In certain cases, such a zone can behave "as a whole". The influence of these zones on the strain behavior and NMC structure evolution will be addressed in Chapter 20 when studying recrystallization of NMC materials.

16.2. LATTICE DISLOCATION EMISSION FROM NONEQUILIBRIUM GRAIN BOUNDARIES

16.2.1. *Preliminary* Remarks

The idea that grain boundaries can act as sources of lattice dislocations was expressed in many works (see, e.g., review (Shtremel et al. 2001)). However, the mechanisms of dislocation emission from boundaries are scarcely studied. This is due to the fact that the dislocations generated from a free surface or emitted by Frank-Read sources cannot be distinguished experimentally from the "out-of-boundary" dislocations. In particular, it is noted that, in several cases, the dislocation-emission mechanism cannot be explained by traditional reasoning about stress concentration in triple junctions under GBS (Styczynska et al. 1985, Johannesson et al. 1972).

Styczynska and Lojkowski (Styczynska et al. 1985) studied emission of dislocations from grain boundaries in polycrystals having precipitation free zones. Dispersed particles inside grains inhibited the propagation of dislocations from the grain body to the boundaries. In this case, dislocations observed near the boundary were nucleated by the boundary. The study shows that the dislocation emission does not necessarily require defects at grain boundaries. The areas that generate dislocations appear to be not connected with any topological features of the boundary, which are usually considered as stress concentration areas under GBS (triple junctions, dispersed particles in the boundary, etc.). It should be emphasized that in this experiment the delocalization of dislocations in grain boundaries was observed. As will be shown below, the latter is necessary for lattice dislocation emission.

The mechanisms of lattice dislocation emission from boundaries are an urgent issue in the theory of NMC materials. The matter is that at a typical size of fine grains $d/b{\sim}10^3$ and small stress $\sigma/G{\sim}10^{-3}$, the functioning of general dislocation sources (like the Frank-Read source) is inhibited. This is due to the fact that a dislocation loop with a large curvature radius $R/b{\sim}G/\sigma$ cannot unfold inside a small grain. Experiments have revealed that dislocation slip in NMC materials occurs uniformly, i.e. dislocation sources in grain boundaries are evenly distributed. This hardly can be attribuited to certain topological features of a boundary, e.g., fractures or facets, because grain boundaries in NMC materials have high migration mobility and can quickly straighten out all the irregularities.

In connection with the above, it is necessary to develop a new approach to the description of dislocation emission from boundaries in NMC materials (Perevezentsev et al. 1989; Chuvil'deev, 2004).

16.2.2. Kinetics of Dislocation Accumulation in Grains

First, let us examine the accumulation kinetics of dislocations inside a grain. A basic equation for the change accumulation rate of the lattice dislocation density can be written as

$$\dot{\rho}_v = I_v^+ - \rho_v/t_s - \chi\rho_v^2 = I_v^+ - I_v^-. \tag{16.1}$$

The first term in the (16.1) describes the emission rate of lattice dislocations. As mentioned above, the functioning of general dislocation sources (like Frank-Read sources) is inhibited due to a fine grain size, and the only effective source of dislocations in these conditions is a grain boundary. A particular mechanism of boundary dislocation emission will be examined later. Here, to evaluate the intensity of such emission from GB, we use a formal description:

$$I_v^+ = \rho_g / dt_g, \tag{16.2}$$

where ρ_g is the linear density of sources in grain boundaries [cm^{-1}], d is the grain size and t_g is a typical time of dislocation emission by the source. The second term in Eq. (16.1), ρ_v / t_s, describes the withdrawal of dislocations from the lattice as a result of their embedding to "ideal sinks". In NMC materials, grain boundaries can act as such sinks. In this case, t_l is the time of dislocation movement through the grain from the "source" at one boundary to the "sink" at another boundary:

$$t_s = d/V_\perp. \tag{16.3}$$

Here, $V_\perp$ is the rate of dislocation slip. Thus, the second term of Eq. (16.1) can be presented as

$$\rho_v / t_s = \rho_v V_\perp / d. \tag{16.4}$$

With the account of Orowan relationship $\dot{\varepsilon}_v = \rho_v V_\perp b$, this expression can be rewritten as:

$$\rho_v / t_s = \dot{\varepsilon}_v / db, \tag{16.5}$$

where $\dot{\varepsilon}_v$ is the intragranular strain rate. The third term of Eq. (16.1) describes the reciprocal capture and annihilation of dislocations opposite in sign. It is known that dislocations having the field of long-range stresses ($\sigma_i \sim Gb/r$) are able to "capture" other dislocations. This process promotes dislocation annihilation and reduces their density. In the coarse-grained material, tThe same process leads to the creation of intragranular dislocation ensembles: while the reciprocally captured dislocations are not yet annihilated, they can move in a correlated manner, and their groups can form complex dislocation configurations.

Parameter χ in Eq. (16.1) is termed as a "capture cross-section" and is defined as (Smirnov, 1981)

$$\chi = \alpha_\perp \frac{G}{\sigma} V_\perp b, \tag{16.6}$$

where $\alpha_\perp$ is a numerical parameter, $\alpha_\perp \sim 10$. Using the Orowan relationship $\dot{\varepsilon}_v = \rho_v b V_\perp$, the expression for χ is rewritten as

$$\chi = \alpha_\perp \frac{G}{\sigma}\frac{\dot{\varepsilon}_v}{\rho_v}. \tag{16.7}$$

Compare the second and the third terms of Eq. (16.1). The critical density of dislocations ρ_v^*, at which these terms are equal, is:

$$\rho_v^* = \left(\frac{1}{\alpha_\perp}\right)\left(\frac{\sigma}{G}\right)\left(\frac{1}{bd}\right). \tag{16.8}$$

Multiplying both parts of this equation by d^2, one obtains the critical number of dislocations in a grain, $N^* = \rho_v^* d^2$:

$$N^* = \left(\frac{1}{\alpha_\perp}\right)\left(\frac{\sigma}{G}\right)\left(\frac{d}{b}\right). \tag{16.9}$$

Now it is possible to evaluate the critical density ρ_v^* and the value of N^*. For typical parameters of NMC materials, viz. $\alpha_\perp=10$, $\sigma/G=10^{-2}$ and $d/b=10^3$, we will have $\rho_v^*=10^{10}$ cm^{-2} and $N^*=1$. At the dislocation density above critical $\rho_v > \rho_v^*$, the third term of Eq. (16.1) dominates, that gives:

$$\dot{\rho}_v = \frac{\rho_g}{t_g d} - \alpha_\perp\left(\frac{G}{\sigma}\right)V_\perp b \rho_v^2. \tag{16.10}$$

In this case, in stationary deformation conditions $\dot{\rho}_v = 0$, the density of dislocations in grains is

$$\rho_v^{st} = \left[\frac{\rho_g}{t_g d}\left(\frac{\sigma}{G}\right)\frac{1}{\alpha_\perp}\frac{1}{V_\perp b}\right]^{1/2}. \tag{16.11}$$

Multiplying both parts of Eq. (16.11) by d^2, the stationary number of dislocations in the grain is

$$N^{st} = \left[\left(\frac{t_s}{t_g}\right)(\rho_g d)\left(\frac{\sigma}{G}\right)\frac{1}{\alpha}\left(\frac{d}{b}\right)\right]^{1/2}, \tag{16.12}$$

where $t_s=d/V_\perp$ is a typical time of dislocation run through a grain. For stationary conditions, the typical time t_s of dislocation "run" through a grain should be equal to the emission time t_g. In this case, for typical parameters $\rho_g d=10$, $\sigma/G=10^{-2}$, $\alpha_\perp=10$, $d=10^3 b$ the stationary value of N is estimated as $N^{st}\sim 10$.

At the dislocation density in grains below critical $\rho_v < \rho_v^*$, the second term in Eq. (16.1) plays the key role. For stationary deformation conditions ($\dot{\rho}_v = 0$) this equation becomes

$$\rho_g / t_g = \rho_v V_\perp \qquad (16.13)$$

Multiplying both parts of Eq, (16.13) by d^2, one finds

$$N^{st} = \rho_v^{st} d^2 = \left(\frac{t_s}{t_g}\right)(\rho_g^{st} d) \qquad (16.14)$$

At $t_s/t_g\sim 1$, Eq. (16.14) gives $N^{st}\sim 10$.

Note that using the Orowan relationship $\dot{\varepsilon}_v = \rho_v b V_\perp$, Eq. (16.13) can be also presented as

$$\rho_g / t_g = \dot{\varepsilon}_v / b \qquad (16.15)$$

To evaluate t_g and ρ_g more accurately, it is necessary to analyze the interaction mechanism of dislocations with grain boundaries.

16.2.3. Kinetics of Defect Accumulation in Grain Boundaries

In accordance with the approach described in (Chuvil'deev, 2004; Perevezentsev et al. 1992), during interaction of dislocations with grain boundaries, orientational misfit dislocations (OMD) and products of their delocalization are accumulated in the boundaries. An evolution of the density of OMD can be described by a kinetic equation

$$\dot{\rho}_b = I_b^+ - \rho_b / t_1 , \qquad (16.16)$$

where the first term relates to the flow of embedded dislocations from the lattice into the grain boundary while the second term defines the absorption of dislocations by grain boundaries. In the first approximation, the intensity of bombardment of grain boundaries by dislocations is proportional to the rate of their withdrawal from the lattice:

$$I_b^+ = I_v^- d \; I_b^+ = I_v^- d \, . \qquad (16.17)$$

In this case, a stationary OMD density ($\dot{\rho}_b = 0$) with the account of Orovan relationship, can be presented as:

$$\rho_b^{st} = \dot{\varepsilon}_v t_1 = V_\perp \rho_v^{st} bt_r \; \rho_b^{st} = \dot{\varepsilon}_v t_1 = V_\perp \rho_v^{st} bt_1 \cdot$$
(16.18)

Eqs. (16.13) and (16.18) show a simple link between the density of dislocation sources in a grain boundary and a typical dislocation emission time on the one hand, and between the OMD density and their absorption rate, on the other hand:

$$\rho_g / t_g = \rho_b^{st} / t_1 \cdot \tag{16.19}$$

The obtained relationship describes allows to understand the dislocation emission mechanism by grain boundaries in NMC materials.

16.2.4. Emission of Lattice Dislocations from Grain Boundaries

As shown in (Ishida et al. 1968; Johannesson et al. 1972; Lojkowski et al. 1981; Pumphrey et al. 1974, Varin, 1982; Varin et al. 1982a,b), a core of a dislocation coming from the grain volume to a grain boundary is unstable and diffusionally spreaded (delocalized) in the boundary plane. In the core area of delocalized dislocations, the atomic structure of the boundary changes and, within typical time t_l termed as "the time of a dislocation effect on diffusion" (Chuvil'deev, 2004), the diffusion processes are accelerated. In the area of radius $r_L a_L$~100b (Varin, 1982; Varin et al. 1982a,b), the grain boundary diffusion coefficient approximates to a value D_b~D_L that is typical for the liquid state. This area in the boundary is termed as "liquid-like", or "L-zone". Another zone of the boundary, where the diffusion coefficient is also increased but does not reach such high value, is termed as "solid-like", or "S-zone". Due to different diffusion coefficients and viscosity, local sliding proceeds with rate U_L in the L-zone and with rate U_s in the S-zone. It is evident that in the liquid-likeL- zone the resistance to viscous intragranular slipding drops sharply, hence $U_L >> U_s$. Local sliding in L-zones under an external stress σ leads to strong stress concentrators at the tips of these zones (Figure 16.2). The concentrated stresses σ_{int} grow with external stresses σ and depend on a ratio of viscosity coefficients of the liquid-likeL- and solid-like S-zones. When viscosity of L-zone is very low, the stress σ_{int} at its tip can be considered as for a shear crack of length $r_L a_L$. At high the substantial value of external stresses σ and large length $r_L a_L$, internal stresses σ_{int} can become sufficiently high to activate plastic sliding in a solid-like zoneS-zone of the boundary as well as slip in the grain volume by emitting lattice dislocations from the tips of the L-zones.

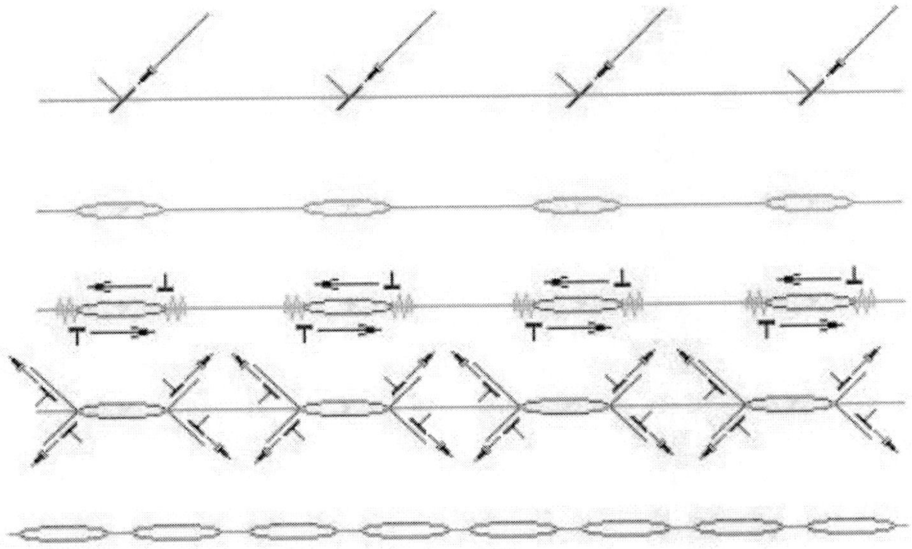

Figure 16.2. Generation of lattice dislocations from nonequilibrium grain boundaries in NMC material:.
a) Lattice dislocations (LD) enter grain boundaries;
b) Delocalization of LD nuclei;
c) Initiation of shears in LD nucleus delocalization area;
d) Emission of LD from the ends of shear areas;
e) Alteration of L- and S-areas in grain boundaries.

Let us determine conditions for the functioning of such source of lattice dislocations. To estimate displacements U_{12} S_{12} at the tip of a shear crack with length $r_L a_L/2$, the Dugdale solution (Dugdale, 1960) or the equivalent Bilby-Cotrell-Swinden solution (Bilby et al. 1963) can be used. For transversal shear crack of type II under plain strain conditions

$$U_{12}(a_L) = \frac{4\tau_y(1-\nu)}{\pi G}\frac{r_L}{2}\ln\left[\sec\left(\frac{\pi\tau}{2\tau_y}\right)\right] \quad U_{12}(r_L) = \frac{4\tau_y(1-\nu)}{\pi G}\frac{r_L}{2}\ln\left[\sec\left(\frac{\pi\tau}{2\tau_y}\right)\right], \tag{16.20}$$

where τ_y is the shear yield stress, τ is the shear stress ($\tau\sim\sigma/2$).

Equating $U_{12}S_{12}(r_L a_L)$ to the Burgers vector of lattice dislocation (b), it is easy to determine the stress at which geometrical condition for the emission of lattice dislocation from nonequilibrium grain boundaries is satisfied:

$$U_{12}S_{12}=b. \tag{16.21}$$

For typical values $\tau_y=10^{-2}G$ and $r_L a_L=10^2 b$, Eq. (16.20) gives for the emission conditions $\tau\sim 0.75\tau_y$. Therefore, the possibility of lattice dislocation emission depends on stress σ and a typical size of the L-zone ($r_L a_L$). In order to link these parameters with the plastic deformation conditions, one has to proceed from a separated liquid-likeL- zone to the whole boundary.

It is important to outline that in activating the above-described mechanism of lattice dislocation emission from grain boundary, the conditions where the emission process can be embedded into the whole boundary surface and become self-sustained, are possible. A dislocation embedded in the boundary after its core has been delocalized can become a source that emits two new lattice dislocations from each tip of the liquid-like core L-zone. Then, two orientational misfit dislocations are formed in the boundary after emission. In turn, the cores of these dislocations will be delocalized, thus creating new liquid-like zoneL-zones. Their tips will also emit lattice dislocations, and the process may recur, thus covering the whole boundary (Figure 16.2). Therefore, the boundary, which acts as a source of lattice dislocations, can be considered as composed of alternating L and S-zones (Figure 16.2). The presence of solid-like zoneS-zones limits the intensity of dislocation flows which are emitted by the boundary or embedded into the boundary, and, consequently, affect the intragranular strain rate.

As shown above, Eq. (16.19), that connects the emission rate of lattice dislocations from grain boundaries with the intensity of their absorption by the boundaries, is true under stationary deformation conditions. Within the developed model, the linear density of sources in the boundary ρ_g is linked to the size of L-zone by an evident expression $\rho_g = r_L^{-1}$ $\rho_g = r_L^{-1}$.
In this case, the characteristic emission time t_g of a lattice dislocation can be presented as:

$$t_g = t_r(a_L\rho_b^{st})\ t_g = t_r(r_L\rho_b^{st}). \tag{16.22}$$

When the average OMD density in a grain boundary is $\rho_b^{st} = a_L^{-1}\ \rho_b^{st} = r_L^{-1}$, the characteristic time of dislocation emission is equal to the time of their absorption. This is the condition of an "ideal" deformation mode.

16.2.5. Stress Conditions of Dislocation Absorption

A grain boundary can absorb dislocations when the external stress field σ exceeds the internal stress field $\sigma_i(\rho_b, w_t)$ created by dislocations embedded in the boundary. It was shown (Chuvil'deev, 2004) that the internal stress field near a boundary is induced by the OMD distributed in the boundary with density ρ_b and by the products of their delocalization with the Burgers vector density w_t. In stationary deformation conditionstate, σ is the condition is that

$$\sigma = \sigma_i. \tag{16.23}$$

In the first approximation, the internal stress field σ_i of created by OMD is proportional to their density [Hirth, 1968]:

$$\sigma_i^{\rho_b} = \varphi_1 G\rho_b^{st}\Delta b. \tag{16.24}$$

The stress field created by tangential components of delocalized dislocationstangential components of Burgers vector of delocalized dislocations has a similar form:

$$\sigma_i^{w_t} = \varphi_2 G w_t . \quad (16.25)$$

The summary stress field:

$$\sigma_i = \varphi_1 G \rho_b^{st} \Delta b + \varphi_2 G w_t \quad (16.26)$$

where φ_1, φ_2 are geometrical coefficients of the order of unity [Hirth, 1968].

When the stress field preventing the lattice dislocations from entering the grain boundary is created only by OMD, the stress condition of dislocation absorption has a general form:

$$\sigma \geq \varphi_1 G \Delta b \rho_b , \quad (16.27)$$

and the stationary OMD density in a grain boundary is:

$$\rho_b^{st} \Delta b = \frac{1}{\varphi_1}\left(\frac{\sigma}{G}\right) . \quad (16.28)$$

A similar relationship can be formulated for w_t:

$$w_t^{st} = \frac{1}{\varphi_2}\left(\frac{\sigma}{G}\right) . \quad (16.29)$$

For typical parameters φ_1, $\varphi_2 \sim 1$ and $\sigma/G \sim 10^{-2}$, the density of defects in a grain boundary is $\rho_b^{st} \Delta b \sim 10^{-2}$ and $w_t \sim 10^{-2}$.

Inserting Eq. (16.28) into Eq. (16.18) gives:

$$\dot{\varepsilon}_v t_1 = \left(\frac{\sigma}{G}\right)\frac{1}{\varphi_1} . \quad (16.30)$$

As noted above, the stationary value of OMD density in a grain boundary is

$$\rho_b^{st} \Delta b = (g_1 \dot{\varepsilon}_v / D_b^*)^{1/4} , \quad (16.31)$$

where $g_1 = \xi_1 (b^3 A_1/\delta)(kT/G\Omega)$. The stationary value of tangential components of delocalized dislocationstangential components of Burgers vector of delocalized dislocations is

$$w_t^{st} = (g_2 d^2 \dot{\varepsilon}_v / D_b^*)^{1/2} . \quad (16.32)$$

where $g_2 = (\xi_2/C_1)(kT/G\Omega)(b/\delta)$. With the account of Eq. (15.122), Eq. (16.28) can be presented as:

$$\dot{\varepsilon}_v \sim \left(\frac{G\Omega}{kT}\right)\left(\frac{D_b^*\delta}{b^3}\right)\left(\frac{\sigma}{G}\right)^4. \quad (16.33)$$

Similarly, combining Eqs. (15.133) and (16.29), one will have

$$\dot{\varepsilon}_v \sim \left(\frac{G\Omega}{kT}\right)\left(\frac{D_b^*\delta}{b^3}\right)\left(\frac{\sigma}{G}\right)^2\left(\frac{b}{d}\right)^2\left(\frac{b}{\Delta b}w_0\alpha_B\right). \quad (16.34)$$

These expressions for the intragranular deformation rate, obtained from stress conditions of dislocation absorption, describe the contribution of intragranular deformation to the rheology of the NMC materials.

At domination of OMD in grain boundaries, Eq. (16.33) is actually a rheological equation. The power dependence of the intragranular deformation rate on stress, $\dot{\varepsilon}_v \sim \sigma^4$ (see Eq. (16.33)), is often used in literature for rheology of NMC alloys at elevated temperatures. At domination of tangential components of delocalized dislocationstangential components of Burgers vector of delocalized dislocations, the dependence $\dot{\varepsilon}_v \sim \sigma^2$ (see Eq. (16.34)) is observed which explains the high strain rate sensitivity coefficient (m~0.5) in a wide range of strain rates for NMC materials.

16.3. Summary

A model for the dislocation emission from grain boundaries under deformation is developed for NMC materials. It is shown that this process is connected with the nucleation of "light shear" L-areas at nonequilibrium grain boundaries; stress concentration at the tips of these areas is sufficient for the dislocation emission. The conditions are derived for uniform distribution of dislocation sources in grain boundaries, which ensures the homogeneity of dislocation flow into grains. The conditions that can prevent the dislocation accumulation in grains are discussed. Expressions for intragranular strain rate of NMC materials at elevated temperatures are obtained. This rate is controlled by the interaction of dislocations moving through grains with the dislocations already embedded in a grain boundary.

Chapter 17

VACANCIES IN NMC METALS

17.1. PROBLEM DESCRIPTION

At present, the data on direct experimental study of the vacancy concentration in materials subjected to ECAE is very limited; accurate results could be obtained by positron annihilation. The concentration of vacancies in NMC materials was obtained by studying heat generation using calorimetry methods and electrical resistance behavior at heating and isothermal annealing. Without describing experimental procedures in detail, we should note that different methods used by different authors gave almost the same value of vacancy concentration: about 10^{-4}, which is very high (Schafler et al. 2005; Setman et al. 2008; Cao et al. 2008). Such a level of vacancy concentration correspond to pre-melting temperatures whereas in NMC materials it is observed at a relatively low (near-room) temperature. Heating of NMC materials leads to intensive relaxation of excess vacancies accompanied by heat generation. The key issue of NMC metals vacancy problem is the role of vacancies in ECA deformation and in NMC material structure evolution.

17.2. NONEQUILIBRIUM VACANCY CONCENTRATION AND VOLUME DIFFUSION COEFFICIENT

It is well-known that nonequilibrium vacancies are generated in a material during plastic deformation (Ruoff, 1963, Ruoff et al. 1967; Mecking et al. 1980; Noskova et al. 2003; Farber, 2002; Shtremel et al. 2002). In some cases, the concentration of vacancies at cold plastic deformation can be above 10^{-4} (Himmel et al. 1964; Cotterill et al. 1965), which significantly exceeds the equilibrium concentration and is comparable with that at the melting point.

For homogeneous deformation, vacancy concentration C_{vac} can be determined using a basic kinetic equation describing the generation of vacancies and their withdrawal to sinks:

$$\dot{C}_{vac} = \dot{C}^{+}_{vac} - C_{vac}/t_{v}, \qquad (17.1)$$

where $\dot{C}^{+}_{vac}$ is the intensity of excess vacancy generation, t_v is a typical time of their withdrawal to sinks.

Several models have been proposed to describe the vacancy generation rate at plastic deformation. The most evident is the "energy" approach (Mecking et al. 1980) within which the vacancy generation rate is presented as

$$\dot{C}^{+}_{vac} = \beta_v \dot{\varepsilon}_v \sigma / U_o \; \dot{C}^{+}_{vac} = \beta_v \dot{\varepsilon}_v \sigma \Omega / U_o , \tag{17.2}$$

where σ is the applied stress, $\dot{\varepsilon}_v$ is intragranular strain rate, β_v is the portion of the plastic deformation energy spent for vacancy generation, and U_0 is the work necessary to generate one vacancy. With typical deformation parameters of ECAE $\dot{\varepsilon}_v = 10^{-1}$ s^{-1}, $\sigma/G \sim 10^{-2}$ and physical properties of metals $G\Omega/kT_m \approx 50$, $U_0 \sim 10$ kT_m, taking $\beta v \sim 0.1$, we obtain $\dot{C}^{+}_{vac} \approx 5 \cdot 10^{-4}$ s^{-1}.

To evaluate the stationary concentration of vacancies in NMC materials, it is necessary to determine the characteristic time of their "withdrawal" to sinks such as lattice dislocations and grain boundaries. First, let us evaluate the effectiveness of lattice dislocations as a vacancy sink. The mechanism of vacancy absorption by dislocations, which is accompanied by a dislocation climb, is well-studied (Hirth, 1969). In this case, a vacancy absorption time is

$$t_{v1} \cong x_v^2 / D_m , \tag{17.3}$$

where x_v is a distance to a sink and D_m is the effective vacancy diffusion coefficient. In the case under examination, x_v is an average distance between dislocations, $x_v = \rho_v^{-1/2}$. Then, the value of D_m is

$$D_m = a^2 \nu_D \exp(S_m / k) \exp(-Q_m / kT), \tag{17.4}$$

where a is the lattice period, ν_D is the Debye frequency, Q_m is the activation energy of vacancy migration, and $S_m \sim k$ is the entropy of vacancy migration (Eds. Cahn, 1983).

The problem of estimating Q_m has not been unambiguously solved (Orlov, 1983). A number of authors suppose that $Q_m \sim 0.5 Q_v$, where Q_v is the activation energy of self-diffusion $Q_v = Q_m + Q_g$, where Q_g is the energy of vacancy generation. In this case, with $Q_v = 18$ kT_m for fcc metals, we have $Q_m \approx 9$ kT_m. Other authors believe that $Q_m \sim 0.4 Q_v$ that gives $Q_m \approx 7.2$ kT_m. Accepting the second point of view and evaluating t_v, we shall insert Eqs. (17.4),) into Eq. (17.3), which gives:

$$t_v = [a^2 \nu_D \exp(S_m / k) \exp(-Q_m / kT) \rho_v]^{-1} . \tag{17.5}$$

The stationary concentration of vacancies from Eqs. (17.2)-(17.4) is

$$C_{vac} = \beta_v \frac{\sigma\dot{\varepsilon}\Omega}{U_o} \frac{1}{D_m \rho_v}. \tag{17.6}$$

In order to estimate t_v using Eq. (17.5), we can use typical values of parameters: $a^2=10^{-15}$ cm^2, $\nu_D=10^{13}$ s^{-1}, $Q_m=7.2$ kT$_m$, $\rho_v=10^{11}$ cm^{-2}. Since the most convincing experimental data on vacancy concentration after ECAE is known for copper at room temperature (Schafler et al. 2005; Setman et al. 2008), we will also examine this material. Taking into account possible heating during ECAE to temperature T=293-323 K, Eq. (17.5) gives $t_v \sim 6 \cdot 10^3 \div 1.25 \cdot 10^5$ s. The calculated time of vacancy absorption by dislocations at these relatively low temperatures appears to be much higher than the typical time of their generation. That means that the mechanism of absorption at room temperature is insufficient for annihilation of excess vacancies generated by deformation.

Let us evaluate the volume self-diffusion coefficient which can be reached for such super-high vacancy concentration. The diffusion coefficient at an arbitrary concentration of vacancies C_{vac} may be described by equation

$$D_v = C_{vac} D_m. \tag{17.7}$$

With tThe equilibrium vacancy concentration, the value C_{vac} is

$$C_{vac}^{st} = \exp(S_g / k) \exp(-Q_g / kT), \tag{17.8}$$

where $S_g \sim k$ is the entropy and $Q_g \sim 0.6 Q_v$ is the enthalpy of vacancy generation at a basic value of $Q_v=18$ kT$_m$, $Q_g=10.8$ kT$_m$. At room temperature, $C_{vac}^{st} \sim 10^{-21}$, and from Eqs. (17.7) and (17.8) the volume self-diffusion coefficient is $D_v \sim 10^{-36}$ cm^2/s.

In case of nonequilibrium vacancy concentration, C_{vac} is determined by Eq. (17.6). Inserting Eq. (17.6) into Eq. (17.7), we obtain

$$D_v^{(1)} = \frac{\beta_v \sigma \Omega \dot{\varepsilon}_v}{U_o} \frac{1}{\rho_v}. \tag{17.9}$$

For typical parameters of ECAE $\beta=0.1$, $\sigma=10^{-2}$G, $G\Omega/kT_m=50$, $\dot{\varepsilon}_v \sim 0.1$, $U_0=10$ kT$_m$ and $\rho_v=10^{11}$ cm^{-2}, D_v is close to $5 \cdot 10^{-15}$ cm^2/s. We should underline that the obtained value of D_v only slightly depends on temperature and is mainly determined by the strain rate and the density of lattice dislocations.

Consider the second possible mechanism of vacancy relaxation, namely the absorption of vacancies by migrating grain boundaries. When a migrating boundary passes over a grain, it is able to effectively absorb lattice vacancies[23]. Diffusion of vacancies in boundaries can

[23] According to the present theory of non-equilibrium grain boundaries, this absobtion increases the free volume, diffusion and degree of nonequilibricity of these boundaries.

provide their effective "withdrawal". A typical time of vacancy "sweepout" by a migrating grain boundary can be evaluated as

$$t_{v2} = d / V_m, \tag{17.10}$$

where V_m is the boundary migration rate. The problem of determining the rate of boundary migration in the ECAE process is complicated (see Chapters 20 and 21). However, a simple approach can be used for a rough estimation. In the collective recrystallization model, V_m is determined by the following formula:

$$V_m = M\gamma_b / d = \left(\frac{\gamma_b}{d}\right)\left(\frac{D_b^*\delta}{b^2}\right)\left(\frac{\Omega}{kT}\right). \tag{17.11}$$

The typical parameters of ECAE are γ_b=Gb/24, $D_b^*=D_{L0}\exp(-Q_L/kT)$, Q_L=3.6 kT_m, $D_{L0}=10^{-4}$ cm^2/s (see Chapter 19), and $d{\sim}10^3b$=250 nm. Then, for ECAE of copper at room temperature (T=293-323 K), from Eqs. (17.10), (17.11) follows t_{v2}=4-20 s.

Combining Eqs. (17.10), (17.11),(17.1), a stationary density of vacancies is

$$C_{vac}^{st} = \frac{(\beta_v \dot{\varepsilon}_v \sigma\Omega / U_o) d^2}{(D_b^*\delta)(\gamma_b / kT)} \quad C_{vac}^{st} = \frac{(\beta\dot{\varepsilon}_v \sigma\Omega / U_o) d^2}{(D_b^*\delta)(\gamma_b b / kT)}, \tag{17.12}$$

i.e C_{vac}^{st} lies within the range of $3\cdot10^{-4}\div1.5\cdot10^{-3}$. In this case, in accordance with Eqs. (17.7) and (17.12), the volume self-diffusion coefficient is expressed as

$$D_v^{(2)} = \frac{D_m}{D_b^*} \frac{\beta_v \dot{\varepsilon}_v (\sigma / G) d^2 kT}{(\delta / b)(\gamma_b / Gb) U_o}. \tag{17.13}$$

Calculations show that for typical characteristics $D_v^{(2)}=10^{-19}$ cm^2/s. Since the activation energy Q_m for vacancy diffusion is close to the activation energy Q_b^* for nonequilibrium grain boundary diffusion, the temperature dependence of the diffusion coefficient is weak and mainly determined by the strain rate $\dot{\varepsilon}_v$ and grain size d.

Obtained coefficients of the volume self-diffusion in both cases are, respectively, by 21 and 17 orders of magnitude higher than the equilibrium diffusion coefficient. Notwithstanding this enormous acceleration, it is provable that volume diffusion cannot significantly influence the deformation processes and structure evolution at ECAE. In fact, for typical distance of diffusion mass transfer by volume diffusion $x \sim \sqrt{D_v t}$ and time of ECAE $t \sim 1/\dot{\varepsilon}_v$~1-10 s, we have $x\approx2\cdot10^{-7}$ cm = 2 nm. This means that vacancies supersaturation cannot significantly influence the total intensity of diffusion mass transfer, but may play a certain role in the processes of "small-scale" diffusion such as local precipitation or dissolution of particles. It should be noted that a weak temperature dependence of $D_v^{(1)}$ and $D_v^{(2)}$ results in an insignificant change of the volume diffusion coefficient with temperature; however, intensive annealing of vacancies occurs due to their withdrawal to dislocations (see Eq. (17.3)) and grain boundaries (see Eq. (17.10)). Thus, in the temperature range where excess vacancies still exist in the crystal lattice, they can manifest themselves only by changing the kinetics of local processes, while at elevated temperature they are quickly annealed and do not have enough time to contribute to the total diffusion mass transfer.

17.3. Comparison with Experimental Data. Discussion

In published works on NMC materials, there is no direct data on vacancy concentration, and only several papers are known where reliable experimental techniques such as differential scanning calorimetry were used for evaluating C_{vac}. In (Cao et al. 2008), heat release was measured at different heating rates for oxygen-free copper subjected to 4, 12 and 16 ECAE passes. The exothermic peak of stored energy was observed at 436 K after 4 and 12 ECAE passes and at 442 K after 16 passes for the heating rate 10 K/min. The stored energy was found to be 0.95 J/g. The activation energy of recovery was determined using a shift in the peak temperature corresponding to the maximum stored energy. towards higher temperatures With increasing heating rate, the shift was toward higher temperatures and attained about 80 kJ/mol ($\approx 7\ kT_m$). It was concluded that the stored energy was connected with contributions of grain boundaries (0.44 J/gr at d~0.2 μm), dislocations (0.035 J/gr) and vacancies (0.52 J/gr). The latter value corresponds to the vacancy concentration of $3.5 \cdot 10^{-4}$. The vacancy concentration of the order of 10^{-4} was also discovered in other materials subjected to severe plastic deformation. These experimental concentrations of vacancies fit well to the theoretical value of $3 \cdot 10^{-4} \div 1.5 \cdot 10^{-3}$ given above. For the experimental recovery temperature of 436 K and the heating rate of 10 K/min, a vacancy withdrawal time corresponds to several minutes (~10^2 s). This information is sufficient to determine the possible mechanisms of vacancy annihilation.

Consider, first, the mechanism of vacancy relaxation by interaction with dislocations. For the temperature of the exothermal peak 436 K, $\rho_V = 10^{10}$ cm^{-2} (Noskova et al. 2003) and typical processing characteristics, Eq. (17.3) gives $t_V \sim 30$ s. That agrees with the experimentally measured value within the order of magnitude.

To evaluate the second mechanism of vacancy absorption by migrating grain boundaries, the activation energy for grain boundary migration at the given temperature should be determined. As noted previously, under ECAE conditions $Q_b^* \approx Q_L$. However, during heating (Chuvil'deev, 2004), Q_b^* rapidly increases due to withdrawal of excess free volume from NGB and recovery of the equilibrium value $Q_b \approx 9\ kT_m$. The value of Q_b^* can be also found from equation (for details see Chapter 20):

$$Q_b^* = T_1 k \ln[D_{b0} \delta G \Omega t_a \ln(\omega_0/\omega^*)/d^3 kTA_1], \qquad (17.16)$$

where T_1 is the critical temperature of recrystallization. For T_1=436 K (0.32Tm), parameter values considered in Chapter 20 and ta~100 sec, we obtain Qb≈6.8kTm. These results confirm that both vacancy absorption mechanisms are effective and high concentration of vacancies revealed experimentally by measuring of thermal properties of NMC materials (Cao et al. 2008) is quite realistic

The role of vacancies in the structure evolution of materials subjected to ECAE is currently under discussion in literature (Shtremel, 2002; Farber, 2002; Noskova et al. 2003; Lotkov et al. 2007). This interest was stimulated by the fact that different experimental techniques detect a high vacancy concentration (about 10^{-4}) in metals subjected to severe plastic deformation and, in particular, ECAE. Such vacancy concentration can influence the behavior of NMC materials. A number of works gave indirect evidence of a change in diffusion properties during (and after) ECAE. In particular, changes in the kinetics of phase dissolution and precipitation were revealed in (Straumal et al. 2004; Gubicza et al. 2007; Stolyarov et al. 2002; Mazilkin et al. 2007) and changes of solid solution solubility limits were reported in (Straumal et al. 2004; Vasil'ev et al. 2006). For example, in (Mazilkin et al. 2007) severe plastic deformation of supersaturated solid solutions in the Al-Zn and Al-Mg alloys, quenched from a high temperature, restored the equilibrium state of the relevant phase diagram. Such unusual behavior can be only explained by assuming that the vacancy concentration in materials under severe plastic deformation is close to that observed near the melting point.

Our observations of structure modification in the Al-12%Si and Al-20%Si alloys subjected to ECAE at T=723÷773 K (Chuvil'deev et al. 2008a), as well as measurements of the scandium solubility in Al-Mg-Sc alloys (Chuvil'deev et al. 2008c) also confirmed an increase in the solubility limits during ECAE.

At the same time, the presented simple estimates, which agree with (Shtremel, 2002), show that a low mobility of vacancies at room temperature does not permit a serious consideration of their role in phase dissolution/separation processes. In our opinion, such contradiction can be explaned by a behavior of nonequilibrium grain boundaries during ECAE. Although reliable quantitative information on the solubility change is not available now, it makes sense to consider some ideas which may direct experimental research. In particular, an analogy can be drawn with a well-studied phenomenon of diffusion-induced grain boundaries migration (DIGM) (Kaur et al. 1989). A rapidly migrating boundary by diffusion distributes the solute atoms and provides effective saturation of the volume. At DIGM, the saturation is accelerated by 2-3 orders of magnitude as compared with diffusion in a monocrystal (Kaur et al. 1989). It can be assumed that strain induced nonequilibrium grain boundaries migrate during ECAE. When these boundaries with high free volume and enhanced grain boundary diffusion pass through grains, they collect vacancies and further increase the boundary free volume and diffusion coefficient. Also, fast transport of soluble atoms along the boundary entails grain homogenization of effective solubility.

17.4. Summary

The performed analysis shows that ECAE at relatively low temperatures ($T \leq 0.3T_m$) can bring a high concentration of vacancies of about $\sim 10^{-4}$. This leads to a significant increase in the coefficient of volume self-diffusion. Its value insignificantly depends on temperature and is mainly determined by the intragranular strain rate, lattice dislocation density and grain size.

However, a high concentration of vacancies at room temperature cannot affect significantly the diffusion and structure evolution because of low vacancy mobility. At elevated temperature, the vacancies are intensively annealed and do not have enough time to significantly contribute to these processes.

Section II. Grain Boundary Diffusion Properties and Structural Stability of NMC Materials

Chapter 18

DIFFUSION PROPERTIES OF GRAIN BOUNDARIES IN NMC METALS

18.1. PROBLEM DESCRIPTION

The theory of NGB developed in the preceeding Chapters links the properties of grain boundaries with their free volume. Hereinafter it will be used to describe diffusion of grain boundaries in NMC materials. The latter are known to differ significantly from coarse grained metals. For NMC material, a low activation energy for grain boundary diffusion (20-40% lower than in equilibrium boundaries) was found along with unusually high values of the grain boundary diffusion coefficient exceeding a typical value by 2-3 orders of magnitude (Grabovetskaya et al. 2008; Kolobov et al. 2001a,b,c; Amirkhanov et al. 2001; Amirkhanov et al. 1998; Amouyal et al. 2007). However, the available experimental data on diffusion properties of NMC materials are controversial (Kolobov et al. 2001c, Amouyal et al. 2007; Fujita et al. 2004; Amouyal et al. 2008) and dufficult to be systematized. In our opinion, it is due to difficulties in measuring D_b^* in NMC materials due to a nonequilibrium statenonequilibricity of grain boundaries.

During ECAE, high-angle grain boundaries are formed. With further deformation, new dislocations are embedded into the boundaries changing their misorientation and structure. The duration of this process is connected with the rate of diffusion restructuring of dislocation cores in a boundary, which strongly depends on temperature. After ECAE, some delocalized defects do not have enough time to withdraw from the boundary, and the boundary can retain a nonequilibrium statenonequilibricity and the excess free volume for a long time. The total number of defects in boundaries and their nonequilibrium statenonequilibricity significantly depend on the processing, post deformation cooling, and the retention interval before measurement of diffusion properties. These factors affect D_b^*, however their role was insufficiently accounted for in experiments.

Another difficulty in the assessment of D_b^* for NMC materials relates to the boundary migration during heating and annealing (Kolobov et al. 2001a,c; Amirkhanov et al. 1998; Amouyal et al. 2008; Molodova et al. 2007; Kim, 2004; Degtyarev, 2005). As D_b^* D_b^* is measured at elevated temperatures, recovery and recrystallizationcan be induced by preheating. When grain boundaries migrate through a deformed matrix, the lattice dislocations within grains are embedded into boundaries changing the free volume and diffusion properties of boundaries. Thus, under certain conditions, the role of boundary migration is opposite to that of recovery: the former can increase D_b^* D_b^* while the latter reduces it. In particular, boundary migration may bring large values of D_b^* D_b^* regardless of high temperature and long annealing time.

In this Chapter, three practically important issues of the NGB diffusion properties in NMC materials are discussed:

1 Dependence of the grain boundary diffusion coefficient in deformed NMC materials on the recovery process (time and temperature of annealing).
2 Change of the NGB diffusion properties during the grain boundary migration.
3 Dependence of the grain boundary diffusion coefficient on strain rate during intragranular deformation.

In order to resolve these problems using the present approach, it is necessary to specify the flow intensity I of lattice dislocations embedded in boundaries. It depends on both processing conditions and structural characteristics of the material. During recovery without intragranular deformation and grain boundary migration,

$$I=0 \tag{18.1}$$

During recrystallization with migration of boundaries but without deformation, the flow intensity I is proportional to the migration rate V_m and density ρ_v of lattice dislocations swept out by the moving boundary:

$$I = \xi_m V_m \rho_v \tag{18.2}$$

Here ξ_m is the relative fraction of like-sign dislocations embedded in a migrating boundary. Depending on a distribution of dislocations in grains, it can vary within a wide range $\xi_m = 10^{-4} \div 10^{-1}$.

For intragranular deformation with strain rate $\dot{\varepsilon}_v$, the flow of dislocations into a boundary is

$$I = \xi_v \dot{\varepsilon}_v / b, \tag{18.3}$$

where ξ_v is a geometrical parameter that, similarly to ξ_m, may vary within a range $\xi_v = 10^{-4} \div 10^{-1}$.

In further, these cases will be considered in more detail.

18.2. Recovery of the Diffusion Properties of Nonequilibrium Grain Boundaries during Annealing

18.2.1. Model

In the recovery processes that occur during heating or isothermal annealing without external stresses, strains and the recrystallization (boundary migration) value I=0. In this case, Eqs. (15.151)-(15.153) for the defect density take forms

$$\dot{\rho}_b \Delta b = -(\rho_b \Delta b)^4 / t_1';, \ \rho_b = \rho_{b0} \text{ at } t=0, \tag{18.4}$$

$$\dot{w}_t = -w_t^2 / t_2';, \ w_t = w_{t0} \text{ at } t=0, \tag{18.5}$$

$$\dot{w}_n = -w_n / t_3;, \ w_n = w_{n0} \text{ at } t=0, \tag{18.6}$$

where

$t_1' = A_1 kT / D_b^* \delta G \Omega$ $t_1' = A_1 kT / D_b^* \delta G$, and $t_2' = (d/b)^2 kT / C_1 D_b^* \delta G$, t_3 – see Eq. (15.156) $t_2' = (d/b)^2 kT / C_1 D_b^* \delta G \Omega$.

In a general case, D_b^* is determined by Eqs. (15.147)–(15.150). The solution of Eqs. (18.4)-(18.6) together with Eqs. (15.148), (15.149) defines the dependence of defect density and grain boundary diffusion on time and, thus, describes the recovery under isothermal annealing.In general, the analytical solutions of Eqs. (18.4)-(18.6) are complicated.

However, in particular cases where prevails either orientational misfit dislocation ($\Delta\alpha = \rho_b \Delta b / w_0$) or tangential components of delocalized dislocationstangential components of Burgers vector of delocalized dislocations ($\Delta\alpha = w_t / w_0$) and $\Delta\alpha \ll \alpha$, Eqs. (18.4)-(18.5) and (15.87) can be resolved analytically. These cases relates to later recovery stages when most of the defects have left the boundary. For the small defect density in grain boundaries ($\alpha + \Delta\alpha \ll \alpha^*$), Eq. (15.87) gives following expressions for the diffusion coefficient in NGB

$$D_b^* = D_b \exp(\Delta\alpha(\rho_b, w_t) / \alpha_B). \tag{18.7}$$

And for the defect densities

$$\rho_b \Delta b(t,T) = \frac{\rho_{b0} \Delta b}{\sqrt[3]{1 + 3(\rho_{b0} \Delta b)^3 (\delta D_b^* / b^3 A_1)(G\Omega / kT)t}}, \tag{18.8}$$

$$w_t(t) = \frac{w_{t0}}{1 + C_1 w_{t0}(\delta D_b^*/b^3)(b/d)^2(G\Omega/kT)t}. \quad (18.9)$$

As shown above, orientational misfit dislocations with density ρ_b dominate in fine grain materials ($d<d_1$) whereas in coarse grain materials ($d>d_1$) the tangential components of delocalized dislocations w_t play the main role. Inserting Eqs. (18.8) and (18.9) into Eqs. (15.143), (15.144), we obtain the dependence of grain boundary diffusion coefficient on annealing time at late stages of recovery with $\rho_b(t,T)<<\rho_{b0}$, $w_t(t,T)<<w_{t0}$:

$$D_b^* = D_b \exp(\zeta_1/t^{1/3}), \quad (18.10)$$

$$D_b^* = D_b \exp(\zeta_2 d^2/t) \quad (18.11)$$

Here $\zeta_1 = \frac{1}{\alpha_B w_0}\left[\frac{b^3 A_1}{\delta D_b^*}\frac{kT}{G\Omega}\right]^{1/3}$ and $\zeta_2 = \frac{1}{\alpha_B w_0}\left[\frac{b}{C_1 \delta D_b^*}\frac{kT}{G\Omega}\right]$.

It is mportant to note that the kinetics of defect recovery depends substantially on the grain size. In fine-grained materials ($d<d_1$), the recovery of diffusion properties is determined by the OMD delocalization with the typical time t_1 while the grain size effect is low. In coarse-grained materials ($d>d_1$), the recovery of diffusion properties is controlled by a withdrawal of tangential components of delocalized dislocationstangential components of Burgers vector of delocalized dislocations with the typical time t_2 which depends on d, and significantly increases with d (Figure 18.1).

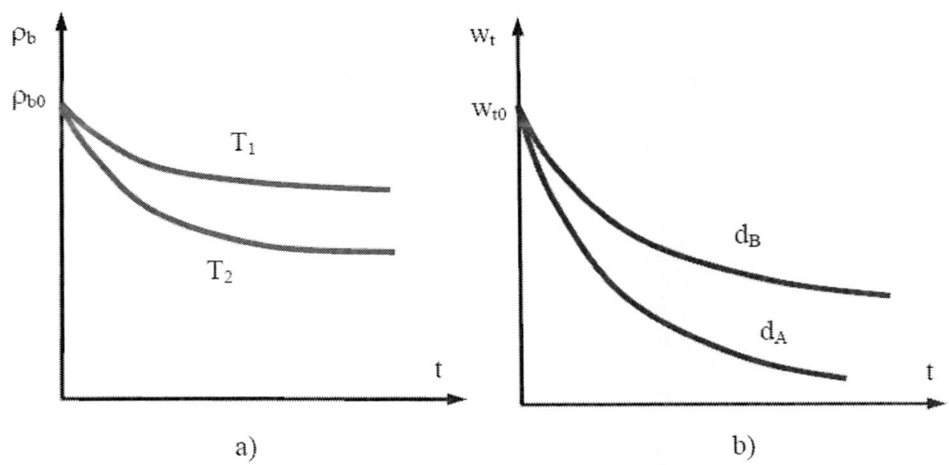

Figure 18.1. Defect recovery kinetics at fine (at $d<d_1$) (a) and coarse (at $d>d_1$) (b) grain sizes in NMC materials for $T_1<T_2$ and $d_1<d_A<d_B$, respectively.

18.2.2. Comparison with Experiments for Fine Grained and Coarse Grained Materials

Currently, no information is available on comparison of the kinetics of recovery processes in microcrystalline materials and NMC materials produced by ECAE. However, we can use results on diffusional recovery during annealing of pre-deformed high-purity aluminum (Kwiecinski et al. 1989; Kwiecinski et al. 1991) and stainless steel (Swiatnicki et al. 1986). In particular, it seems important to verify the theoretical predictions that (i) the recovery kinetics is different for materials with fine and coarse grains, and (ii) the activation energy Q_b for grain boundary diffusion changes with annealing time much slower in coarse-grained materials than in fine-grained ones.

In papers (Kwiecinski et al. 1989; Kwiecinski et al. 1991; Swiatnicki et al. 1986), the kinetics of grain boundary recovery was analyzed using data on the delocalization of lattice dislocations embedded in grain boundaries during preliminary plastic deformation. As it is known (Lojkowski et al. 1981; Varin, 1982), parameters of the delocalization provide the diffusion characteristic of grain boundaries and, in particular, a change in the boundary diffusion coefficient during postdeformation annealing. Experiments in these works were performed for aluminum and steels.

Two types of high-purity aluminum designated as *S* and *G* with the average grain size of 4 and 22 μm, respectively, were subjected to preliminary deformation ε=1.5% (Kwiecinski et al. 1989; Kwiecinski et al. 1991). The dependence of a relative number of grain boundaries (1-*x*) containing non-delocalized dislocations on the annealing time t was analyzed at temperatures 293, 343, 373 and 403 K for polycrystals *G* and temperatures 293 and 373 K for samples *S*. The characteristic time was found from formula

$$t_{diff} = A_1 kTS_d^3/G\Omega\delta D_b \tag{18.12}$$

where A_l is a numerical constants of A_l=10 in (Kuzhidlovsky, 1998) and A_l=20 in (Kwiecinski et al. 1991), r_d is the dislocation core width at which the dislocation diffraction contrast can no longer be observed by electron microscopy. In (Kwiecinski et al. 1989; Kwiecinski et al. 1991), it was accepted that $r_d=2.25\cdot10^{-5}$ cm. Their results on the activation energy Q_b for grain boundary diffusion in function of (1-*x*) (fraction of grain boundaries) are listed in Table 18.2.

In papers (Kwiecinski et al. 1989; Kwiecinski et al. 1991; Swiatnicki et al. 1986), the kinetics of grain boundary recovery was analyzed using data on the delocalization of lattice dislocations embedded in grain boundaries during preliminary plastic deformation. As it is known (Lojkowski et al. 1981; Varin, 1982), parameters of the delocalization provide the diffusion characteristic of grain boundaries and, in particular, a change in the boundary diffusion coefficient during postdeformation annealing. Experiments in these works were performed for aluminum and steels.

Table 18.1. Parameter values used in calculations

Parameters	Notation	Units	Value			
			Al 2 μm	Al 22 μm	steel A 2.1 μm	steel B 30 μm
INTERFACE PARAMETERS						
Enthalpy of solid/liquid interface (kT_m/b^2)	$\gamma^0_{S/L}$	$\gamma^0_{S/L}\ b^2/kT_m$	1.16		0.8	
Entropy of solid/liquid interface (k/b^2)	$S_{S/L}$	$S_{S/L}b^2/k$	0.92	0.92	1.4	1.35
Free surface energy of the boundary S-phase areas (kT_m/b^2)	γ_0	□	1.5		1.6	
Critical free volume of a boundary	α^*	□	0.5		0.5	
Free volume of a boundary	□	□	0.38		0.273	
DIFFUSION PARAMETERS						
Activation energy of self-diffusion in a melt (kT_m)	Q_L	Q_L/kT_m	3.7		3.5	
Pre-exponent in the expression for diffusion coefficient in a melt (cm^2/s)	D_{L0}	□	$1.4 \cdot 10^{-5}$		$8 \cdot 10^{-4}$	
ADJUSTABLE PARAMETERS Numerical factor in the expression for delocalization time	A_1	□	20		10	
Numerical factor in the expression for the time of "withdrawal" of tangential components of delocalized dislocationstangential components of Burgers vector of delocalized dislocations from boundaries	C_1	□	0.15		140	
Initial OMD density	$\Delta b\rho_{b0}$	□	0.005	0.004	0.01	0
Initial density of tangential components of delocalized dislocationstangential components of Burgers vector of delocalized dislocations	w_{t0}	□	0.006	0.006	0	0.01
CRYSTALLOGRAPHIC PARAMETERS						
Burgers vector (cm)	b	□	$2.86 \cdot 10^{-8}$		$2.58 \cdot 10^{-8}$	
Grain boundary width	δ	2b	2b			
Atomic volume (cm^3)	Ω	□	$1.66 \cdot 10^{-23}$		$1.16 \cdot 10^{-23}$	
THERMODYNAMIC AND ELASTIC CONSTANTS						
Specific heat of fusion	λ	$\lambda\rho b^3/kT_m$	$\lambda\rho b^3=1.96\ kT_m$		$\lambda\rho b^3=1.5\ kT_m$	
Density	ρ					
Melting point	T_m	$G\Omega/kT_m$	$G\Omega/kT_m=32$		$G\Omega/kT_m=31$	
Shear modulus (MPa)	G		$2.5 \cdot 10^4$		$6.0 \cdot 10^4$	

Two types of high-purity aluminum designated as *S* and *G* with the average grain size of 4 and 22 μm, respectively, were subjected to preliminary deformation $\varepsilon=1.5\%$ (Kwiecinski et al. 1989; Kwiecinski et al. 1991). The dependence of a relative number of grain boundaries (1-*x*) containing non-delocalized dislocations on the annealing time t was analyzed at temperatures 293, 343, 373 and 403 K for polycrystals *G* and temperatures 293 and 373 K for samples *S*. The characteristic time was found from formula

$$t_{\mathrm{diff}} = A_1 kTS_d^3 / G\Omega\delta D_b \qquad (18.12)$$

where A_1 is a numerical constants of $A_1=10$ in (Kuzhidlovsky, 1998) and $A_1=20$ in (Kwiecinski et al. 1991), S_d is the dislocation core width at which the dislocation diffraction contrast can no longer be observed by electron microscopy. In (Kwiecinski et al. 1989; Kwiecinski et al. 1991), it was accepted that $S_d=2.25\cdot10^{-5}$ cm. Their results on the activation energy Q_b for grain boundary diffusion in function of (1-*x*) (fraction of grain boundaries) are listed in Table 18.2.

Table 18.2. Comparison between experimental and theoretical values of the activation energy for grain boundary diffusion in aluminum

Grain size, μm	Fraction of grain boundaries containing non-delocalized dislocations	Activation energy Q_b^*, kJ/mol	
		Experiment (Kwiecinski et al. 1991)	theory
4 (samples *S*)	0.2	84.6 (extrapolation)	74.0
	0.5	64.4	64.4
	0.8	44.3	52.7
22 (samples *G*)	0.2	49.3	52.7
	0.5	46.2	46.2
	0.8	42.2	41.0

It has been revealed (Kwiecinski et al. 1989; Kwiecinski et al. 1991) that at the same level of recovery, the activation energy for grain boundary diffusion is significantly higher in coarse-grained samples *G* than in fine-grained samples *S*. In samples *G* recovery proceeded much slower than in samples *S*. The difference in Q_b^* is more noticeable at later stages of recovery. As seen in Table 18.2, for a 20% fraction of boundaries with non-delocalized dislocations, the activation energies at later stages of recovery for *S* and *G* samples differ by 70%; for a 50% fraction they differ by 30%, and at a 80% fraction – by only 5%.

Swiatnicki et al (Swiatnicki et al. 1986) used the same method to investigate the annealing of dislocations embedded into grain boundaries for stainless steels *A* and *B* having different chemical compositions (Table 18.3) and grain sizes 2.1μm and 30 μm, respectively. The dependences of D_b^* on annealing time were obtained at different temperatures. These results are presented in Table 18.4.

Usually, in phenomenological models, the number and nature of adjustable parameters are subjected to discussion. On this reason, the comparison process is further described in detail together with discussion of adjustable parameters at every step of comparison. The values of the determined parameters for the studied materials are listed in Table 18.17.

For aluminum:

1 To find the value of the grain-boundary free volume α, we use thermodynamic constants for Al (Frost et al. 1982; Chuvil'deev, 2004) and experimental parameters of grain boundary diffusion Q_b and D_{b0} for equilibrium grain boundaries (Kwiecinski et al. 1991) (see Table 18.1). The free volume α is calculated from Eqs. (15.147)-(15.150) without any adjustable parameters;

2 Then, the excess free volume $\Delta\alpha$ is found at the beginning of the recovery process (t=0). In this case, the activation energy for grain boundary diffusion should be less than or equal to the minimal experimental value of Q_b = 49.3 kJ/mol (Kwiecinski, 1991). That corresponds to an 80% fraction of boundaries with non-delocalized dislocations. The value of $\Delta\alpha$ at the initial instant of time, $\Delta\alpha_0$, is calculated from Eqs. (15.147)-(15.150) without adjustable parameters;

3 The total defect density in grain boundaries is estimated. For values $\Delta\alpha_0$, the initial total OMD density ρ_{b0} and the density of tangential components of delocalized dislocationstangential components of Burgers vector of delocalized dislocations ($\rho_{b0}\Delta b+w_{t0}$) are found from Eq. (15.15) using only one adjustable factor w_0 = 0.1 (Chuvil'deev, 2004);

4 Now, the fractions of different-type defects, namely the orientational misfit dislocations and the tangential components of delocalized dislocationstangential components of Burgers vector of delocalized dislocations , can be estimated. For this purpose, an adjustable parameter of the initial (at t=0) density of tangential components of delocalized dislocationstangential components of Burgers vector of delocalized dislocations w_{t0}=0.006 is introduced (the first adjustable parameter);

5 From Eqs. (15.151)-(15.152) and (18.4)-(18.5), the time change $\rho_b(t)\Delta b$ of the density of OMD and that w_t of tangential components of delocalized dislocationstangential components of Burgers vector of delocalized dislocations is calculated for the experimental temperature of (Kwiecinski, 1991). These calculations require two adjustable parameters A_1 and C_1 in Eqs. (18.4), (18.5). We use A_1=20, which corresponds to (Kwiecinski et al. 1991), and choose C_1=0.15. It should be noted that for the Zn-20% Al alloy different values of C_1=3 and C_1 = 50 were used in (Petryaev et al. 2000), and (Perevezentsev, 1987), respectively;

6 Activation energy Q_b^* is directly calculated from Eqs. (15.148), (15.150) for different stages of recovery and temperatures without adjustable parameters.

The obtained theoretical values of Q_b^* were used in Table 18.2 for comparison with experimental data.

For Stainless steel, experimental results of (Swiatnicki et al. 1986) show that the recovery process is in many ways similar to that in aluminum, and the procedure remains the same. However, many characteristics of steels are unknown and a number of adjustable parameters increase. The thermodynamic characteristics $\gamma_{S/L}$, $S_{S/L}$ and γ_0 are listed in Table 18.1. To evaluate $\alpha+\Delta\alpha$, the initial value of the activation energy for grain boundary diffusion is taken

$Q_b^* \approx 120$ kJ/mol, that corresponds to an 80% fraction of non-delocalized dislocations together with chosen adjustable parameters $A_1=10$ and $C_1=140$. The experimental and theoretical values of Q_b for steels are listed in Table 18.4.

Table 18.3. Chemical composition (wt.%) of stainless steels (Swiatnicki et al. 1986)

Steel	C	Mn	Si	P	S	Cr	Ni	Cu	Mo
A	0.12	1.23	0.47	0.028	0.008	22.3	17.0	-	0.28
B	0.06	0.002	0.002	0.002	0.007	19.0	18.6	0.02	0.01

Table 18.4. Comparison between experimental (Swiatnicki et al. 1986) and theoretical values of the activation energy for grain boundary diffusion in stainless steels

Grain size, μm	Fraction of grain boundaries containing non-delocalized dislocations	Activation energy, kJ/mol	
		Experiment (Swiatnicki et al. 1986)	Theory
2.1 steel A	0.2	191.2	185.7
	0.5	190.9	163.2
	0.8	153.4	153.4
30 steel B	0.2	142.4	142.4
	0.5	163.8	124.1
	0.8	120.6	117.1

Aluminum and Stainless Steel

For aluminum, experimental and theoretical values of activation energy show a good agreement. The discrepancy does not exceed 6.6% for *G* samples and 15% for *S* samples (unfortunately, there are only two experimental points for the S samples) (see Table 18.2). A good agreement for steels (Table 18.4) is less meaningful due to a large number of adjustable parameters. The developed model explaines well the significant difference in diffusion parameters between fine-grained and coarse-grained polycrystals both for aluminum and for steels. As seen from Eqs. (15.147)-(15.150), diffusion properties of grain boundaries depend on excess free volume $\Delta\alpha$. A change in the value of $\Delta\alpha$ with time during annealing is determined by the kinetics of diffusion "withdrawal" of OMD and the products of their delocalization w_t from grain boundaries. Eq. (15.155) and Table 18.5 imply that the evolution of w_t depends on the grain size d: the larger d is, the slower the tangential components of delocalized dislocationstangential components of Burgers vector of delocalized dislocations depart from grain boundaries.

That means that even with the same value of (1-*x*), i.e. with the same level of OMD delocalization, the excess free volume in grain boundaries for fine-grained and coarse-grained materials will differ significantly due to a difference in w_t. Respectively, the activation energy for grain boundary self-diffusion and the parameters characterizing the intensity of recovery will also differ.

For aluminum samples *G* (grain size d=22 μm), the activation energy returns to its equilibrium value Q_b=77.5 kJ/mol within a $4{\cdot}10^6$ s at temperature T=403 K while for aluminum samples *S* (d=4 μm) this time at the same temperature is $2{\cdot}10^5$ s that is less by the factor of 20. For stainless steel samples B with d=30 μm, the annealing time $2{\cdot}10^4$ s at T=653T is required to reach the equilibrium value of Q_b=190 kJ/mol of the activation energy; at T=673 K, this time is about 10^3 s. For stainless steel samples A with d=2.1 μm, the annealing time is about 800 s and 300 s, respectively.

Summarizing these results, it should be noted that the theoretical prediction concerning differences in the recovery kinetics for fine-grained and coarse-grained materials appears to be true. Thus, the developed approach can also be used to describe the recovery processes in NMC materials produced by ECAE.

18.2.3. Comparison with Experiments for NMC Materials

Since the dependence of diffusion coefficient D_b^* or activation energy Q_b^* on the isothermal annealing time in NMC materials is not available in literature, we use indirect data to analyze the recovery kinetics in such materials.

1. The influence of a low-temperature (pre-recrystallization) annealing on internal stresses in a Cu-0.5% Al_2O_3 NMC composite produced by high-pressure torsion (HPT) at room temperature was studied in (Amirkhanov et al. 2001). The average size of copper grains and Al_2O_3 particles after HPT was 80 nm and 10 nm, respectively. Microdeformation (Δε, %) after a 30 min isothermal anneal was measured using XRD analysis (Gorelik et al. 2002). A dependence of Δε on annealing temperature in the range from 293 K to 673 K (the latter value corresponds to the temperature of recrystallization) are presented in Table 18.5 and Figure 18.2.

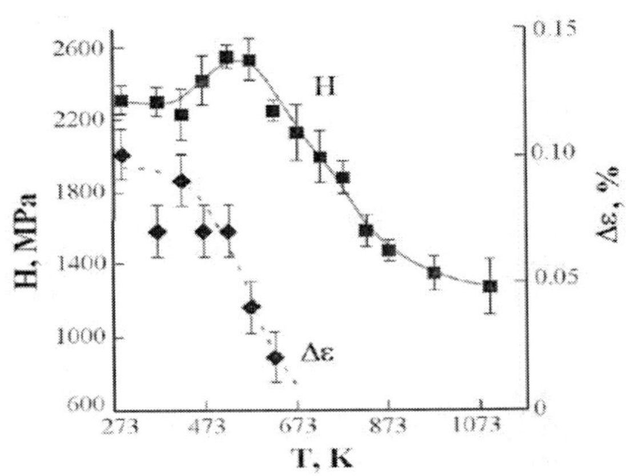

Figure 18.2. Microhardness (H) and microdistortions (Δε) vs. annealing temperature for a Cu+0.5%Al_2O_3 NMC composite (Amirkhanov et al. 2001).

Table 18.5. Dependence of microdeformation Δε, %, on annealing temperature in a Cu-0.5% Al_2O_3 NMC composite

T, K	293	373	423	473	523	573	623	673
Δε, %	$0.10^{\pm 0.02}$	$0.07^{\pm 0.02}$	$0.09^{\pm 0.02}$	$0.07^{\pm 0.02}$	$0.07^{\pm 0.02}$	$0.04^{\pm 0.02}$	$0.02^{\pm 0.02}$	0

Then, using formula $\sigma_c = E \cdot \Delta\varepsilon$ $\sigma_c = E \cdot \Delta\varepsilon$, where E is the elastic modulus of copper, the dependence of elastic stresses on the annealing temperature was determined. An increase of temperature from 293 K to 623 K reduces the internal elastic stresses from 120 MPa to 24 MPa. Assuming that recovery of NMC copper is controlled by the kinetics of diffusional withdrawal of orientational misfit dislocations from grain boundaries, the value of elastic stresses σ_c can be connected with the OMD density by a relationship $\rho_b\Delta b = \varphi_1\sigma_c/G$ $\rho_b\Delta b = \varphi_1\sigma_c/G$, where φ_1~1. Calculations show that an increase of the annealing temperature from 293 K to 623 K reduces the OMD density ($\rho_b\Delta b$ $\rho_b\Delta b$) from $2.9\cdot10^{-3}$ to $5.7\cdot10^{-4}$. In accordance with Eqs. (15.148)-(15.150), such a change in OMD density corresponds to an increase in the activation energy for grain boundary self-diffusion from 7.8 kT_m to 9.2 kT_m and a buildup of pre-exponent D^*_{b0} from $1.69\cdot10^{-6}$ to $8.83\cdot10^{-6}$ m^2/s. Related parameters are listed in Table 18.6.

Now, the dependence of OMD density on the annealing temperature at t=30 min can be determined from Eq. (18.8) for the calculated value of $\rho_{b0}\Delta b$ = $\rho_{b0}\Delta b = 3\cdot10^{-3}$ and assumed value A_1=100. The comparison of calculated and experimental values is presented in Figure 18.3.

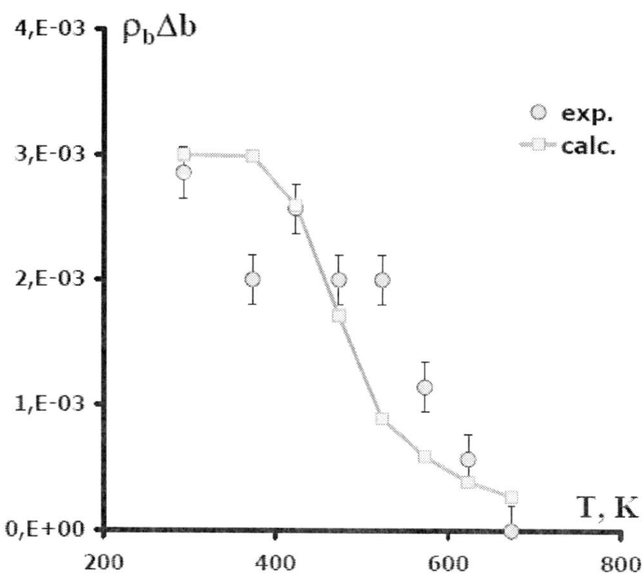

Figure 18.3. Numerical calculation of the OMD density vs. temperature for a 30-minute isothermal anneal of a Cu-0.5%Al_2O_3 NMC composite (Amirkhanov et al. 2001).

Table 18.6. Parameters used for calculations (copper)

Parameters	Notation	Units	Value
Crystallographic and geometrical parameters			
Burgers vector	b	m	$2.56\cdot10^{-10}$
Atomic volume	Ω	m^3	$1.18\cdot10^{-29}$
Typical grain size	d	b	$2\cdot10^{5}$
Grain boundary width	δ	b	2
Thermodynamic and elastic constants			
Shear modulus	G	$G\Omega/kT_m$	50
Melting point	T_m		
Specific heat of fusion	λ	$\lambda\rho b^3/kT_m$ $\lambda\rho b^3/kT_m$	1.15
Density	ρ		
Volume jump at melting	ΔV_m	-	$5\cdot10^{-2}$
Interface parameters			
Solid/liquid interface enthalpy	$\gamma^0_{S/L}$	$\gamma^0_{S/L}$ $\gamma^0_{S/L}b^2/kT_m$ b^2/kT_m	0.9
Solid/liquid interface entropy	$S_{S/L}$	$S_{S/L}b^2/k$ $S_{S/L}b^2/k$	0.767
Relative free volume of grain boundary	α	-	0.36
Critical free volume of grain boundary	α^*	-	0.5
Excess free volume of grain boundary	$\Delta\alpha$	-	0.01, 0.1
Grain boundary free energy ($T=0.5T_m$)	γ_b	$\gamma_b b^2/kTm$ $\gamma_b b^2/kT_m$	1.93
Free surface energy of boundary S-phase areas	γ_0	$\gamma_0 b^2/kTm$ $\gamma_0 b^2/kT_m$	1.95
Grain boundary enthalpy (at $\alpha=0.36$)	γ^0_b	γ^0_b $\gamma^0_b b^2/kT_m$ b^2/kT_m	4.32
Grain boundary entropy (at $\alpha=0.36$)	S_b	$S_b b^2/k$ $S_b b^2/k$	2.30
Diffusion properties			
Preexponential factor for grain boundary diffusion coefficient ($\alpha=0.36$)	δD_{b0}	m^3/s	$5.0\cdot10^{-15}$
Activation energy for grain boundary self-diffusion ($\alpha=0.36$)	Q_b	Q_b/kT_m	9.2
Preexponential factor for diffusion coefficient in melt	D_{L0}	m^2/s	$14.6\cdot10^{-8}$
Activation energy for diffusion in melt	Q_L	Q_L/kT_m	3.6

2. NMC nickel formed by ECAE with average grain size 0.13 μm was investigated by XRD analysis in (Kolobov et al. 2001c). The measured microdeformation $\Delta\varepsilon$ of the samples after annealing in the pre-recrystallization temperature range are listed in Table 18.7. Using the above-described procedure, it was found that reduction of σ_c from 156 MPa to 51 MPa during the recovery corresponds to a decrease in the OMD density from $2.0 \cdot 10^{-3}$ to $6.7 \cdot 10^{-4}$. According to Eqs. (15.148)-(15.150), this change shows an increase in the activation energy for grain boundary diffusion from 7.9 kT_m to 8.7 kT_m and the preexponential factor D^*_{b0} from $1.53 \cdot 10^{-6}$ m^2/s to $5.28 \cdot 10^{-6}$ m^2/s.

Table 18.7. Dependence of microdeformation Δε, %, in NMC nickel on the annealing temperature (Kolobov et al. 2001c)

T, K	293	323	373	423	443
Δε, %	0.076	0.045	0.05	0.049	0.025

Comparison of experimental and calculated data on the OMD density vs. annealing temperature for NMC ECAE nickel demonstrates a satisfactory agreement (Figure 18.4).

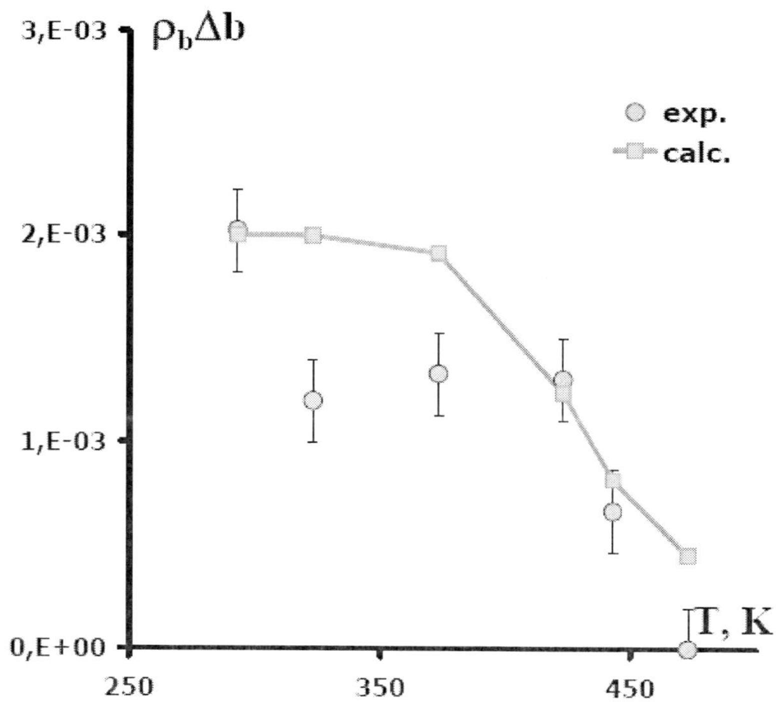

Figure 18.4. Numerical modeling of the OMD density vs. temperature for a 30-minute isothermal anneal of NMC nickel (experiment (Kolobov et al., 2001c)).

18.3. DIFFUSION PROPERTIES OF NONEQUILIBRIUM GRAIN BOUNDARIES DURING MIGRATION

18.3.1. Model

When the recrystallization processes is accompanied by grain boundary migration without deformation, the flow of dislocations embedded in grain boundary I is proportional to the migration rate V_m and the density of lattice dislocations ρ_v swept out by the moving boundary (see Eq. (18.2)). In this case, equations for stationary density of OMD and the tangential components of delocalized dislocationstangential components of Burgers vector of delocalized dislocations are:

$$\rho_b^{st}\Delta b = (\varphi_{1m}\rho_v V_m)^{1/4}, \tag{18.13}$$

$$w_t^{st} = d(\varphi_{2m}\rho_v V_m)^{1/2}, \tag{18.14}$$

where

$$\varphi_{1m} = \xi_m A_1 (kT/G\Omega)(b^3/D_b^*)$$

and

$$\varphi_{2m} = (\xi_m/C_1)(kT/G\Omega)(b/D_b^*).$$

Combining Eqs. (18.13), (15.87) and (15.150), we obtain an approximate expression for the grain boundary diffusion coefficient D_b^* under the conditions of boundary migration with (provided $\Delta\alpha << \alpha$):

$$D_b^* = D_b \exp\left\{\frac{(\varphi_{1m}\rho_v V_m)^{1/4} + d(\varphi_{2m}\rho_v V_m)^{1/2}}{w_0\alpha_B}\right\} \tag{18.15}$$

The acceleration parameter of grain boundary diffusion $\eta = D_b^*/D_b$ during migration with rate V_m is

$$\eta = \frac{D_b^*}{D_b} = \exp\left\{\frac{(\varphi_{1m}\rho_v V_m)^{1/4} + d(\varphi_{2m}\rho_v V_m)^{1/2}}{w_0\alpha_B}\right\}. \tag{18.16}$$

Eq. (18.15) shows that the grain boundary diffusion depends exponentially both on the boundary migration rate V_m and the density of lattice dislocations ρ_v ρ_v, as well as on the thermodynamic characteristics of a material.

It follows from Eqs. (15.151)-(15.152) and (15.154)-(15.155) that grain boundaries accumulate OMD ($\rho_b\Delta b >> w_t$) for small grain sizes $d<d_1$ or low temperatures $T<T^*$. At

temperatures $T>T^*$ or coarse grain sizes $d>d_1$, the tangential components of delocalized dislocationstangential components of Burgers vector of delocalized dislocations $w_t>>\rho_b\Delta b$ prevail in boundaries. The effect of diffusion acceleration is possible when the dislocation flow through the boundary $I^+=\xi_m\rho_v V_m$ $I^+ = \xi_m \rho_v V_m$ exceeds the intensity of defect withdrawal due to recovery ($I^-=\rho_b/t_1$ $I^- = \rho_b / t_1$ or $I^-=w_t/t_1 b$ $I^- = w_t / t_2 b$). Equating these intensities allows to estimate the migration rate V_m^* at which D_b^* can increase:

$$V_m^* = \rho_b^{st} / \rho_v t_1 \xi_m \tag{18.17}$$

$$V_m^* = w_t^{st} / \rho_v t_2 b \xi_m \tag{18.18}$$

According to Eq. (15.154), for NMC copper at $T=0.4T_m$ (523 K) and typical values $\rho_b\Delta b=10^{-2}$ and $w_t=10^{-2}$ the characteristic time is $t_1\approx 2.01\cdot 10^{-2}$ s. Similarly, Eq. (15.15) gives $t_2=2.46\cdot 10^{-3}$ s at the same temperature and $d_0=0.2$ μm. Then, using Eqs. (18.17)-(18.18) and values $\rho_v=10^{15}$ m^{-2} and $\xi_m=0.01$, we obtain $V_m^* \sim 10^{-10}$ m/s. Therefore, when the boundary migration rate V_m in NMC materials exceeds the value of V_m^*, the grain boundary diffusion coefficient increases significantly; when $V_m< V_m^*$, the grain boundary diffusion coefficient remains identical to the equilibrium value.

18.3.2. Comparison with Experimental Data

To verify this model, we shall examine the experimental data on recrystallization processes in NMC materials. Figure 18.5 displays our results on the average grain size d and the volume fraction f_v^R of the recrystallized material vs. annealing temperature for NMC copper M0b (99.995%) and M1f (99.98%). The dependence of average grain size on annealing temperature in NMC detects three stages. At the first stage ($T\le T_1$), only insignificant homogeneous grain growth is observed. At the second stage, abnormal grain growth at temperature $T= T_1$ (Chuvil'deev et al. 2003b, 2004a; Nokhrin et al. 2004) results in bimodal structure with coarse grains which size d_c exceeds the average grain size of a homogeneous NMC matrix by an order of magnitude. With the increase of annealing temperature, the volume fraction f_v^R of the abnormal grains increases and $f_v^R \to 1$ at temperature T_2. At temperatures $T\ge T_2$, collective recrystallization begins. The recrystallization mechanisms in NMC materials will be described in detail in Chapter 20. During the abnormal grain growth, the dependence of the average grain size on time and annealing temperature can be calculated from Eq. (20.21) (see Chapter 20):

$$\bar{d} = d_c(1 - \exp\{-t/t_3\}), \tag{18.19}$$

where t_3 is determined by Eq. (15.156). The third stage ($T\ge T_2$) can be described by kinetics of collective recrystallization (Gorelik et al. 2005):

$$\dot{d} = \gamma_b \delta b D_b^* / dkT \tag{18.20}$$

In this case, the grain size dependence on the annealing time is

$$d^2 - d_0^2 = 2\gamma_b bt\delta D_b^* / kT, \tag{18.21}$$

where d_0 is the initial grain size (at the beginning of that stage).

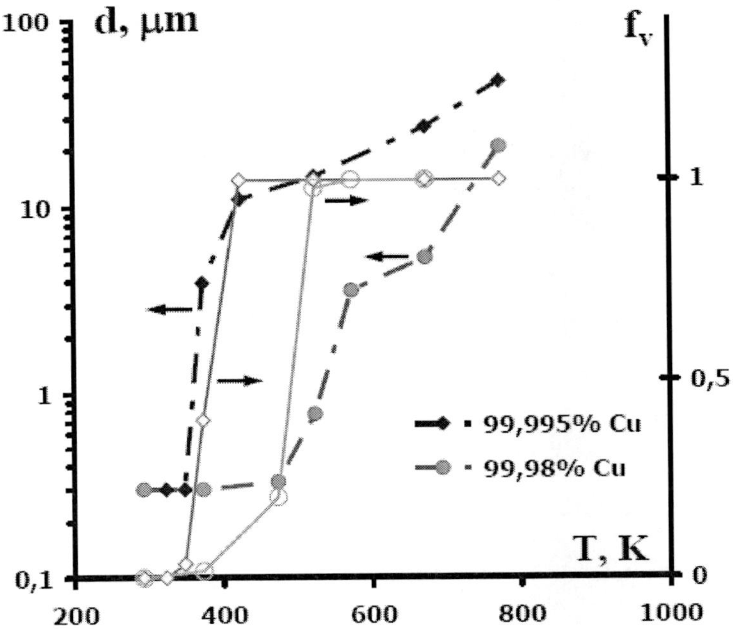

Figure 18.5. Average grain size d and the volume fraction of material involved in the abnormal grain growth process f_v^R vs. temperature of a 30-minute isothermal anneal for M0b (99.995%) and M1f (99.98%) NMC copper (8 ECAE passes, T=293 K, Route D).

Eqs. (18.20)-(18.21) give the grain boundary diffusion coefficient at each stage of recrystallization, and the corresponding activation energy can be found from formula $Q_b^* / kT_m = -\ln(D_b^* / D_{b0})/(T_m / T)$ where D_b^* may be determined at every stage using the appropriate formula. The results of analysis are presented in Figure 18.6. With transition to the abnormal grain growth, the diffusion coefficient D_b^* increases sharply from $9.6 \cdot 10^{-29}$ m²/s to $5.5 \cdot 10^{-20}$ m²/s for NMC copper M1f (99.98%) NMC copper and from $6.1 \cdot 10^{-28}$ m²/s to $3.4 \cdot 10^{-15}$ m²/s for NMC copper M0b (99.995%). The corresponding activation energy Q_b^* reduces from (9.0-9.2)kT_m to 7.6kT_m in NMC copper M1f (99.98%) that is by about 20% lower than that for equilibrium grain boundaries; in NMC copper M0b (99.995%) the activation energy decreases to 4.6-5.0kT_m. During the abnormal grain growth, the diffusion coefficient decreases non-monotonically while the activation energy Q_b^* approaches the typical value for equilibrium grain boundary diffusion (≈9.1kT_m). A similar effect is found in other materials such as iron, nickel and some magnesium alloys where the abnormal grain growth is observed. A reduction of the activation energy for grain boundary diffusion during their migration was also observed in NMC magnesium alloy AZ31 (Kim et al. 2004). At annealing temperatures below $0.58T_m$, the activation energy is 70.1 kJ/mol; in the temperature range from $0.58T_m$ to $0.74T_m$, Q_b^*=24.5 kJ/mol; with a further increase of the annealing temperature, Q_b^* increases to 114 kJ/mol. Similar results were reported in (Kim, 2004).

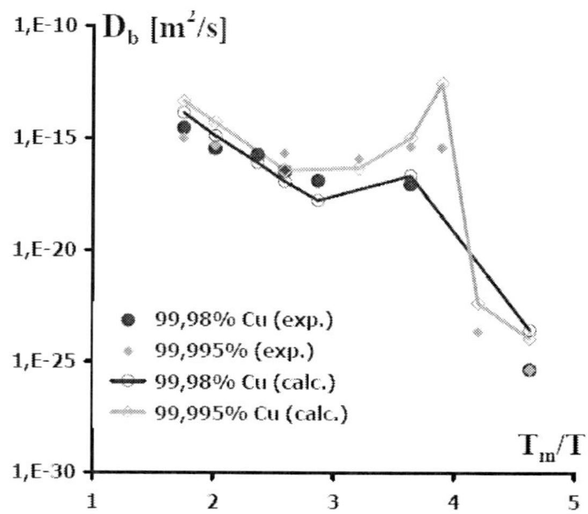

a)

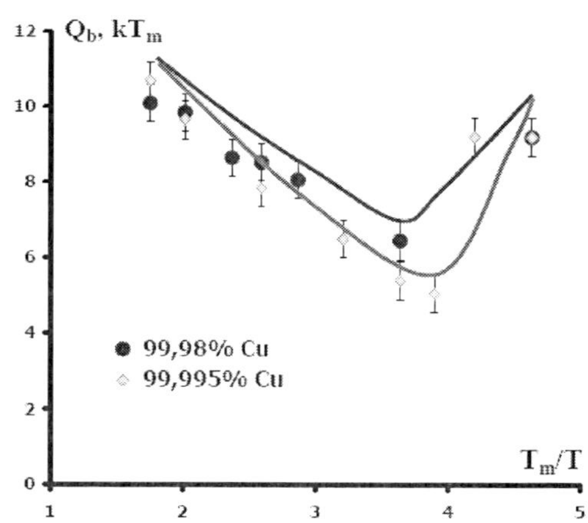

b)

Figure 18.6. Diffusion coefficient (a) and activation energy of grain boundary diffusion (b) vs. annealing temperature for M0b (99.995%) and M1f (99.98%) NMC copper.

It should be mentioned that Fujita et al (Fujita et al. 2004) attributed the enhanced diffusion coefficient during grain boundary migration of Al-0.1%Sc NMC alloy (Figure 18.7) to heterodiffusion of zinc along boundaries of NMC metals what is different from the present theorymodel.

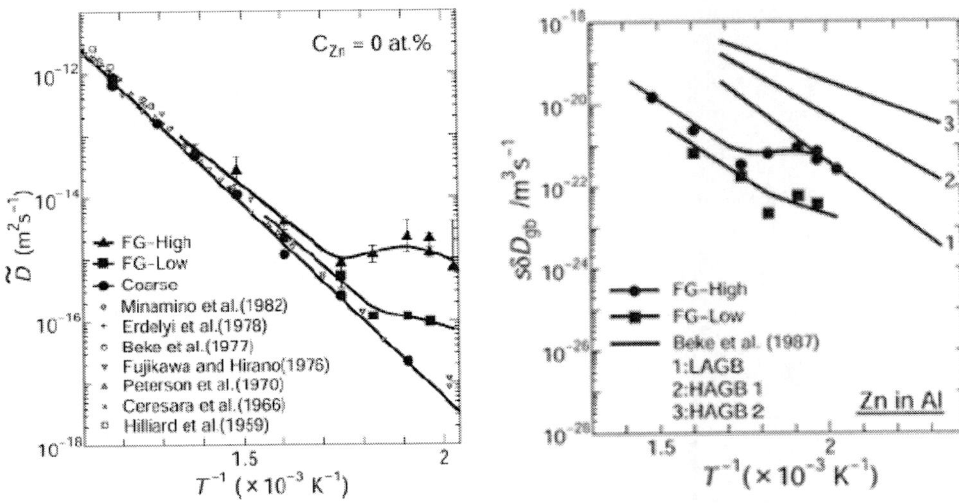

Figure 18.7. Temperature dependence of the grain boundary diffusion coefficient in NMC alloy Al-0.1%Sc/Al-2.1%Zn-0.1%Sc (Fujita et al. 2004).

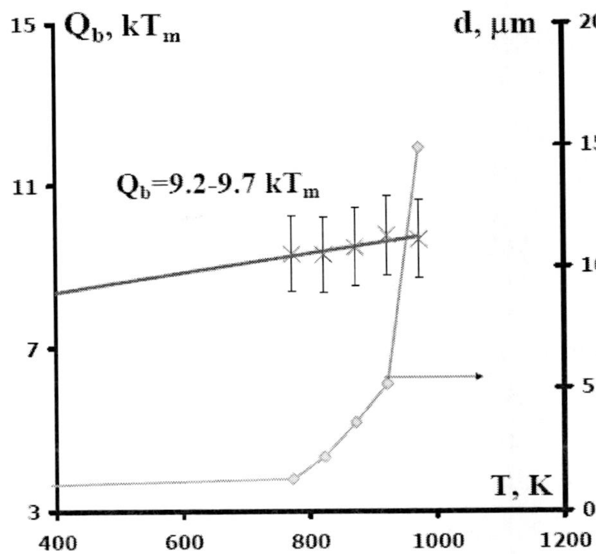

Figure 18.8. Grain size d and activation energy of grain boundary diffusion Q_b vs. annealing temperature for VT1-00 NMC titanium alloys.

In the materials subjected to ECAE at elevated temperature, abnormal grain growth does not always take place. Instead, collective recrystallization is often observed. For collective recrystallization, the effect of the grain boundary diffusion acceleration is not as strong as in the case of abnormal grain growth, and a change in the value of Q_b^* is small in a wide temperature range. Figure 18.8 exemplifies a dependence of the activation energy of grain boundary diffusion on the annealing temperature for VT1-00 NMC titanium (d=0.8 μm) subjected toproduced by ECAE at T=653÷673 K. As seen from Figure 18.8, value Q_b^* is close to its equilibrium value of 9.7 kT_m. In this material the grain boundary migration rate

V_m at T=773 K (about $0.4T_m$) does not reach its critical value $V_m < V_m^* = 10^{-10}$ m/s while in the NMC copper at a similar homologous temperature ($T=0.4T_m \approx 523$ K), a typical migration rate V_m lies in the range of $10^{-8} \div 10^{-9}$ m/s, that is by 1-2 orders of magnitude above the critical V_m^*. As follows from Eq. (18.15), V_m significantly influences the diffusion properties of migrating grain boundaries in NMC materials: at abnormal grain growth and high migration rates, the activation energy for grain boundary diffusion decreases while during collective recrystallization with low migration rate Q_b^* is almost constant.

Using Eqs. (18.13)-(18.15), the diffusion coefficient in NMC copper can be calculated as a function of annealing temperature during grain boundary migration. A difficulty presents evaluation of V_m together with the lattice dislocation density. In order to do so, it is necessary to introduce several assumptions: (i) the diffusion coefficient depends on the annealing temperature and migration rate according to Eq. (18.15); (ii) the dependence of the grain boundary migration rate on annealing temperature for stages of abnormal grain growth and collective recrystallization are described by Eqs. (18.21) and (18.20), respectively; (iii) evolution of the lattice dislocation density during grain boundary migration is

$$\dot{\rho}_v = -\rho_v / t_4 , \qquad (18.22)$$

where t_4 is the characteristic time of absorbtion of dislocations by migrating grain boundaries. It can be presented as $t_4 \sim r_\perp / V_m$, where $r_\perp$ is the average distance between dislocations. Fot a uniform distribution of dislocations, $r_\perp \sim \rho_v^{-1/2}$, and

$$t_4^{-1} = V_m \sqrt{\rho_v} \,. \qquad (18.23)$$

The solution of Eqs. (18.22), (18.23) gives a general expression for the density of lattice dislocations swept out by migrating boundaries:

$$\rho_v(t,T) = \frac{\rho_{v0}}{1 + 2V_m t \sqrt{\rho_{v0}}} \quad \rho_v = \frac{\rho_{v0}}{1 + 2V_m t \sqrt{\rho_{v0}}} \,. \qquad (18.24)$$

Then, Eq. (18.24) and (18.15) defines the grain boundary self-diffusion coefficient D_b^* on annealing temperature. The calculated values of D_b^* versus temperature T_m/T are presented in Figure 18.6a for NMC copper M1f (99.98%) and M0b (99.995%) which characteristics are listed in Table 18.6. As seen from Figure 18.6, the proposed model shows a satisfactory agreement with experimental data.

18.4. Diffusion Properties of Nonequilibrium Grain Boundaries under Intragranular Deformation

18.4.1. Model

A dislocation flow through a grain boundary caused by intragranular deformation with the strain rate $\dot{\varepsilon}_v$ can be described by Eq. (18.3). In this case, the density of OMD and the products of their delocalization in a grain boundary depends on $\dot{\varepsilon}_v$. According to Eqs. (15.147) and (15.150)-(15.153) this means that the grain boundary diffusion coefficient depends substantially on $\dot{\varepsilon}_v$. So, we shall find an explicit form of $D_b^*(\dot{\varepsilon})$. In accordance with Eq. (15.87), at small changes in the free volume $D_b^* = D_b \exp(\Delta\alpha/\alpha_B)$. In the first approximation, $\Delta\alpha$ is assumed to be proportional to a stationary density of defects distributed in a grain boundary: $\Delta\alpha = (\rho_b^{st}\Delta b + w_t^{st})/w_0$. The values of $\rho_b^{st}\Delta b$ and W_t^{st} can be determined from Eqs. (15.151) and (15.152) with the account of $I = \xi\dot{\varepsilon}_v/b$:

$$\rho_b^{st}\Delta b = \chi_1\dot{\varepsilon}_v^{1/4}, \qquad (18.25)$$

$$w_t^{st} = \chi_2\dot{\varepsilon}_v^{1/2}, \qquad (18.26)$$

where $\chi_1 = (g_1/D_b^*)^{1/4}$ and $\chi_2 = d(g_2/D_b^*)^{1/2}$. Inserting these expressions into Eq. (15.87), in a rough approximation we obtain:

$$D_b^* = D_b \exp(\chi_1'\dot{\varepsilon}_v^{1/4} + \chi_2'\dot{\varepsilon}_v^{1/2}).\ (\Delta\alpha \ll \alpha) \qquad (18.27)$$

where $\chi_1' = \chi_1/w_0\alpha_B$ and $\chi_2' = \chi_2/w_0\alpha_B$. At high strain rates, the density of intragrain defects is high, and for $\alpha+\Delta\alpha=\alpha^*$ the value of diffusion coefficient D_b^* is close to that in a melt, D_L:

$$D_b^* = D_L\ (\alpha+\Delta\alpha=\alpha^*). \qquad (18.28)$$

The diagram of $D_b^*(\dot{\varepsilon}_v)$ corresponding to Eqs. (18.27), (18.28) is shown in Figure 18.9.

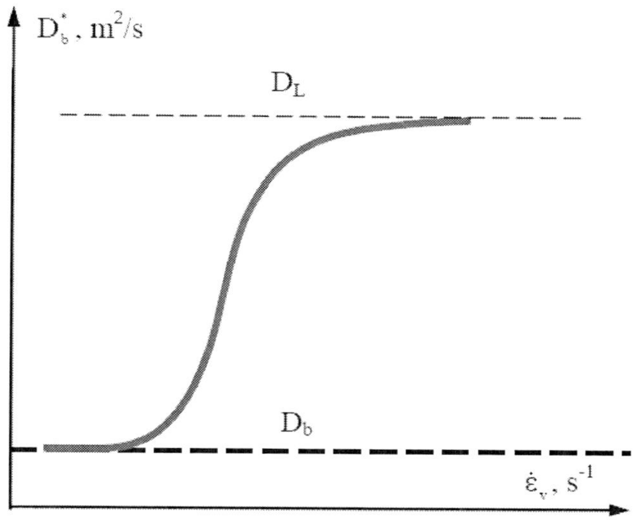

Figure 18.9. Grain boundary diffusion coefficient D_b^* vs. strain rate $\dot{\varepsilon}_v$.

18.4.2. Comparison with Experimental Data

Direct experimental data on the effect of strain rate on the grain boundary diffusion coefficient in NMC materials is scarce. However, the indirect evidence of enhanced diffusion under active intragranular deformation is the well established effect of high-strain rate superplasticity in NMC structures (Nieh et al. 1997; Mukherjee, 2002; Valiev et al. 1997; Horita et al. 2000; Perevezentsev et al. 2002b; Furukawa et al. 2001; Musin et al. 2004; Valiev et al. 2006; Garcia-Infanta et al. 2008).

For superplastic conditions, a rheological equation between strain rate $\dot{\varepsilon}\dot{\varepsilon}$, yield stress σ and grain boundary diffusion coefficient D_b^* can be presented as (Nieh, 1997):

$$\dot{\varepsilon} = A_b \left(\frac{\sigma^*}{G}\right)^{1/m} \left(\frac{b}{d}\right)^p D_b^* \left(\frac{G\delta}{kT}\right), \qquad (18.29)$$

where m is the strain-rate sensitivity coefficient of flow stress, $A_b=10^2$ (Perevesentsev et al. 1992). Experimental data show that for optimum conditions m=0.5 and p=2. Then, It follows from Eq. (18.29) becomesthat in optimum conditions D_b^* may be evaluated using the following formula:

$$D_b^* = \frac{\dot{\varepsilon}b^2}{A_b}\left(\frac{G}{\sigma}\right)^2 \left(\frac{d}{b}\right)^2 \left(\frac{kT}{G\Omega}\right), \qquad (18.30)$$

and then, knowing D_b^*, the activation energy for grain boundary diffusion may be determined:

$$\frac{Q_b^*}{kT_m} = -\frac{\ln(D_b^* / D_{b0})}{T_m / T}. \tag{18.31}$$

For comparison of these equations with experimental results, we used data on high-strain rate superplasticity in NMC Al-Mg-Sc alloys at 673 K (Furukawa et al. 2001) shown in Figure 18.10. Related theoretical calculations are presented in Table 18.8 and Figure 18.11. It is seen that grain boundary diffusion of NMC aluminum alloys at T=673K is enhanced: an increase of strain rate from 10^{-3} s^{-1} to 10^0 s^{-1} accelerates the grain boundary diffusion coefficient by about two orders of magnitude; whereas in this case, the activation energy Q_b^* and the preexponential factor D_{b0}^* drop from 10.8 kT_m to 6.1 kT_m and from $8.74 \cdot 10^{-5}$ m^2/s to $1.11 \cdot 10^{-6}$ m^2/s, respectively ($G\Omega/kT_m$=32, b=$2.86 \cdot 10^{-10}$ m, d=$4 \cdot 10^{-6}$ m).

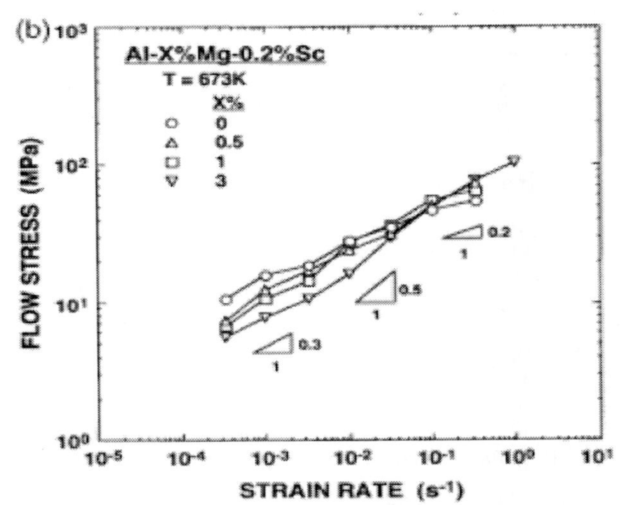

Figure 18.10. Flow stress vs. strain rate for NMC Al-Mg-Sc alloys with a different magnesium content (Furukawa et al. 2001).

Table 18.8 Influence of strain rate $\dot{\varepsilon}$ on the grain boundary diffusion parameters in Al-3%Mg-0.2%Sc NMC alloy at T=673 K (Furukawa et al. 2001)

$\dot{\varepsilon}$, s^{-1}	σ^S, MPa	D_b^*, m^2/s	D_{b0}^*, m^2/s	Q_b^*, kT_m
0	-	$2.8 \cdot 10^{-11}$	$8.74 \cdot 10^{-5}$	10.8
$1 \cdot 10^{-3}$	8	$3.5 \cdot 10^{-11}$	$6.76 \cdot 10^{-6}$	8.8
$3 \cdot 10^{-3}$	13	$4.0 \cdot 10^{-11}$	$6.37 \cdot 10^{-6}$	8.6
$1 \cdot 10^{-2}$	18	$6.7 \cdot 10^{-11}$	$3.49 \cdot 10^{-6}$	7.6
$3 \cdot 10^{-2}$	25	$1.0 \cdot 10^{-10}$	$2.83 \cdot 10^{-6}$	7.3
$1 \cdot 10^{-1}$	45	$1.1 \cdot 10^{-10}$	$2.73 \cdot 10^{-6}$	7.3
$3 \cdot 10^{-1}$	80	$1.2 \cdot 10^{-10}$	$2.51 \cdot 10^{-6}$	7.2
$1 \cdot 10^{0}$	100	$2.3 \cdot 10^{-10}$	$1.11 \cdot 10^{-6}$	6.1

Table 18.9 comprises experimental values of σ and $\dot{\varepsilon}$ and the calculated values of D_b^* and Q_b^* for superplastic 1570 NMC alloy (Musin et al. 2004). It is seen that when strain rate at temperature T=723 K increases from $1.4{\cdot}10^{-3}$ s^{-1} to 1.4 s^{-1}, the grain boundary diffusion coefficient grows by the factor of 60 with a corresponding reduction of the activation energy from the equilibrium value $Q_b^* \approx 10.8$ kT_m (Frost et al. 1982) to 5.8 kT_m.

Table 18.9. Influence of superplastic strain rate on the grain boundary diffusion properties in 1570 NMC alloy (d_0^a~1 μm, d_{sp}=3.1-5.8 μm) (Musin et al. 2004)

$\dot{\varepsilon}$, s^{-1}	σ^S, MPa	D_b^*, m^2/s	D_{b0}^*, m^2/s	Q_b^*, kT_m
-	-	$7.74{\cdot}10^{-11}$	$8.74{\cdot}10^{-5}$	10.8
$1.4{\cdot}10^{-3}$	10	$2.03{\cdot}10^{-11}$	$7.79{\cdot}10^{-6}$	9.9
$1.4{\cdot}10^{-2}$	13	$1.21{\cdot}10^{-10}$	$5.27{\cdot}10^{-6}$	8.3
$5.6{\cdot}10^{-2}$	18	$2.51{\cdot}10^{-10}$	$2.34{\cdot}10^{-6}$	7.1
$1.4{\cdot}10^{-1}$	27	$2.79{\cdot}10^{-10}$	$2.01{\cdot}10^{-6}$	6.9
$5.6{\cdot}10^{-1}$	43	$4.41{\cdot}10^{-10}$	$9.59{\cdot}10^{-7}$	6.0
$1.4{\cdot}10^{0}$	65	$4.82{\cdot}10^{-10}$	$8.14{\cdot}10^{-7}$	5.8

The effect of strain rate on the grain boundary diffusion coefficients calculated from Eqs. (18.25)-(18.27) and determined experimentally for Al-3%Mg-0.25%Sc NMC aluminum alloy at T=673 K is presented in Figure 18.11. An increase of the grain boundary diffusion coefficient and reduction of the activation energy at high-strain rate superplasticity were also observed in aluminum (Park, 2003), magnesium (Chuvil'deev et al. 2004c) and titanium (Ko et al. 2005) alloys.

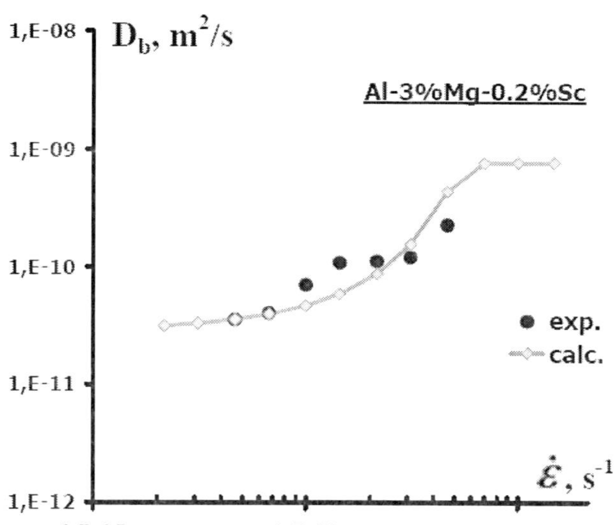

Figure 18.11. Calculation of the grain boundary diffusion coefficient vs. strain rate for Al-3%Mg-0.2%Sc NMC alloy (experimental data of (Furukawa et al. 2001)).

Detailed analysis of D_b^* during superplastic deformation is complicated for a few reasons. First, the temperature of superplastic deformation is usually high. Hence, the recovery rate of the grain-boundary defect structure is also high. In these conditions, a large density of excess defects that affects D_b^* can be accumulated in grain boundaries only at very high strain rates. Second, at high temperatures of superplastic deformation most of the NMC materials are recrystallized. This process also contributes to D_b^* D_b^* and hinders unambiguous evaluation of its dependence on $\dot{\varepsilon}$.

18.5. SUMMARY

1 The diffusion coefficient in NGB depends on the annealing temperature and time. In the course of annealing, nonequilibrium diffusion coefficient D_b^* D_b^* decreases and approaches its equilibrium value D_b. The recovery time depends on temperature and the material grain size.

2 Under recrystallization conditions, the diffusion coefficient in NGB depends on the grain boundary migration rate and the density of lattice dislocations swept out by migrating boundaries.

3 In the conditions of active intragranular deformation, the diffusion coefficient in NGB increases with the increase of strain rate.

Chapter 19

LIMIT OF GRAIN REFINEMENT DURING ECAE

19.1. PROBLEM DESCRIPTION

To date, there is vast literature on structure refinement in metals under severe plastic deformation. It is well-known that, during severe plastic deformation, misorientated microareas or fragments are generated in the material. As SPD proceeds, the misorientation of fragments increases and their sizes are gradually reduced, reaching a certain value d^* (Figure 19.1) which is termed as the grain refinement limit. The grain refinement limit d^* depends on the structure and properties of a material as well as on the processing conditions (strain, strain rate, temperature, deformation mode and history)

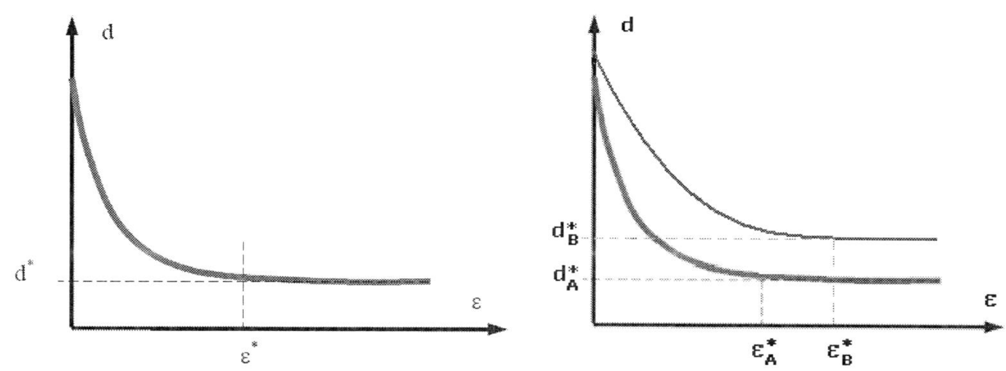

Figure 19.1. Grain size vs. strain under SPD at different temperatures $T_B^{def} > T_A^{def}$.

The temperature of deformation T^{def} is one of the most important factors. Its influence on the grain refinement limit under ECAE was studied for aluminum alloys (Islamgaliev, 2003; Wang, 2004), magnesium alloys (Kim et al. 2005; Matsubara et al. 2003); copper (Huang et al. 2004), iron (Huang et al. 2008) and nickel (Pilyugin et al. 2008). In all cases, the minimum grain size d^* increases with the deformation temperature. A typical dependence of d^* on T^{def} exhibits two stages (Figure 19.2). Above a certain value T_1, which can be called the ECAE recrystallization temperature, d^* rises intensively with temperature.

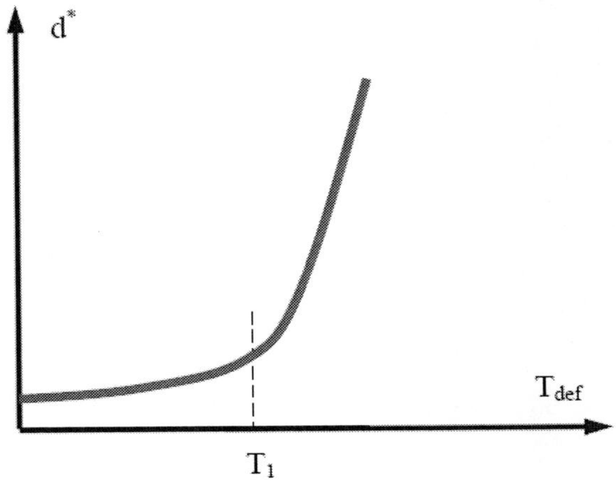

Figure 19.2. Grain refinement limit vs. deformation temperature.

In commercially pure metals, the refinement limit for materials with high melting temperatures (e.g. iron, nickel, copper) is less than for materials with low melting temperatures materials (e.g., magnesium, aluminum). For technically pure Al processed at a room or near-room temperature, $d^* \geq 1$ μm (Valiev et al. 2006; Kopylov et al. 2004) whereas similar processing of nickel and copper produces the average grain size $d^*=0.2$ μm (Segal et al. 1994; Chuvil'deev et al. 2004b; Hebesberger et al. 2005). Therefore, the minimum grain size d^* reduces with a decrease of the homologous deformation temperature T^{def}/T_m (Figure 19.3).

The influence of strain rate on the grain refinement limit of ECAE has not been adequately studied. Available experimental results obtained by different techniques of SPD often contradict each other. For example, analysis of structure refinement during ECAE in (Valiev et al. 2006; Brodova et al. 2008) did not reveal any changes in the grain refinement limit d^* for aluminum and its alloys after increasing the pressing rate by more than two orders of magnitude. Similarly, ECAE of titanium and copper (Khomskaya et al. 2008; Zel'dovich et al. 2008) did not detect any significant changes of d^* after one ECAE pass for quasi-static and dynamic processing with speed 92 m/s and strain rate $10^3 \div 10^5$ s^{-1}. However, adiabatic heating in the last case was significant up to temperatures 770-870K that exceeded the recrystallization temperature (Zeldovich et al. 2008) and should led to grains sized from 1μm to 20μm.

Despite of extensive research on characterization of NMC structures, the mechanisms of grain refinement and structure formation during SPD remain unknown. A model described below allows one to evaluate the grain refinement limit d* and links it to the deformation rate and temperature and strain rate. The model is based on our experimental research on the structure and properties of a wide range of metals and alloys subjected to ECAE as well as on ideas of the theory of severe plastic deformation (Rybin, 1986) and the theory of nonequilibrium grain boundaries (Chuvil'deev, 2004).

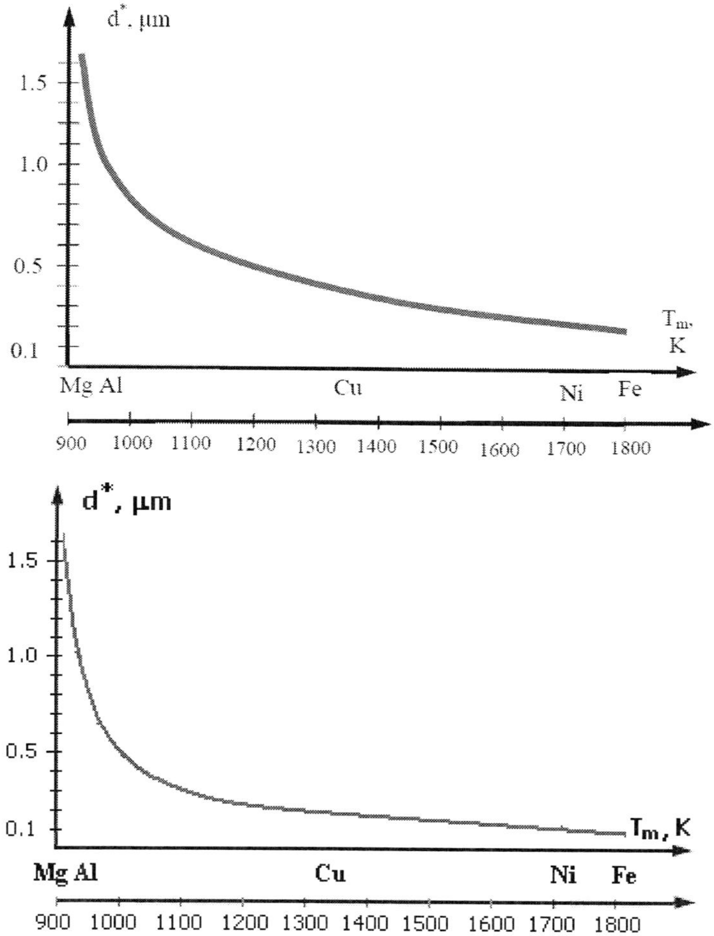

Figure 19.3. Grain refinement limit at room temperature vs. melting point of a material.

19.2. Grain Refinement Limit

19.2.1. Model

During intragranular plastic deformation, defects are accumulated at grain boundaries (Figure 19.4) thus producing layers of complex systems of dislocation and disclination defects. In the first approximation, these defect layers can be described as systems of non-self-consistent flat dislocation arrays and a system of triple junction disclinations (Rybin, 1986; Chuvil'deev, 2004; Perevezentsev et al. 1983) (Figure 19.5). The defects formed at boundaries and, first of all, triple junction disclinations, generate strong internal stresses σ_i in grains which trigger intragranular accommodation slip. It is important that the stress fields created by triple junction disclinations during SPD are so large that the accommodation movement of dislocations in grains becomes collective. The collective movement was described in (Rybin, 1986; Rybin, 2002) using the idea of generation of a special type of defects, viz. broken dislocation boundaries, and their movement through a crystal. When deformation proceeds, the broken dislocation boundaries intersect each other and gradually fragment the crystal (Figure 19.6).

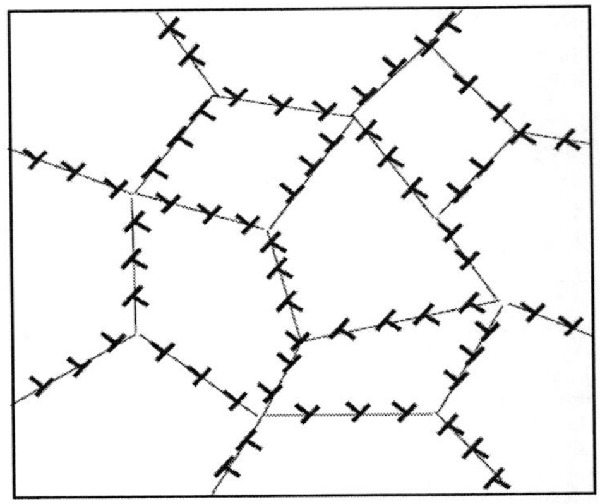

Figure 19.4. Defects at grain boundaries.

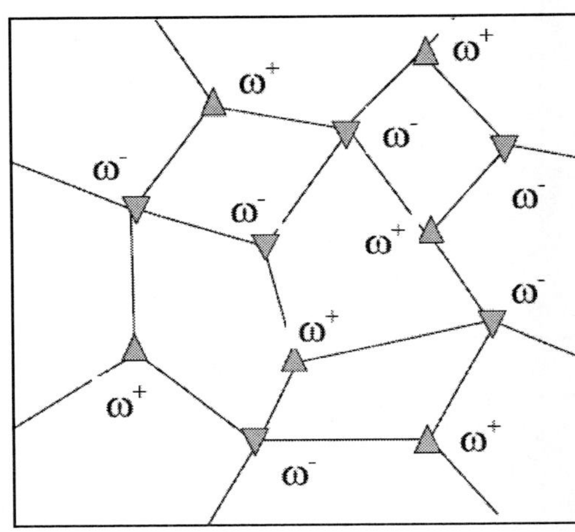

Figure 19.5. Disclinations in a deformed metal.

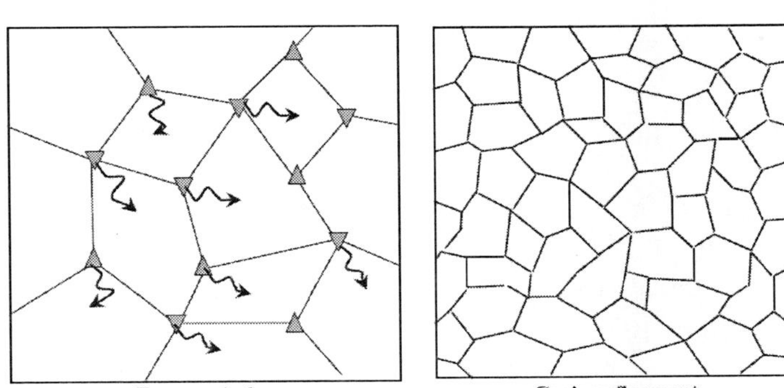

Figure 19.6. Metal fragmentation under severe plastic deformation.

In view of the aforesaid, irrespective of the specific fragmentation mechanism, the condition for fragmentation can be presented in terms of the intensity of internal stresses σ_i

generated by triple junction disclinations: the stresses must be sufficient to trigger plastic deformation in grains, i.e. $\sigma_i \geq \sigma_1$. Stress σ_1 is assumed as being proportional to the material yield stress σ_y, that is $\sigma_1 \approx \psi_2 \sigma_y$. Using the Hall-Petch formula $\sigma_y = \sigma_0 + K/\sqrt{d}$, where σ_0 is the stress at which a gliding dislocation can overcome obstacles in the crystal lattice, K is the Hall-Petch coefficient and d is a grain size, an expression for σ_1 can be written as $\sigma_1 = \psi_2(\sigma_0 + K/\sqrt{d})$. In the first approximation, σ_i is proportional to the power ω of triple junction disclinations $\sigma_i = \varphi_3 G\omega$ (Likhachev et al. 1975), where $\varphi_3 \sim 1$ is a geometrical factor, G is shear modulus. Then the fragmentation condition can be presented as $\omega \geq \sigma_1/\varphi_3 G$, or

$$\omega \geq \omega^* = \chi_\omega(\sigma_0/G + K/G\sqrt{d}) \approx 1/\sqrt{d}, \qquad (19.1)$$

where $\chi_\omega = \psi_2/\varphi_3$ is a geometrical factor and ω^* is a critical power of triple junction disclinations at which they begin emitting broken dislocation boundaries; the latter denotes the onset of fragmentation. In NMC materials, the term $K/\sqrt{d}$ significantly exceeds σ_0, and the fragmentation condition can be written in a more simpform $\omega \geq \omega^* = \chi_\omega K/G\sqrt{d} \sim 1/\sqrt{d}$. If the dislocation pile-ups are not formed because of a fine grain size, the Hall-Petch relation for σ_y is not valid. In such cases, σ_y becomes independent of d (see Chapter 24) and the fragmentation condition becomes $\omega \geq \omega^* = \chi_\omega \sigma_y$.

In the above description, fragmentation is considered as an accommodation process, which is driven by internal stresses σ_I and provides relaxation of the elastic energy stored during deformation. In this model, it is mainly connected with triple junction disclinations. However, depending on deformation conditions, other accommodation processes can occur and be more effective than fragmentation, i.e. proceed at a higher rate and terminate grain refinement. Such an alternative (with regard to fragmentation) mechanism is diffusion mass transfer, which is possible at the certain conditions of ECAE. In Chapter 15, it was shown that the kinetics of a diffusion-induced change in the power of triple junction disclinations can be approximately described as $\dot{\omega} = \xi_3 \dot{\varepsilon}_v - \omega/t_3$, where ξ_3 is the geometrical factor characterizing the homogeneity of deformation, $\dot{\varepsilon}_v$ is the intragranular strain rate and t_3 is the characteristic time of diffusion accommodation. In the dipole approximation (at d=const), t_3 is determined by Eq. (15.156). The related parameters are listed in Table 19.1. As the first approximation at $D_b^* = \text{const}$, the function $\omega(t)$ is given by

$$\omega(t) = \omega^{st}(1 - \exp(-t/t_3)), \qquad (19.2)$$

$$\omega^{st} = \xi_3 \dot{\varepsilon}_v t_3 \approx A_1(\xi_3 \dot{\varepsilon}_v d^3/\delta D_b^*)(kT/G\Omega). \qquad (19.3)$$

The time dependence of the power of a triple junction disclination dipole at different grain sizes d is shown in Figure 19.7. From Figure 19.7 and Eq. (19.2) it is seen that at a large time $t \geq t_3$ the dipole power reaches its stationary value ω^{st}, which depends on the dipole arm d (i.e.

grain or fragment size), deformation temperature, strain rate, and the parameters entering the grain boundary diffusion coefficient D_b^*, viz. the activation energy Q_b^* and preexponential factor D_{bo}^*.

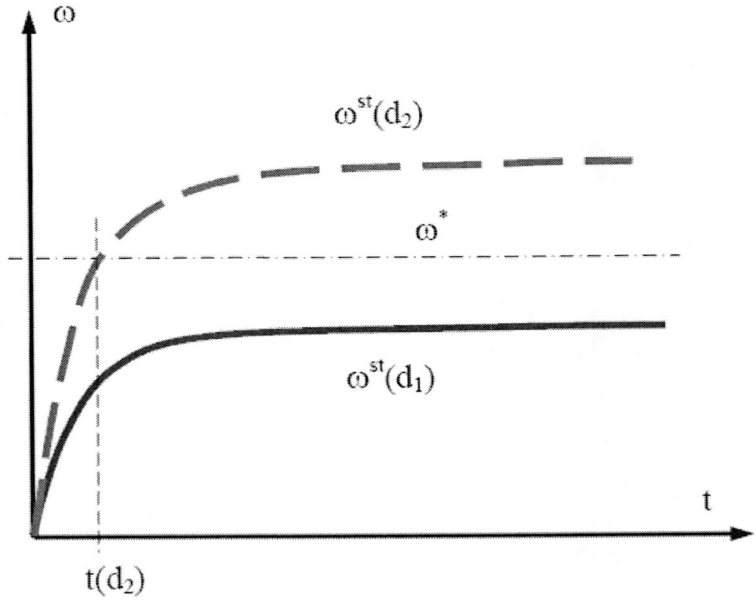

Figure 19.7. Power of a disclination dipole vs. deformation time at different grain sizes ($d_2 > d_1$).

A horizontal line in Figure 19.7 denotes the critical power of dipole ω^* corresponding to the fragmentation condition. At grain size d_2 the dipole power does not reaches its stationary value $\omega^{st}(d_2)$, and fragmentation starts at $t=t(d_2)$ leading to the grain size reduction. At a smaller grain size d_1, the stationary value of disclination power appears to be below the level necessary for fragmentation, $\omega^{st}(d_1)<\omega^*$, and at $d \leq d^*$ fragmentation becomes impossible. The fragmentation condition is $\omega^{st} \geq \omega^*$. Using Eqs. (19.1) and (19.3), it can be also represented as $d \geq d^*$. Here, d^* is the minimum grain size when fragmentation is terminated:

$$(d^*)^{3.5} \geq \chi_\omega (K/G)(\delta D_b^* / A_1 \xi_3 \dot{\varepsilon}_v)(G\Omega/kT). \quad (19.4)$$

The termination of fragmentation at $d \leq d^*$ is due to the fact that at finer sizes of fragments the diffusion accommodation of triple junction disclinations becomes very intensive. Hence, the power of triple junction disclinations cannot reach the critical value necessary for fragmentation, and the latter is terminated at $d=d^*$. It should be noted that at small d the diffusion accommodation is a more effective mechanism for reducing the free energy of the system than fragmentation. Thus, a grain refinement limit, i.e. the minimum grain size that cannot be reduced under the given conditions ($\dot{\varepsilon}$ and T) of plastic deformation, appears due to the development of diffusion accommodation of triple junction disclinations. This limit is determined by Eq. (19.4).

Table 19.1. Parameter values used in calculations

Parameter (units)	Notation	Value		
		Cu	Mg	Al
Activation energy for self-diffusion in melt (kT_m)	Q_L	3.6	3.8	3.8
Activation energy for self-diffusion in equilibrium grain boundaries (kT_m)	Q_b	9.2	11.0	10.8
Preexponential factor for the grain boundary diffusion coefficient ($10^{15}\cdot m^3/s$)	δD_{b0}	5.0	$5\cdot 10^3$	50
Preexponential factor for the diffusion coefficient in a melt (cm^2/s)	D_{L0}	$8\cdot 10^{-4}$		
Burgers vector ($m\cdot 10^{10}$)	b	2.56	3.21	2.86
Boundary width	□	2b		
Atomic volume ($cm^3\cdot 10^{23}$)	Ω	1.18	2.33	1.66
Melting point (K)	T_m	1356	924	933
Shear modulus (GPa)	G	42	17	25
Relative free volume of the boundary	α	0.35		
Critical free volume of the boundary	α^*	0.5		
Excess free volume of the boundary	$\Delta\alpha$	0.02; 0.1		
Specific heat of fusion Density	λ ρ	$\lambda\rho b^3=1.5\ kT_m$		
Enthalpy of the solid/liquid interface	$\gamma^0_{S/L}$	$\gamma^0_{S/L}b^2 = 1\ kT_m$		
Entropy of the solid/liquid interface	$S_{S/L}$	$S_{S/L}b^2=0.8k$		
Free energy of the grain-boundary S-phase	γ_0	$\gamma_0 b^2 \sim 1.4\ kT_m$		
Hall-Petch coefficient ($MPa\cdot mm^{1/2}$)	K	1.5	3.0	3.0
Homogeneity coefficient of plastic deformation	ξ_3	10^{-4}		
Numerical parameter	A_1	10		

19.2.2. Temperature Dependence of Grain Refinement Limit

Let us examine the influence of temperature on the parameters entering Eq. (19.4).

(a) Temperature dependence of strain rate. The rate of intragranular deformation $\dot{\varepsilon}_v$ depends on the local internal stresses in the deformation zone σ_1^* and the deformation temperature. At relatively low temperatures and high stresses, the intragranular strain rate is determined by the slip of dislocations and their interaction with obstacles (Frost et al. 1982):

$$\dot{\varepsilon}_v^I = \dot{\varepsilon}_0 \exp[-(\Delta F/kT)(1-\sigma_1^*/\sigma^0)], \tag{19.5}$$

where $\dot{\varepsilon}_0 = \rho_v b V_\perp \dot{\varepsilon}_0 = \rho_v b V_0$, ΔF is the free energy necessary for overcoming obstacles, σ^0 is the related necessary stress at zero temperature, ρ_v is density of lattice dislocations, b is the Burgers vector, and $V_{\perp 0}$ is a typical velocity of dislocations. Values of ΔF and σ^0 depend on the nature of obstacles. In pure metals, they are mainly determined by lattice type and are very different for fcc and transition bcc metals. It should be noted that this difference is one of the main reasons why fcc and bcc metals are assigned to different isomechanical groups (Frost et al. 1982). In fcc metals $\Delta F/kT_m<10^{-2}$ and $\sigma^0/G<10^{-4}$, in transition bcc metals $\Delta F/kT_m=3.2\pm0.7$ and $\sigma^0/G=(1\pm0.4)\cdot10^{-2}$ (Frost at al, 1982).

At elevated temperature, the intragranular deformation rate $\dot{\varepsilon}_V$ is controlled by the rate of diffusion accommodation:

$$\dot{\varepsilon}_v^{II} = A_v\left(\frac{G\Omega}{kT}\right)\left(\frac{\sigma_1^*}{G}\right)^{n_v}\left[D_v+\rho_v S_d^2 D_c\right]/b^2\ \dot{\varepsilon}_v^{II} = A_v\left(\frac{G\Omega}{kT}\right)\left(\frac{\sigma_1^*}{G}\right)^{n}[D_v+\rho_v r_d^2 D_c]/b^2\cdot \tag{19.6}$$

where $D_v = D_{v0}\exp(-Q_v/kT)$, $D_c = D_{c0}\exp(-Q_c/kT)$. Here, D_{v0} and D_{c0}, Q_v and Q_c are the preexponential factors and activation energies for diffusion in the volume and along dislocation cores, respectively, n_v is the dimensionless constant, which lies within the range of 3 to 10 (depending on the material), and A_v is the Dorn constant (Frost et al. 1982).

The temperature T_v of a transition from intragranular strain rate determined by the dislocation slip (see Eq. (19.5) for $\dot{\varepsilon}_v^I$) to the intragranular strain rate determined by diffusion accommodation (see Eq. (19.6) for $\dot{\varepsilon}_v^{II}$) for pure metals can be estimated from equation $\dot{\varepsilon}_v^I = \dot{\varepsilon}_v^{II}$. Assuming that $S_d^2\rho_v D_c$ $r_d^2\rho_v D_c >> D_v$, the transition temperature is

$$T_v/T_m=Q_c-\Delta F(1-\sigma_1^*/\sigma^0)/\ln(Z_v)kT_m,$$

where $Z_v=A_v(G\Omega/kT)(\sigma_1^*/G)^{n_v}(D_{c0}/V_0 b)$. For typical parameters $A_v=10^6$ (fcc metals) or $A_v=10^8$ (bcc metals), $G\Omega/kT_m=50$, $n_v=5$, $D_{c0}=10^{-2}$ cm^2/s, $V_0=10^{-2}$ cm/s, $Q_c=10\ kT_m$ and a high ratio $\sigma_1^*/G=10^{-2}$ during ECAE, for fcc metals $\Delta F(1-\sigma_1^*/\sigma^0)<<Q_c$ and calculations give $T_v=0.65T_m$ while for bcc metals $\Delta F(1-\sigma_1^*/\sigma^0)\approx3$ and $T_v = 0.33T_m$.

(b) Temperature dependence of coefficient K. The dependence of the Hall-Petch coefficient K on temperature has not been adequately studied. There are evidences that in the range of $(0.1\div0.3)T_m$ it varies insignificantly with temperature. Hence, this coefficient is assumed to be constant.

(c) Influence of strain-enhanced grain growth. When temperature increases, strain-enhanced grain growth becomes possible during deformation. It can significantly influence the value of d^* measured after deformation and corrupt the dependence $d^*(T)$. To eliminate this effect, the deformation temperature should be kept below the recrystallization temperature of a NMC material. Approaches to creating mModels for estimating this value can be found described in (Chuvil'deev et al. 2003c). Hereinafter, when comparing the results of calculations with experiment, we will consider only deformation temperatures which are below the recrystallization temperature.

19.3. Comparison with Experimental Data

19.3.1. Temperature Effect

Pure Metals

Experimental information on the effect of deformation temperature on d^* is limited. Most data relate to severe plastic deformation (SPD) of pure metals at room temperature (T = 293 K). Since the melting temperature T_m of these materials varies, the dependence of the grain refinement limit on the inverse homologous temperature T_m/T can be determined. Experimental results for N passes ECAE are shown in Table 19.2. In accordance with the present model [see Eq.(19.4)], refinement data for fcc metals after ECAE as well as those for bcc metals after HPT deformation (Valiev et al 1996; Ivanisenko et al. 1999; Chashchukhina et al. 2001; Krasil'nikov et al. 2 001; Korznikov et al. 1998) are presented in Figure 19.8 as a diagram of $3.5\ln(d^*/b)$ versus inverse homologous temperature T_m/T. For each group of materials this dependence is well-described by straight lines.The slope μ of these lines is $\mu_{1fcc}\approx 3.1$ for fcc metals and $\mu_{2bcc}\approx 1.1$ for bcc metals. The similar theoretical slope is

$$\mu = Q_b^*/kT_m - (\Delta F/kT_m)(1-\sigma_1^*/\sigma^0) \qquad (19.7)$$

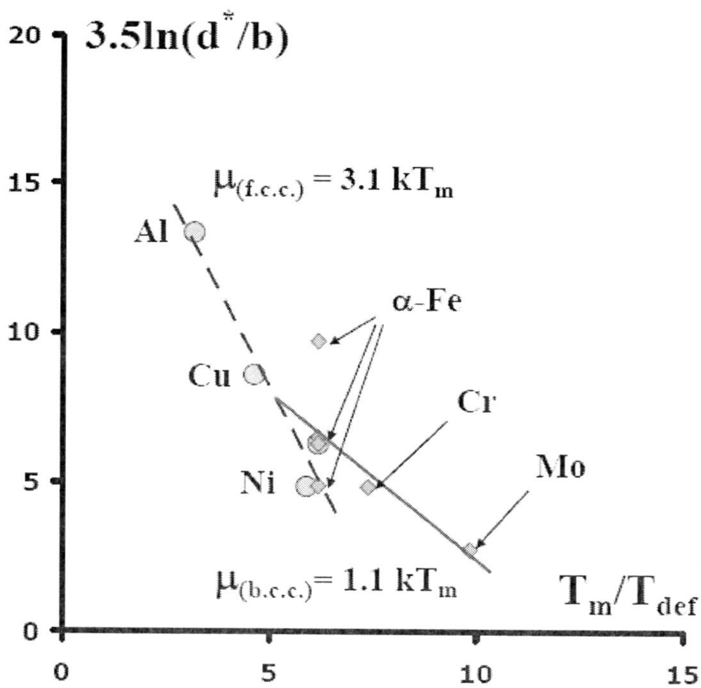

Figure 19.8. Minimum grain size vs. deformation temperature for pure fcc (ECAE) (experimental data of (Chuvil'deev, 2004bsee Table 21.2) and bcc metals (HPT), T=293 K (experimental data of (Valiev et al. 1996; Ivanisenko et al. 1999; Chashchukhina et al. 2001; Krasil'nikov, 2001; Korznikov et al. 1998)), T=293 K.

Table 19.2. Effect of temperature on the minimum grain size of pure metals during ECAE (Chuvil'deev et al. 2004b)

Metal	T_{def}, K	T_m, K	T_{def}/T_m	N	d^*, μm
Al	293 K	933	3.18	8	1.3
Cu		1356	4.63	12	0.3
Ni		1726	5.89	12	0.1
Fe		1808	6.17	10	0.1

At room temperature $T_{def} < T_v$, and $\dot{\varepsilon}^I_v \sim \dot{\varepsilon}_v$ should be used in Eq. (19.4). As for fcc metals the second term in Eq. (19.7) is small, $\mu_1 \approx Q_b^*$ and $\mu_{fcc} \approx Q_b^*$ $\mu_1 \approx Q_b^*$; in this case, Q_b^* = 3.1 kT_m. For bcc metals, $\Delta F/kT_m \approx 3$, $(1-\sigma_1^*/\sigma^0) \approx 1$, and $\mu_2 \approx Q_b^* - 3$ $\mu_{fcc} \approx Q_b^*$, In this case, Q_b^* $Q_b^* \approx 4.1$ kT_m. Thus, for both groups of metals Q_b^* $Q_b^* = (3.1 \div 4.1)kT_m$ that is close to the activation energy for diffusion in melts $Q_L \approx (3 \div 4)kT_m$. This surprising result can be interpreted within the theory of NGB: due to a high content of excess defects, the free volume of boundaries in metals subjected to ECAE reaches the limiting value of α^*. In such boundaries, the activation energy for diffusion becomes comparable with that in melts (see Chapter 15).

Alloys

The theoretical and experimental results on the influence of deformation temperature on grain refinement limit d* are presented in Table 19.3 for a few aluminum and magnesium alloys.

For Al-X%Mg-0.22%Sc-0.15%Zr alloys (X=0, 0.15 and 3%) (Chuvil'deev et al. 2004b), 6 passes ECAE were performed at temperatures 373, 433 and 573 K. Experimental values of μ are 4÷5.6 and depend on the magnesium content. Note that the activation energy for grain boundary diffusion Q_b^* Q_b^* in the alloy without magnesium (X=0) is close to $Q_L = 3.8kT_m$ for pure Al; in alloys containing 1.5 and 3% Mg, Q_b^* Q_b^* is $5.5kT_m$. The theoretical values of d^* calculated for parameters of Table 19.1 agree well with experimental data (see Table 19.3).

For 5052 (Al-2.65%Mg) aluminum alloy (Chen et al. 2003), ECAE was performed at temperatures from 323 to 573 K with an increment of 50K. In the range T=323÷523 K, μ≈4.9 (Figure 19.9a). The corresonding value of Q_b^* Q_b^* differs from that given above by only 0.5kT for 1.5-3% Mg. However, at T>573 K, d* and $Q_b^*(T)$ $Q_b^*(T)$ are significantly higher. This is due to the fact that at high temperatures the pattern of grain refinement is distorted by strain-enhanced grain growth. Experimental and calculated sizes d^* at T=323÷573 K are shown in Table 19.3. A good agreement between both data except temperature T=573 can be seen again.

Table 19.3. Effect of deformation temperature on the grain refinement limit for NMC Al-Mg and Mg-Al alloys at ECAE

Alloy	Experiment				Calculation		
	T^{def}, K	T_m/T^{def}	d^*, μm	Q^*_b, Q_b^*/kT_m (see Eq. 19.7)	$\dot{\varepsilon}_v$, s^{-1}	Q^*_b, Q_b^*/kT_m	d^*, μm
Al-0.22%Sc-0.15%Zr (Chuvil'deev et al. 2004b)	373	2.50	1.2	4.0	$5 \cdot 10^{-2}$	3.8	1.04
	433	2.15	1.8				1.46
Al-1.5%Mg--0.22%Sc-0.15%Zr (Chuvil'deev et al. 2004b)	373	2.50	0.3	5.5	$5 \cdot 10^{-2}$	5.5	0.33
	473	1.97	0.7				0.72
Al-3%Mg--0.22%Sc-0.15%Zr (Chuvil'deev et al. 2004b)	373	2.50	0.35	5.6	$5 \cdot 10^{-2}$	5.5	0.38
	473	1.97	0.8				0.81
5052 (Chen et al. 2003)	323	2.89	0.25	4.9	$2 \cdot 10^{-1}$	5.5	0.14
	373	2.50	0.30				0.25
	423	2.21	0.40				0.39
	473	1.97	0.50				0.55
	523	1.78	0.75				0.71
	573	1.63	2.0 (*)				0.88
MA2-1 (Chuvil'deev et al. 2004b)	473	1.95	1.7	4.4	$3 \cdot 10^{1}$	4.5	1.78
	523	1.77	2.4				2.38
	553	1.67	2.8				2.77
	653	1.40	4.0				4.07
AZ91 (Chuvil'deev et al. 2004b)	423	2.18	1.3	4.9	$1 \cdot 10^{2}$	4.5	1.03
	653	1.42	3.4 (*)				2.45
AZ91 (Kubota et al. 1999)	293	3.15	1.0	4.9	$1 \cdot 10^{2}$	4.5	1.23
	573	1.61	7.6				7.37
	673	1.37	15.4 (*)				9.57
	753	1.22	66.1 (*)				11.18

(*) strain-enhanced grain growth was observed.

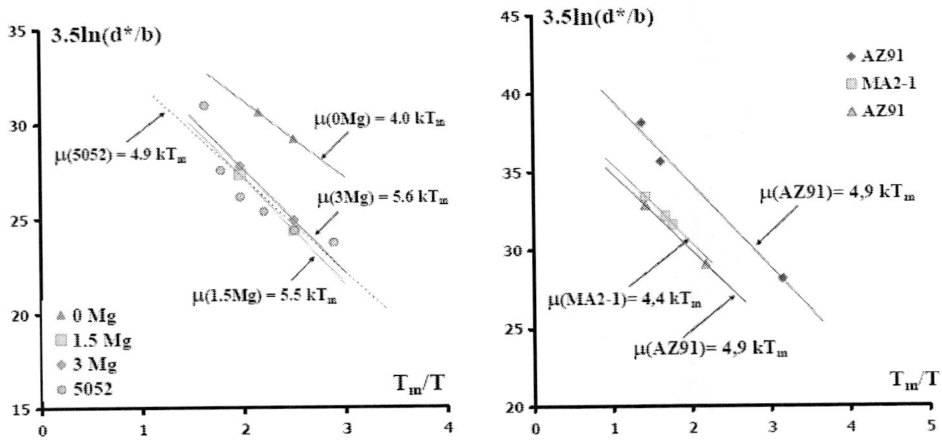

a) b)

Figure 19.9. Minimum grain size d^* vs. deformation temperature for Al-Mg (a), Mg-Al (b) alloys. Data for 5052 (Chen et al. 2003) for AZ91 (Kubota et al. 1999).

For mMagnesium alloys, Table 19.3 contains results related to: (a) MA2-1 alloy after ECAE at temperatures 473, 523, 553 and 653 K, (Chuvil'deev et al. 2004b); (b) AZ-91 alloy after ECAE at temperatures 423 and 653 K (Chuvil'deev et al. 2004b); (c) AZ-91 alloy after ECAE at temperatures 293, 573, 673 and 753 K (Kubota et al. 1999). Average experimental values of the activation energy for grain boundary diffusion Q_b^* are also presented. For the sake of comparison, here it is assumed that Q_L=4.5kT_m. The values of d^* and Q_b*calculated using the parameters of Table 19.1 and Q_L =4.5kT_m are presented in Table 19.3 and Figure 19.9b. A good correlation between the experimental and theoretical data is evident.

19.3.2. Effect of Chemical Composition

(A) Diluted Copper Alloys

Some results for copper alloys with small amounts of alloying elements are presented in Table 19.4. In these cases, alloying elements change insignificantly the value of d^*. This result can be explained in terms of approaches developed in (Chuvil'deev et al. 1999; Chuvil'deev et al. 2001b). It is known that even low dopings change noticeable diffusion and, thus, free volume of grain boundaries. At the same time, if the excess free volume of boundaries is large due to embedded dislocations, the role of free volume connected with foreign atomssmall alloying additives becomes insignificant. This means that in nonequilibrium grain boundaries of copper strained at the room temperature $Q_b^* \sim Q_L$, and influence of alloying atoms is minor. Such conclusion agrees with experimental observations. However, alloying elements can significantly influence d^* in the case of high-temperature deformation when recovery processes in grain boundaries lead to an increase in the activation energy for grain boundary diffusion up to $Q_b^* \sim Q_b$. The theoretical estimation of the grain refinement limit for copper from Eq. (19.4) with parameters of Table 19.1 gives d^*=0.22 μm which is close to experimental data (see Table 19.4).

Table 19.4. Effect of chemical composition on the grain refinement limit for copper and copper alloys at ECAE (Chuvil'deev et al. 2004b)

	Experiment			Calculation		
Material	T^{def}, K	T^{def}/T_m	d^*, d^*, μm	$\dot{\varepsilon}_v$, s^{-1}	Q_b^*, kT_m	d^*, d^*, μm
Technical grade copper						
Cu M0b	293	4.63	0.25	$5\cdot10^0$	3.6	0.22
Cu M1			0.25			
Cu M3			0.20			
Cu M1f			0.30			
Cu-Cr alloys						
Cu-0.3%Cr	293	4.63	0.30	$5\cdot10^0$	3.6	0.22
Cu-0.4%Cr			0.30			
Cu-0.5%Cr			0.30			
Cu-0.8%Cr-0.1%Zr			0.30			

(B) Solid Solutions in Al-Mg Alloys

Table 19.5 displays the results obtained for Al-X%Mg-0.22%Sc-0.15%Zr alloys with a magnesium content X=0, 1.5 and 3% after 6 ECAE passes at temperatures 373 and 473 K. Similar data of (Furukawa et al. 2001) for Al-X%Mg-0.3%Sc alloys with X=0, 0.5, 1 and 3% and (Semiatin et al. 2001) for Al-X%Mg alloys with X= 0, 1 and 3% are shown in tables 19.6, 19.7, respectively.

In order to explain the influence of magnesium on the grain refinement limit, it is necessary to introduce additional assumptions. According to (Chuvil'deev et al. 1999), alloying of aluminum with magnesium decreases the grain boundary self-diffusion. As there is no reliable quantitative data on the effect of magnesium on the activation energy for grain boundary self-diffusion, it is assumed that with increasing the magnesium concentration C_v to its solubility limit in the solid state $C_v^*(T)$, the activation energy for grain boundary self-diffusion grows linearly. It is also assumed that at $C_v \geq C_v^*$ the value of $Q_b^* = Q_b^*(C_v^*)$ is constant. Then,

$$Q_b^*(Mg) = Q_b^* + \Delta Q_b \cdot C_v \qquad \text{at } C_v < C_v^* \tag{19.8}$$

$$Q_b^*(Mg) = Q_b^* + \Delta Q_b \cdot C_v^* \qquad \text{at } C_v \geq C_v^* \tag{19.9}$$

Logarithmation of Eq. (19.4) for two magnesium concentration C_1 and C_2 (C_1, $C_2 < C_v^*$) at T=const gives:

$$3.5\ln\left[\frac{d^*(C_1)}{d^*(C_2)}\right] = \ln\left[\frac{K(C_1)}{K(C_2)}\right] + \frac{\Delta Q_b}{kT_m}(C_2 - C_1)\frac{T_m}{T} \quad \text{at } C_v < C_v^* \tag{19.10}$$

$$3.5\ln\left[\frac{d^*(C_1)}{d^*(C_2)}\right] = \ln\left[\frac{K(C_1)}{K(C_2)}\right] + \frac{\Delta Q_b}{kT_m}(C_v^{max})\frac{T_m}{T} \quad \text{at } C_v \geq C_v^* \tag{19.11}$$

Thus, in order to determine the grain refinement limit at arbitrary magnesium concentration, it is necessary to know: (i) the grain refinement limit of an alloy with a known magnesium content; (ii) Hall-Petch coefficients K for alloys with a different content of alloying elements (for Al-Mg alloys, the K values were found in (Furukawa et al. 1996)); (iii) the solubility limit at a processing temperature (for Mg dissolved in Al at 293 K, C_v^*=1.5% (Ed. Lyakishev, 1996)); (iv) the value of ΔQ_b which characterizes the effect of alloying element on the activation energy of grain boundary diffusion. The parameter ΔQ_b can be found experimentally or theoretically (Chuvil'deev et al. 1999). Here, we will use the experimental results for d^* versus the magnesium concentration (Furukawa et al. 2001). Calculations using data of tables 19.5, 19.3 show that for Mg dissolved in Al $\Delta Q_b \approx 1.13\ kT_m$ per one percent of magnesium within the concentration range from 0 to 1.5% Mg. For C≥1.5% $\Delta Q_b \approx 1.7kT_m$.

Table 19.5. Effect of chemical composition on the grain refinement limit of Al-X%Mg-0.22%Sc-0.15%Zr alloys at ECAE (Chuvil'deev et al. 2004b)

	Experiment			Calculation		
Magnesium content	T^{def}, KT_{def}, K	T^{def}/T_m T_{def}/T_m	d^*, μm d^*, μm	$\dot{\varepsilon}_v$, s^- $_1\dot{\varepsilon}_v$, s^{-1}	Q_b^*, kT_m Q_b^*, kT_m	d^*, μm d^*, μm
X = 0	393	2.50	1.0	$5 \cdot 10^{-2}$	3.8	1.04
X = 1.5			0.3		5.5	0.33
X = 3.0			0.35		5.5	0.35
X = 0	493	1.97	2.0	$5 \cdot 10^{-2}$	3.8	1.73
X = 1.5			1.0		5.5	0.72
X = 3.0			1.0		5.5	0.76

Using this approach and coefficients K(C) from Table 19.8, the grain refinement limits were calculated from Eq. (19.4) for some alloys and presented in tables 19.5-19.7. A fair agreement with experimental data supports the validity of the developed models to capture the effect of alloying on the grain refinement limit.

Table 19.6. Effect of chemical composition on the grain refinement limit of Al-X%Mg-0.2%Sc alloys during ECAE (Furukawa et al. 2001)

	Experiment			Calculation		
Magnesium contentMagnesium content	T^{def}, KT_{def}, K	$T^{def}/T_mT_{def}/T_m$	d^*, μm d^*, μm	$\dot{\varepsilon}_v$, s^- $_1\dot{\varepsilon}_v$, s^{-1}	Q^*_b, kT_m Q^*_b, kT_m	d^*, μm d^*, μm
X = 0	293	3.18	0.70	$1\cdot10^{-2}$	3.8	0.84
X = 0.5			0.48		4.6	0.44
X = 1.0			0.36		5.5	0.21
X = 3.0			0.20		5.5	0.22

Table 19.7. . Effect of chemical composition on the grain refinement limit for Al-X% Mg alloys during ECAE (Semiatin et al. 2001)

	Experiment			Calculation		
Magnesium contentMagnesium content	T^{def}, KT_{def}, K	$T^{def}/T_mT_{def}/T_m$	d^*, μm d^*, μm	$\dot{\varepsilon}_v$, s^- $_1\dot{\varepsilon}_v$, s^{-1}	Q^*_b, kT_m Q^*_b, kT_m	d^*, μm d^*, μm
X = 0	293	3.18	1.30	$5\cdot10^{-3}$	3.8	1.03
X = 1.0			0.45		5.2	0.33
X = 3.0			0.27		5.5	0.27

Table 19.8. The Hall-Petch parameter K for Al-Mg alloys at different magnesium content (Furukawa et al. 1996)

Mg, %	K, MPa·mm$^{1/2}$
0	3.0
0.5	3.5 (*)
1.0	4.0 (*)
2.25	5.3
3.0	4.67

(*) extrapolation of data (Furukawa et al. 1996).

19.3.3 Effect of Strain Rate

The strain rate effect on the grain refinement limit during severe plastic deformation relates to a similar function $D^*_b(\dot{\varepsilon})$ for grain boundary diffucion briefly described in Chapter 18. The function $D^*_b(\dot{\varepsilon})$ can be divided into three stages (Petryaev et al. 2000). At a low strain rate ($\dot{\varepsilon} < \dot{\varepsilon}_1$), the diffusion coefficient insignificantly depends on $\dot{\varepsilon}$ and can approximately be considered as constant, $D^*_b \sim D_b$. At medium strain rates ($\dot{\varepsilon}_1 < \dot{\varepsilon} < \dot{\varepsilon}_2$), grain boundary diffusion significantly depends on strain rate. The function $D_b^*(\dot{\varepsilon})$ is close to linear and well described by Eq. (18.27) within the range $\dot{\varepsilon}=10^{-2}\div10^{2}$ s^{-1}. At higher strain rates, $\dot{\varepsilon}>\dot{\varepsilon}_2$ Db(ε) is flattensed and approximates to diffusion in melt $D_b^*(\dot{\varepsilon})\rightarrow D_L$ when the density of excess defects in boundaries is very high and free volume α→α*. According to the concepts described in Chapter 18, the transition strain rates $\dot{\varepsilon}_1$ and $\dot{\varepsilon}_2$ significantly depend on the deformation temperature and thermodynamic properties of a material. Therefore, the values of $\dot{\varepsilon}_1$ and $\dot{\varepsilon}_2$ should be found for corresponding conditions.

In order to evaluate the effect of $D_b^*(\dot{\varepsilon})$ on d^*, Eq. (19. 4) should be presented as $d^* \sim [D_b^*(\dot{\varepsilon})/\dot{\varepsilon}]^{1/3.5}$. This description of $D_b^*(\dot{\varepsilon})$ implies that the influence of strain rate on d^* is ambiguous. At low strain rates when $D_b^*(\dot{\varepsilon}) \sim D_b = \mathrm{const}$, the strain rate dependence of the grain refinement limit is $d^* \sim \dot{\varepsilon}^{-1/3.5}$. This means that an increase of strain rate by the factor of 10 reduces d^* by approximately 2 times. At medium strain rates, which are typical for ECAE at a room temperature, D_b^* increases near linearly with $\dot{\varepsilon}$, and strain rate affects d^* insignificantly. In such cases, an increase of strain rate during ECAE by one order of magnitude does not change the grain size d* (Semiatin et al. 2001). At very high strain rates, when $D_b^* \sim D_L = \mathrm{const}$, it can be expected that an increase in the strain rate should result in smaller d^*. However, because of adiabatic heating, the final grain size remains ambiguous.

19.4. SUMMARY

A model of structure refinement limit during severe plastic deformation was suggested. The model shows that the minimum grain size d^* depends on strain rate, temperature, composition and diffusion properties of the material. Comparison of the calculated values of d^* with experimental data for different alloys confirms their good correlation and the effectiveness of the model in describing of the limit of grain refinement.

Chapter 20

RECRYSTALLIZATION AND GRAIN GROWTH IN PURE NMC METALS

20.1. PROBLEM DESCRIPTION

The peculiarities of recrystallization in a number of NMC metals produced by severe plastic deformation were established experimentally. First, it is reduction (Valiev et al. 1996; Ivanisenko et al. 1999; Chashchukhina et al. 2001; Krasil'nikov, 2001; Korznikov et al. 1998; Neishi et al. 2002; Morris et al. 2002; Mao et al. 2005; Kang et al. 2008; Ferreasse et al. 2008) of the critical temperature of recrystallization (CTR) by $(0.1\div0.2)T_m$ as compared with coarse-grain metals. It is necessary to note that this unexpected effect was discovered when studying the CTR dependence on strain in NMC metals: with increasing strain, CTR was reducing (as usually) but then, upon the attainment of a certain value, CTR was observed to increase (Nokhrin et al. 2003; Wei et al. 2008) (Figure 20.1). Similar effects were revealed in NMC metals produced by high-pressure torsion (HPT) and rolling (Gagen-Torn, 1979). In particular, the grain boundary migration rate in NMC α-Fe was observed to slow down with an increase of strains (Degtyarev, 2002).

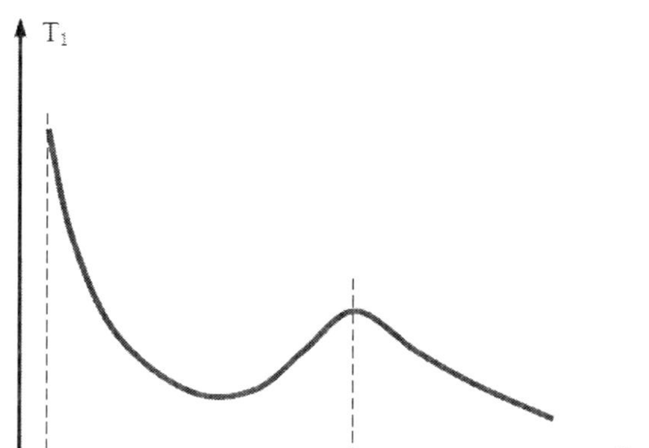

Figure 20.1. Critical recrystallization temperature vs.strain in NMC metals.

Second, it is unusual grain growth behavior in NMC metals under annealing. It has been shown experimentally that recrystallization in NMC copper is a two-stage process. At the first stage, at a near-CTR temperature, abnormal grain growth is observed: some grains begin to grow reaching the size of about 1÷5 μm, which by an order of magnitude exceeds the average grain size of stable NMC matrix (about 0.2 μm) (Chuvil'deev et al. 2004a; Neishi et al. 2003; Eddahbi et al. 2005; Liu et al. 2004). In this case, bimodal grain size distribution is observed (Chuvil'deev, 2004) (see Figure 20.2). The volume fraction of coarse grains increases exponentially (Molodova, 2007; Huang, 2004). After coarse grains cover the whole volume of material, the abnormal growth changes into collective recrystallization.

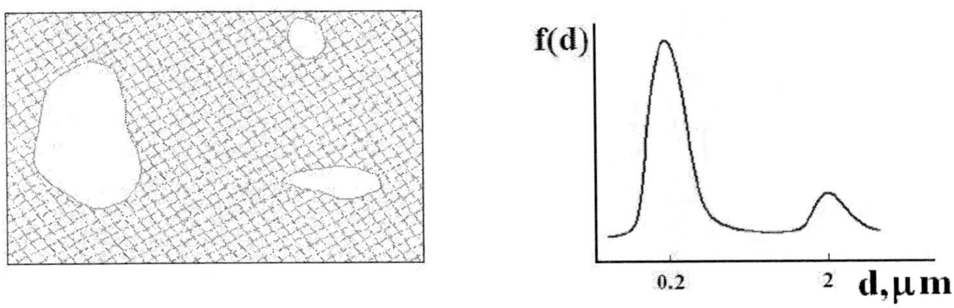

a) b)

Figure 20.2. Abnormal grain growth in a NMC material at annealing: coarse grains in NMC matrix (a) and bimodal grain size distribution (b).

The above features of recrystallization process in NMC metals produced by different SPD methods cannot be described within traditional models. This necessitates the development of novel approaches accounting for the specific structure of nano- and microcrystalline metals produced by ECAE. In this chapter, we present the experimental results on the thermal stability of NMC structures and describe the models for recrystallization in microcrystalline metals.

The experimental data on the average grain size during annealing were analyzed for NMC copper after ECAE passes N=12 (figures 20.3a and c) and N=16 (Figure 20.3b and d) via Route D at temperature T=293 K.

In Figure 20.3, a and b, it is seen that the dependence of the average grain size on annealing temperature has three stages. At the first stage, in the temperature range from 293 to 423 K, the grain size of NMC copper (N=16) does not change. At the second stage at $T > T_1 = 423$ K for N=16 and $T_1 = 453$ K for N=12, intensive grain growth occurs. At the third stage, at $T > T_2 = 453$ K for N=16 and $T_2 = 553$ K for N=12, a slower grain growth is observed during fixed annealing times. The values of characteristic temperatures T_1 and T_2 of the recrystallization process for NMC Cu (N=8) and NMC Ni (N=8) for different isothermal annealing times are listed in Table 20.1. Similar results for NMC copper subjected to different numbers N of ECAE passes are presented in Table 20.2. Temperature dependences of the average grain size for NMC copper show that for N=16 an increase in the annealing time from 1 to 4 h leads to the reduction of temperature T_1 by 25K from 423 to 398 K, for N=12 an increase of the annealing time from 10 to 60 min reduces T_1 by 40 K from 493 to 453 K (Figure 20.3), and for N=8 a decrease of T_1 is 20K from 393 to 373 K (Figure 20.3). A similar phenomenon was observed for NMC Ni. The characteristic recrystallization temperatures T_1 and T_2 (tables 20.1-20.2) can be presented in a homologous form. Their values for both NMC copper and Ni are close to T_1/T_m~0.29, T_2/T_m~0.33 after 1h anneal, and T_1/T_m~0.28, T_2/T_m~0.32 after 5h anneal. The average size of recrystallized grains reduces with a number of passes untill some constant value.

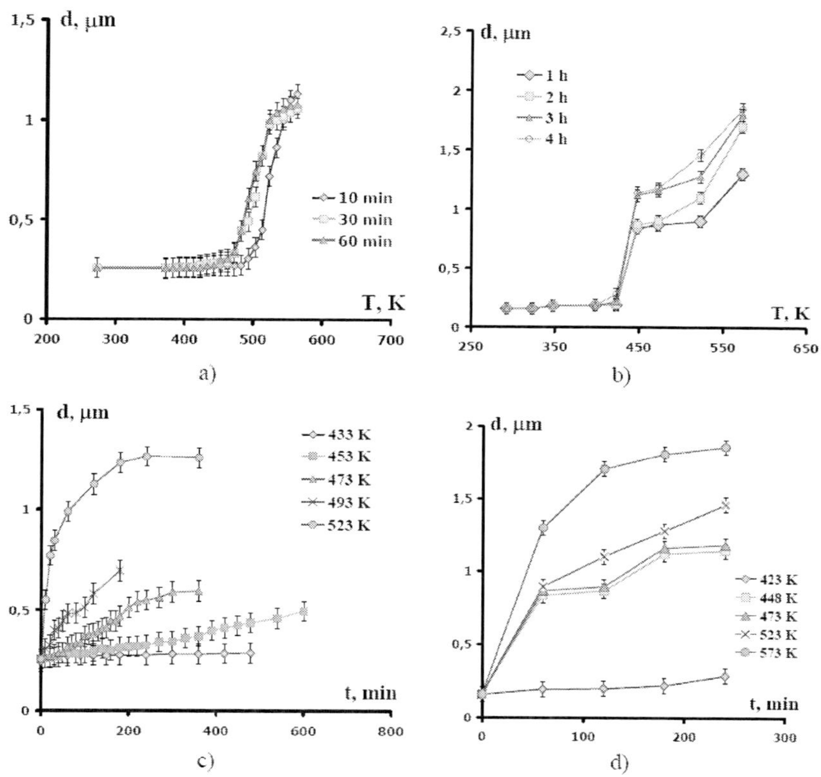

Figure 20.3. Average grain size vs. temperature (a, b) and time (c, d) after isothermal annealing of M1 (99.98%) NMC copper subjected to 12 (a, c) and 16 (b, d) ECAE passes.

Table 20.1. Experimental values of characteristic temperatures T_1 and T_2 for abnormalous grain growth in NMC copper and nickel

Material		Cu (N=8)		Ni (N=8)	
Temperature \ Annealing time		1 hour	5 hours	1 hour	5 hours
T_1	K	393	373	523	473
	T_1/T_m	0.29	0.28	0.30	0.27
T_2	K	443	423	573	553
	T_2/T_m	0.33	0.31	0.33	0.32

Table 20.2. Characteristic temperatures T_1 and T_2 of NMC copper for a one-hour anneal

Temperature \ Deformation		N=4	N=8	N=12	N=16
T_1	K	383	393	453	423
	T_1/T_m	0.28	0.29	0.33	0.31
T_2	K	433	443	553	453
	T_2/T_m	0.32	0.33	0.41	0.33

Diagrams of the average grain size in NMC nickel (at N=8) versus the anneal temperature are presented in Figure 20.4 for two heating rates V_{h1}=1 K/s and V_{h2}=100 K/s. In both cases, curves $\overline{d}$ (T) feature three stages. At the first stage ($T<T_1$), no grain growth is observed. At the second stage ($T_1<T<T_2$), intensive grain growth takes place. At the third stage ($T>T_2$), the dependence of $\overline{d}$ on temperature becomes flatter. The temperature T_1 of a fast grain growth depends on the heating rate: T_1=573 K at V_{h1}=1 K/s and T_1=623 K for V_{h2}=100 K/s. Also, the grain growth for various heating rates differs substantially. For V_{h1} =1 K/s, the average grain growth rate at the second stage is $1.0\cdot10^{-6}$ cm/s, while for V_{h2} =100 K/s it constitutes $1.4\cdot10^{-4}$ cm/s.

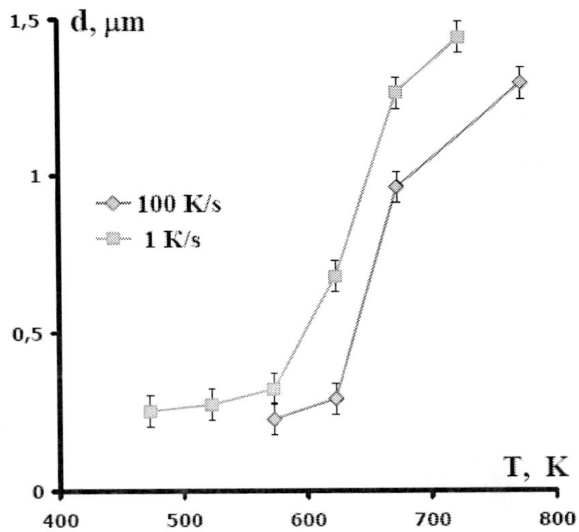

Figure 20.4. Average grain size of NMC nickel (N=8) vs. temperature for different heating rates.

Observations of the microstructures formed in NMC copper M1 (99.98%) for N=12 and N=16 after annealing at T=423÷673 K show that within the temperature range $T_1<T<T_2$, the pattern of recrystallization is quite unusual with some large grains of 1÷5 μm on the background of a stable NMC matrix of the average grain size 0.2÷0.5 μm. Dispersion in grain sizes is 5÷10 times. A relative area of large grains is 10% at $T\sim T_1$ and increases to 80% at $T\rightarrow T_2$. At $T>T_2$, collective recrystallization begins. The examples of abnormal grain growth in NMC Cu and Ni after one-hour annealing at T=412÷623 K are shown in Figure 20.5. When the annealing temperature approaches 473 K (Cu, N=16), a fraction of grain boundaries with characteristic striated contrast increases, while the average grain size in NMC copper remains unchanged. Annealing at T=448÷453 K for 1 h results in the occurrence of abnormally growing grains, whose volume fraction increases rapidly with raising the annealing temperature. After a one-hour anneal at 523 K the volume fraction of coarse grains is about 80%.

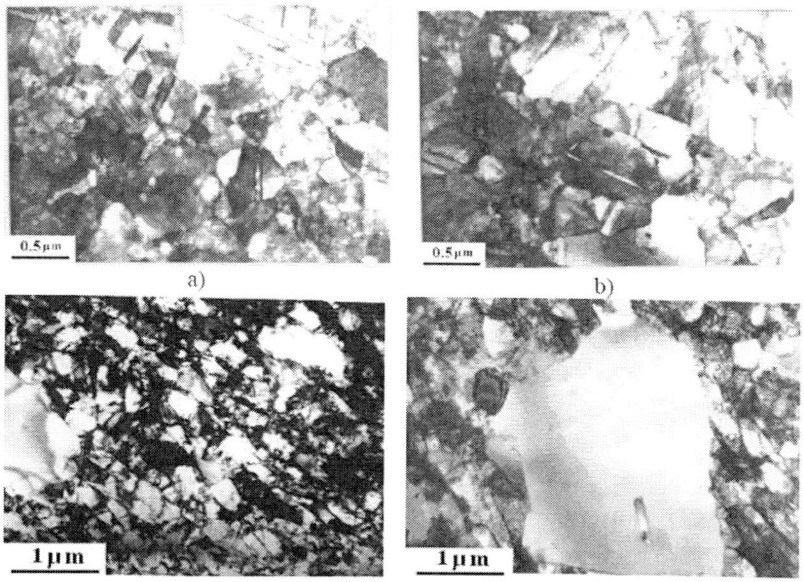

Figure 20.5. TEM-microstructures after one-hour annealing of NMC copper (N=16) at 412 K (a) and 523 K (b), and of NMC nickel (N=8) at 548 K (c) and 623 K (d).

20.2. Recrystallization of Pure NMC Metals

20.2.1. Basic Notions of Recrystallization

Before proceeding to recrystallization of NMC metals, we shall briefly review the basic notions of recrystallization in metallic materials (Gorelik et al. 2005; Ed. Haessner, 1978):

1 *Threshold deformation* is the minimum strain after which recrystallization becomes possible;
2 *Critical temperature of recrystallization,* T_1, is the temperature corresponding to the onset of recrystallization after certain annealing time (t_R), which is termed the latent period;
3 *Latent period of recrystallization,* t_R, is a characteristic time after which the recrystallization process is initiated in the material provided that the critical annealing temperature T_1 is maintained;
4 *Volume fraction* of the recrystallized material, f_v^R (t, T), is a relative volume of material which has undergone recrystallization at a given temperature and annealing time;

5 *Recrystallization nuclei* are the domains of a defect-free, or "good", material surrounded by high-angle grain boundaries whose growth leads to a gradual transition of the whole volume into a recrystallized condition.

It is necessary to outline that a general theory of recrystallization has not been developed so far. In this connection, when creating particular models, it seems necessary to address all the basic problems inherent in recrystallization. Then, a model for recrystallization of NMC materials is supposed to consider the nature and the value of prestrain ε_1, the critical temperature of recrystallization as a function of annealing time, the latent period versus annealing temperature, f_v^R as a function of strain and temperature and the mechanism of formation of the recrystallization nuclei. The following model to a certain extent addresses most of these challenges. It should be noted that the formation mechanism of recrystallization nuclei is still an open issue (Gorelik et al. 2005). Two essentially different standpoints are being discussed in literature. According to them, the nuclei are formed either during deformation or at heating after the deformation, e.g., during the latent period.

20.2.2. Description of the Approach

The present approach to the grain boundary migration during recrystallization of polycrystalline NMC materials is based on assumptions:

1. Nonequilibrium grain boundaries, i.e. high-angle boundaries with embedded defects are considered;

2. As earlier, three basic types of defects in grain boundaries comprise: orientational misfit dislocations with density ρ_b, tangential components of delocalized dislocationstangential components of Burgers vector of delocalized dislocations with Burgers vector density w_t, and normal components of delocalized dislocations with Burgers vector density w_n, which can be described as triple junction disclinations with power ω;

3. The migration rate, V_m, of grain boundaries is determined by general expression (Gorelik et al. 2005)

$$V_m = MP_m, \tag{20.1}$$

where M is the grain boundary mobility, P_m is a driving force for migration;

4. The mobility of NGB, M^*, considerably differs from that of equilibrium grain boundaries M_b. It is known that displacement of an equilibrium boundary does not require long-distance diffusion mass transfer, and the diffusional rearrangement of atoms takes place only in a layer whose thickness is close to the grain boundary width δ. In this case, the grain-boundary mobility is determined as

$$M_b = D_b\delta/kT. \tag{20.2}$$

Migration of a nonequilibrium grain boundary requires diffusion flows which are necessary to displace the defects distributed in the boundary. In this case, the grain boundary mobility is determined by the expression (Perevezentsev et al. 1985a; Perevezentsev et al. 1991):

$$M^{-1} = M_b^{-1} + M_\omega^{-1} + M_\rho^{-1} + M_w^{-1}, \tag{20.3}$$

where M_ω is the mobility of disclinations at grain junctions connected with additional misorientation of boundaries due to normal components of delocalized dislocations, M_ρ is the mobility of orientational misfit dislocations (OMD) and M_w is the mobility of tangential components of delocalized dislocationstangential components of Burgers vector of delocalized dislocations .These mobilities are (Chuvil'deev, 2004):

$$M_\omega = M_b \left(\frac{b}{d}\right)^2 \frac{1}{\omega^2}, \tag{20.4}$$

$$M_\rho = M_b \left(\frac{b}{d}\right) \frac{1}{\rho_b \Delta b}, \tag{20.5}$$

$$M_w = M_b \left(\frac{b}{d}\right) \frac{1}{w_t}, \tag{20.6}$$

where ω is the power of a disclination dipole and d is the average grain size;

5. The presence of distributed defects in grain boundaries results in an additional driving force necessary for migration, which is connected with the interaction of internal and external stress fields with the stress fields generated by the aforesaid defects. The force acting on a grain boundary in the field of external σ and internal σ_i stresses can be expressed as (Perevezentsev et al. 1991)

$$P = (\sigma + \sigma_i)(\varphi_{p1}\rho_b \Delta b + \varphi_{p2} w_t + \varphi_{p3}\omega)\cdot \tag{20.7}$$

In rough approximation, σ_i can be determined by formula:

$$\sigma_i = \varphi_1 G(\rho_b \Delta b) + \varphi_2 G w_t + \varphi_3 G\omega, \tag{20.8}$$

where $\varphi_{p1,p2,p3}$ and φ_1, φ_2, φ_3 are numerical factors of the order of unity;

6. For typical conditions of severe plastic deformation such as ECAE, recrystallization nuclei are formed during deformation. In this situation, the high-angle boundaries surrounding a nucleus have deformation origin, i.e. they are non-equilibrium because of a high content of embedded excess defects. Then, the above definition of recrystallization nuclei will read as "a domain of a "good" material surrounded by *nonequilibrium* high-angle

boundaries which contain defects absorbed during deformation". Hence the behavior of a nucleus will substantially depend on the defects formed in its boundaries at prestrain;

7. The fraction of the total volume occupied by recrystallization nuclei, $f_{fr}(\varepsilon)$, is proportional to the portion of high-angle grain boundaries in the material. This is because the formation and further evolution of the nuclei is connected with the presence and motion of high-angle boundaries. From this standpoint, the misorientations of grains seems to be the main structural parameter of a material that determines its recrystallization behavior. Unfortunately, a detailed (pass-by-pass) analysis of the evolution of grain misorientations during ECAE has not to been performed. Thus, we have to use a schematic diagram of the volume fraction occupied by nuclei versus strain (Rybin, 2002; Goloborodko et al. 2004; Prangnell et al. 2004; Sitdikov et al. 2007) (Figure 20.6). The diagram can be subdivided into three areas. At a small (pre-threshold) strain $\varepsilon<\varepsilon_1$, the material does not recrystallize during heating after deformation. In the area of moderate strains $\varepsilon_1<\varepsilon<\varepsilon_2$, separate recrystallization nuclei appear and their number increases with strain. At large plastic deformation $\varepsilon>\varepsilon_2$, a continuous lattice of high-angle boundaries is formed, and nearly all fragments (subgrains) of fragmented microstructure are actually individual grains. Then, at $\varepsilon\geq\varepsilon_2$ all fragments become the recrystallization nuclei.

The main objective of a recrystallization model is quantitative evaluation of function $f_{fr}(\varepsilon)$. At present, this is possible only on the basis of experimental data (Kopylov et al. 2002; Rybin, 2002; Sitdikov et al. 2007; Goloborodko et al. 2004; Prangnell et al. 2004) (Figure 20.6). According to (Kopylov et al. 2002), after 12 passes of ECAE, in particular in Cu, the misorientation distribution of grain boundaries differs insignificantly from the random distribution, i.e. f_{fr} is close to unity. After 8 passes of ECAE, f_{fr}=0.5 (Goloborodko et al. 2004; Prangnell et al. 2004).

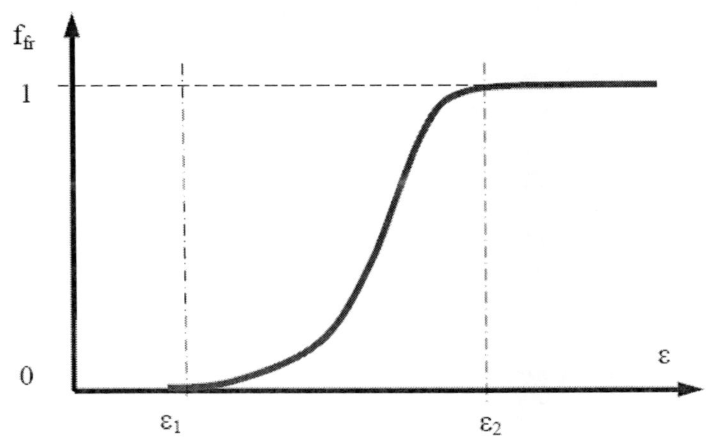

a)

Figure 20.6. Continued on next page.

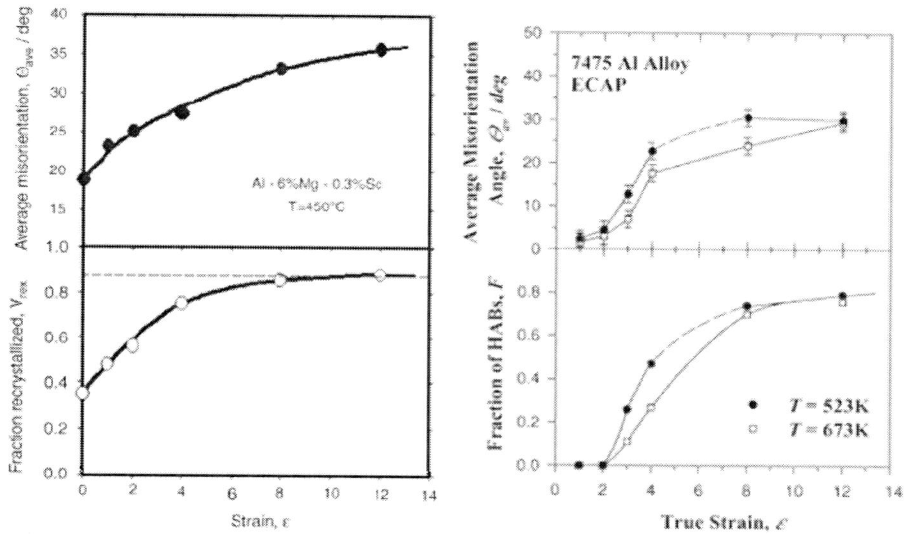

b)

Figure 20.6. Strain dependence of the volume fraction of a material captured by recrystallization nuclei: diagram (a) (Rybin, 2002) and experiment (b) (Sitdikov et al. 2007; Goloborodko et al. 2004).

20.2.3. Model

The mobility of high-angle boundaries with embedded defects, which surround the recrystallization nuclei, depends on density of defects. Generally, this mobility is considerably lower than that of equilibrium boundaries. Thus, it can be assumed that non-equilibrium grain boundaries cannot migrate fast until the concentration of defects decreases. This will correspond to the onset of recrystallization, i.e. migration of the nuclei boundaries. For the boundaries to become mobile, certain time should pass during which defects will depart from the boundary due to recovery processes. This time corresponds to the latent period of recrystallization (CTR), t_R. In our model, the latent period will largely depend on the annealing temperature. Reciprocally, for a given annealing time, there will be a certain temperature at which the defects can depart from grain boundaries to start recrystallization. This temperature corresponds to the critical temperature of recrystallization T_1.

We suppose that the boundary will migrate when the mobility of a least mobile defect becomes comparable with the mobility of a defect-free grain boundary. According to Eqs. (20.4)-(20.6), triple junction disclinations possess the least mobility. Then, the condition for the onset of migration is

$$M_{\omega} \approx M_{b} \cdot \quad (20.9)$$

As a first approximation with the account of Eqs. (20.4)-(20.6), this condition can be rewritten as

$$\omega \approx \omega_{R} \approx b/d \cdot \quad (20.10)$$

When the power of triple junction disclinations is high ($\omega >> \omega_R$), the boundary is immobile while at $\omega \le \omega_R$ fast migration of boundaries is possible. Reduction of the power of a triple junction disclination during annealing can be described by formula (see Eq. 15.153 and 19.2)

$$\omega(t) = \omega_0 \exp(-t/t_3), \qquad (20.11)$$

where ω_0 is the initial power (before heating). Equating $\omega(t)$ to ω_R, one determines the time t_R at which the boundary start to migrate:

$$t_R = t_3 \ln\left(\frac{\omega_0}{\omega_R}\right) = \left[\frac{G\Omega}{kT}\frac{\delta D^*_{b0}\exp(-Q^*_b/kT)}{d_R^3}\frac{1}{A_1}\right]^{-1} \ln\left(\frac{\omega_0}{\omega_R}\right). \qquad (20.12)$$

This time can be interpreted as the latent period of recrystallization. From Eq. (20.12), the time t_R exponentially depends on the annealing temperature. Parameter d_R is the characteristic distance of diffusion mass transfer at diffusion accommodation of triple junction disclinations.

The critical temperature of recrystallization (CTR) T_1 can be determined from Eq. (20.12) by setting the annealing time t_a as equal to t_R:

$$T_1 = Q^*_b \Big/ k \ln\left(\frac{D^*_{b0}\delta}{d_R^3}\frac{G\Omega}{kT}\frac{t_a}{A_1}\ln\frac{\omega_0}{\omega_R}\right). \qquad (20.13)$$

20.3. Comparison with Experimental Data

20.3.1. Prestrain Effect

The defects embedded in the nuclei boundaries during deformation demonstrate dual influence on CTR. On the one hand, embedded defects increase the free volume of grain boundaries ($\alpha+\Delta\alpha$) and reduce the diffusion activation energy Q^*_b Q^*_b in Eq. (20.13). Thus, the value of CTR should decrease with increasing the density of embedded defects. The calculations from Eq. (15.148) show that an increase of $\Delta\alpha$ by an order of magnitude from 0.01 to 0.1 that corresponds to an increase in the OMD density ρ_b from ~ 10^5 to 10^6 cm^{-1} reduces Q^*_b Q^*_b from $9.1kT_m$ to $5.2kT_m$ and CTR from $0.45T_m$ to $0.23T_m$ at $d_R = 0.1$ μm. This is in a good agreement with experimental data for Ni (see Table 20.1).

On the other hand, embedded defects affect the initial power of a disclination dipole ω_0: the higher the defect density, the larger ω_0. According to Eqs. (20.12) and (20.13), an increase of the initial power of a dipole ω_0 increases the latent period t_R and reduces CTR. However, due to double logarithmical dependence of CTR on ω_0 (see Eq. (20.13), the effect of ω_0 on T_1 is weak.

As seen from Eqs. (20.12) and (20.13), an important parameter influencing CTR is the distance of diffusion mass transfer d_R. When d_R changes from 0.1 to 1 μm at $\Delta\alpha=0.1$ and $Q^*_b \approx 0.6Q_b$, CTR increases from $0.18T_m$ to $0.31T_m$ for copper and from $0.19T_m$ to $0.33T_m$

for iron. That conforms to experimental data on recrystallization in Cu and α-Fe (Akhmadeev et al. 1992; Islamgaliev et al. 1998; Valiev et al. 2000; Pyshmintsev et al. 2001; Chuvil'deev et al. 2003a; Mughrabi et al. 2003; Noskova et al. 2003).

Let us examine the changes in d_R during deformation in more detail. As noted above, at a moderate strain $\varepsilon_1<\varepsilon<\varepsilon_2$ the recrystallization nuclei are located far from each other. With increasing strain, the distance between nucleuses is reduced and their number grows. However, they do not contact each other, and the characteristic length scale at which the dislocation charge distributed at nucleus boundaries is "nulled" can be assumed as equal to the nucleus size d_R. As is seen in Figure 20.7, the dislocation charge and, respectively, the power of a disclination dipole is effectively reduced only via mass transfer over characteristic distance d_R equal to the fragment size d^*.

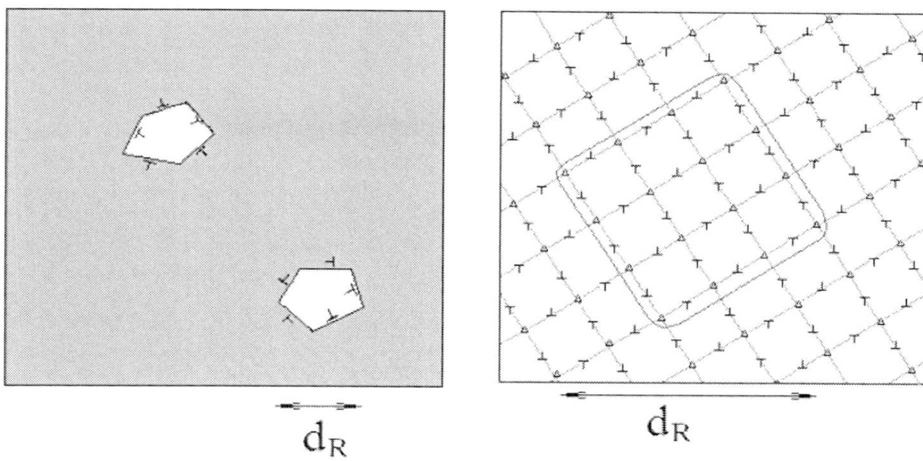

$\varepsilon_1<\varepsilon<\varepsilon_2$ $\varepsilon\geq\varepsilon_2$

Figure 20.7. Change of d_R during NMC structure formation by ECAE.

At large deformations ($\varepsilon>\varepsilon_2$) the situation changes considerably: a continuous lattice of high-angle boundaries is formed in the material and each fragment of the structure becoming a recrystallization nucleus. Then triple junction disclinations are formed which cannot be presented as an ideal system of disclination dipoles with arm d^*. In this system, the dislocation charge is nulled at a length scale d_R which exceeds the fragment size d^*. It is estimated that $d_R\approx(2\div3)d^*$. Such a change in d_R at $\varepsilon\geq\varepsilon_2$ can lead to an increase of CTR. For typical parameters values listed in tables 19.1 and 20.5, a change d_R from d^* to $3d^*$ increases the value of CTR for copper by about $0.05T_m$, i.e. by 60÷70 K (see Eq. (20.13)).

Therefore, the considered model can describe three areas recrystallization shown in Figure 20.1.

20.3.2. Effect of Heating Rate

For a constant heating rate V_h, the heating time is $t_h=T_1/V_h$. Assume that heating time t_h to the temperature T_1 is equal to annealing time t_a to some temperature $T_\eta = T_1/\eta_1$, where $0<\eta_1<1$. Then, the effect of heating rate V_hc can be estimated from Eq. (20.13) using $t_a=\eta_1 T_\eta/V_h$ and accepting that for high heating rate and short annealing time Q_b^*~const. Within this approach, CTR depends logarithmically on the heating rate. For parameters of Table 20.5, that gives that a change of the heating rate by two orders of magnitude increases CTR by $0.15T_m$. This result is in a good agreement with experimental data for Cu, Ni, Fe and Al (see Table 20.3).

20.3.3. The Critical Temperature of Recrystallization

For numerical calculations of CTR, it is necessary to determine the activation energy for grain boundary diffusion in NGB and the characteristic length scale of diffusion mass transfer d_R versus the preliminary deformation. Other parameters for NMC metals are listed in Table 20.5. As noted above, Q_b^* depends on density of defects embedded in boundaries and changes during early stage of annealing due to recovery processes. In each particular case, this value should be calculated numerically. For a qualitative assessment, the limit value Q_b^*~Q_b can be used for equilibrium boundaries and $Q_b^* \approx 0.6Q_b$ can be used for boundaries with excess density of defects. Correspondingly, two temperatures T_1^{max} and T_1^{min} are found from Eq. (20.13). At strains $\varepsilon<\varepsilon_2$, the grain size d_{R1} is equal to the average size of nuclei d^*. For strains $\varepsilon\geq\varepsilon_2$, d_{R2} is several times larger: $d_R\approx n_R d^*$, where n_R depends on the strain homogeneity and other factors, and cannot be calculated rigorously. For further estimates we assume $n_R=3$. Strain ε_2 , at which forming of new high-angle all grain boundaries completes become high-angle, should be determined experimentally. For ECAE of copper at room temperature via Route D, ε_2 corresponds to 12 passes.

Table 20.3 presents calculated values of CTR for a few materials and processing conditions of ECAE. Their comparison with identical experimental results shows a good agreement and confirms the adecuacy of the theoretical model.

20.4. Abnormal Grain Growth

With abnormal grain growth, the structure comprizes substantially different sizes of fine grains $d_f(t)$ and coarse grains $d_c(t)$ (Chuvil'deev et al. 2003c; Neishi et al. 2003; Eddahbi et al. 2005; Liu et al. 2004). Using the simple rule of mixtures, the average grain size $\bar{d}(t)$ is

$$\bar{d}(t) = f_f(t)d_f(t) + (1-f_f(t))d_c(t), \tag{20.15}$$

Table 20.3. Experimental and theoretical values of the critical recrystallization temperature for NMC metals

Metal	Condition		Experiment			Calculations		
	deformation	heat treatment	T_1^{exp}	T_1^{exp}/T_m	d^*, μm	d_R, μm	T_1^{min}/T_m	T_1^{max}/T_m
Ni (NP-1)	ECAE (293 K, N=8)	annealing 1 h	523 K	0.303	0.30	0.30	0.261	0.326
		annealing 5 h	473 K	0.274			0.246	0.308
		heating 1 K/s	573 K	0.332			0.286	0.358
		heating 100 K/s	623 K	0.361			0.348	0.435
Cu (M1)	ECAE (293 K, N=4)	annealing 1 h	383 K	0.282	0.20	0.20	0.232	0.324
Cu (M1)	ECAE (293 K, N=8)	annealing 1 h	393 K	0.290	0.20	0.20	0.232	0.324
		annealing 5 h	373 K	0.275			0.217	0.307
Cu (M1)	ECAE (293 K, N=12)	annealing 10 min	473 K	0.349	0.25	0.75	0.261	0.353
		annealing 0.5 h	473 K	0.349			0.247	0.34
		annealing 1 h	453 K	0.334			0.239	0.332
Cu (M1)	ECAE (293 K, N=16)	annealing 1 h	423 K	0.312	0.16	0.48	0.225	0.317
		annealing 4 h	398 K	0.294			0.213	0.303
α-Fe (Islamgaliev et al. 1998)	ECAE (773 K)	annealing 1 h	773 K	0.428	1.0	1.0	0.421	0.566
	ECAE+HPT (1.5 GPa)		723 K	0.399	0.40	1.2	0.385	0.518
Cu (M1) (Mughrabiet al. 2003)	ECAE (293 K, N=12)	heating 0.1 K/min + annealing 2 h	443 K	0.327	0.30	0.90	0.301	0.408
Al (Mughrabi et al. 2003)	ECAE (293 K, N=12)	heating 0.5 K/min + annealing 2 h	493 K	0.528	0.8÷0.9	2.7	0.419	0.568
α-Fe (Mughrabi et al. 2003)	ECAE (623 K, N=8)	heating 0.5 K/min + annealing 2 h	623 K	0.345	0.75	0.75	0.355	0.481
Cu (M1) (Mughrabi et al. 2003)	ECAE (293 K, N=16)	annealing 1 h	423 K	0.312	0.16	0.48	0.225	0.317
Cu (M1) (Pyshmintsev et al. 2001)	ECAE (293 K)	annealing 3 min	463 K	0.341	0.30	0.90	0.364	0.529
		annealing 0.5 h					0.326	0.474
Ni (Kolobov et al. 2001b)	ECAE (293 K)	annealing 1 h	473 K	0.274	0.26	0.78	0.307	0.417

where $f_f(t)$ is the volume fraction of fine grains. Eq. (20.15) implies that parameters $d_f(t)$, $d_c(t)$ and $f_f(t)$ change with time t. The critical volume fraction f_f^* at which coarse grains dominate in structure is:

$$f_f^* = d_c(t)/(d_f(t)+d_c(t)). \tag{20.16}$$

In NMC materials, the size of coarse grains is significantly larger than that of fine grains $d_c(t)>>d_f(t)$. In this case, the second term of Eq. (20.16) determines $\bar{d}(t)$ if $f_f<0.8$, and the average grain size is:

$$\bar{d}(t) = (1-f_f(t))d_c(t). \tag{20.17}$$

To describe the time dependence of the average grain size $\bar{d}(t) = (1-f_f(t))d_c(t)$ $\bar{d}(t)$, the volume fraction of fine grains $f_f(t)$ and maximum grain size $d_c(t)$ should be determined.

Time Dependence of Volume Fraction $f_f(t)$

In describing the process of collective recrystallization, it is assumed that migration of the grain boundaries starts as soon as the sample is heated up to a prescribed temperature. Thus, all the grain boundaries should have the same mobility. In NMC materials, the situation is substantially different. As shown above, boundary migration in NMC metals at a given temperature cannot start immediately because accumulated defects during preliminary deformation inhibit their motion. Migration becomes possible only after significant reduction of the density of boundary defects due to the development of recovery processes, i.e. after a certain time t_R. The main defects inhibiting grain boundary mobility are triple junction disclinations. The kinetics of power reduction of these disclinations during isothermal annealing is described by Eq. (20.11). Boundaries can migrate when their power decreases from initial ω_0 to ω_R=b/d,. In this approximation, $t_R \sim t_3$.

In a metal deformed to strain ε_2, each grain becomes a recrystallization nucleus. However, the process of defect relaxation departure in boundaries of different grains proceeds at dissimilar rates and accomplishes in different periods of time. The value of t_3 for different boundaries depends on the local values of ω_0, D_b^*, d_f and ω_R. At present, there are insufficient data on the distribution of these parameters over an ensemble of grain boundaries. Thus, it seems impossible to determine rigorously the distribution of function t_3 and a time dependence of the number of boundaries with $\omega<\omega_R$ that are able to migrate. Hence, we shall limit ourselves to simple estimations.

Let's suppose that function $\omega(t)$ is identical for all the boundaries, i.e. t_3=const, and that the initial power ω_0 is evenly distributed (from 0 to ω_{max}) over the grain-boundary ensemble. Then, due to exponential function $\omega(t)=\omega_0\exp(-t/t_3)$ (see Eq. (20.11)), the number of boundaries able to migrate will also exponentially change with time. Assuming further that the volume fraction of growing fine grains is proportional to the number of boundaries that can migrate, function f(t) can be presented as

$$f_f(t) = f_0 \exp(-t/t_3), \qquad (20.18)$$

where f_0 can is found from the normalization condition $\int f_f(t)dt=1$. Thus, the volume fraction of fine grains $f_f(t)$ in an ensemble of NMC grains exponentially decreases with time whereas the volume fraction of coarse grains $(1-f_f(t))$ exponentially grows; the characteristic time of that process t_3 can be found using Eq. (15.156).

Grain Growth Rate in NMC Metals

Within the developed approach, there are at least three stages of grain growth. The first stage probably corresponds to a large volume fraction of fine grains, $f_f(t)>0.8$. In this case, the average grain size and its growth rate are evaluated as

$$\bar{d}(t) \approx f_f(t)d_f(t);, \ \dot{\bar{d}}(t) \approx \dot{f}_f(t)d_f(t) + \dot{d}_f(t)f_f(t). \qquad (20.19)$$

The function $\bar{d}(t)$ displays a complex behavior because $f_f(t)$ exponentially decreases with time while $d_f(t)$ is an increasing function. As reliable experimental data on the behavior of NMC metals during the first stage are not available, it is impossible to compare the obtained expression with experiment.

The second stage commences when the volume fraction of coarse grains reaches ~0.2. Then, the growth kinetics of an average grain size is mainly determined by the change of volume fraction of fine grains:

$$\bar{d}(t) = (1 - f_f(t))d_c(t). \qquad (20.20)$$

Combining Eqs. (20.18) and (20.20) gives

$$\bar{d}(t) = d_c(t)(1 - \exp(-t/t_3)). \qquad (20.21)$$

In the first approximation at $d_c(t)=d_c=\text{const}$, the grain growth rate $\dot{\bar{d}}(t) = \dot{f}_f(t)d_c$ can be calculated by the following formula:

$$\dot{\bar{d}} = (d_c/t_3)\exp(-t/t_3). \qquad (20.22)$$

Inserting Eq. (15.156) in Eq. (20.22) yields

$$\dot{\bar{d}} \approx d_c\left[\frac{D_b^*}{A_1}\frac{G\Omega}{kT}\frac{\delta}{(d^*)^3}\right]\exp(-t/t_3). \qquad (20.23)$$

The third stage of grain growth takes place when boundaries of coarse grains become free from defects accumulated during migration owing to the completion of recovery processes. The further grain growth follows a collective recrystallization law.

Growth Rate of Coarse Grains

Coarse grains with sizes $d_c >> d_f$ are formed in NMC metals as a result of fast growth of some fine grains whose boundaries have undergone recovery and become defect-free. Then a question arises: why does this growth slow down and finally terminate? This issue should be addressed in the theoretical model.

A migrating grain boundary absorbs dislocations distributed in the matrix. The density of these dislocations even after partial recovery during time t_3 is high $\rho_v \sim 10^9 \div 10^{10}$ cm^{-2}. This absorption results in (i) changes in the free volume of grain boundaries, (ii) increase of their diffusion coefficient, and (iii) new accumulation of orientational misfit dislocations and products of their delocalization. These defects can diffusionally depart from boundaries with the characteristic withdrawal time as a cubic function of the initial grain size (see Eq. (15.156) for t_3). At the onset of the grain growth, the diffusion accommodation is fast due to small grain sizes. Hence, the density of defects in grain boundaries is low and defects do not inhibit migration. However, as $\bar{d}(t)$ increases, the rate of defect withdrawal from the boundaries declines. Then, the defect density in boundaries increases and their mobility reduces substantially. After a certain size d_c, the grain ceases to grow, i.e. $d_c(t)$=const. Further growth of coarse grains becomes possible only after the boundaries become devoid of defects once again due to diffusion withdrawal in accordance with the above mechanism.

Now let us evaluate d_c. As noted above, the grain boundary mobility is mainly influenced by triple junction disclinations. For a boundary migration at a rate V_m through the matrix with the lattice dislocation density ρ_v, the accumulation kinetics of the triple-junction-disclination power is:

$$\dot{\omega} = \xi_m \rho_v^+ V_m b - \omega / t_3. \tag{20.24}$$

Here ξ_m is a geometrical factor, ρ_v^+ is the excess density of like-sign lattice dislocations (because embedding of such dislocations into a grain boundary increases the power ω_m of a triple junction disclination). Typically, ρ_v^+ is a small fraction of the total dislocation density ρ_v: $\rho_v^+ \sim 10^{-2}\rho_v$.

We suppose that the boundary migration kinetics can be described by a general expression (Gorelik et al. 2005) $V_m = M_b P_\rho$, where $M_b = D_b\delta(\Omega/kT)/b^2$ is the boundary mobility, $P_\rho = G\rho_v b^2$ is the driving force for migration; P_ρ is connected with the reduction of the system free energy due to absorption of lattice dislocations by boundaries.

Combining Eq. (20.11) with Eq. (15.156) gives a relation between the stationary power of triple junction disclinations generated during migration and the grain or nucleus size:

$$\omega_m^{st} = \xi_m A_1 (\rho_v b^2)(\rho_v^+ b^2)\left(\frac{d}{b}\right)^3 . \quad (20.25)$$

The critical power at which triple junction disclinations begin considerably influencing the boundary mobility is evaluated from condition $\omega_m^{st} >> \omega_R = b/d$. Assuming $\omega_m^{st} = A_\omega \omega_R$ where A_ω=10, from Eq. (20.25) we obtain the grain size at which the mobility of grain boundaries declines and the grain growth is arrested:

$$\frac{d_c}{b} = \left[A_\omega A_1 \xi_m (\rho_v^+ b^2)(\rho_v b^2)/A\omega\right]^{-1/4} . \quad (20.26)$$

For the typical values A_ω=10, A_1=10, $\xi_m=10^{-4}$, $\rho_v^+ \sim 10^{-2}\rho_v$ and $\rho_v=10^{14}$ m^{-2}, Eq. (20.26), gives $d_c \sim 1$ μm that is in a good agreement with experimental observations of abnormal grain growth (Chuvil'deev et al. 2003a; Chuvil'deev et al. 2003b).

20.5. COMPARISON WITH EXPERIMENTAL DATA

The experimental values of the characteristic time t_3 at different annealing temperatures for Cu after N=12 and N=16 passes of ECAE are presented in Table 20.4.

Table 20.4. Experimental and theoretical values of characteristic time t_3 at different annealing temperatures for NMC copper 99.98%

	experiment	theoretical calculations	
T_{ann}, K	t_3^{exp}, s	$t_3^{th(max)}$, s	$t_3^{th(min)}$, s
NMC Cu (N=12)			
433	$1.0\cdot10^6$	$4.6\cdot10^7$	$6.9\cdot10^1$
453	$2.5\cdot10^5$	$1.1\cdot10^7$	$3.3\cdot10^1$
473	$1.0\cdot10^5$	$3.2\cdot10^6$	$1.7\cdot10^1$
493	$3.3\cdot10^4$	$1.0\cdot10^6$	$9.5\cdot10^0$
523	$1.0\cdot10^4$	$2.2\cdot10^5$	$4.2\cdot10^0$
NMC Cu (N=16)			
243	$2.5\cdot10^5$	$2.2\cdot10^7$	$2.6\cdot10^1$
448	$1.7\cdot10^4$	$3.9\cdot10^6$	$1.1\cdot10^1$
473	$1.6\cdot10^4$	$8.2\cdot10^5$	$4.5\cdot10^0$
523	$3.3\cdot10^3$	$5.8\cdot10^4$	$1.1\cdot10^0$

The theoretical estimation of t_3 was performed for two limiting cases when the activation energy for grain boundary diffusion (a) corresponds to the equilibrium value ($Q_b^* = Q_b$) and (b) maximally deviates from the equilibrium value (Q_b^* Q_b^*=0.6Q_b). As in intermediate cases, $0.6Q_b < Q_b^* < Q_b$, the experimental value t_3^{exp} should lie between the maximum value $t_3^{th(max)}$ corresponding to the first case and the minimum value $t_3^{th(min)}$ corresponding to the second case. These theoretical values calculated from Eq. (15.156) for parameters of Table 20.5 are also presented in Table 20.4. Comparison of theoretical and experimental results with experimental reveals that for all annealing temperatures t_3^{exp} meets the condition $t_3^{th(min)} < t_3^{exp} < t_3^{th(max)}$.

The dependence of the average grain size on annealing time and temperature during abnormal grain growth in NMC materials is described by Eqs. (20.18)-(20.23). Numerical results of their solution for NMC copper after N=12 passes ECAE are shown in Figure 20.8. For parameters listed in Table 20.5, there are very good qualitative agreement between experimental and modeling data for the second stage abnormal grain growth at temperatures $T>T_1$.

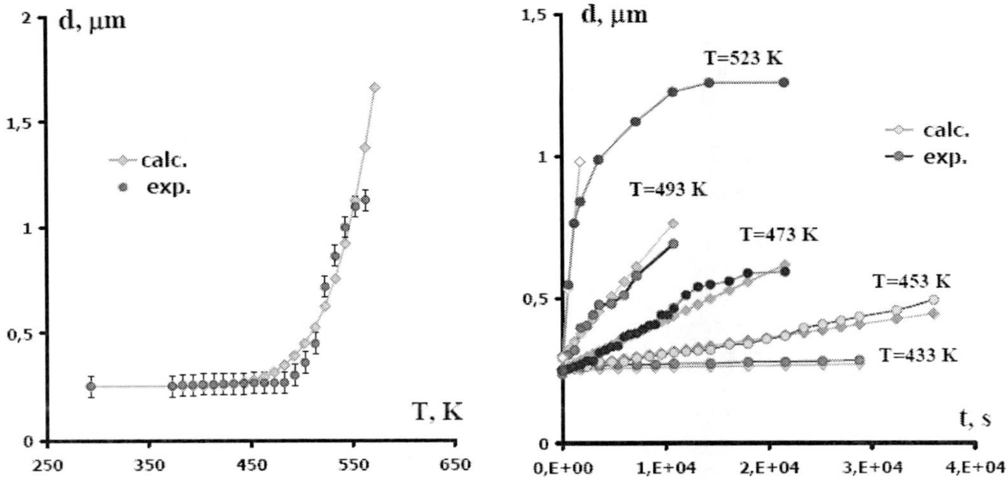

a) b)

Figure 20.8. Numerical simulation of abnormal grain growth in NMC copper M1 99.98% (N=12): (a) experimental and calculated dependence of the average grain size on temperature at a 10-minute anneal, and (b) experimental and calculated dependence of the average grain size on the isothermal annealing time at different temperatures.

Table 20.5. Parameter values used in calculations

Parameter	Notation	Value				
		Ni	Cu	Fe	Mo	Al
Activation energy for self-diffusion in melt *	Q_L/kT_m	3.3	3.6	3.65	5.6	3.4
Activation energy for diffusion in equilibrium grain boundaries **	Q_b/kT_m	9	9.2	11.6	15.7	10.8
Preexponential factor for grain boundary diffusion coefficient ($10^{15}\cdot m^3/s$) **	δD_{b0}	3.5	5.0	1.1	$55\cdot10^{-3}$	75
Preexponential factor for diffusion coefficient in melt (cm^2/s)*	D_{L0}	$8\cdot10^{-4}$				
Burgers vector (10^{10}m)**	b	2.49	2.56	2.48	2.73	2.86
Boundary width	δ	2b				
Atomic volume ($10^{29}m^3$)**	Ω	1.09	1.18	1.18	1.53	1.21
Melting temperature (K)**	T_m	1726	1356	1808	2895	934
Shear module (GPa)**	G	79	42	78	120	25
Initial power of triple junction disclinations*	ω_0	$2\cdot10^{-2}$				
Critical power of triple junction disclinations*	ω^*	10^{-3}				
Relative free volume *	α	0.35				
Critical free volume*	α^*	0.5				
Excess free volume of the boundary*	$\Delta\alpha$	0.02; 0.1				
Specific heat of fusion* Density	λ ρ	$\lambda\rho b^3=1.5\ kT_m$				
Enthalpy of solid/liquid interface*	$\gamma^0_{S/L}$	$\gamma^0_{S/L}b^2=1\ kT_m$				
Entropy of solid/liquid interface*	$S_{S/L}$	$S_{S/L}b^2=0.8k$				
Free energy of grain-boundary S-phase*	γ_0	$\gamma_0 b^2=1.4\ kT_m$				
Numerical parameters	A_0	10.0				
	C_1	50.0				

* Chuvil'deev, 2004

** Frost et al

20.6. Summary

The basic features of recrystallization in NMC materials are reviewed. A model for estimating the critical temperature of recrystallization in pure NMC metals is developed. The model is based on the assumption that the boundaries of a recrystallization nucleus can start moving only when the density of excess defects embedded in the boundaries (orientational misfit dislocations and the products of their delocalization) decreases to a certain threshold value. This reduction of the defect density is controlled by diffusion of the nucleus in nonequilibrium boundaries, and at preset heating times is attained at certain temperatures.

Also, a model is described for abnormal grain growth in NMC metals at the initial stage of recrystallization. It is shown that the onset of recrystallization in NMC materials is controlled by the processes of defect recovery in NGB. During recovery, the number of migrating boundaries increases exponentially with time and, respectively, the average grain size commences depending on time exponentially.

Chapter 21

RECRYSTALLIZATION AND GRAIN GROWTH IN NMC METALS WITH SMALL AMOUNTS OF ADDITIVES

21.2. PROBLEM DESCRIPTION

It is known that even a small amount of alloying elements can substantially influence the recrystallization temperature of a metal (Gorelik et al. 2005; Ed. Haessner, 1978, Gleiter, 1972). In order to reveal the specific features of recrystallization in NMC materials, it is necessary to compare the effect of small amounts of additives on the recrystallization temperature in cast (i.e. coarse-grained) and NMC materials. However, despite a vast body of experimental data, there are only few works studying pure and dilute NMC materials or comparing the recrystallization behavior of cast and NMC alloys of the same chemical composition. Here, we will concentrate on pure Al and dilute Al-Sc and Al-Zr alloys (Lee ert al, 2002; Furukawa et al. 2002; Chuvil'deev et al. 2009). The analysis of experimental results obtained for different alloys, e.g., Cu-0.5%Sn (Ferrase et al. 2008), Cu-0.5%Al_2O_3 (Ivanov, 2001; Amirkhanov et al. 2001) and Fe-0.009%C (Degtyarev, 2005), have shown that the effect of alloying elements on the recrystallization temperature T_r in these systems is similar to the influence of scandium and zirconium on T_r, of aluminum. The analysis of published experimental data has revealed that the effect of alloying on T_r is very similar for cast and NMC materials of the same composition: the recrystallization temperature increases with the addition of alloying elements to a pure metal. However, this increase for NMC materials is smaller than for cast ones. For example, in a cast Al-0.2%Sc alloy an increase of T_r in comparison with pure Al was $0.23T_m$ (Elagin et al. 1991) whereas in an Al-3%Mg-0.2%Sc NMC alloy the similar increase was only $0.16T_m$ (Lee et al. 2002). Also, 0.2% Zr in an Al-3%Mg-0.2%Zr NMC alloy (Lee et al. 2003) did not change significantly the recrystallization temperature, while in cast Al-0.2%Zr aluminum alloy (Elagin, 1975) the value of T_r was increased by $0.1T_m$.

Thus, the objective of this Chapter is to develop a theoretical model for estimating the recrystallization temperature in a material containing small amounts of alloying elements and to reveal the peculiarities of recrystallization in NMC alloys.

21.2. RECRYSTALLIZATION TEMPERATURE

21.2.1. Model

As prevoiusly, we will suppose that the recrystallization nuclei are generated during ECAE. These nuclei are domains of a defect-free material surrounded by high-angle boundaries with embedded dislocations. It is implied that the beginning of recrystallization corresponds to the onset of intensive migration of their boundaries. In pure metals, the migration begins after the completion of boundary recovery and diffusion withdrawal of embedded defects which inhibit the boundary displacement. That way, the critical temperature of recrystallization in a pure metal is described in terms of initiation of a migration process of the nucleus boundaries. The same approach can be extended to dilute alloys.

An expression for the migration rate of grain boundaries (including boundaries of a recrystallization nucleus) in materials with alloying elements is (Gorelik et al. 2005):

$$\dot{d} = M(C,\omega) \cdot P_m(C,\rho_v) \quad (21.1)$$

Here, $M(C, \omega)$ is the mobility of boundaries which is proportional to the grain boundary diffusion coefficient $D_b^*(C)$ (see Eq. (20.2)), $P_m(C, \rho_v)$ is an effective driving force for migration, C is the volume concentration of solutes, ρ_v is the density of lattice dislocations and ω is the power of triple junction disclinations that influence the grain boundary mobility. Within this approach, recrystallization in alloys containing small amounts of solutes commences when the two independent conditions are satisfied.

First, the mobility of the boundaries of a recrystallization nucleus should be high. According to (Chuvil'deev, 2004), the mobility $M_b(C)$ of a boundary with a zero density of embedded dislocations can be assumed as a criterion of high mobility

$$M(C, \omega) \approx M_b(C). \quad (21.2)$$

Second, the effective driving force P_m for the boundary migration should be positive $P_m \geq 0$. It can be expressed as (Martin, 1976)

$$P_m = P_\rho - P_z \quad (21.3)$$

Here P_ρ is a driving force component proportional to the volume density ρ_v of lattice dislocations (Gorelik, 2005):

$$P_\rho = \beta_1 G b^2 \rho_v, \quad (21.4)$$

where β_1 is a numeric factor. Parameter P_z in Eq. (21.3) is the pinning or Zener force of dispersed particles (Martin, 1980):

$$P_z = f_v(C)\frac{\gamma_b(C)}{R_v(C)}, \tag{21.5}$$

where $f_v(C)$ is the volume fraction of second-phase particles, $R_v(C)$ is their characteristic size and $\gamma_b(C)$ is the grain-boundary energy.

The combination of Eqs. (21.1)-(21.5) gives the following expression for the grain boundary migration rate:

$$\dot{d} = M(C)\left(\beta_1 G b^2 \rho_v - f_v(C)\frac{\gamma_b(C)}{R_v(C)}\right). \tag{21.6}$$

21.2.2. Influence of Solute Atoms on Grain Boundary Mobility

If a grain boundary is free from solute atoms, heating to the critical recrystallization temperature T_1 withdraws excess defects from boundaries and they start to migrate. According to (Chuvil'deev, 2004), tThe solute atoms at a grain boundaries change their free volume and the activation energy for diffusion (Chuvil'deev, 2004):

$$Q_b(C) = W_1\left(\frac{\alpha^*}{\alpha + \Delta\alpha(C)} - 1\right) + W_2\left(\sqrt{\frac{\alpha^*}{\alpha + \Delta\alpha(C)}} - 1\right) + Q_L, \tag{21.7}$$

where W_1 and W_2 are the constants which depend on thermodynamic parameters of a material (see Eq. (15.48)). At a small concentration of solute atoms, a change in the diffusion coefficient can be described by the general formula

$$D_b(C) = D_b \exp(\Delta\alpha(C)/\alpha_B), \tag{21.8}$$

where $\Delta\alpha$ is a change in the free volume of a grain boundary at solute concentration C:

$$\Delta\alpha = (\Delta V/\Delta V_m)(\alpha C/K_c). \tag{21.9}$$

Here, ΔV is the difference in atomic volumes between the solute and matrix atoms, ΔV_m is the volume jump at melting, α is the boundary free volume and K_c is the solute distribution ratiocoefficient. Thus, in the presence of solute atoms, the grain boundary mobility depends on a number of parameters. In this case, the critical temperature of boundary migration in an alloy $T_1(C)$ differs from that in a pure metal, T_1. The value of $T_1(C)$ can be estimated by a formula similar to Eq. (20.13) wherein the parameters for a pure metal should be replaced by of those for an alloy and Q_b^* should be substituted by $Q_b^*(C)$. In more complex cases, when solute atoms change D_b and influence the kinetics of dislocation withdrawal from the grain boundaries, an expression for the critical temperature of migration becomes more sophisticated (Chuvil'deev, 2004).

It should be emphasized that for alloys $T_1(C)$ is not the recrystallization temperature. As will be shown below, boundaries in alloys travel over a short distance close to d_0 and then stop. For the onset of recrystallization in these cases, the withdrawal of defects from boundaries is a necessary but not a sufficient condition because for migration boundaries have to overcome obstacles of precipitated particles.

21.2.3. Influence of Alloying Elements on Migration Driving Force

Consider the effect of alloying elements on the driving force of migration for boundaries of a recrystallization nucleus. We will suppose that alloying atoms are in a solid solution. This corresponds either to supersaturated solid solutions after quenching or to dopant concentrations below the solubility limit. In these cases, there are not second-phase particles ($f_v(C)=0$) and Zener forces (see Eq. (21.5)) that can inhibit the grain boundary migration, and Zener force is equal to zero (see Eq. (21.5)). Upon heating to temperature $T_1(C)$, such boundaries start to migrate and encounter solute atoms on its way. If the distribution coefficient of a solute K_c is less than unity ($K_c<1$), foreign atoms are attracted by the grain boundary. After some time, sweeping out solute atoms produce local supersaturated areas with subsequent precipitation of dispersed particles at grain boundaries (Figure 21.1a). These particles can pin boundaries in accordance with Eqs. (21.5), (21.6). In another case, when the distribution coefficient is above unity $K_c>1$, the moving boundaries repel solute atoms and develop supersaturated areas with precipitated particles in front of boundaries (Figure 21.2b). These particles also can pin grain boundaries and inhibit their migration.

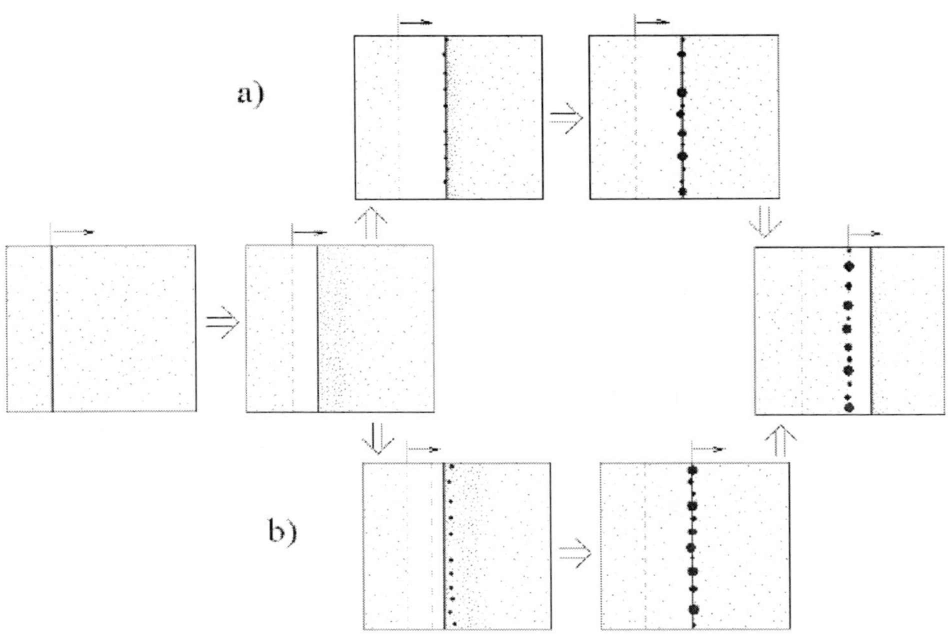

Figure 21.1. Influence of dispersed particles on boundary migration: (a) attracted dopants ($K_c<1$) and (b) repelled dopants ($K_c>1$).

For each concentration of alloying elements, depending on the particle fraction and size necessary for pinning, the migration distance can be estimated from geometrical consideration. However, if the initial concentration is close to the solubility limit, the distance traveled by a boundary to its stop in both cases is small ($x<d_0$) and difficult to be measured experimentally. Such boundaries are pinned by precipitates very soon. Thus, the critical temperature of recrystallization is connected with the depinning condition. In accordance with Eq. (21.6), this condition can be presented as:

$$\beta_1 G\rho_v b^2 \geq f_v(C)\gamma_b(C)/R_v(C). \tag{21.10}$$

For simplicity, we assume that formation of precipitations has been accomplished and function $f_v(C)$ has reached its peak value $f_v(C)=f_v^* = C(at.\%)/n_c$, where n_c is the fraction of alloying atoms in a precipitating phase. Then, the particle size $R_v(C)$ is the only variable parameter with time in Eq. (21.10). Related condition of recrystallization is depinning of boundaries from particles that can be presented in terms of the particle size:

$$R_v(C) > R_1(C) = f_v^*(C)\gamma_b(C)/\beta_1 G\rho_v b^2. \tag{21.11}$$

It is known that the growth of precipitates at f_v=const can be described by the Liefshitz-Wagner-Slezov model (see, for example, (Martin et al. 1976)). Under domination of grain boundary diffusion, an equation for the particle radius R_v is

$$R_v^4(C) - R_0^4(C) = 9\delta D_b^*(C)\gamma C_v^*\Omega t/32 A_R B_R kT. \tag{21.12}$$

Here, $A_R = 2/3 - \gamma_b/2\gamma + \gamma_b/6\gamma$, $B_R=1/2\ln(1/F_b)$ (Martin et al. 1976), F_b is the fraction of a grain boundary occupied by the precipitating phase, C_v^* is the solubility limit, and R_0 is the initial particle size. In general, a radius R_0 depends on many factors and should be experimentally determined for each alloy. It affects the value of surface energy γ of a growing particle. For $R_0<10\div20$ nm, the particle/matrix interphase is coherent and its surface energy γ is small, comprising about 10% of the energy of a non-coherent boundary (Martin et al. 1976). Hence, after a particle reaches its critical size (about 16 nm for Al_3Sc (Elagin et al. 1991)), the interface energy increases and, according to Eq. (21.12), the particle growth rate rises substantially.

Eqs. (21.11) and (21.12) determine the temperature when particles attain the necessary size R_1 for depinning of grain boundaries after certain time t. Apparently, this temperature corresponds to the recrystallization temperature T_1 of the alloy:

$$T_1(C) = \frac{Q_b^*(C)}{\ln\left[\dfrac{A_1 b^5\rho_v^5 C_{vp}(\gamma/Gb)\delta D_{b0}^*(C)t}{(f_v^*\gamma_b^*/Gb)^4 - \alpha_1^4 b^4\rho_v^4 R_0^4}\right]} \qquad T_1(C) = \frac{Q_b^*(C)}{\ln\left[\dfrac{A_c b^5\rho_v^5 C_{vp}(\gamma/Gb)\delta D_{b0}^*(C)t}{(f_v^*\gamma_b^*/Gb)^4 - \alpha_1^4 b^4\rho_v^4 R_0^4}\right]}, \tag{21.13}$$

where $A_c = 9\alpha_1^4(G\Omega/kT)/(32A_RB_R)$ $A_1 = 9\alpha_1^4(G\Omega/kT)/(32A_RB_R)$. For typical parameters listed in Table 21.1, A_1=2.2·10^2. Eq. (21.13) shows that the effect of alloying elements on the recrystallization temperature is mainly connected with their influence on the grain-boundary diffusion parameters $Q_b^*(C)$ and $D_{b0}^*(C)$, and the particle characteristics $f_v^*(C)$, $R_0(C)$ and $\gamma_b(C)$.

Depending on alloying element concentration C_v and solubility C_{vp} in the volume, the alloying element concentration C_b and the solubility C_{bp} in grain boundaries, and on the distribution coefficient K_c, different precipitation scenarios can take place (see Table 21.2). Regardless of the complexity and variety of scenarios, the above-described model for the recrystallization temperature appears to be the same because the onset of recrystallization is controlled by the precipitate growth until depinning of the boundaries of recrystallization nuclei can occur.

Table 21.1. Parameter values used in calculations for aluminum

Parameter	Symbol	Value
Activation energy for self-diffusion in equilibrium grain boundaries	Q_b/kT_m	10.8[24]
Preexponential factor for grain boundary diffusion coefficient (10^{-14} m^3/s)	δD_{b0}	5.0[25]
Burgers vector (10^{-10} m)	b	2.86[26]
Grain boundary width	δ	2b[27]
Atomic volume (10^{-29} m^3)	Ω	1.18[28]
Melting point (K)	T_m	933[29]
Shear modulus	$G\Omega/kT_m$	50[30]
Annealing time (s)	t	3,600
Surface energy of random grain boundaries	γ_b/Gb	0.042[31]
Surface energy of noncoherent Al_3Sc and Al_3Zr particles in aluminum matrix	γ/Gb	0.042[32]
Surface energy of coherent Al_3Sc and Al_3Zr particles in aluminum matrix	γ/Gb	0.0042[33]
Numerical factors	β_1	0.5
	A_R	0.38
	B_R	0.35

24 Frost et al. 1982
25 Frost et al. 1982
26 Frost et al. 1982
27 Frost et al. 1982
28 Frost et al. 1982
29 Frost et al. 1982
30 Orlov et al. 1980
31 Orlov et al. 1980
32 Martin et al. 1976
33 Chuvil'deev, 2004

Parameter	Symbol	Value
Thermodynamic coefficients	W_1/kT_m	14.87
	W_2/kT_m	14.50
Fraction of Sc (Zr) atoms in Al_3Sc (Al_3Zr) particle	n_c	1/4
Free volume of a grain boundary	α	0.38[34]
Aluminum ion radius (10^9, m)	r_{Al}	0.143[35]
Scandium ion radius (10^9, m)	r_{Sc}	0.162[36]
Zirconium ion radius (10^9, m)	r_{Zr}	0.160[37]
Distribution coefficient of scandium atoms in aluminum	K_c^{Sc}	0.48[38]
Distribution coefficient of zirconium atoms in aluminum	K_c^{Zr}	2.70[39]

Table 21.2. Scenarios for particle precipitation in alloys versus the solute concentration and distribution coefficient

No.	Concentration condition	Scenario for precipitation of particles
Distribution coefficient $K_c<1$		
1	$C_v<C_{pv}$ $C_b<C_{pb}$	Alloying agent in solid solution in grain volume and boundary
2	$C_v<C_{pv}$ $C_b>C_{pb}$	a) alloying element in solid solution in grain volume; b) particle precipitation in grain boundary; c) particle growth in grain boundary;
3	$C_v>C_{pv}$ $C_b<C_{pb}$	a) alloying element in solid solution in grain boundary; b) particle precipitation in grain volume; c) particle growth in grain boundaries;
4	$C_v>C_{pv}$ $C_b>C_{pb}$	a) particle precipitation in grain boundary and volume b) particle growth in grain boundary
Distribution coefficient $K_c>1$		
5	$C_v<C_{pv}$ $C_b<C_{pb}$	a) particle precipitation in boundary area; b) particle growth in grain boundary;
6	$C_v>C_{pv}$ $C_b<C_{pb}$	a) particle precipitation in grain volume; b) particle growth in grain boundary;
7	$C_v>C_{pv}$ $C_b>C_{pb}$	a) particle precipitation in both grain boundary and volume; b) particle growth in grain boundary;

21.3. Comparison with Experimental Data

We will compare the recrystallization behavior of Al-3%Mg-0.2%Sc and Al-3%Mg-0.2%Zr NMC alloys (Lee et al. 2002, 2003) and Al-0.2%Sc (Elagin et al. 1991) and Al-0.2%Zr (Elagin, 1975) cast alloys (see Table 21.3). NMC structure in Al-3%Mg-0.2%Sc and

34 Chuvil'deev, 2004
35 Ed. Smithells, 1976
36 Ed. Smithells, 1976
37 Ed. Smithells, 1976
38 Ed. Hein, 1983
39 Ed. Hein, 1983.

Al-3%Mg-0.2%Zr alloys with an average grain size $d^* \approx 0.2$ μm was produced by ECAE (N=8) at room temperature. At annealing, Al-3%Mg-0.2%Sc and Al-3%Mg-0.2%Zr alloys retained their fine-grained structure with d<1μm up to temperatures T=650 K and 600 K, respectively. Rapid grain growth in these alloys was observed at 773 K and 573 K, respectively. Main thermodynamic and diffusion parameters of cast and NMC alloys are presented in Table 21.4. Both Sc and Zr have a low solubility limit in Al and therefore during heating after quenching precipitation of second-phase particles occurs in these alloys ($C > C^*$). The distribution coefficient of Sc atoms in aluminum is less than unity ($K_c^{Sc} < 1$), therefore Al_3Sc particles tend to precipitate mainly at grain boundaries (Figure 21.1a). The distribution coefficient of Zr atoms in the Al matrix is above unity ($K_c^{Zr} > 1$), hence they are repelled from boundaries and Al_3Zr particles precipitate mainly inside grains.

When estimating T_r, it is necessary to take into account that the solute atoms in grain boundaries can noticeably influence the value of Q_b (Chuvil'deev, 2004). For instance, the solute atoms that are larger than the matrix atoms, e.g., Mg, Sc and Zr in Al, can increase Q_b (Chuvil'deev, 2004). In this model, it was assumed that the contribution of different alloying atoms to the activation energy of grain-boundary diffusion is additive. For 0.2 wt.% Sc in aluminum and distribution coefficient $K_c^{Sc}=0.48$, the value of Q_b increased by $0.2kT_m$ as compared with an equilibrium value for pure Al, while alloying aluminum with 3 wt.% Mg raised the value of Q_b by $3.2kT_m$. Since Zr atoms are repelled from grain boundaries in aluminum and are mainly distributed in the lattice, they do not influence the grain boundary diffusivity.

The maximum volume fraction of precipitates in an alloy with 0.2 wt.% Sc is $f_v^*=0.48\%$ while for an alloy containing 0.2 wt% Zr f_v^* $f_v^*=0.24\%$. In estimating the recrystallization temperature, it is also necessary to account for the fact that the dislocation density ρ_v in NMC alloys is substantially higher than in cast alloys (Noskova et al. 2003; Valiev et al. 2006). For examined systems, in NMC materials is by an order of magnitude higher than $\rho_v \sim 10^{13}$ m^{-2} (see Table 21.4) in cast alloys.

Let us consider the initial size of precipitates R_0. As shown Table 21.4, the size of Al_3Sc and Al_3Zr particles in Al-Sc(Zr) and Al-Mg-Sc(Zr) NMC alloys is, respectively, 3 nm and 6 nm. That is about three times smaller than their size of 9 nm and 18 nm in cast alloys. In our view, this is because the number of nuclei in alloys is proportional to average spacing between dislocations. Then, the ratio of interparticle distances in NMC and coarse grain (CG) materials is $\sqrt{\rho_v^{CG} / \rho_v^{NMC}}$. Since the density of dislocations in NMC alloys is by an order of magnitude larger than in cast alloys, the number of nuclei in NMC alloys will be three times higher and the particle size, correspondingly, lower.

The recrystallization temperature $T_1(C)$ for NMC and cast alloys calculated from Eq. (21.14) for parameters of Table 21.1 are presented in Table 21.4 along with experimental data. A fair agreement between both data confirms the validility of the theoretical model. From the obtained results (see Table 21.4), the following conclusions can be drawn:

1 The process that controls recrystallization nucleus growth is identical in both materials, viz.coalescence of second-phase particles at nucleus boundaries;

2 Recrystallization temperature for NMC alloys is lower than for cast alloys of the similar chemical composition. This is mainly due to the fact that the dislocation density and the driving force for the nucleus boundary migration are substantially higher in NMC alloys;

3 Notwithstanding the similarity of Sc and Zr atoms (close ion sizes, low solubility limits in Al and similar structures of precipitating Al_3Sc and Al_3Zr particles), scandium atoms in aluminum exhibit a more pronounced stabilizing effect on the microstructure than Zr atoms for two reasons. First, at the same *weight* concentration of Sc and Zr, the number of Sc atoms in Al is two times higher due to a scandium's smaller atomic mass. Correspondingly, the volume fraction of second-phase particles and the boundary pinning force are higher for Al_3Sc particles than for Al_3Zr particles. Second, during decomposition of a solid solution, Sc atoms in aluminum form coherent Al_3Sc particles whose size is smaller than that of Al_3Zr particles. Therefore, the growth time of an Al_3Sc particle to the critical size, at which the coherency is lost, is much longer than the corresponding time for an Al_3Zr particle.

Table 21.3. Experimental values of recrystallization temperature for cast and NMC aluminum alloys

Cast alloys		NMC alloys	
Alloy	T_r/T_m	Alloy	T_r/T_m
Al	0.45-0.5 (Gorelik et al. 1967)	Al	0.43 (Lee et al. 2002)
		Al-3%Mg	0.4 (Lee et al. 2002) 0.45(Furukawa et al. 2002)
Al-0.2%Sc	0.84÷0.91 (Elagin et al. 1991)	Al-3%Mg-0.2%Sc	0.51-0.56 (Lee et al. 2002); 0.56-0.61 (Furukawa et al. 2002)
Al-0.2%Zr	0.6÷0.62 (Elagin, 1975)	Al-3%Mg-0.2%Zr	0.4-0.45 (Lee et al. 2002)

Table 21.4. Theoretical values of recrystallization temperature for cast and NMC aluminum alloys

Alloy	Structural state	R_0, nm	$\gamma/Gb \cdot 10^{-3}$	ρ_V, m^{-2}	$\frac{Q_b^*(C)}{kT_m}$	f_v^*,%	T_r/T_m
Al-3%Mg-0.2%Sc	NMC	3	4.2	$6 \cdot 10^{13}$	14.2	0.48	0.51
Al-0.2%Sc	Cast	9	42.0	$4 \cdot 10^{12}$	11.0	0.48	0.85
Al-3%Mg-0.2%Zr	NMC	6	4.2	$6 \cdot 10^{13}$	14.0	0.24	0.44
Al-0.2%Zr	Cast	20	42.0	$4 \cdot 10^{12}$	10.8	0.24	0.60

21.4. SUMMARY

A model describing the influence of a small amount of alloying elements on the recrystallization temperature of NMC alloys has been developed.

It has been shown that grain boundary pinning mechanisms in cast and NMC alloys can be described within a unified approach. A difference in the behavior of recrystallization temperature and its dependence on concentration for NMC materials , which reveals, in particular, a lower thermal stability of NMC materials, relate to a nonequilibrium statenonequilibricity of grain boundaries and to a substantially larger driving force for grain boundary migration due to a higher density of lattice dislocations. An expression for the dependence of critical temperature of recrystallization in alloys containing small amounts of additives on the grain boundary diffusion and on the concentration of alloying elements has been obtained.

SECTION III. MECHANICAL PROPERTIES OF NMC MATERIALS

Chapter 22

OPTIMUM GRAIN SIZE FOR STRUCTURAL SUPERPLASTICITY

22.1. PROBLEM DESCRIPTION

Classical structural superplasticity is known as an ability of polycrystalline ultrafine grain materials (with a grain size less than 10 μm) to display high plasticity with elongations reaching hundreds and thousands percent during tensile tests. The conditions for classical superplasticity include an elevated temperature $T \geq 0.7T_m$, (where T_m is a melting point) and relatively low optimum strain rate $\dot{\varepsilon} \sim 10^{-4}\ s^{-1}$ which is inversely proportional to the square of grain size, $\dot{\varepsilon} \sim d^{-2}$ (Edington, 1976; Kaibyshev, 1987; Nieh et al. 1997; Novikov et al. 1981). Along with a small size of grains, their about equaxed shape and a substantial fraction of high-angle grain boundaries in the material (Todd, 1997; Watanabe, 1997) are also important structural conditions for superplasticity. A significant role of the grain size, shape and boundary state is connected with the fact that the main mechanism of superplasticity is grain boundary sliding (GBS) (Edington, 1976; Perevezentsev et al. 1992; Valiev et al. 2006).

In order to ensure GBS, it is necessary to create conditions for grain boundary "shearing" (the so-called "intrinsic GBS") and accommodation of this "shearing" in triple junctions (Edington, 1976; Kaibyshev et al. 1987). As shown in works (Orlov et al. 1982; Perevezentsevet al. 1992; Chuvil'deev, 2004), the condition for intrinsic GBS is the transition of grain boundaries into a nonequilibrium statenonequilibricity due to their interaction with flows of lattice dislocations bombarding the boundaries during intragranular deformation.

Along with GBS, intragranular dislocation slip is also important for superplastic deformation (Edington, 1976; Kaibyshev, 1987; Perevezentsev et al. 1992; Chuvil'deev, 2004).

The rheological equation of superplastic flow is usually presented as (Nieh et al. 1997):

$$\dot{\varepsilon} = A_b \left(\frac{\sigma^*}{G}\right)^{1/m} \left(\frac{b}{d}\right)^p \frac{D_o^{eff} \exp(-Q^{eff}/kT)}{b^2} \left(\frac{G\Omega}{kT}\right), \qquad (22.1)$$

where m is the strain rate sensitivity coefficient of the flow stress whose value depends on strain rate $\dot{\varepsilon}$ (Figure 22.1) (for the optimum superplastic strain rate $\dot{\varepsilon} = \dot{\varepsilon}^{opt}$, m≥0.5), p=2÷3 is a numerical parameter, Q^{eff} is the activation energy for superplastic flow which is usually close to that for grain boundary diffusion Q_b, σ^* is the effective superplastic flow stress (most works suppose that $\sigma^* = \sigma$). This form of rheological equation for superplastic deformation is not the only one. Some authors (Li et al. 1999; Huang et al. 1998; Padmanabhan, 1981; Padmanabhan et al. 1983) believe that it is more correct to define the effective stress in Eq. (22.1) as

$$\sigma^* = \sigma - \sigma_0^{sp}(T), \tag{22.2}$$

where $\sigma_0^{sp}(T)$ is the superplastic threshold stress. In a number of works (for example, (Ashby etal, 1973; Perevezentsev et al. 1992)), the rheological equation is presented in a more rigorous manner separating the contribution of intergranular (connected with grain boundary sliding) and intragranular (connected with dislocation slip) deformation rates:

$$\dot{\varepsilon} = \dot{\varepsilon}_b + \dot{\varepsilon}_v. \tag{22.3}$$

Intergranular strain rate $\dot{\varepsilon}_b$ is usually described as

$$\dot{\varepsilon}_b = A_b \left(\frac{\sigma}{G}\right)^2 \left(\frac{b}{d}\right)^2 \left(\frac{G\Omega}{kT}\right) \left(\frac{D_b\delta}{b^3}\right), \tag{22.4}$$

where A_b~100 is a numerical factor (Perevezentsev et al. 1985b). For calculating the intragranular strain rate $\dot{\varepsilon}_v$, a power-law creep equation (Frost et al. 1982) has been proposed (Larin et al. 1990, Huang et al. 1998, Ridley et al. 2007):

$$\dot{\varepsilon}_v = A_v \left(\frac{\sigma}{G}\right)^n \frac{D_v}{b^2} \left(\frac{G\Omega}{kT}\right), \tag{22.5}$$

where A_v is the Dorn constant and D_v is the volume self-diffusion coefficient.

The range of problems considered in the superplasticity theory traditionally includes the strain-enhanced grain growth and cavity nucleation and growth resulting in the material failure.

In NMC materials, superplasticity was discovered in aluminum, magnesium, titanium and other alloys (Lee et al. 2002; Matsubara et al. 2003; Valiev et al. 2007; Chen, H.-H. et al. 2008; Shammazov et al. 2000; Komura et al. 1998; Perevezentsev et al. 2007; Figueiredo et al. 2008; Sergueeva et al. 2002; Kubota et al. 1999; Chuvi'deev et al. 2004d). When studying the superplastic deformation mechanisms in NMC materials, there are several problems which cannot be solved within a traditional theory of superplasticity. First, it is the issue of high-strain rate superplasticity mechanisms and, in particular, the mechanisms of low-temperature superplasticity in NMC materials. According to the key superplastic Eq. (22.1)), the refinement of grain structure from a micron (d_0) to submicron grain size (d_1) should increase the optimum superplastic strain rate by the factor of $(d_1/d_0)^2$. However, the optimum strain rate found experimentally sometimes appeared to be several times higher than the theoretical value calculated with the account of factor $(d_1/d_0)^2$. At a given strain rate, this effect often reveals itself as a shift of the optimum temperature of superplastic deformation towards lower values, which in some cases reaches 50÷100 K (Kubota et al. 1999; Chuvil'deev et al. 2004d; Kang et al. 2008; Mukherjee, 2002).

Second, experimental studies of superplasticity in NMC materials (Kang et al. 2008; Mukherjee, 2002; Islamgaliev et al. 2003, 2006; Higashi et al. 1995; del Valle et al. 2007; Lapovok et al. 2005; Yunusova et al. 2007) revealed an unusual effect: the highest superplasticity parameters were reached not in materials with the smallest grain size as it follows from a traditional approach, but in materials with a near-micron grain size (Figure 22.2). This phenomenon, viz. "the optimum grain size for superplastic deformation", cannot be interpreted within the traditional theory of superplasticity and requires special explanation.

The first of these problems has partially been addressed in Chapter 18 where the acceleration of grain boundary diffusion under active deformation was briefly discussed. Another aspect of the problem related to low-temperature deformation will be touched upon in Chapter 24. In this Chapter, we consider the problem of an optimum grain size for superplasticity.

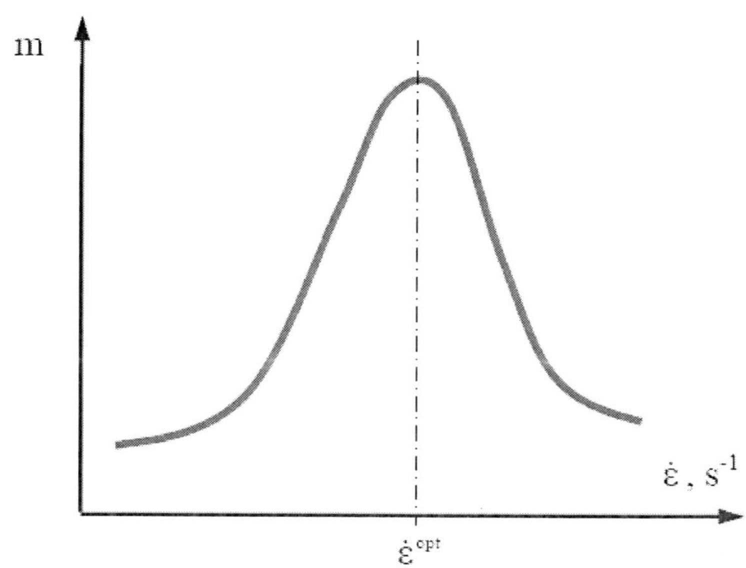

Figure 22.1. Strain rate sensitivity coefficient of flow stress m vs. strain rate $\dot{\varepsilon}$ under superplasticity ($\dot{\varepsilon}^{opt}$ is the optimum superplastic strain rate).

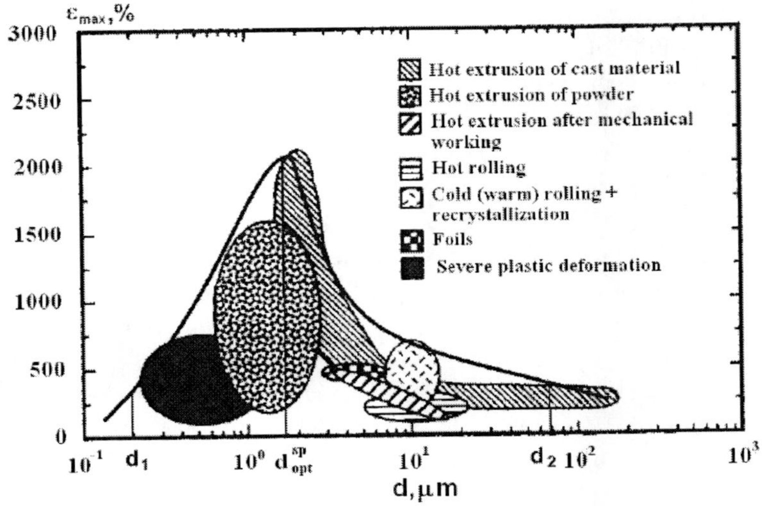

Figure 22.2. Elongation to failure vs. grain size in superplastic materials (Higashi et al. 1995).

22.2. MODEL OF OPTIMUM GRAIN SIZE FOR SUPERPLASTIC DEFORMATION

As noted above, the basic mechanism of superplastic deformation is the grain boundary sliding (GBS). Intrinsic GBS is conditioned by a transition of grain boundaries to a nonequilibrium statenonequilibricity due to their interaction with lattice dislocation flows bombarding the boundaries during intragranular deformation. Delocalization of lattice dislocation cores leads to the occurrence of zones with high local grain boundary diffusion coefficient and low effective viscosity. In these zones, grain boundary sliding is substantially facilitated and provides easy propagation of intrinsic GBS. GBS accommodation at grain junctions is conditioned by small grain sizes and a high value of grain boundary diffusion coefficient, which is connected with the transition of a grain boundary to the nonequilibrium statenonequilibricity (Orlov et al. 1982; Perevezentsev et al. 1992; Chuvil'deev, 2004).

Thus, the conditions for the onset and development of the superplastic flow substantially depend on the possibility of creating and maintaining the nonequilibrium statenonequilibricity of grain boundaries. The interaction of grain boundaries with lattice dislocations brings about two types of defects in grain boundaries: (i) orientational misfit dislocation (OMD) and (ii) the products of OMD delocalization. In accordance with (Chuvil'deev, 2004), the stationary density of OMD, ρ_b^{st}, does not depend on the grain size d while the stationary density of the Burgers vector of tangential components of delocalized dislocationstangential components of Burgers vector of delocalized dislocations , w_t^{st}, is proportional to d (see Eqs. (18.25)-(18.26)). The grain size at which the domination of OMD in the boundaries changes into the prevalence of tangential components of delocalized dislocationstangential components of Burgers vector of delocalized dislocations is determined by condition $\rho_b^{st}\Delta b = w_t^{st}$ and can be calculated using the formula:

$$\left(\frac{d_1}{b}\right)=\left(C_1^2 A_1 \frac{D_b^* \delta G}{kT}\right)^{1/4}\left(\frac{\xi_1}{\xi_2^2}\frac{1}{\dot{\varepsilon}_v}\right)^{1/4}. \tag{22.6}$$

For the parameters listed in Table 22.1, $d_1/b{\sim}10^3$.

Table 22.1. Typical values of parameters used in calculations

Parameter	Symbol	Value
Crystal parameters		
Burgers vector	b	$2.86\cdot 10^{-10}$ m
Grain boundary width	δ	2b
Grain size	d	$(10^3\div 10^5)b$
Atomic volume	Ω	$1.5\cdot 10^{-29}$ m^3
Diffusion parameters		
Pre-exponential factor for grain boundary diffusion coefficient	D_{b0}	10^{-6} m^2/s
Activation energy for grain boundary diffusion	Q_b	10 kT_m
Pre-exponential factor for diffusion coefficient in a melt	D_{L0}	10^{-8} m^2/s
Activation energy for diffusion in a melt	Q_L	3 kT_m
Thermodynamic and elastic constants		
Relative volume jump at crystal melting	ΔV_m	5%
Specific heat of fusion	λ	$\lambda\rho b^3/kT_m$=1.5
Density	ρ	
Shear modulus	G	$G\Omega/kT_m$ = 50
Melting point	T_m	
Grain boundary parameters		
Free volume of grain boundary	V_b	2%
Relative free volume of grain boundary	α	0.4
Critical free volume	α^*	0.5
Excess free volume of grain boundary	$\Delta\alpha$	0.1; 0.01
Enthalpy of solid/liquid interface	$\gamma^0_{S/L}$	$\gamma^0_{S/L}\ b^2/kT_m$ =1
Entropy of solid/liquid interface	$S_{S/L}$	$S_{S/L} b^2/k$ =0.8
Grain boundary enthalpy (α=0.35)	γ^0_b	$\gamma^0_b\ b^2/kT_m$ =3.8
Grain boundary entropy (α=0.35)	S_b	$S_b b^2/k$ =2.3
Free (surface) energy of grain-boundary S-phase areas	γ_0	$\gamma_0 b^2/kT_m$ =1.8
Grain boundary free energy	$\square_b$	$\gamma_b b^2/kT_m$ =2.5

Dislocations delocalized in a grain boundary change its GB's free volume α which increaseby $\Delta\alpha$, which is proportional to the density of these defects: $\Delta\alpha=(w_t+\rho_b\Delta b)/w_0$. Accumulation of the additional free volume changes the grain boundary diffusion coefficient. For small $\Delta\alpha$ ($\Delta\alpha<<\alpha$), the latter is estimated by Eq. (15.87) that can be rearranged using Eqs. (15.150), (18.25)-(18.26) as

$$D_b^* = D_b \exp(\varphi_{d1} d/b + \varphi_{d2}), \quad (22.7)$$

where

$$\varphi_{d1} = A_k / w_0\alpha_B, \ \varphi_{d2} = \rho_b \Delta b / w_0\alpha_B$$

and

$$A_k = (\xi_2 \dot{\varepsilon}_v kT / C_1 \delta G D_b^*)^{1/2}.$$

At large $\Delta\alpha$ ($\Delta\alpha+\alpha\geq\alpha^*$), Eq. (15.87) is not valid, and $D_b^* = D_L = D_{L0}\exp(-Q_L/kT)$, where Q_L, D_L and D_{L0} are the activation energy, diffusion coefficient and the preexponential factor for diffusion in a melt, respectively. The grain size, at which relation $\Delta\alpha + \alpha = \alpha^*$ is true, in the first approximation can be estimated by the formula obtained from Eqs. (15.129)-(15.130) and (18.27):

$$\frac{d_L}{b} = (\alpha^* - \alpha) w_o \left[\frac{D_L \delta}{\Omega} \frac{G\Omega}{kT} \frac{1}{\xi_2 \dot{\varepsilon}_v} \right]^{1/2} \quad (22.8)$$

For typical parameters presented in Table 22.1, $d_2/b{\sim}10^5$.

Eq. (22.7) shows that D_b^* depends on the grain size. At small grain sizes $d<d_1$, D_b^* weakly depends on d ($\varphi_{d1}d<<\varphi_{d2}$) while in the range of medium grain sizes, $d_1<d<d_L$, D_b^* grows exponentially with d. At coarse grain sizes $d\to d_L$, the approximation $\Delta\alpha+\alpha<<\alpha^*$ is not correct, and more complex expressions should be used for estimating $D_b^*(d)$ (Chuvil'deev, 2004). At $d\geq d_L$, D_b^* reaches the value of D_L and becomes independent of the grain size.

The physical meaning of Eq. (22.7) is apparent. Diffusion coefficient in NGB depends on the density of defects distributed in the boundaries. These defects are embedded in a grain boundary due to intragranular slip and can be withdrawn from it due to the development of diffusion accommodation mass transfer. A typical distance for this diffusion mass transfer depends on the grain size. The larger the grain is, the slower the recovery of diffusion properties of grain boundaries becomes. Thus, a grain boundary in a more coarse-grained material will possess a higher level of nonequilibrium and a larger diffusion coefficient than a boundary in ultrafine grain material.

In view of these results, we shall study the behavior of materials in structural superplastic conditions. Combination of Eq. (22.7) for $D_b^*(d)$ with Eq. (22.4) for intergranular strain rate $\dot{\varepsilon}_b$ gives

$$\dot{\varepsilon}_b = Z_d D_b \exp(\varphi_{d1} d + \varphi_{d2}) / d^2, \quad (22.9)$$

where

$Z_d = A_b(\sigma/G)(G\Omega/kT)(\delta/b)$.

The obtained Eq. (22.9) and Figure 22.3 show that the dependence of $\dot{\varepsilon}_b$ on d is substantially non-monotonic due to non-monotonic function $D_b^*(d)$. For small grains $d<d_1$, the grain boundary strain rate drops with increasing d because in these conditions the grain boundary diffusion coefficient is independent of d ($D_b^* \approx D_b \exp(\varphi_{d2})$), and the behavior of $\dot{\varepsilon}_b$ is determined by factor d^{-2}. For the medium grains $d_1<d<d_L$, an increase of d leads to a growth of $\dot{\varepsilon}_b(d)$ as $D_b^*(d) \approx D_b \exp(\varphi_{d1} d)$. For coarse grains $d \geq d_L$, the maximum value of the grain boundary diffusion coefficient $D_b^* = D_L$ is reached, D_b^* becomes independent of the grain size, $\dot{\varepsilon}_b(d)$ is proportional to d^{-2} and decreases with d.

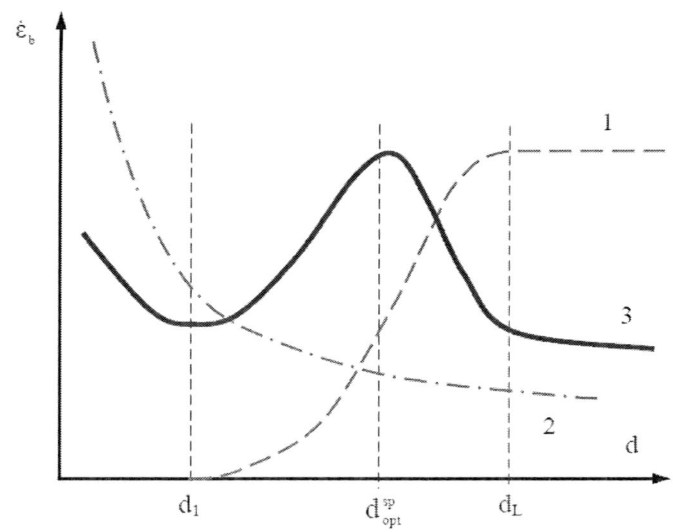

Figure 22.3. Intragranular strain rate $\dot{\varepsilon}_b$ of superplastic materials vs. grain size d. Curve 1 is the dependence $D_b^*(d)$, curve 2 is the dependence $d^{-2}(d)$, and curve 3 is $\dot{\varepsilon}_b \sim D_b^*/d^2$.

During superplasticity, the maximum tensile elongation ε_{max} is observed when the total deformation is dominated by grain boundary deformation, i.e. $\varepsilon_{max} \sim \dot{\varepsilon}_b$. In this case, the maximum ε_{max} in elongation curves corresponds to the maximum of the grain-boundary strain rate $\dot{\varepsilon}_b$, i.e. to the point at the d axis coinciding with the optimal grain size for superplastic deformation $d = d^{sp}_{opt}$ (Figure 22.3). Let us estimate this parameter. Apparently, the maximum of $\dot{\varepsilon}_b(d)$ lies within the $d_1<d<d_L$ range and the condition $\rho_b \Delta b << w_t$ should be satisfied in this vicinity. Then, a formula for d^{sp}_{opt} determined from the condition $\partial\dot{\varepsilon}_b/\partial d = 0$ is:

$$d^{sp}_{opt} / b = Z_{opt}(D^*_b)^{1/2} / \dot{\varepsilon}^{1/2}_v, \qquad (22.10)$$

where

$$Z_{opt} = 2w_0\alpha_B\left(\frac{G\Omega}{kT}\frac{C_1}{\xi_2}\frac{\delta}{b^3}\right)^{1/2}.$$

For the parameters of Table 22.1, Eq. (22.10) gives $d^{sp}_{opt}/b \sim 10^4$. As seen from Eq. (22.10), the value of d^{sp}_{opt} follows a power-law dependence on strain rate, changes exponentially with temperature and is connected with the thermodynamic properties of a material.

22.3. Comparison with Experimental Data

(a) Magnesium Alloy AZ31

The grain size effect on superplastic properties of AZ31 magnesium alloy was studied in (Kang et al. 2008). The grain size range from 3 μm to 6.5 μm was induced by ECAE with different temperatures and passes. Superplasticity tests were performed at temperatures 473, 523, 573, and 623 K and strain rates from 10^{-4} to 10^{-1} s^{-1}. The dependence of relative elongation on the grain size is presented in Figure 22.4. At the deformation temperature 623 K The maximum plasticity at temperature 623 K is observed for the initial grain size 3.8 μm.

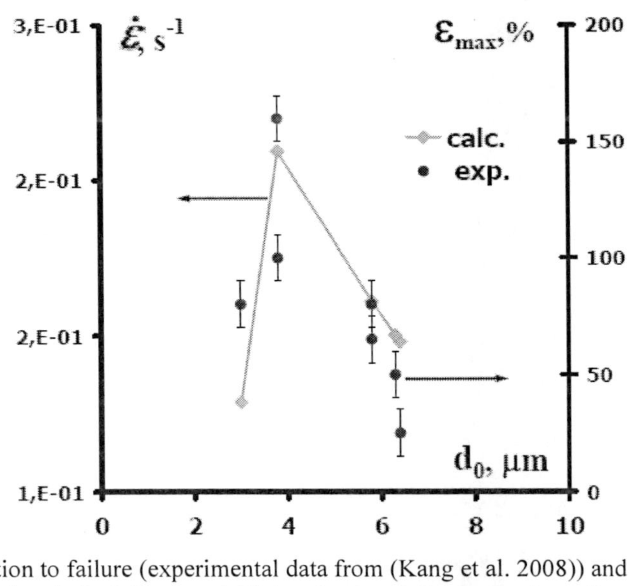

Figure 22.4. Elongation to failure (experimental data from (Kang et al. 2008)) and the optimum rate of superplastic deformation (calculation) vs. initial grain size d_0 for magnesium alloy AZ31 (T=623 K).

When comparing the theory with experiment, it should be remembered that heating to superplastic testing results in recrystallization of an NMC alloy. In particular, heating of AZ31 NMC alloy to 623 K increases the grain size to about 8-9 μm (Kang et al. 2008).

Using Eq. (22.10), we can evaluate the grain boundary diffusion coefficient corresponding to superplasticity with the optimum grain size d^{sp}_{opt}:

$$D^*_b = \left(\left[\frac{d^{sp}_{opt}}{b} \right] \frac{\dot{\varepsilon}_v^{1/2}}{Z_{opt}} \right)^2 . \qquad (22.11)$$

For $\dot{\varepsilon} = 10^{-2}$ s^{-1}, $d^{sp}_{opt} = 8\cdot 10^{-6}$ m, $b = 3.21\cdot 10^{-10}$ m, $C_1 = 5$, $\xi_2 = 1$ and T=623 K ($T_m/T = 1.49$), the grain boundary diffusion coefficient is $D^*_b = 2.37\cdot 10^{-9}$ m^2/s. At $D_{b0} = 7.8\cdot 10^{-3}$ m^2/s, this value of D^*_b corresponds to the activation energy for grain boundary diffusion $Q^*_b \approx 10.1kT_m$ (see Eq. (18.31)) that is typical for superplastic deformation of NMC magnesium alloys (Kubota et al. 1999; Chuvil'deev et al. 2004d).

In order to determine $\dot{\varepsilon}_b(d)$ and $\varepsilon_{max}(d)$, it is necessary to find the current grain size accounting for a grain growth from d_0 to $d(t,T,d_0)$ during heating time t to temperature T. In (Kang et al. 2008), the heating time to temperatures 573-623 K was 10 min. The current size $d(t,T,d_0)$ was found from Eq. (18.20) describing collective recrystallization. Assuming that $2\gamma_b bt\delta D^*_b / kT$=const and d_0=3.0, 3.8, 5.8, 6.3, 6.4 μm, the calculated values of $d(t,T,d_0)$ for these conditions are 6.4, 7.6, 8.0, 9.1 and 9.5 μm, respectively. Then, the dependence of superplastic strain rate on the grain size of NMC magnesium alloy at T=623 K were calculated from Eqs. (22.1) and (22.4) using the actual grain sizes d, the flow stress σ^*=25-45 MPa (Kang, 2008) and parameters $\alpha_B = 0.02$, $\xi_2 = 1$, G=16.6 GPa, $A_b = 50$, $Q_b = 10.1kT_m$. The final results $\dot{\varepsilon}_b(d)$ are presented in Figure 22.4 and exhibit a satisfactory agreement with the experimental dependence data.

(B) Aluminum Alloys

The superplasticity of AA 1421 with grain sizes from 0.3μm to about 2μm produced by ECAE at temperatures from 613 K to 683 K was studied in papers (Islamgaliev et al. 2003; 2006). For the alloy with the grain size d =0.3-0.4 μm after ECAE at T=643 K, tensile tests at temperature T=673 K and strain rate $\dot{\varepsilon} = 10^{-2}$ s^{-1} showed the largest elongations to failure of about 1,500%. Other grain sizes detected lower ductility (Figure 22.5). Also, the superplasticity of this alloy was studied after high-pressure torsion (Islamgaliev et al. 2006). By varying the HPT temperature, the average grain size was changed from 40 nm to 100 nm. Tensile tests at temperatures 648 K and 673 K and with strain rates $10^{-1} \div 10^{-2}$ s^{-1} have revealed a monotonic decrease of elongation to failure with reducing d (Figure 22.6). These results demonstrate that the optimal original grain size for superplastic deformation of AA 1421 at

T=648÷673 K and $\dot{\varepsilon}=10^{-1}\div10^{-2}$ s^{-1} is d^{sp}_{opt}=0.3-0.4 μm. However, it was found (Islamgaliev, 2006) that heating AA1421 to testing temperature of 673 K led to an increase of the average grain size to about 1 μm.

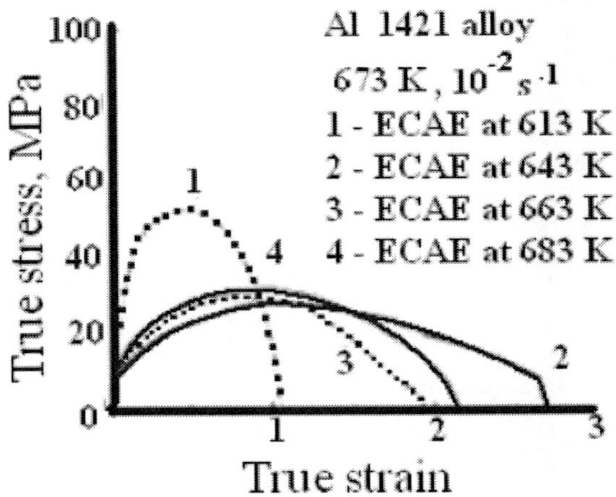

Figure 22.5. Stress-strain curves for NMC alloy 1421 after ECAE (T=673 K, $\dot{\varepsilon}=10^{-2}$ s^{-1}) (Islamgaliev et al. 2003, 2006).

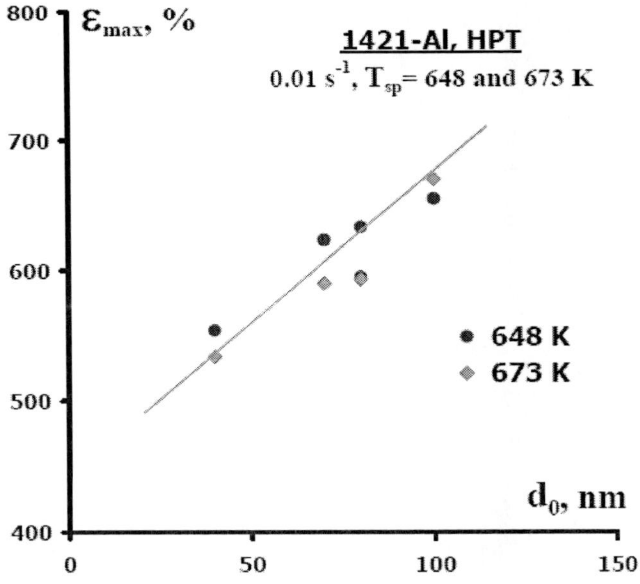

Figure 22.6. Elongation to failure of NMC alloy 1421 vs. grain size (data generalization (Islamgaliev et al. 2006)).

Following the procedure described above, we estimated the grain boundary diffusion coefficient D_b^* and the activation energy Q_b^*=10.9kT_m corresponding to the optimum grain size of AA 1421. These values are in good agreement with the tabular data for Al. Then, using the collective recrystallization Eq. (18.20), the current grain size $d(t,T,d_0)$ was determined under testing conditions of (Islamgaliev et al. 2003; 2006). Finally, the grain size dependence on the optimum rate of superplastic deformation was calculated from Eqs. (22.1) and (22.4). The results are presented in Figure 22.7. A satisfactory agreement with experimental dependence $\varepsilon_{max}(d)$ is obvious.

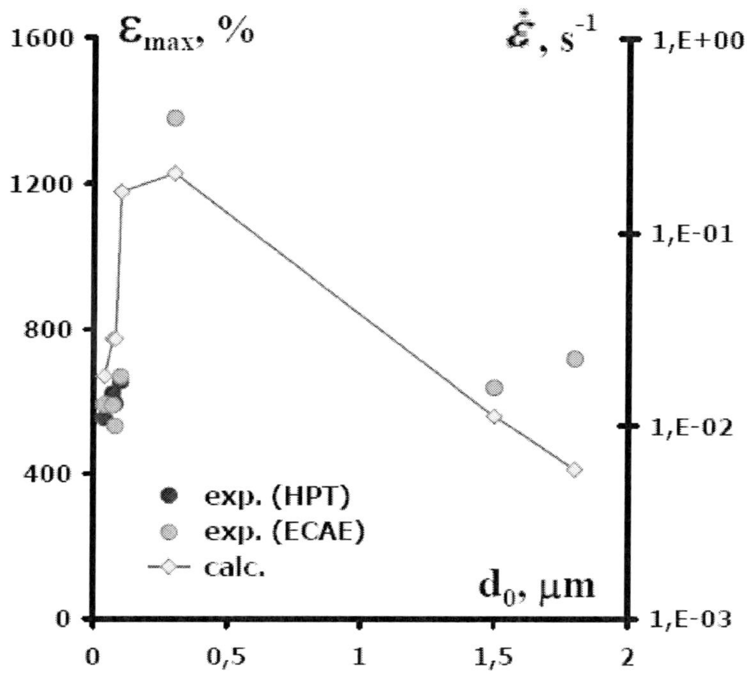

Figure 22.7. Experimental dependence of deformation to failure on initial grain size $\varepsilon_{max}(d_0)$ and theoretical dependence of superplastic strain rate $\dot{\varepsilon}_b$ on d_0 for NMC alloy 1421 at T_{def}=673 K.

The analysis of vast experimental data on superplasticity (Sergueeva et al. 2005) has shown that reducing the grain size of aluminum and other NMC alloys below d^{sp}_{opt}=1 μm results in a decrease of maximum elongation to failure. Using the above-described procedure, Eqs. (22.9)-(22.10) and assuming $d^{sp}_{opt}\approx 1$ μm, the effective grain boundary diffusion coefficient D^*_b and the activation energy Q^*_b were calculated for aluminum alloy 1421. At $\dot{\varepsilon}=10^{-1}$ s^{-1} (Lee et al. 2002; Komura et al. 2005; Islamgaliev et al. 2003, 2006; Perevezentsev et al. 2002a), d^{sp}_{opt}=1.5 μm, b=2.86·10^{-10} m and T=0.77T_m (i.e. T=723 K), we obtain D^*_b=3.2·10^{-11} m^2/s that corresponds to the activation energy of about 10.4kT_m.

22.4. SUMMARY

A new model for determining the optimum grain size for structural superplasticity has been proposed. As is known, in superplastic materials a fine grain size is necessary to ensure intensive accommodation of grain boundary sliding at triple junctions of grains. Therefore, it is usually assumed that the smaller the grain size is, the higher superplastic properties are. However, this approach does not take into account the fact that fine-grained materials are characterized by fast recovery of diffusion properties of nonequilibrium grain boundaries, and their effective coefficient of grain boundary diffusion is lower than that of more coarse-grain materials where the recovery rate is lower. From this standpoint, the optimum grain size for superplastic deformation can be calculated as a size at which a high level of grain boundary nonequilibrium is combined with sufficient intensity of grain boundary sliding accommodation. An expression for estimating the dependence of the optimum grain size for superplastic deformation on strain rate, temperature and material thermodynamic parameters is formulated.

Chapter 23

STRAIN HARDENING UNDER SUPERPLASTICITY

23.1. PROBLEM DESCRIPTION

Preliminary Remarks

A typical dependence of flow stress σ on strain ε for NMC alloys under superplastic conditions is shown in Figure 23.1a (Mishra et al. 1999). In a general case, the curve $\sigma(\varepsilon)$ is divided into three stages: (a) stage I of hardening which is sometimes subdivided into substages Ia of severe hardening and Ib of weak hardening), (b) stage II of stationary flow, and (c) stage III of softening. The length of stages depends on the deformation temperature and strain rate. At low strain rates, only the hardening stage is usually observed. At strain rates corresponding to optimum superplasticity, the hardening stage is followed by the stage of stationary flow. At high strain rates stage I is immediately followed by stage III (softening) (see Figure 23.1).

It is typically assumed that strain hardening during superplastic deformation is connected with the processes that inhibit grain boundary sliding. This standpoint is used in some models (Larin et al. 1990; Likhachev et al. 1987) where the grain growth under superplasticity is considered as a basic reason for strain hardening. Indeed, the rheological equation Eq. (22.1) describes the steady superplastic flow as a function between grain boundary strain rate $\dot{\varepsilon}$, flow stress σ^* and grain size d. When the grain size increases, a corresponding increase in the flow stress σ is necessary to provide the constancy of strain rate $\dot{\varepsilon}$.

A grain growth under superplasticity is described by the power-law dependence (Wilkinson et al. 1984; Perevezentsev et al. 1992; Chuvil'deev, 2004):

$$d^z - d_0^z = g(\dot{\varepsilon})t \quad d^Z - d_0^Z = \alpha(\dot{\varepsilon})t\,, \tag{23.1}$$

where d_0 and d are the initial grain size and the current grain size, respectively, z=2 or 3 is the power exponent and $g(\dot{\varepsilon})$ $\alpha(\dot{\varepsilon})$ is a factor depending on strain rate $\dot{\varepsilon}$, diffusion constants of the material, structural parameters and strain conditions. Inserting Eq. (22.1) in Eq. (23.1) and assuming that at $\dot{\varepsilon} = \text{const}$ $t = \varepsilon / \dot{\varepsilon}$, we obtain:

$$\sigma^* \sim A'\dot{\varepsilon}^m \left[d_0^z + g(\dot{\varepsilon})\varepsilon/\dot{\varepsilon}\right]^{mp/z} \sigma^* \sim A'\dot{\varepsilon}^m \left[d_0^z + \alpha(\dot{\varepsilon})\varepsilon/\dot{\varepsilon}\right]^{mp/z}. \quad (23.2)$$

In this case, the strain hardening coefficient $n = \partial \lg \sigma / \partial \lg \varepsilon$ is expressed as

$$n = \frac{mp}{z} \cdot \frac{\alpha(\dot{\varepsilon}) \cdot \varepsilon/\dot{\varepsilon}}{(d_0^z + \alpha(\dot{\varepsilon}) \cdot \varepsilon/\dot{\varepsilon})} = \frac{mp}{z} \cdot \left(\frac{d^z - d_0^z}{d^z}\right) n = \frac{mp}{z} \cdot \frac{\alpha(\dot{\varepsilon}) \cdot \varepsilon/\dot{\varepsilon}}{(d_0^z + g(\dot{\varepsilon}) \cdot \varepsilon/\dot{\varepsilon})} = \frac{mp}{z} \cdot \left(\frac{d^z - d_0^z}{d^z}\right) \quad (23.3)$$

From Eqs. (23.1)-(23.3) it is seen that at optimal parameter values p=2, m=0.5 and z=2 for structural superplasticity, and for substantial changes in the grain sizes ($d^z >> d_0^z$) typical in NMC materials, the value of n will be equal to 0.5. Thus, according to Eq. (23.1), the grain growth can provide only relatively weak hardening. Therefore, strong hardening with $n \approx 1$observed in most superplastic NMC alloys cannot be explained by the grain growth.

The objective of this Chapter is a new model to explain substantial strain hardening at stage I of superplastic flow in NMC materials depending on strain rate, temperature and microstructural characteristics.

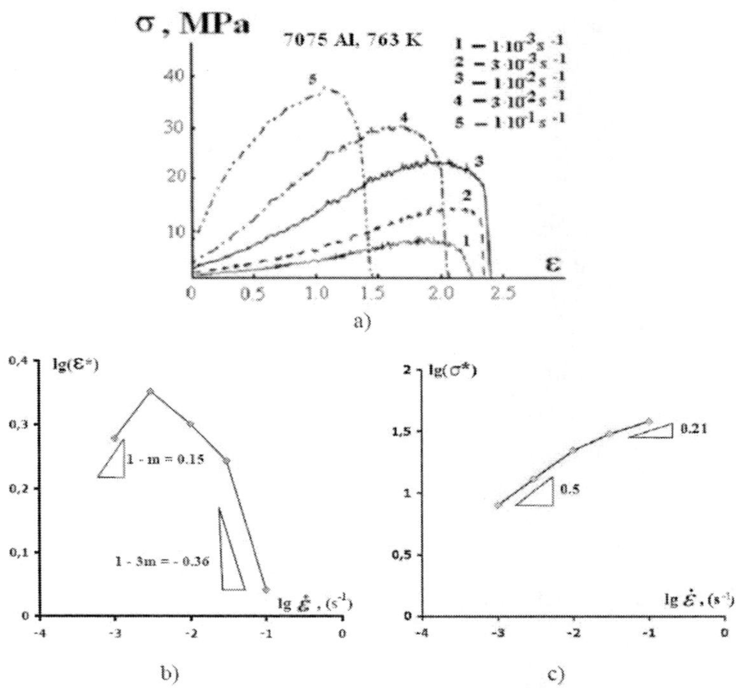

Figure 23.1. Typical dependences of flow stress on strain at different superplastic strain rates. As an example, dependences for Al-7075 alloy at 763 K (a) (Mishra, 1999),; are shown: line $\lg\varepsilon^*$ vs. $\lg\dot{\varepsilon}$ (b) (plotted using the data from Figure a), and dependence of $\lg\sigma^*$ on $\lg\dot{\varepsilon}$ (c) (built using the data from Figure a).

Experimental Data

The stress-strain dependence $\sigma(\varepsilon)$ for different superplastic MC and NMC materials has been extensively studied in literature. Typical experimental data for MC and NMC alloys can be found in (Mishra et al. 2005; Kaibyshev et al. 2000, 2001; Watanabe et al. 1999b; Piatti, 1983) and other. We will describe curves $\sigma(\varepsilon)$ at the hardening stage by parameters σ^* and ε^* denoting the stress and strain at the end of this stage followed by either stationary flow or softening. Since σ^* and ε^* depend on the strain rate $\dot{\varepsilon}$, for further analysis $\lg \varepsilon^*$ and $\lg \sigma^*$ are plotted versus $\lg \dot{\varepsilon}$ in Figure 23.1(b and c). It is seen that these functions are non-monotonic and contain at least two stages. At low superplastic strain rates $\dot{\varepsilon} < \dot{\varepsilon}^{\varepsilon}$, a rise in ε^* is observed with increasing $\dot{\varepsilon}$. At high strain rates $\dot{\varepsilon} > \dot{\varepsilon}^{\varepsilon}$, ε^* reduces with increasing strain rate. However, two-stage behavior is not a characteristic of all alloys. In some cases, only the descending branch of curve "$\lg \varepsilon^*$ vs. $\lg \dot{\varepsilon}$" is observed. For a quantitative description, an additional parameter $K^{\varepsilon} = \partial \lg \varepsilon^* / \partial \lg \dot{\varepsilon}$ is introduced that is positive at small strain rates ($\dot{\varepsilon} < \dot{\varepsilon}^{\varepsilon}$) and negative at high strain rates ($\dot{\varepsilon} > \dot{\varepsilon}^{\varepsilon}$). The values of K^{ε} for the alloys under examination are listed in Table 23.1.

The dependence of $\lg \sigma^*$ on $\lg \dot{\varepsilon}$ is shown in Figure 23.1c. This curve can also be subdivided into two stages. At small strain rates ($\dot{\varepsilon} < \dot{\varepsilon}^{\sigma}$), its tangent $K^{\sigma} = \partial \lg \sigma^* / \partial \lg \dot{\varepsilon}$ is higher than at high strain rates ($\dot{\varepsilon} > \dot{\varepsilon}^{\sigma}$). The values of K^{σ} in the corresponding ranges of strain rates are presented in Table 23.1.

In the preceeding text, $\dot{\varepsilon}^{\varepsilon}$ and $\dot{\varepsilon}^{\sigma}$ are certain "transient" strain rates whose meaning will be clarified below. The values of $\dot{\varepsilon}^{\sigma}$ and $\dot{\varepsilon}^{\varepsilon}$ are different for dissimilar materials and depend on the deformation temperature. Typically, at T=const the tangent of curves "$\lg \varepsilon^*$ vs. $\lg \dot{\varepsilon}$" and "$\lg \sigma^*$ vs. $\lg \dot{\varepsilon}$" is changed at close strain rates, i.e. $\dot{\varepsilon}^{\varepsilon} \approx \dot{\varepsilon}^{\sigma}$.

23.2. Hardening under Superplastic Deformation

23.2.1. Model

The phenomenon of structural superplasticity is attributed to the transition of grain boundaries to a nonequilibrium statenonequilibricity due to lattice dislocation flow (Perevezentsev et al. 1992; Chuvil'deev, 2004). These dislocations emitt from grain boundaries and junctions, quickly glide across grains and are embedded into boundaries. If the generation rate of dislocations is close to their absorption rate by the boundaries, the defects are not accumulated inside grains. In this case, the stress necessary for intragranular slip is conditioned either by dislocation emission from grain boundaries and junctions or by dislocation embedment into grain boundaries. Both of these processes are closely interconnected and can be described within a single approach in many cases (see Chapter 16).

Interaction of lattice dislocations with grain boundaries during the intragranular slip changes boundary diffusion properties due to transition to a nonequilibrium statenonequilibricity and leads to defect accumulation in the boundaries. Accumulated defects create internal stresses σ_i which inhibit the lattice glide and affect intragranular deformation. Stresses σ_i depend on the density ρ_b of orientational misfit dislocations distributed in boundaries and the density w_t of the Burgers vector of tangential components of delocalized dislocationstangential components of Burgers vector of delocalized dislocations , that is $\sigma_i=\sigma_i(\rho_b, w_t)$ (Chuvil'deev, 2004). It should be noted that normal components w_n of the Burgers vector of delocalized dislocations form additional misorientation in the boundary and create triple junction disclinations, which can be described as a system of disclination dipoles with the arm d and power $\omega \sim w_n$ (Perevezentsev et al. 1983). The effective stress fields of such dipoles act over a distance of the order of the grain size d and, thus, just slightly affect the internal stresses σ_i near grain boundaries. It is supposed that internal stresses $\sigma_i(\rho_b, w_t)$ determine the level of external stresses which is required to perform superplastic deformation (from Eqs. (16.24)-(16.26)), $\sigma=\sigma_i(\rho_b,w_t)$). In a first approximation, σ_i is proportional to the density of defects distributed in the grain boundary and is given by Eq. (16.26) (Chuvil'deev, 2004).

Within this approach, the nature and mechanism of strain hardening is determined by the kinetics of defect accumulation at grain boundaries during superplastic deformation.

23.2.2. Kinetics of Defect Accumulation in Grain Boundaries

The accumulation kinetics of orientational misfit dislocations ρ_b and tangential components of the Burgers vector w_t in grain boundaries due to the flow of lattice dislocation $I=\xi_v \dot{\varepsilon}_v/b$ is described by Eqs. (18.25)-(18.26). The set of Eqs. (15.147)–(15.150), (15.154)–(15.155) for ρ_b and w_t can be solved in cases when the dominating defects in the boundary are the orientational misfit dislocation (OMD), i.e. $\rho_b\Delta b>>w_t$, and when the boundary is dominated by the tangential components of delocalized dislocationstangential components of Burgers vector of delocalized dislocations , i.e. $w_t>>\rho_b\Delta b$. The former case refers to ultrafine grain materials where the relationship $d<d_1$ is true while the latter case takes place for $d>d_1$. The value of d_1 is determined from the relation $\rho_b\Delta b = w_t$ as:

$$d_1/b \approx (C_1A_1)^{1/2}/\rho_b\Delta b. \tag{23.4}$$

Taking ito account a connection between $\rho_b\Delta b$ and σ_i, Eq. (23.4) becomes

$$(d_1/b) = (C_1A_1)^{1/2}(G/\sigma)(1/\varphi_1). \tag{23.5}$$

When $\rho_b\Delta b>>w_t$ (i.e. $d<d_1$), Eqs. (15.147)-(15.151), (15.154) and (18.3) gives for the OMD density in grain boundaries:

$$\dot{\rho}_b \Delta b = \xi_1 \dot{\varepsilon}_v - (\rho_b \Delta b)^4 (G D_b \delta / A_1 kT) \exp(\rho_b \Delta b / \alpha_B w_0). \tag{23.6}$$

In the case $\rho_b \Delta b << w_t$ (i.e. $d > d_1$), the tangential components of delocalized dislocationstangential components of Burgers vector of delocalized dislocations dominate in the boundary, and the kinetic equation for the defect density follows from Eqs. (15.147)-(15.150) and (15.155):

$$\dot{w}_t = \xi_2 \dot{\varepsilon}_v - w_t^2 (b/d)^2 (C_1 G D_b \delta / kT) \exp(w_t / \alpha_B w_0). \tag{23.7}$$

23.2.3. Description of Strain Hardening

With the account of Eqs. (16.24)-(16.26) which link the external and internal stresses to the density of defects distributed in grain boundaries, Eqs. (23.6) and (23.7) can be resolved for $\dot{\sigma}$. When OMD dominate in the grain boundary, Eq. (16.26) becomes $\sigma_\rho = \varphi_1 G \Delta b \rho_b$. Inserting this expression into Eq. (23.6), gives

$$\dot{\bar{\sigma}}_\rho = \xi_1 \dot{\varepsilon}_v - \bar{\sigma}_\rho^4 \left[\frac{D_b \delta G}{A_1 kT}\right] \exp\left(\frac{\varphi_1 \bar{\sigma}_\rho}{\alpha_B w_0}\right) \quad \text{(at } \rho_b \Delta b >> w_t\text{)}, \tag{23.8}$$

where $\bar{\sigma}_\rho = \sigma_\rho / G \varphi_1$.

When the tangential components of delocalized dislocationstangential components of Burgers vector of delocalized dislocations dominate in the grain boundary, $\sigma_w = \varphi_2 G w_t$, and Eq. (23.7) can be rewritten as

$$\dot{\bar{\sigma}}_w = \xi_2 \dot{\varepsilon}_v - \bar{\sigma}_w^2 \left[\left(\frac{b}{d}\right)^2 \frac{D_b \delta G}{kT}\right] \exp\left(\frac{\varphi_2 \bar{\sigma}_w}{\alpha_B w_0}\right) \quad \text{(at } w_t << \rho_b \Delta b\text{)}, \tag{23.9}$$

where $\bar{\sigma}_w = \sigma_w / G \varphi_2$.

The obtained equations for $\dot{\sigma}$ can be integrated; however, analytical expressions for $\sigma(t)$ are lengthy. From Eqs. (23.8) and (23.9) it is seen that at certain instants of time t_1 and t_2 stresses σ_ρ and σ_w, respectively, reach their stationary values σ_ρ^{st} and σ_w^{st} when $\dot{\sigma}_\rho = 0$ and $\dot{\sigma}_w = 0$. For a small density of defects distributed in the boundary, $\exp(\varphi_1 \bar{\sigma}_\rho / \alpha_B w_0) \sim 1$ and $\exp(\varphi_2 \bar{\sigma}_w / \alpha_B w_0) \sim 1$, stationary stresses $\bar{\sigma}_\rho^{st}$ and $\bar{\sigma}_w^{st}$ are:

$$\bar{\sigma}_\rho^{st} = \left[\xi_1 \dot{\varepsilon}_v \frac{kTA_1}{D_b^* \delta G}\right]^{1/4}, \tag{23.10}$$

$$\overline{\sigma}_w^{st} \approx \left[\xi_2 \dot{\varepsilon}_v \frac{kT(d/b)^2}{D_b^* \delta G C_1}\right]^{1/2} \tag{23.11}$$

In a first approximation, their stationary values are reached at $t=t_1$ (see Eq. (15.154)) and $t=t_2$ (see Eq. (15.155)), respectively. With the account of Eqs. (16.28) and (16.29), Eqs. (15.154) and (15.155) can be presented as:

$$t_\rho = G_1/(\overline{\sigma}_\rho^{st})^3 ,\ t_w = G_2/\overline{\sigma}_w^{st} \quad , \tag{23.12}$$

where parameters G_1 and G_2 are:

$$G_1 = A_1 (kT/G\Omega)/(\Omega/\delta D_b^*), \tag{23.13}$$

$$G_2 = (d/b)^2 (kT/G\Omega)/(\delta C_1 D_b^*/\Omega). \tag{23.14}$$

Stationary stresses σ_ρ^{st} and σ_w^{st} correspond to certain strains which are $\varepsilon_\rho^{st} = \dot{\varepsilon} t_\rho$ and $\varepsilon_w^{st} = \dot{\varepsilon} t_w$. Then, using Eq. (23.12), we obtain

$$\varepsilon_\rho^{st} = G_1 \dot{\varepsilon}/(\overline{\sigma}_\rho^{st})^3 , \tag{23.15}$$

$$\varepsilon_w^{st} = G_2 \dot{\varepsilon}/\overline{\sigma}_w^{st} . \tag{23.16}$$

In accordance with Eqs. (23.4)-(23.5) and (23.8)-(23.11), an equation (23.15) is applicable for small grains $d<d_1$ and an equation (23.16) - for coarse grains $d>d_1$. Taking into account the ordinary relation $\sigma \sim \dot{\varepsilon}^m$ for the superplastic flow (see also Eq. (22.1)), Eqs. (23.15)-(23.16) take forms:

$$\varepsilon_\rho^{st} \sim \dot{\varepsilon}^{1-3m} \text{ for } d<d_1 \tag{23.17}$$

$$\varepsilon_w^{st} \sim \dot{\varepsilon}^{1-m} \text{ for } d>d_1 \tag{23.18}$$

These equations show that the stationary strains depend on the strain-rate sensitivity coefficient m and grain size d. For $m<0.3$, both ε_ρ^{st} and ε_w^{st} increase with $\dot{\varepsilon}$. For $m\geq 0.4$), in the case when OMD dominate, ε_ρ^{st} is reduced with strain rate increase, whereas, when w_t dominate, ε_ρ^{st} ε_w^{st} is increased with strain rate increase when w_t dominate.

23.3. Comparison with Experimental Data

Direct calculation of curves $\sigma(\varepsilon)$ using Eqs. (23.8) and (23.9) requires a large number of poorly determined fitting parameters. Therefore, we will use another procedure that does not involve complex calculations.

At first, we shall compare the experimental (ε^* and σ^*) and theoretical (ε^{st} and σ^{st}) values of deformation and stress that correspond to the end of the hardening stage and then study the strain-rate dependence of these parameters. The experimental dependences of ε^* and σ^* on strain rate $\dot{\varepsilon}$ have been described above using parameters $K^{\varepsilon} = \partial \lg \varepsilon^* / \partial \lg \dot{\varepsilon}$ and $K^{\sigma} = \partial \lg \sigma^* / \partial \lg \dot{\varepsilon}$ whose values are listed in Table 23.1.

Theoretical dependences $\sigma^{st}(\dot{\varepsilon})$ and $\varepsilon^{st}(\dot{\varepsilon})$ are described by Eqs. (23.10)-(23.11) and Eqs. (23.15)-(23.16), respectively. As noted previously, at small stresses (i.e. small strain rates) Eqs. (23.11) and (23.16) should be used while at large stresses and strain rates Eqs. (23.10) and (23.15) are valid. With these formulae, the theoretical values of $K^{\sigma}_{\rho(th)} = \partial \lg \sigma^{st}_{\rho} / \partial \lg \dot{\varepsilon}$, $K^{\varepsilon}_{\rho(th)} = \partial \lg \varepsilon^{st}_{\rho} / \partial \lg \dot{\varepsilon}$, $K^{\sigma}_{w(th)} = \partial \lg \sigma^{st}_{w} / \partial \lg \dot{\varepsilon}$ and $K^{\varepsilon}_{w(th)} = \partial \lg \varepsilon^{st}_{w} / \partial \lg \dot{\varepsilon}$ can be determined. At small strain rates they are

$$K^{\varepsilon}_{w(th)} = 1 - m, \; K^{\sigma}_{w(th)} = 1/2 , \tag{23.19}$$

while at high strain rates

$$K^{\varepsilon}_{\rho(th)} = 1 - 3m, \quad K^{\sigma}_{\rho(th)} = 1/4 \tag{23.20}$$

It should be noted that at small strain rates $K^{\varepsilon}_{w(th)}$ is always positive, while at high $\dot{\varepsilon}$ and m>0.33, the value of $K^{\varepsilon}_{\rho(th)}$ may be negative. Also, it seems interesting that the values of $K^{\sigma}_{w(th)}$ at low strain rates are twice as high as the values of $K^{\sigma}_{\rho(th)}$ at high strain rates.

The comparison of theoretical (K_{th}) and experimental (K_{exp}) values of K^{ε} and K^{σ} listed in Table 23.1 shows that the developed model captures main peculiarities of strain hardening of superplastic alloys.

It should be noted that superplastic flows are usually presented either as "stress σ–strain ε" diagrams or "stress σ–elongation δ" diagrams. In the last situation, the true strains are $\varepsilon = \ln(1+\delta)$. However, the strain-rate sensitivity coefficient m was not specified in many papers. In such cases, we estimated m from "flow stress σ–strain rate $\dot{\varepsilon}$" curves using the formula $m = \partial \lg \sigma / \partial \lg \dot{\varepsilon}$; these values listed in Table 23.2 were marked by asterisk$^{(*)}$. The experimental and theoretical values of K^{σ} and K^{ε} for aluminum and magnesium NMC

Table 23.1. Experimental and theoretical values of hardening parameters for ultrafine grain materials under superplastic deformation

Alloy 7055 T_{def}=723 K (d=2 μm) (Kaibyshev et al. 2000)	$\dot{\varepsilon}$, s^{-1}	$8.3 \cdot 10^{-5}$	$3.3 \cdot 10^{-4}$	$1.7 \cdot 10^{-3}$	$8.3 \cdot 10^{-3}$	$3.3 \cdot 10^{-2}$	$8.3 \cdot 10^{-2}$
	σ^*, MPa	6.8	7.1	13.4	22.1	38.6	42.7
	ε^*	2.03	2.30	0.37	0.35	0.18	0.06
	m_{exp}	0.50					
	K^{ε}	$K^{\varepsilon}_{(exp)}$ = -0.56, $K^{\varepsilon}_{(th)}$ = -0.50					
	K^{σ}	$K^{\sigma}_{(exp)}$ = 0.36, $K^{\sigma}_{(th)}$ = 0.25					

Alloy 2219 T_{def}=773 K (d=12 μm) (Kaibyshev et al. 2001)	$\dot{\varepsilon}$, s^{-1}	$2.2 \cdot 10^{-4}$	$5.5 \cdot 10^{-4}$	$2.2 \cdot 10^{-3}$	$5.5 \cdot 10^{-3}$	$2.2 \cdot 10^{-2}$	$1.1 \cdot 10^{-1}$
	σ^*, MPa	5.3	7.7	12.8	14.9	22.7	31.8
	ε^*	0.83	0.65	0.37	0.19	0.20	0.20
	m_{exp}	0.43÷0.48					
	K^{ε}	$K^{\varepsilon}_{(exp)}$ = -0.48, $K^{\varepsilon}_{(th)}$ = -0.29÷0.44					
	K^{σ}	$K^{\sigma}_{(exp)}$ = 0.25, $K^{\sigma}_{(th)}$ = 0.25					

Alloy ZK60 T_{def}=473 K (d=6 μm) (Watanabe et al. 1999a)	$\dot{\varepsilon}$, s^{-1}	$2.6 \cdot 10^{-6}$	$5.0 \cdot 10^{-6}$	$1.0 \cdot 10^{-5}$	$3.0 \cdot 10^{-5}$
	σ^*, MPa	18.9	24.8	30.2	44.7
	ε^*	0.79÷0.89	0.67÷0.74	0.36	0.15÷0.30
	m_{exp}	0.50			
	K^{ε}	$K^{\varepsilon}_{(exp)}$ = -0.70, $K^{\varepsilon}_{(th)}$ = -0.60			
	K^{σ}	$K^{\sigma}_{(exp)}$ = 0.35, $K^{\sigma}_{(th)}$ = 0.25			

Alloy Al-Ca T_{def}=723 K (Piatti, 1983)	$\dot{\varepsilon}$, s^{-1}	$4.16 \cdot 10^{-5}$	$2.08 \cdot 10^{-4}$	$4.16 \cdot 10^{-4}$	$1.04 \cdot 10^{-3}$	$2.08 \cdot 10^{-3}$	$4.16 \cdot 10^{-3}$	$1.04 \cdot 10^{-2}$	$4.16 \cdot 10^{-2}$	$1.04 \cdot 10^{-1}$	$2.08 \cdot 10^{-1}$	$4.16. \cdot 10^{-1}$
	σ^*, MPa	1.8	2.7	4.5	5.1	6.5	8.5	10.7	15.5	19.9	24.1	24.7
	ε^*	0.32	0.46	0.60	0.71	0.78	0.79	0.79	0.71	0.74	0.63	0.33
	m_{exp}	0.72				-						
	K^{ε}	$K^{\varepsilon}_{(exp)}$ = 0.23, $K^{\varepsilon}_{(th)}$ = 0.28				$K^{\varepsilon}_{(exp)}$ = -0.90, $K^{\varepsilon}_{(th)}$ = -1.10						
	K^{σ}	$K^{\sigma}_{(exp)}$ = 0.34, $K^{\sigma}_{(th)}$ = 0.50				$K^{\sigma}_{(exp)}$ = 0.27, $K^{\sigma}_{(th)}$ = 0.25						

alloys are presented in Table 23.2. Their analysis has revealed that a fair agreement is attained under the assumption that strain hardening in NMC alloys is controlled by the process of OMD accumulation. The situation when strain hardening is controlled by the accumulation of tangential components of delocalized dislocationstangential components of Burgers vector of delocalized dislocations takes place only in relatively coarse-grained materials deformed either at elevated temperatures or at high strain rates like for ZK60 alloy.

23.4. SUMMARY

A model for strain hardening of MC and NMC materials under structural and high-rate superplasticity has been developed. It is shown that in superplastic conditions the dependence of flow stress on strain is determined by the internal stress fields created by defects embedded in grain boundaries during intragranular slip. Expressions are obtained to describe effects of strain rate, temperature and structural parameters on the flow stress of NMC materials during superplastic deformation.

Table 23.2. Experimental and theoretical values of hardening parameters for NMC materials in superplastic conditions

ZK60 alloy T_{def}=473 K (d= 0.8 μm) (Figueiredo et al. 2006)	$\dot{\varepsilon}$, s^{-1}	$3.3\cdot10^{-5}$	$1\cdot10^{-4}$	$2\cdot10^{-4}$	$3.3\cdot10^{-4}$	$1\cdot10^{-3}$	$1\cdot10^{-2}$
	σ^*, MPa	6	10	14	16	28	52
	ε^*	0.59	0.79	0.96	0.74	0.41	0.26
	m_{exp}	0.42			0.38		
	K^ε	$K^\varepsilon_{(exp)}$ = 0.27, $K^\varepsilon_{(th)}$ = 0.58			$K^\varepsilon_{(exp)}$ = -0.28, $K^\varepsilon_{(th)}$ = -0.14		
	K^σ	$K^\sigma_{(exp)}$ = 0.47, $K^\sigma_{(th)}$ = 0.50			$K^\sigma_{(exp)}$ = 0.33, $K^\sigma_{(th)}$ =0.28		

Al-3.3%Mg-0.2%Sc-0.2%Zr alloy (d= 0.2 μm) (Geng et al. 2004)	T_{def}=523 K	$\dot{\varepsilon}$, s^{-1}	$5\cdot10^{-4}$	$1\cdot10^{-3}$	$1\cdot10^{-2}$	$1\cdot10^{-1}$
		σ^*, MPa	85	120	195	260
		ε^*	0.14	0.18	0.26	0.26
		m_{exp}	0.20			
		K^ε	$K^\varepsilon_{(exp)}$ = 0.20, $K^\varepsilon_{(th)}$ =0.40			
		K^σ	$K^\sigma_{(exp)}$ = 0.21, $K^\sigma_{(th)}$ =0.25			
	T_{def}=623 K	$\dot{\varepsilon}$, s^{-1}	$5\cdot10^{-4}$	$1\cdot10^{-3}$	$1\cdot10^{-2}$	$1\cdot10^{-1}$
		σ^*, MPa	25	30	50	90
		ε^*	0.14	0.18	0.22	0.23
		m_{exp}	0.24			
		K^ε	$K^\varepsilon_{(exp)}$ = 0.14, $K^\varepsilon_{(th)}$ =0.28			
		K^σ	$K^\sigma_{(exp)}$ = 0.24, $K^\sigma_{(th)}$ =0.25			
	T_{def}=773 K	$\dot{\varepsilon}$, s^{-1}	$5\cdot10^{-4}$	$1\cdot10^{-3}$	$1\cdot10^{-2}$	$1\cdot10^{-1}$
		σ^*, MPa	4	7	16	24
		ε^*	0.14	0.18	0.22	0.26
		m_{exp}	0.31			
		K^ε	$K^\varepsilon_{(exp)}$ = 0.10, $K^\varepsilon_{(th)}$ =0.07			
		K^σ	$K^\sigma_{(exp)}$ = 0.33, $K^\sigma_{(th)}$ =0.25			

Table 23.2. (Continued)

Alloy Mg-9%Al T_{def}=423 K (d= 0.15 μm) (Kai et al. 2008)	$\dot{\varepsilon}$, s^{-1}	$5 \cdot 10^{-4}$	$1 \cdot 10^{-3}$	$5 \cdot 10^{-3}$		
	σ^*, MPa	25	29	45		
	ε^*	0.34	0.37	0.69		
	m_{exp}	0.26(*)				
	K^{ε}	$K^{\varepsilon}_{(exp)}$ = 0.33, $K^{\varepsilon}_{(th)}$ =0.22				
	K^{σ}	$K^{\sigma}_{(exp)}$ = 0.26, $K^{\sigma}_{(th)}$ =0.25				
Alloy 1570 T_{def}=723 K (d= 1 μm) (Musin et al. 2004)	$\dot{\varepsilon}$, s^{-1}	$1.4 \cdot 10^{-3}$	$1.4 \cdot 10^{-2}$	$5.6 \cdot 10^{-2}$	$1.4 \cdot 10^{-1}$	
	σ^*, MPa	9	11	17	27	
	ε^*	1.4	1.6	1.8	1.8	
	m_{exp}	0.31				
	K^{ε}	$K^{\varepsilon}_{(exp)}$ = 0.06, $K^{\varepsilon}_{(th)}$ =0.07				
	K^{σ}	$K^{\sigma}_{(exp)}$ = 0.23, $K^{\sigma}_{(th)}$ =0.25				
Alloy Al-7034 T_{def}=623 K (d= 0.3 μm) (Xu et al. 2003)	$\dot{\varepsilon}$, s^{-1}	$1.0 \cdot 10^{-3}$	$1.0 \cdot 10^{-2}$	$3.3 \cdot 10^{-2}$	$1.0 \cdot 10^{-1}$	
	σ^*, MPa	18	26	34	55	
	ε^*	0.26	0.34	0.47	0.34	
	m_{exp}	0.33				
	K^{ε}	$K^{\varepsilon}_{(exp)}$ = 0.08, $K^{\varepsilon}_{(th)}$ =0.01				
	K^{σ}	$K^{\sigma}_{(exp)}$ = 0.23, $K^{\sigma}_{(th)}$ =0.25				
Alloy ZK61 T_{def}=473 K (d= 0.5 μm) (Watanabe et al. 1999b)	$\dot{\varepsilon}$, s^{-1}	$1 \cdot 10^{-5}$	$1 \cdot 10^{-4}$	1.10^{-3}	$1 \cdot 10^{-2}$	$1 \cdot 10^{-1}$
	σ^*, MPa	16	27	50	105	165
	ε^*	0.35	1.10	1.50	0.95	0.40
	m_{exp}	0.20				
	K^{ε}	$K^{\varepsilon}_{(exp)}$ = 0.31, $K^{\varepsilon}_{(th)}$ =0.40				
	K^{σ}	$K^{\sigma}_{(exp)}$ = 0.26, $K^{\sigma}_{(th)}$ =0.25				

Chapter 24

YIELD STRESS OF NMC METALS AT ROOM TEMPERATURE. HALL-PETCH RELATION

24.1. PROBLEM DESCRIPTION

In polycrystalline metals, the influence of the average grain size d on the yield stress σ_y is conventionally described by the Hall-Petch relation (Eds. Cahn, 1983):

$$\sigma_y = \sigma_0 + K/\sqrt{d}, \quad (24.1)$$

where σ_0 is the macroelastic limit of the yield stress, K is the Hall-Petch coefficient characterizing the contribution of grain boundaries to hardening. The macroelastic limit σ_0 is the minimum stress necessary for the onset of microplastic deformation inside individual polycrystal grains (Dudarev et al. 1988; Ed. McMahon, 1967; Golovin et al. 1980; Goldstein et al. 1986). To start microplastic deformation, it is necessary to apply external stress σ exceeding the total amount of internal stresses created by intragranular obstacles that inhibit dislocation glide:

$$\sigma \geq \sigma_0 = \sigma_{PN} + \sigma_{ss} + \sigma_d + \sigma_p, \quad (24.2)$$

where σ_{PN} is the lattice friction (the Paierls-Nabarro stress), σ_{ss} is the stress connected with alloying elements dissolved in the lattice or "solid-solution hardening", σ_d is the stress created by dislocations or "dislocation hardening", σ_p is the stress created by second-phase dispersed particles. The above-mentioned values can be evaluated from the well-known relationships (Goldstein et al. 1986).

For the theoretical justification of Eq. (24.1) and evaluation of coefficient K, a number of models were suggested: (i) a "pile-up" model attributing the grain size effect to the stress concentration in dislocation pile-ups; (ii) a strain hardening model accounting for the dislocation density and their run distance, (iii) a model of surface or grain boundary dislocation sources in transferring a slip from grain-to grain; (iv) a model involving accumulation of incidental and geometrically necessary dislocations, and others (Shtremel et al. 1997; Noskova et al. 2003). All these models provide a qualitative explanation for the power-law dependence of yield stress on grain size under the assumption that parameter K is constant and does not depend the deformation conditions such as temperature T, strain ε and strain rate $\dot{\varepsilon}$.

A traditional procedure for the determination of σ_0 and K values consists in plotting σ_y vs. $d^{-1/2}$, which gives nearly a straight line. Its inclination angle corresponds to the value of K while the intersection with the ordinate axis gives σ_0 (Figure 24.1). For obtaining materials with different grain sizes d, annealing of an ultrafine grain material to different temperatures is usuall performed. Such procedure assumes that both σ_0 and K are independent of the grain size and annealing conditions. However, it is well-proven for the case of fully annealed polycrystals where the hardening coefficient K is mainly connected with the presence of grain boundaries acting as "crystallographic" obstacles for dislocation glide. By now, extensive experimental data (including those for coarse-grained materials) have been accumulated which cannot be interpreted within the traditional assumption about constancy of K and σ_0. In particular, a substantial dependence of K on the prestrain value, strain rate, temperature, pre-recrystallization annealing time and other factors was revealed (Kashyap et al. 1997; Meyers et al. 1995; Panin et al. 1990).

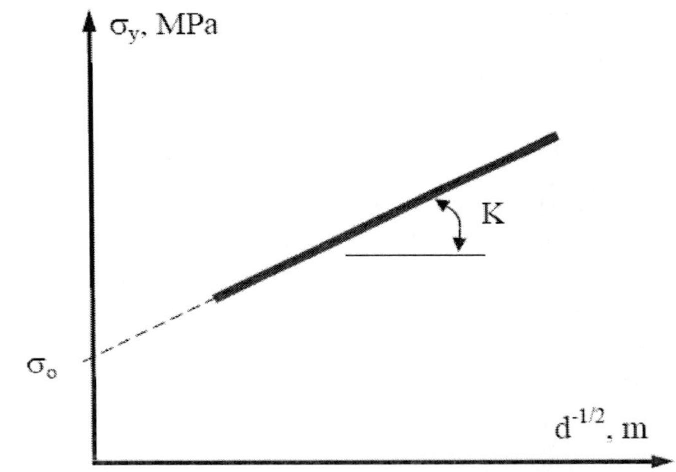

Figure 24.1. Determination of Hall-Petch parameters using a traditional approach.

In regard to NMC metals, the existing data are contradictory. In some works (del Valle et al. 2006; Sergueeva et al. 2001; Furukawa et al. 1997; Firstov et al. 2004; Kim et al. 2004), the classic linear dependence of σ_y on $d^{-1/2}$ was observed while other works reported deviations from such dependence (Conrad, 2003; Shafler et al. 2007; Pozdnyakov, 2003; Su et al. 2005; Kim et al. 2002). Moreover, a decrease of the yield stress in a range of very small grain sizes was observed in some works (Pozdnyakov, 2003; Su et al. 2005; Kim et al. 2002). This phenomenon was termed as the "inverse" or "anomalous" Hall-Petch relation. A non-monotonic dependence of K on the temperature of recrystallization annealing was also noted (Munoz-Morris et al. 2003; Valiev et al. 2000). The "anomalous" behaviour of σ_y with $d^{-1/2}$ was observed in metals with a grain size of 0.1-0.2 μm and less as well as in a number of NMC metals produced by ECAE with relatively coarse micron-sized grains. A transition point to the anomalous behavior occurs at a grain size which has not been uniquely determined and depends on many factors. Consequently, the transition point in curves $\sigma_y \sim d^{-1/2}$ in areas of small grain sizes (d<1 μm) should be used very carefully when determining the Hall-Petch parameters.

It should be noted that determination of the Hall-Petch parameters for NMC metals is connected with a one more problem. For obtaining of different grain sizes, these metals should be annealed at low temperatures. It has been revealed experimentally that a number of NMC metals exhibit the anomalous grain growth and bimodal microstructures during annealing (Nokhrin et al , 2004; Molodova et al. 2007; Wang et al. 1996) when an evaluation of the average grain size in the Hall-Petch equation is complicated. Additionally, a hardening effect detected in many NMC metals after annealing (Nokhrin et al. 2003) should be taken into account. This effect was observed in the pre-recrystallization range of temperatures in Cu (Amirkhanov et al. 1998; Chuvil'deev et al. 2003c), Mo (Korznikov et al. 1998), Fe (Ivanisenko et al. 1999; Korznikov et al. 1993), Ti (Valiev et al. 1998), Ni (Chuvil'deev et al. 2003c), and some other metals. Such hardening is substantial. For example, annealing of NMC Armco iron at temperatures 673÷773 K increases the yield stress by 50% (Ivanisenko net al. 1999; Korznikov et al. 1993).

In order to overcome these difficulties and verify the Hall-Petch relation for NMC metals, new theoretical and experimental approaches should be developed. For the theoretical analysis of this problem, we will use the theory of nonequilibrium grain boundaries. In accordance with its concepts, parameters K and σ_0 in NMC metals are not constants and depend on the density of defects embedded in grain boundaries, i.e. on the level of their nonequilibrium (Chuvil'deev, 2004). Therefore, the annealing to control the grain size also changes the density of defects embedded in grain boundaries. This means that K and σ_0 may be different at each point of a curve $\sigma_y(d^{-1/2})$, and a standard procedure for evaluation of Hall-Petch parameters seems to be inapplicable for NMC metals.

In this Chapter, a related model is developed and compared with experimental results that takes into account effects of diffusion-controlled recovery and recrystallization on the macroelastic limit σ_0, yield stress σ_y and grain boundary hardening coefficient K.

24.2. Yield Stress-Grain Size Relation in NMC Metals

24.2.1. Model

Plastic deformation in metals occurs if the stress condition for dislocation slip is satisfied. In the case of a monocrystal, this condition can be presented as

$$\sigma \geq \sigma_0, \qquad (24.3)$$

where the macroelastic limit σ_0 depends on the nature of obstacles during a dislocation glide through the lattice (Golovin et al. 1980; Dudarev et al. 1988; Goldstein et al. 1986), σ is the externally applied stress. In polycrystals, along with obstacles distributed inside grains, an additional stresses should be applied to cross grain boundaries:

$$\sigma \geq \sigma_0 + \sigma_i^{(b)}. \tag{24.4}$$

Here $\sigma_i^{(b)}$ can be termed as a grain boundary barrier to plastic deformation. This parameter can be presented as the sum of long-range stress fields developed by defects embedded in grain boundaries and a short-range or "geometrical" barrier. If dislocation pile-ups are not formed near obstacles, yield stress σ_y is:

$$\sigma_y = \sigma_0 + \sigma_i^{(b)}. \tag{24.5}$$

For pile-ups adjacent to grain boundaries, the stress generated at the head of a pile-up is (Hirth et al. 1969)

$$\sigma_i^* = n_\perp(\sigma)\sigma, \tag{24.6}$$

where $n_\perp(\sigma) = \pi(1-\nu)(\sigma/G)(L_\perp/b)$. When this stress exceeds some critical value σ^* in any of the adjacent grains (Figure 24.2), plastic deformation in the whole crystal begins:

$$\sigma_i^* = \sigma^* \tag{24.7}$$

Within a traditional approach (Hirth et al. 1969), $n_\perp$ and σ^* are determined under the assumption that grain boundaries are purely geometrical obstacles and do not create stress fields that could inhibit the pile-up generation and propagation of plastic deformation to adjacent grains. This is true for well-annealed coarse-grained materials.

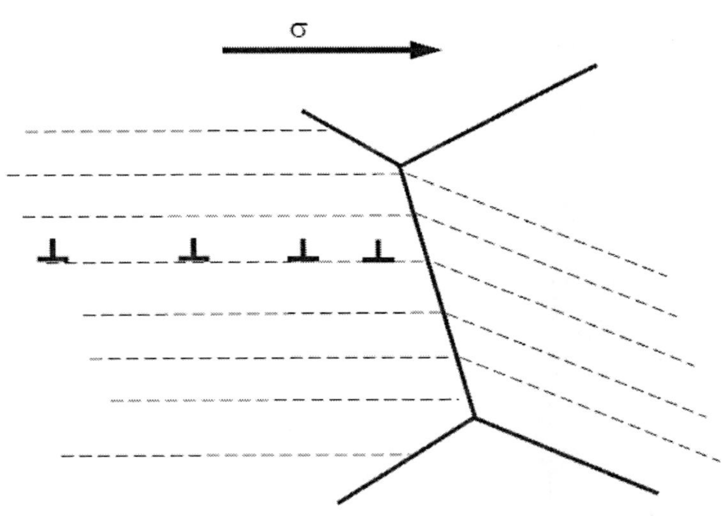

Figure 24.2. Generation of a dislocation pile-up within the traditional Hall-Petch model.

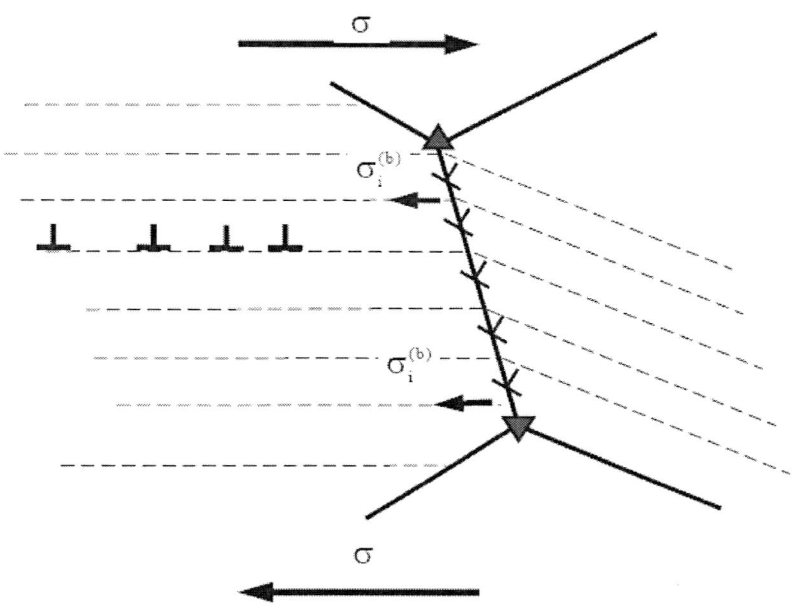

Figure 24.3. Generation of a dislocation pile-up in NMC materials.

For NMC materials, the situation is different. Grain boundaries in NMC metals after ECAE have an excess density of embedded defects which create long-range stress fields $\sigma_i^{(b)}$ inhibiting the glide of lattice dislocations and generation of dislocation pile-ups. Then, the stress applied to the head dislocation in the pile-up can be presented as

$$\sigma_i^* = n_\perp(\sigma, \sigma_i^{(b)}) \cdot (\sigma - \sigma_i^{(b)}). \tag{24.8}$$

where $n_\perp(\sigma, \sigma_i^{(b)})$ is the number of dislocations in a pile-up which depends on σ and $\sigma_i^{(b)}$:

$$n_\perp(\sigma, \sigma_i^{(b)}) = \pi(1-\nu)([\sigma - \sigma_i^{(b)}]/G)(L_\perp/b). \tag{24.9}$$

The condition to propagate deformation into adjacent grains (Eq. (24.7)) should include internal stress fields $\sigma_i^{(b)}$, and can be written as

$$\sigma_i^* = \sigma^* + \sigma_i^{(b)}. \tag{24.10}$$

Inserting a general relationship $L_\perp = d/2$ into Eq. (24.9) gives the relation between applied stress σ and grain size d:

$$\sigma = \sigma_i^{(b)} + \sqrt{\frac{b}{d}\frac{2G(\sigma^* + \sigma_i^{(b)})}{\pi(1-\nu)}}. \tag{24.11}$$

Therefore, with generation of dislocation pile-ups, Eq. (24.5) for yield stress in NMC metals is

$$\sigma_y = \sigma_0 + \sigma_i^{(b)} + \sqrt{\frac{2G(\sigma^* + \sigma_i^{(b)})}{\pi(1-\nu)}\left(\frac{b}{d}\right)}. \tag{24.12}$$

Similarly to the classic theory (Hirsh et al. 1969), the macroelastic limit σ_0 for polycrystals in Eq. (24.12) depends on obstacles for a dislocation glide such as lattice dislocations, solute atoms, etc.

In well-annealed coarse-grained (CG) metals, the effect of long-range stress fields on grain boundaries is insignificant $\sigma_i^{(b)} = 0$, and Eq. (24.12) may be rearranged to the classical Hall-Petch relation:

$$\sigma_y^{cg} = \sigma_0^{cg} + \sqrt{\frac{2G\sigma^*}{\pi(1-\nu)}\left(\frac{b}{d}\right)} = \sigma_0^{cg} + \frac{K_0}{\sqrt{d}} \tag{24.13}$$

Here σ_y^{cg} and σ_0^{cg} are, respectively, yield stress and macroelastic limit in coarse-grained metals, and K_0 is the of Hall-Petch coefficient:

$$K_0 = \sqrt{\frac{2Gb\sigma^*}{\pi(1-\nu)}}. \tag{24.14}$$

In NMC materials, the contribution of $\sigma_i^{(b)}$ becomes significant due to defects distributed in the boundaries $\sigma_i^{(b)}(\rho_b, w_t)$(Chuvil'deev, 2004). As demonstrated above, a complex of defects distributed in grain boundaries can be reduced to systems of orientation misfit dislocations with density ρ_b and tangential components of delocalized dislocationstangential components of Burgers vector of delocalized dislocations with the density of Burgers w_t. If grain boundaries contain such defects, $\sigma_i^{(b)}$ can be described by Eq. (16.26). Inserting Eq. (16.26) in Eq. (24.7), we obtain a formula for yield stress of NMC metals without pile-ups at grain boundaries:

$$\sigma_y = \sigma_0 + \varphi_1 G\Delta b\rho_b + \varphi_2 Gw_t. \tag{24.15}$$

When there are dislocation pile-ups at near grain boundaries, the combination of Eq. (16.26) and Eq. (23.11) gives similar expression for yield stress:

$$\sigma_y = \sigma_0 + \varphi_1 G\rho_b \Delta b + \varphi_2 G w_t + \sqrt{\frac{2G(\varphi_1 G\rho_b \Delta b + \varphi_2 G w_t + \sigma^*)}{\pi(1-\nu)}\left(\frac{b}{d}\right)} \qquad (24.16)$$

In its turn, the density of defects distributed in the boundaries (ρ_b and w_t) depends on the rate and the temperature of prestrain (Chuvil'deev, 2004), as well as on the time and temperature of after-strain annealing (Chuvil'deev et al. 2001a). An analysis of defect evolution during and after ECAE is necessary to evaluate the yield stress under different conditions. But first, we shall consider the generation of dislocation pile-ups in NMC materials.

Let us estimate the minimum grain size d_{min} at which the formation of dislocation pile-ups is still possible. A traditional approach is based on the classical formula (24.9) for the amount of dislocations in a pile-up where $L_\perp = d/2$. With the lowest number of dislocations n=2, the minimum grain size is

$$d_{min} = \frac{4b}{\pi(1-\nu)}\left(\frac{G}{\sigma}\right). \qquad (24.17)$$

For the characteristic yield stress $\sigma=\sigma_y \sim 10^{-2}G$ in NMC metals, the last equation gives $d_{min}=10\div 50$ nm (see, for example, (Andrievskii et al. 2000; Estrin et al. 2007)).

In our opinion, for NMC materials this estimation is not correct. Unlike coarse-grained material where grain boundaries act only as geometrical obstacles for dislocations, in NMC materials defects distributed in the grain boundaries create long-range stress fields $\sigma_i^{(b)}$. Accordingly, Eq. (24.17) becomes

$$d_{min}^{(*)} = \frac{4b}{\pi(1-\nu)}\left(\frac{G}{\sigma - \sigma_i^{(b)}}\right). \qquad (24.18)$$

As the value of $\sigma_i^{(b)}$ calculated from Eq. (16.26) is about $(1/2\div 2/3)\sigma_y$ (see Table 24.1), Eq. (24.18) show that $d_{min}^{(*)}$ by the factor of 2-3 exceeds d_{min} obtained within a traditional approach. The numerical estimates of $d_{min}^{(*)}$ for different NMC materials using the data on σ_y and $\sigma_i^{(b)}$ are presented in Table 24.1. It is seen that the minimum grain size $d_{min}^{(*)}$ approximates to 0.5 μm and appears to be higher than the limit of grain refinement d^* of the same material under ECAE (see Chapter 19).

Table 24.1. Calculated minimum grain size d_{min} for the formation of dislocation pile-ups in NMC metals

Metal	$\rho_b^{st}\Delta b$	G, MPa	b, m	$\sigma_i^{(b)}$, MPa	σ_y, MPa	$d_{min}^{(*)}$, μm
Mg	$1\cdot10^{-2}$	16600	$3.21\cdot10^{-10}$	166	180	0.72
Al	$1\cdot10^{-2}$	25000	$2.86\cdot10^{-10}$	250	275	0.54
Cu	$1\cdot10^{-2}$	42000	$2.56\cdot10^{-10}$	420	450	0.68
Ni	$1\cdot10^{-2}$	78900	$2.49\cdot10^{-10}$	790	850	0.61

24.2.2. Hall-Petch Parameters

We suppose that the grain boundaries after ECAE contain excess density of defects ρ_b and w_t which correspond to the stationary values ρ_b^{st} and w_t^{st} of the density accumulated in the course of ECAE. This is possible when the recovery processes in grain boundaries after deformation have not accomplished yet, e.g., in the case of fast cooling (quenching) of samples from the ECAE temperature and a small aging time at room temperature. Stationary values $\Delta b\rho_b^{st}$ and w_t^{st} are determined from Eqs. (15.151)-(15.152) at $I = \xi\dot{\varepsilon}_v / b$:

$$\rho_b^{st}\Delta b = (g_1\dot{\varepsilon}_v / D_b^*)^{1/4}, \tag{24.19}$$

$$w_t^{st} = d(g_2\dot{\varepsilon}_v / D_b^*)^{1/2} \quad . \tag{24.20}$$

where g_1 and g_2 are numerical parameters.

Consider, first, the case when the dislocation pile-ups are not formed during deformation of NMC materials. Combination of Eqs. (24.19), (24.20) and (24.15) gives

$$\sigma_y^{(1)} = \sigma_o^{(1)} + K_1 d, \tag{24.21}$$

where $\sigma_0^{(1)}$ and K_1 are:

$$\sigma_0^{(1)} = \sigma_0 + \varphi_1 G(g_1\dot{\varepsilon}_v / D_b^*)^{1/4}, \tag{24.22}$$

$$K_1 = \varphi_2 (g_2\dot{\varepsilon}_v / D_b^*)^{1/2}. \tag{24.23}$$

A qualitative form of dependence $\sigma_y^{(1)}$ on d for NMC metals is compared with that for coarse-grain materials in Figure 24.4. It is seen that in the domain of small grain sizes where the dislocation pile-ups at grain boundaries are not formed in NMC metals, the yield stress dependence on grain size is described as $\sigma_y = \sigma_0 + K_1 d$. In coordinates "$\sigma_y$ vs. $d^{-1/2}$", this dependence is termed as "inverse Hall-Patch" (Pozdnyakov, 2003; Su et al. 2005; Kim et al. 2002). In this case, the grain boundary hardening coefficient K_1 is connected with the grain boundary diffusion coefficient D_b^*, which, in turn, depends on the aging time and temperature after ECAE (i.e., before σ_y measurement) due to the development of recovery processes (Chuvil'deev et al. 2001a; Grabovetskaya, 2008; Kolobov et. al., 2001c).

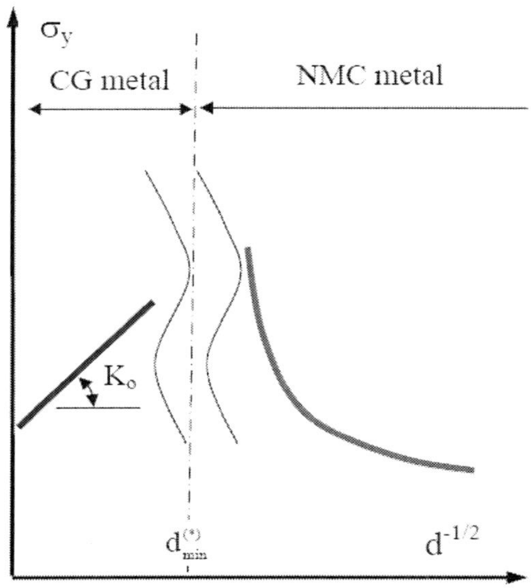

Figure 24.4. Qualitative form of "yield stress–grain size $d^{-1/2}$" dependence with no pile-ups: d^* is the grain refinement limit (see Chapter 21), $d_{min}^{(*)}$ is the minimum grain size at which the pile-up generation is possible.

For generation of dislocation pile-ups, the dependence of σ_y on grain size d is described by Eq. (24.16). If the long-range stress fields associated with grain boundaries are small ($\sigma_i^{(b)}$~0), the dependence of σ_y on d will have the following form (see Eq.(24.5)). On the contrary, if stress fields of defects distributed in grain boundaries play the main role (σ_y~ $\sigma_i^{(b)}$), Eq. (24.16) for the yield stress can be analyzed for two limiting cases. In the first, when $\rho_b^{st}\Delta b >> w_t^{st}$ and OMD dominate in grain boundaries, Eq. (24.16) can be presented as

$$\sigma_y^{(2)} = \sigma_0 + \sigma_i^{\rho} + \frac{K_2}{\sqrt{d}}, \qquad (24.24)$$

where

$$\sigma_i^{\rho} = \varphi_1 G(g_1\dot{\varepsilon}_v / D_b^*)^{1/4}, \qquad (24.25)$$

$$K_2 = \sqrt{\frac{2Gb}{\pi(1-\nu)}\left[\sigma_0 + \varphi_1 G\left(\frac{g_1\dot{\varepsilon}_v}{D_b^*}\right)^{1/2}\right]}. \qquad (24.26)$$

In this case, the dependence $\sigma_y \sim d^{-1/2}$ is given by a classical Hall-Petch relation, but the values of coefficient K and macroelastic limit σ_0 differ from those for coarse-grained materials by a value proportional to the density of orientational misfit dislocation embedded in the grain boundaries during ECAE (Figure 24.5). In the second, when $\rho_b^{st}\Delta b << w_t^{st}$ and tangential components of delocazed dislocation dominate in grain boundaries, Eq. (24.16) can be represented as

$$\sigma_y^{(3)} = \sigma_0 + \sigma_i^w(d) + K_3(d), \tag{24.27}$$

where

$$\sigma_i^w(d) = \varphi_2 G d\sqrt{g_2\dot{\varepsilon}_v / D_b^*(d)}, \tag{24.28}$$

$$K_3(d) = \sqrt{\frac{2Gb}{\pi(1-\nu)}\left[\frac{\sigma_o}{d} + \varphi_2 G\left(\frac{g_2\dot{\varepsilon}_v}{D_b^*(d)}\right)^{1/2}\right]}. \tag{24.29}$$

In this case, the dependence $\sigma_y \sim d^{-1/2}$" have a complex form determined by the interplay of parameters $\sigma_y \sim K_0/\sqrt{d}$, $\sigma_i^w \sim d/\exp(\varphi_{2d}d)$ and $K_3(d)$.

For small grain sizes $d<d_1$, yield stress is described by Eq. (24.24) (Figure 24.5). For intermediate grain sizes $d_2>d>d_1$ and under domination of tangential components of delocalized dislocationstangential components of Burgers vector of delocalized dislocations in grain boundaries, the yield stress dependence on grain size is linear and the inverse Hall-Petch effect is observed in curve $\sigma_y \sim d$. For coarse grain sizes $d>d_2$, the classical dependence $\sigma_y \sim d^{-1/2}$ is valid (Figure 24.5).

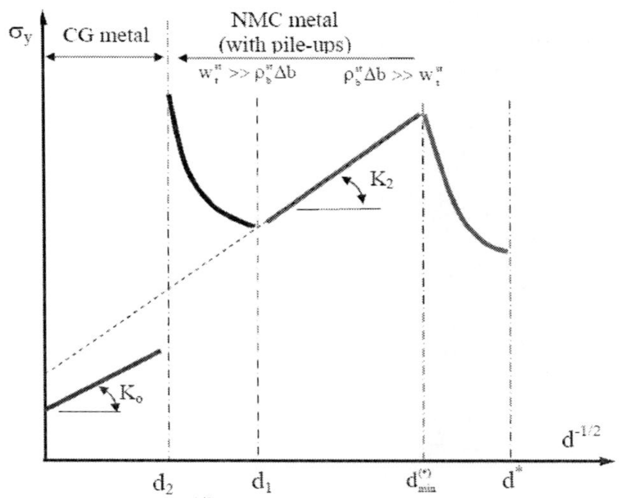

Figure 24.5. Yield stress vs. grain size $d^{-1/2}$ for NMC metals.

24.3. Comparison with Experimental Data

When choosing particular experimental data for analysis and comparison, it is necessary to take into account that, within the developed model, at least three types of dependences $\sigma_y(d)$ can exist in an NMC material in different ranges of grain sizes. This imposes stringent requirements on studied grain sizes because the curve $\sigma_y(d)$ should have a sufficient number of experimental points. That substantially narrows the number of works suitable for analysis. Another limitation relates to the parameters used in the model. They include (i) grain size and mechanical properties measured in the same material with an NMC coarse-grained structures, (ii) temperature, strain, strain rate and deformation history during SPD processing, and (iv) strain dependence of the density of lattice dislocations. It should also be noted that in literature there is a vast body of experimental results on the dependence of the ultimate strength, σ_s, on grain size. However, these data are not relevant here because the mechanisms determinimg the yield stress and the ultimate strength are substantially different (see Chapter 25).

According to the present model, three types of relationships between σ_y and d are possible in NMC materials:

1 The Hall-Petch relation in the range of submicron grain sizes $d^{(*)}_{min} < d < d_l$ when dislocation pile-ups can be formed in NMC grains, orientational misfit dislocations dominate within grain boundaries ($\rho^{st}_b \Delta b >> w^{st}_t$), and the yield stress dependence on grain size is determined by Eqs. (24.24)-(24.26) (see Table 24.2);

2 Deviation from the Hall-Petch relation in the range of relatively coarse (micron-sized) grains $d \geq d_l$ when dislocation pile-ups are generated in NMC grains, tangential components of delocalized dislocationstangential components of Burgers vector of delocalized dislocations dominate in grain boundaries ($w^{st}_t >> \rho^{st}_b \Delta b$), and dependence σ_y~d is described by Eqs. (24.27)-(24.29) (see Table 24.2);

3 The "inverse Hall-Petch" relation in the range of small (nano and submicron) grain sizes $d \leq d^{(*)}_{min}$ when dislocation pile-ups are not generated in NMC grains, the OMD density exceeds the density of tangential components of delocalized dislocationstangential components of Burgers vector of delocalized dislocations ($\rho^{st}_b \Delta b >> w^{st}_t$), and the yield stress dependence on grain size is described by Eqs. (24.21)-(24.23) (see Table 24.2).

In order to correctly calculate the dependence of σ_y on d for particular materials, it is necessary to determine the critical grain sizes d_l and $d^{(*)}_{min}$ using Eqs. (15.142) and (24.18). The choice of an appropriate formula according to a relation between d_l, $d^{(*)}_{min}$ and d is presented in Table 24.2.

Table 24.2. Dependence of yield stress on grain size for NMC metals

Type of dependence $\sigma_y(d)$ in NMC metals	Grain size range	Defect density in grain boundaries	Equations
Hall-Petch relation is valid	$d_1<d<d_{min}^{(*)}$	$\rho_b^{st}\Delta b >> w_t^{st}$	(24.24)-(24.26)
Deviation from Hall-Petch relation in the case of pile-up generation	$d \geq d_1$	$w_t^{st} >> \rho_b^{st}\Delta b$	(24.27)-(24.29)
Deviation from Hall-Petch relation due to the absence of pile-ups	$d \leq d_{min}^{(*)}$	$\rho_b^{st}\Delta b >> w_t^{st}$	(24.21)-(24.23)

Eqs. (24.21) and (24.24) include material parameters σ_0, σ^*, K_0 and characteristics of grain boundary diffusion, in particular Q_b^*. For rigorous calculation of σ_0, Eq. (24.2) should be used. However, for approximate evaluations, it can be assumed that in pure NMC metals the main contribution to σ_0 is due to dislocation hardening $\sigma_0 \cong \sigma_d = \alpha_\rho Gb\sqrt{\rho_v}$, where α_ρ=0.3. The density of lattice dislocations ρ used in calculations is often presented in experimental papers. If this value is not available, it can be estimated using the Orowan's relationship $\varepsilon=\rho b\lambda_\perp$ assuming that the free run of dislocations is $\lambda_\perp$=d. The value of parameter σ^* is connected with K_0 (Eds. Burke, 1973) for coarse-grained materials; which the latter can also be easily calculated from Eq. (24.14).

The comparison of experimental results and theoretical predictions set forth below demonstrates that different $\sigma_y(d)$ relations may be valid for particular materials and processing and testing conditions.

24.3.1. Hall-Petch Relation in NMC Materials

Interstitials Steel

The structure and properties of an interstitial free (IF) NMC steel produced by ECAE with strain rate $\dot{\varepsilon}_v = 10^{-3}$ s^{-1} were examined in (Messemaeker et al. 2004). The grain size varied from 55 μm to about 0.5 μm; corresponding yield stresses were measured by compression tests and changed from 150 MPa to 560 MPa. Values of σ_y versus d for parameters of Table 24.3 were calculated from Eqs. (24.24)-(24.26). The experimental and theoretical data fotr $\sigma_y \sim d^{-1/2}$ are displayed in Figure 24.6 and show a satisfactory agreement between.

Al-7%Si Aluminum Alloy

The influence of a number of ECAE passes on the structure and mechanical properties of an Al-7%Si NMC alloy was investigated in (Gutierrez-Urrutia et al. 2007). A cast alloy was subjected to 1–10 ECAE passes at room temperature with strain rate 10^{-1} s^{-1}. With an increase of passes, the average grain size reduces from about 500 μm to 0.17-0.42 μm while the yield stress increases from 90 to 245 MPa.

It has been accepted that strain for each ECAE pass was about unity, and the density of lattice dislocations with changing the number of ECAE passes from 1 to 10 increased from $6.5 \cdot 10^{13}$ m^{-2} to $1.2 \cdot 10^{15}$ m^{-2}. Results of calculations similar to the interstitial steel show close correlation with experimental data (Figure 24.7).

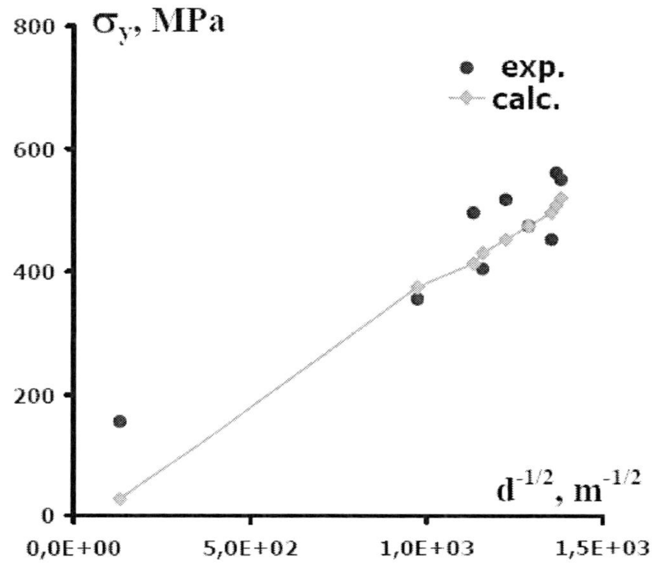

Figure 24.6. Experimental (Messemeker et al. 2004) and theoretical dependences of yield stress on grain size for a IF steel.

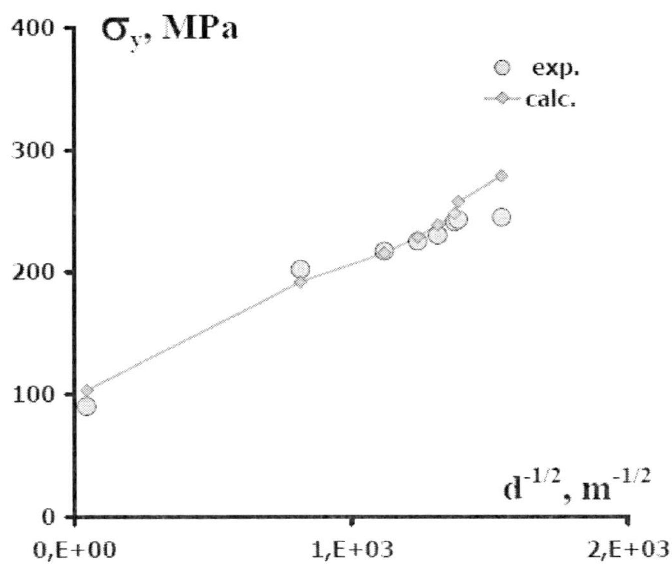

Figure 24.7. Experimental (Gutierrez-Urrutia et al. 2007) and theoretical dependence of yield stress on grain size for Al-7%Si NMC alloy.

24.3.2. Deviation from the Hall-Petch Relation in Micron Grain Size Materials

Magnesium Alloy AZ61

Experimental research of magnesium alloy AZ61 processed by ECAE at temperature 578K and strain rate $4\cdot10^{-2}$ s^{-1} (Kim, Hong, 2003) showed that the Hall-Petch relation is true for grain sizes larger than 10-16 μm. However, a systematic decrease of yield stress from the Hall-Petch formula was observed for smaller grains. Corresponding theorethical calculations were obtained using Eqs. (24.27)-(24.29) and Table 24.3 and assuming that the density of lattice dislocations increases from $1.8\cdot10^{13}$ m^{-2} to $2.8\cdot10^{15}$ m^{-2} with changing the number of ECAE passes from 1 to 10 (Kim, Hong, 2003). Again, a good agreement between both data can be seen in Figure 24.8.

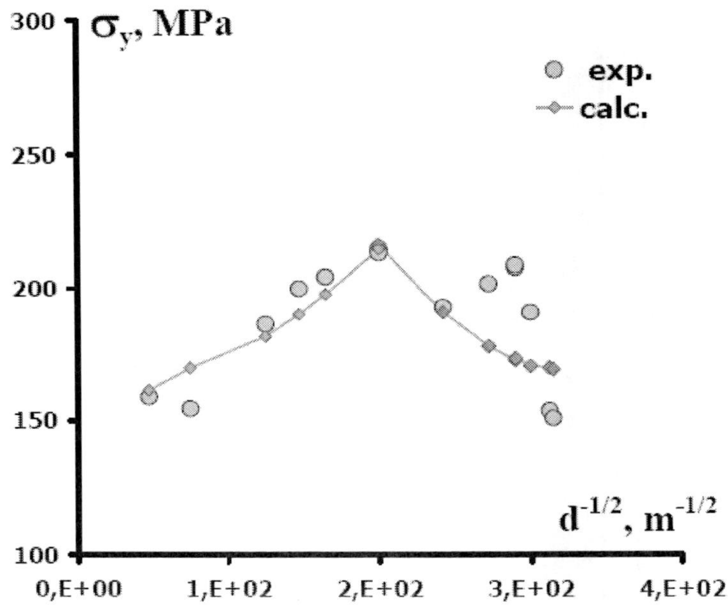

Figure 24.8. Experimental (Kim et al. 2003) and theoretical dependence of yield stress on grain size for AZ61 NMC alloy.

Pure magnesium and Mg-0.9%Al Alloy

Pure (99.9%) magnesium (99.9%) and a model magnesium alloy Mg-0.9%Al were subjected to N=1-4 ECAE passes at temperatures 473, 573 and 673 K with strain rate $\dot{\varepsilon}_v=2\cdot10^{-1}$ s^{-1} (Yamashita et al. 2001). The average grain size was changed from about 400 μm to 17 μm resulting in an increase of the yield stress from 60 to 140 MPa. A related dependence σ_y~$d^{-1/2}$ was nonmonotonic (Figure 24.9, blue cicrcles) and followed to the Hall-Petch formula for grain sizes more than 100 μm. In the grain size range from 50 to about 100 μm, the yield stress reduced insignificantly. With a further reduction of grains to 17 μm, the yield stress increases again. A similar result was reported for a Mg-3%Al-1%Zn NMC alloy at temperatures 423-473 K by (Barnet et al. 2004). Experimental data are quite identical to theoretical calculations shown for Mg-0.9%Al alloy in Figure 24.9 (it was assumed that the lattice dislocationdensity increased from $1.4\cdot10^{13}$ m^{-2} to $1.6\cdot10^{14}$ m^{-2} during four passes ECAE).

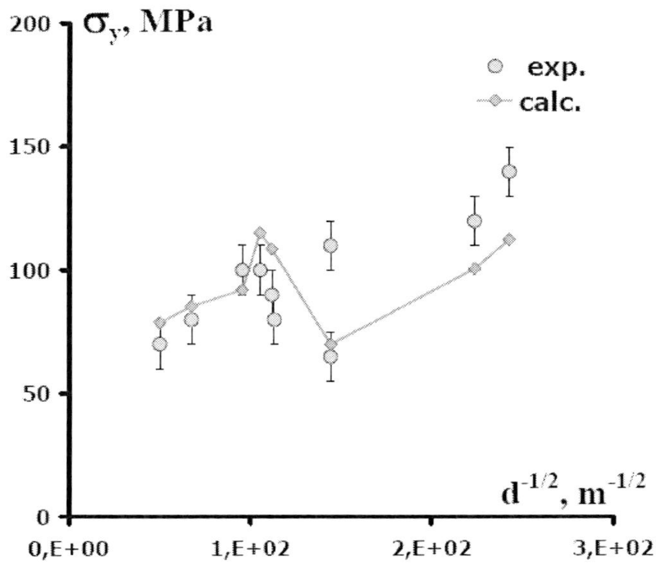

Figure 24.9. Experimental (Yamashita et al. 2001) and theoretical dependence of yield stress on grain size for ECAE of pure magnesium and Mg-0.9%Al magnesium alloy.

24.3.3. Deviation from the Hall-Petch Relation in NMC Materials

NMC Copper

The structure and properties of high-purity NMC copper produced by high-pressure torsion (HPT) were studied in (Degtyarev et al. 2007). The obtained NMC metal exhibited a non-monotonic dependence of hardness on the grain size (Figure 24.10). With reduction of grain size from 0.6 μm to 0.3 μm, the material hardness H increased from 1.63 to 1.77 GPa. However, further grain refinement during HPT resulted in a decrease of hardness to 1.58 GPa. These results coincide with related theoretical data of Figure 24.10 assuming that hardness and the yield stress are connected by the known empirical equation $H\approx3\sigma_y$.

NMC copper (99.98%) produced by N= 1, 2, 4, 8 and 12 passes of ECAE at room temperature with strain rate $1.25\cdot10^{-2}$ s^{-1} was studied in (Wei et al. 2006). This processing results in structure refinement from 106 μm to 0.15-0.2 μm. The yield stress increased from its initial value of about 40 MPa to the maximal value of 380 MPa at N=4 (d=0.2 μm), then remained constant up to N=8, and reduced to about 350 MPa with a further increase of ECAE passes to N=12 (d=0.14÷0.15 μm) (Figure 24.11). The theoretical diagram of σ_y~$d^{-1/2}$ for copper displays a similar behaviour (Figure 24.11) (in the calculations, the density of lattice dislocations increases with passes from $1.2\cdot10^{14}$ m^{-2} to $2.7\cdot10^{15}$ m^{-2} that conforms to the estimates of (Wei et al. 2006)).

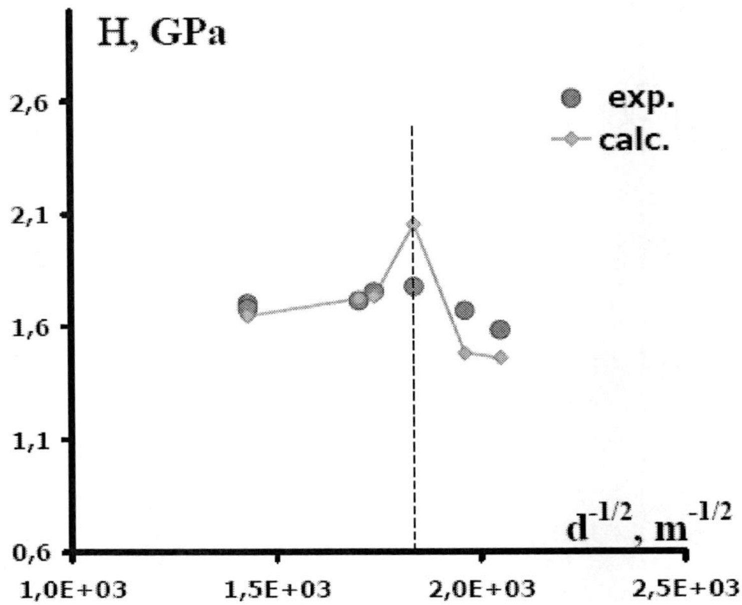

Figure 24.10. Experimental (Degtyarev et al. 2007) and theoretical dependence of the hardness of NMC copper after HPT on grain size.

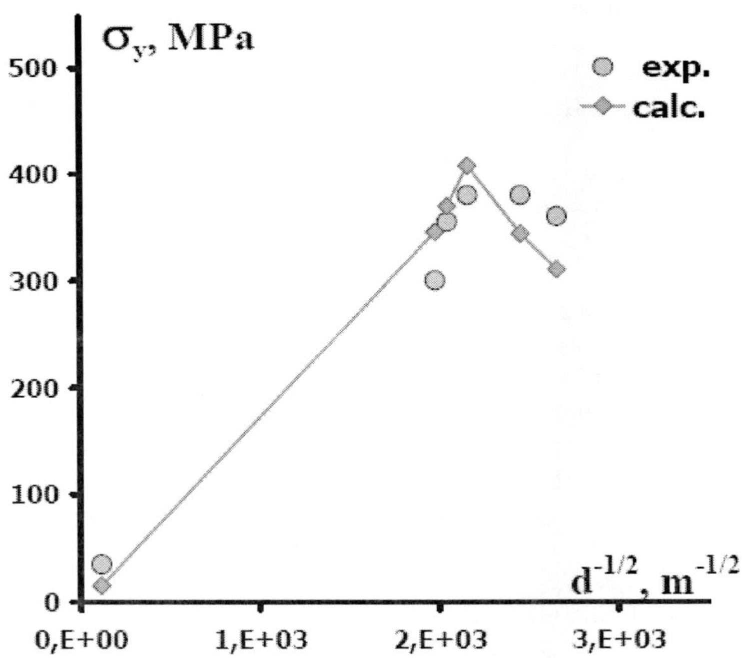

Figure 24.11. Experimental (Wei et al. 2006) and theoretical dependence of yield stress on grain size of NMC ECAE copper.

Table 24.3. Parameter values used for calculating σ_y vs. $d^{-1/2}$

Parameter	No pile-ups		Generation of pile-ups			
			$\rho_b^{st}\Delta b >> w_t^{st}$		$w_t^{st} >> \rho_b^{st}\Delta b$	
	Cu (Degtyarev et al. 2007)	Cu (Wei et al. 2006)	AZ61 (Kim et al. 2003)	Mg (Yamashita et al. 2001)	IF-steel (Messmaeker et al. 2004)	Al-7%Si (Gutierez et al. 2006)
Testing temperature (T_m/T)	4.63	4.63	3.15	3.15	6.18	3.18
Elastic modulus ($G\Omega/kT_m$) (Frost et al. 1982)	50	50	32	32	50	32
Solid/liquid interface enthalpy ($\gamma_{S/L}^0 b^2/kT_m$) (Chuvil'deev, 2004)	0.900		1.158		0.971	1.158
Solid/liquid interface entropy ($S_{S/L} b^2/k$) (Chuvil'deev, 2004)	0.767		0.920		0.833	0.920
Specific heat of fusion (λ); Density(ρ) (Ed.Grigor'ev, 1991): $\frac{\lambda \rho b^3}{kT_m}$	1.15		1.38		1.25	1.38
Surface energy of grain-boundary S-phase areasEntalpy of GB ($\gamma_b^0 b^2/kT_m$) (Chuvil'deev, 2004)	4.32		4.98		4.63	4.98
Macroelastic limit for a CG metal (σ_0^{CG}, MPa)	43÷204	79	2÷9	7÷34	17÷139	5÷81
Theoretic sStress value (σ^*, MPa)	100	450	50	50	450	50
Hall-Petch coefficient for a CG metal (K_0, $MPa \cdot m^{1/2}$)	3.3	7	1.6		8	1.8
Burgers vector ($b \cdot 10^{10}$, m) (Frost et al. 1982)	2.56		3.21		2.48	2.86

Table 24.3. (Continued)

Parameter		No pile-ups		Generation of pile-ups			
				$\rho_b^{st}\Delta b >> w_t^{st}$		$w_t^{st} >> \rho_b^{st}\Delta b$	
		Cu (Degtyarev et al. 2007)	Cu (Wei et al. 2006)	AZ61 (Kim et al. 2003)	Mg (Yamashita et al. 2001)	IF-steel (Messmaeker et al. 2004)	Al-7%Si (Gutierez et al. 2006)
Preexponential factor for the grain boundary diffusion coefficient ($\delta D_{b0}\cdot 10^{15}$, m^3/s) (Frost et al. 1982)		5.0		$5.0\cdot 10^3$		$1.1\cdot 10^3$	50
Activation energy for grain boundary diffusion (Q_b^*, kT_m)		5.6	5.8	7.9	7.0	5.1	7.5
Strain rate ($\dot{\varepsilon}_v$, s^{-1})		$2\cdot 10^{-3}$	$1.3\cdot 10^{-2}$	$4\cdot 10^{-2}$	$2\cdot 10^{-1}$	$1\cdot 10^{-2}$	$1\cdot 10^{-1}$
Numerical factors	ξ_1	$8\cdot 10^{-3}$	0.01	1	1	0.01	0.01
	ξ_2	$2\cdot 10^{-1}$	$2\cdot 10^{-3}$	$2\cdot 10^{-3}$	$1\cdot 10^{-3}$	-	-
	C_1	50	50	50	100	50	50
	A_1	10	10	10	10	10	10
The density of orientational misfit dislocation ($\rho_b^{st}\Delta b$)		$4.5\cdot 10^{-3}$	$5\cdot 10^{-3}$	$6.5\cdot 10^{-3}$	$4.7\cdot 10^{-3}$	$3.8\cdot 10^{-3}$	$5.7\cdot 10^{-3}$
The density of tangential components of delocalized dislocationstangential components of Burgers vector of delocalized dislocations (w_t^{st})		$2.5\cdot 10^{-3}$	$3\cdot 10^{-3}$	$3\cdot 10^{-3}$	$4.8\cdot 10^{-3}$	-	-
Minimum grain size (d^*_{min}, μm)		0.25	0.4	3.3	1.8	0.14	0.16

24.4. SUMMARY

A model for estimating yield stress and the parameters appearing in the Hall-Petch relation (macroelastic limit σ_0 and grain boundary strengthening factor K) is proposed with regard to NMC materials. It is shown that a specific type of the yield stress dependence on grain size in NMC materials is connected with (i) a possibility of the dislocation pile-up formation, (ii) grain size, and (iii) structural condition and diffusion properties of nonequilibrium grain boundaries. If there are no dislocation pile-ups (i.e. the grain sizes are very small), an inverse Hall-Petch effect is observed in NMC materials. In the case when dislocation pile-ups are formed and the orientational misfit dislocations dominate in the grain boundaries of NMC materials (small grain sizes), a regular Hall-Petch relation is observed. However, the values of Hall-Petch parameters for such NMC materials differ from those typical of coarse grained metals. In the case of both pile-up generation and domination of tangential components of delocalized dislocationstangential components of Burgers vector of delocalized dislocations , which corresponds to micron-sized grains, a deviation from the Hall-Petch relation takes place. The formulas are derived for the dependence of Hall-Petch parameters on the strain rate, temperature and the grain-boundary diffusion coefficient during ECAE, as well as on structural characteristics of materials.

Chapter 25

STRENGTH AND DUCTILITY OF NMC METALS AT ROOM TEMPERATURE

25.1. PROBLEM DESCRIPTION

It is known that severe plastic deformation leads to intensive material hardening. In the first works dating back to 1980s we revealed it was found that ECAE brought about a threefold-fourfold increase in strength at room temperature as compared with annealed or cast condition (Segal et al. 1981; Segal et al. 1994). As for room temperature plasticity of NMC materials, it seems to be an intricate and intriguing issue. Typically, an increase in strength is associated with a reduction of ductility. This feature is exemplified in Figure 25.1, which is often given in reviews (see, for example, (Valiev et al. 2006)). The higher the tensile strength σ_S, the lower elongation to failure ε_{max} ; it is true for most any individual materials in its different structural states (see Figure 25.2 for aluminum grade AA1100). However, a number of NMC materials are exceptions from that rule: they retain very high strength and at the same time exhibit a sufficiently high ductility comparable with initial coarse-grained (CG) counterparts. Such behavior was observed in aluminum (Sabirov et al. 2008b, Sivaraman et al. 2008; Kim et al. 2005; Rhee et al. 2005) and magnesium alloys (Cheng et al. 2007; Yamashita et al. 2001; Minger et al. 2007; Zuberova et al. 2007; Estrin et al. 2006), as well as in pure copper (Wang et al. 2002; Valiev et al. 2001, 2002; Valiev et al. 2001, 2002). This Chapter addresses strength and ductility and their unusual behavior in NMC materials.

Characteristic "stress-strain" diagrams of NMC materials at room temperature are shown in Figure 25.3. Depending on the strain rate and grain size, three types of curves are observed. For type A, ductility is restricted by material softening and flow localization. The type B exhibits a pronounced yield stress plateau with a good combination of high strength and ductility. For curves C, a prolonged stage of hardening and stable flow is followed by an abrupt failure. When describing the plasticity of NMC materials during tensile tests, homogeneous deformation ε^* and the total deformation to failure ε_{max} should be clearly distinguished because they are controlled by various physical mechanisms and present a different practical interest. Examples of real stress-strain curves $\sigma(\varepsilon)$ for aluminum (Tsuji et al. 2002), aluminum alloy (Wang et al. 2008) and steel (Tsuji et al. 2002) are shown in figures 25.4, 25.5. In these cases, the yield stress $\sigma_{0.2}$ and the ultimate tensile strength σ_S differ substantially. As shown in Chapter 24, the dependence of $\sigma_{0.2}$ on a graind size d in NMC materials can be described in terms of the Hall-Petch relation while σ_B versus d may demonstrate a different behavior (Yamashita et al. 2001; Perez-Prado et al. 2005; Amirkhanov et al. 2001).

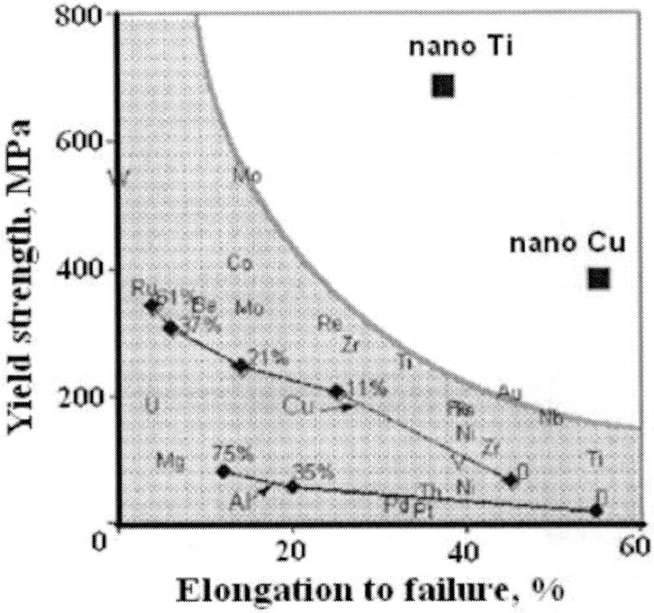

Figure 25.1. Relation between plasticity and strength of metallic materials (Valiev et al. 2006).

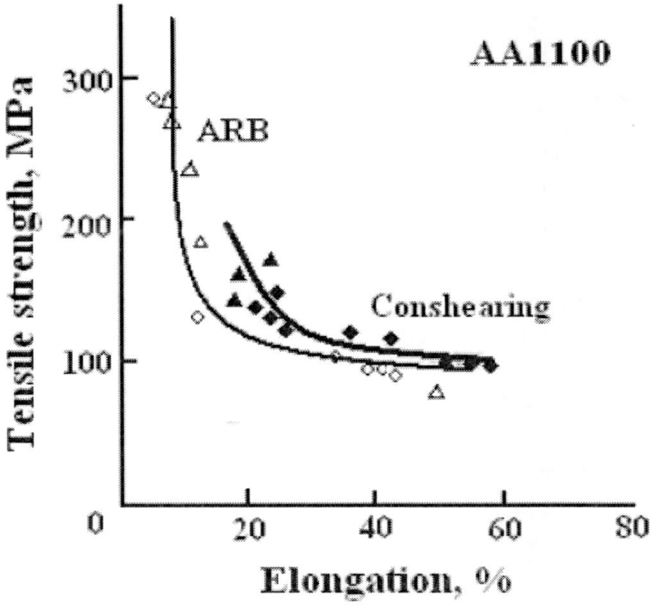

Figure 25.2. Strength vs. elongation for AA1100 aluminum (Utsunomiya et al. 2004).

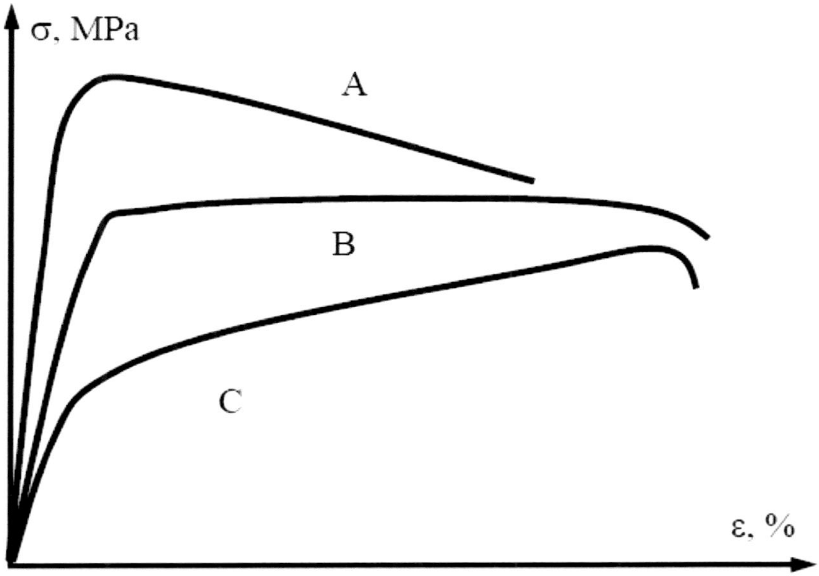

Figure 25.3. Typical stress- strain diagrams for NMC materials at room temperature.

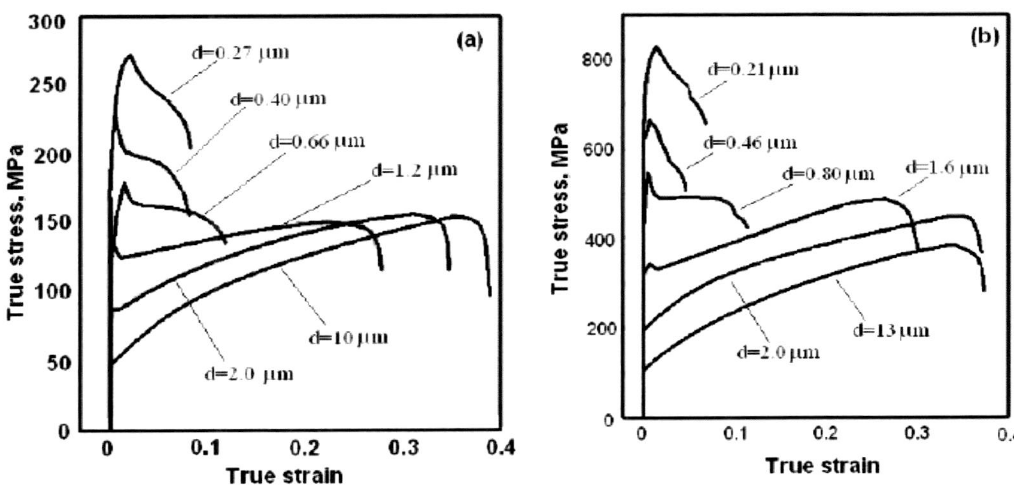

Figure 25.4. Dependences of σ on ε for NMC aluminum (a) and steel (b) (Tsuji et al. 2002).

Hereinafter, when discussing the mechanical properties of NMC materials and determining the conditions for attaining a combination of high strength and high ductility, the values of σ_S and ε^* will be used as measures of related properties. Detailed experimental studies of NMC at different $\dot{\varepsilon}$ and T are infrequent in literature. Therefore, a brief overview of their specifics can shed light on the nature of observed effects.

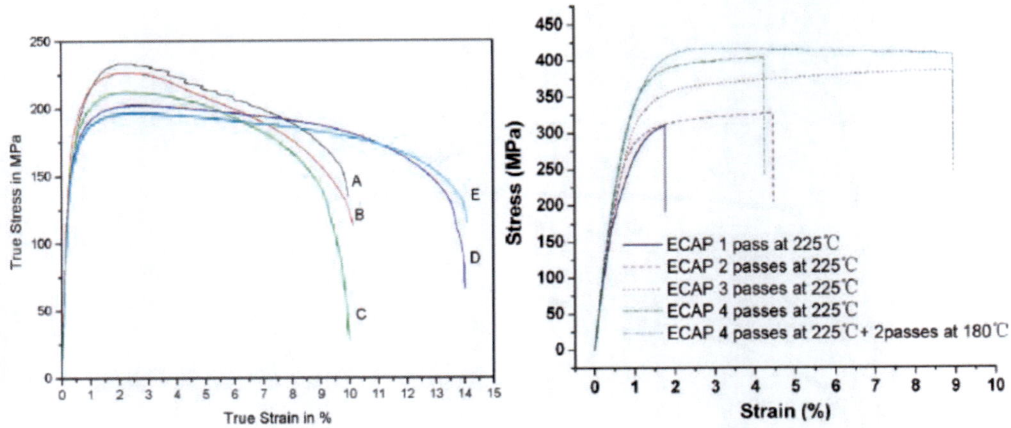

(a) aluminum alloy (b) AZ31 magnesium alloy

Figure 25.5. Dependence of σ on ε for Al-1050 NMC aluminum (a) (strain rates: 10^{-1} s^{-1} (A), 10^{-2} s^{-1} (B), 10^{-3} s^{-1} (C), 10^{-4} s^{-1} (E) and $5{\cdot}10^{-5}$ s^{-1} (D)) (Wang et al. [2008]) and AZ91 magnesium alloy (Chen et al. 2008) (b) strained at room temperature.

Yield Stress Plateau

An important peculiarity of $\sigma(\varepsilon)$ curves for some aluminum (Sabirov et al. 2008b; Tsuji et al. 2002; Wang et al. 2008), magnesium (Cheng et al. 2007; Chen, B. et al. 2008) and copper (Wang et al. 2002; Amirkhanov et al. 2001) alloys is a pronounced yield stress plateau: after a certain value of deformation ε^*, intensive hardening ceases and yield stress σ^* reaches a stationary value.

Strain Rate Sensitivity

The yield stress plateau usually takes place at small strain rates $\dot{\varepsilon} \leq 10^{-4}$ s^{-1} for aluminum alloy 6082 (Wang et al. 2008; Sabirov et al. 2008b) (Figure 25.5). In NMC magnesium alloys, the yield stress plateau and a high ductility are observed only at strain rates of $5{\cdot}10^{-4} \div 3{\cdot}10^{-3}$ s^{-1} (Cheng et al. 2007; Kim et al. 2005; Rhee et al. 2005; Minger et al. 2007; Zuberova et al. 2007; Estrin et al. 2006; Chen, B. et al. 2008). For Cu+0.5%Al_2O_3 NMC alloy, high elongations of 81% was detected at $\dot{\varepsilon} = 10^{-3}$ s^{-1} while at larger (10^{-2} s^{-1}) and smaller (10^{-4} s^{-1}) strain rates it was 11% and 22%, respectively (Amirkhanov et al. 2001).

A strain rate dependence of the ultimate tensile strength for commercial NMC magnesium alloys are exemplified in Figure 25.6 (Perez-Prado et al. 2005). A non-monotonic dependence of the ultimate strength on strain rate was also observed in aluminum (Wang et al. 2008) and copper alloys (Amirkhanov et al. 2001).

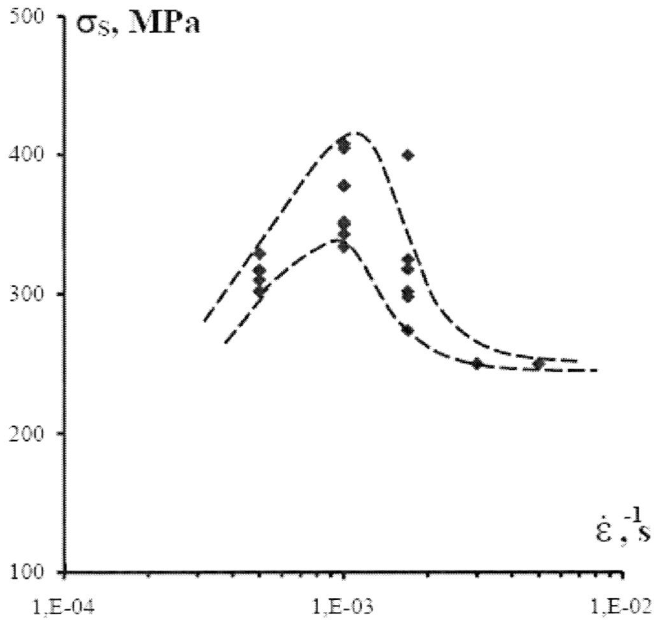

Figure 25.6. Ultimate strength of NMC magnesium alloys vs. strain rate at room temperature (Perez-Prado et al. 2005).

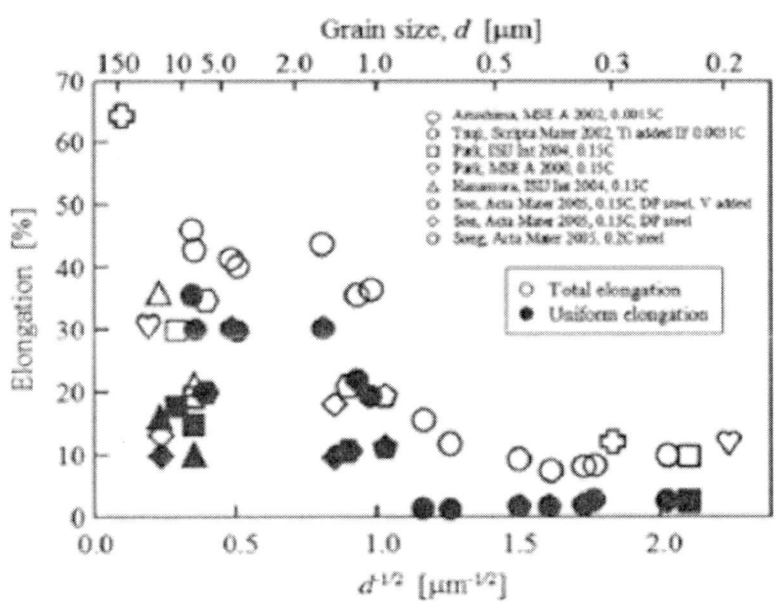

Figure 25.7. Grain size influence on relative elongation at room temperature for low-alloyed steels (Song et al. 2006).

Grain Size Effect on Elongation

An unusual dependence of maximum elongation on the grain size is an important peculiarity of strain behavior of NMC materials at room temperature. Their ductility reaches its maximum at grain sizes of a few microns but not for finer (submicron) or coarser (~10 μm) structures. A notable example of this effect is low-alloyed steels (Song et al. 2006) (Figure 25.7). A similar dependence was also observed in copper (Pyshmintsev et al. 2001) and titanium (Valiev et al. 2003). However, this effect is not obvious from diagrams presenting the maximum elongation ε_{max} versus the annealing temperature where grain growth may occur (Figure 25.8).

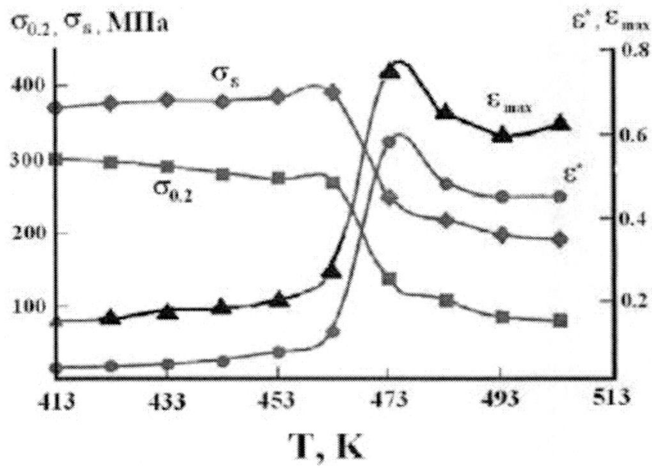

a

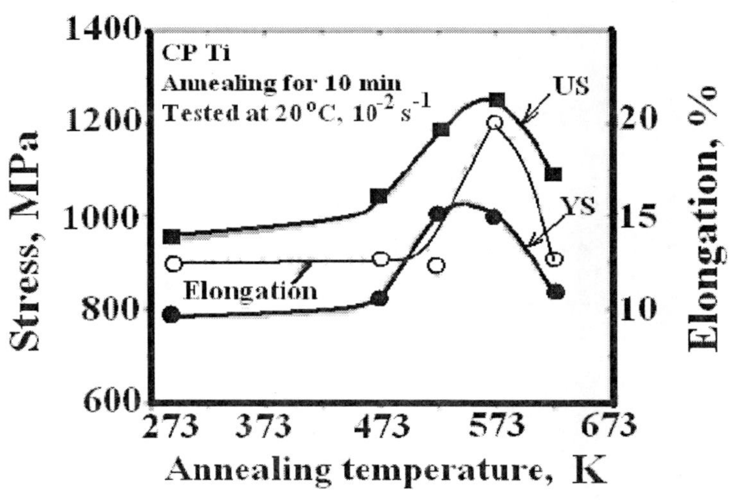

b

Figure 25.8. Elongation to failure and strength vs. annealing temperature for NMC copper (Pyshmintsev et al. 2001) (a) and titanium (Valiev et al. 2003) (b).

25.2. Ultimate Strength Model

The structure of NMC metals after SPD comprises almost dislocation free grains which boundaries retain substantial density of absorbed defects. These defects include orientational misfit dislocations (OMD) of density ρ_{b0} and the products of their delocalization such as (i) the tangential components of delocalized dislocationstangential components of Burgers vector of delocalized dislocations with the Burgers vector density w_t^0 and (ii) normal components of delocalized dislocations of with the density w_n^0 which generate triple junctions disclination dipoles with power ω_0. The defects embedded in grain boundaries create internal stress fields that inhibit the dislocation movement. As shown in Chapter 24, a high strength of NMC metals is mainly conditioned by internal stress fields generated by defects.

During a tensile test nonequilibrium grain boundaries emit lattice dislocations under the action of the external stress σ. When a sufficient level of external stress is attained, the emitted dislocations glide across grains and again embedd in grain boundaries. Otherwise, if the internal stress σ_i generated by a boundary is high, the dislocations are accumulated inside grains. In the case when the dislocations performing intragranular deformation reach the grain boundary, two alternative scenarios are possible. In the first, the dislocations arriving at grain boundaries cannot be absorbed by the latter and, thus, cannot be effectively accommodated. Then, the OMD density in boundaries continuously increases in accordance with the expression:

$$\dot{\rho}_b = \xi_1 \dot{\varepsilon}_v / b. \qquad (25.1)$$

At a constant intragranular strain rate $\dot{\varepsilon}_v$, i.e. when $t = \varepsilon_v / \dot{\varepsilon}_v$, the OMD density will increases in proportion with strain:

$$\rho_b = \xi_1 \varepsilon_v / b. \qquad (25.2)$$

According to Eq. (16.24), the internal stresses are also proportional to density of OMD and continuously increase with strain. They inhibit the movement of other dislocations and their embedding into boundaries. At a certain defect density ρ_b^m in grain boundaries or, what is equivalent, at the ultimate level of internal stress σ_m, the dislocation movement inside grains entirely ceases that denotes the depletion of the material's ductility. The corresponding ultimate strain ρ_b^m can be found from Eqs. (25.2) and (16.24) (see also Figure 25.9, curve A):

$$\varepsilon_m^0 = \frac{1}{\xi_1 \varphi_1}\left(\frac{\sigma_m}{G}\right)\left(\frac{b}{\Delta b}\right). \qquad (25.3)$$

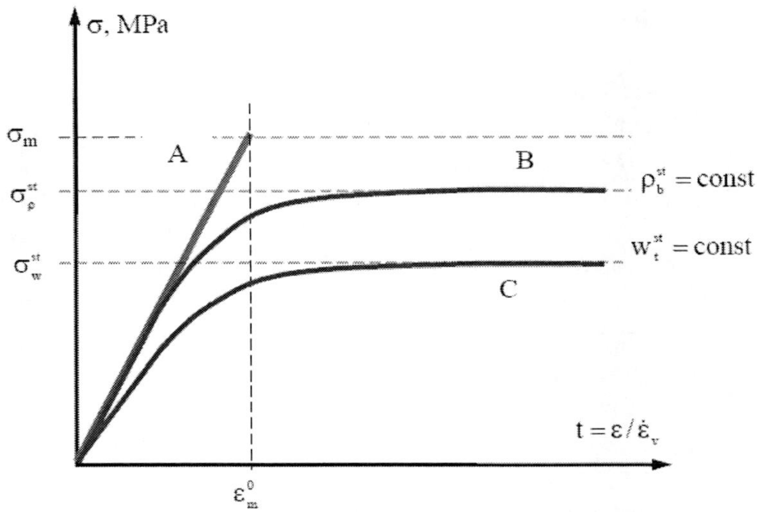

Figure 25.9. Internal stresses generated by defects distributed in grain boundaries vs. strain ε.

In the second scenario, the grain boundaries can effectively accommodate gliding dislocations. For the sake of clarity, we suppose that the flux of all dislocations impinging upon a grain boundary is compensated by their diffusion withdrawal from the boundary. Then, the densities of orientational misfit dislocations (OMD) and the products of their delocalization in grain boundaries reach stationary values ρ_b^{st} and w_t^{st}, respectively. Since σ_i is proportional to the defect density, the straining stress also becomes constant and forms the "yield stress plateau". In the case of OMD domination in grain boundaries, the stress value corresponding to this plateau can be determined using Eqs. (16.24), (15.151) and (15.154):

$$\overline{\sigma}_\rho^{st} = \left[\xi_1 \dot{\varepsilon}_v A_1 \left(\frac{kT}{G\Omega}\right)\frac{b^3}{D_b^* \delta}\right]^{1/4} \text{ at } \rho_b^{st}\Delta b >> w_t^{st}, \tag{25.4}$$

while in the case of domination of tangential components of delocalized dislocationstangential components of Burgers vector of delocalized dislocations this stress is

$$\overline{\sigma}_w^{st} \approx \left[\frac{\xi_2 \dot{\varepsilon}_v}{C_1}\left(\frac{d}{b}\right)^2 \left(\frac{kT}{G\Omega}\right)\frac{b^3}{D_b^* \delta}\right]^{1/2} \text{ at } w_t^{st} >> \rho_b^{st}\Delta b, \tag{25.5}$$

where $\overline{\sigma}_w^{st} = \sigma_w / G\varphi_2$ and $\overline{\sigma}_\rho^{st} = \sigma_\rho / G\varphi_1$. Eq. (25.4) is true for small grain sizes $d \le d_1$ while Eq. (25.5) is valid for coarse grains $d > d_1$, where d_1 is determined by Eqs. (15.142) and (22.6). As can be seen from the above expressions that the stationary stress substantially depends on strain rate $\dot{\varepsilon}_v$ and the grain boundary diffusion coefficient D_b^* D_b^*. Eq. (25.5) shows that for coarse grains $d > d_1$ the stationary stress increases with d, and at a certain grain size d_2 can reach its ultimate value σ_m. The condition gives:

$$\frac{d_2}{b}=\left(\frac{\sigma_m}{G}\left(\frac{\delta D_b^*}{b^3}\right)\left(\frac{G\Omega}{kT}\right)\frac{C_1}{\dot{\varepsilon}_v\xi_2}\right)^{1/2} \tag{25.6}$$

For typical values $\sigma_m/G=10^{-1}$, $D_b^*=1.3\cdot10^{-18}$ m^2/s (at $D_{b0}=8.74\cdot10^{-5}$ m^2/s, $Q_b=10kT_m$ and $T_m/T=3.18$), $b=2.86\cdot10^{-10}$ m, $\delta=2b$, $C_1=50$, $\xi_2=1$ and $\dot{\varepsilon}_v=10^{-3}$ s^{-1},we obtain $d_2\approx1.2$ μm. Noteworthy is a substantial dependence of d_2 on strain rate (the lower $\dot{\varepsilon}_v$ is, the higher d_2 is) and grain boundary diffusion. In particular, decreasing the activation energy for grain boundary diffusion from the equilibrium value of $10.8kT_m$ to a nonequilibrium value of about $9kT_m$ increases d_2 from 0.3 μm to about 5.7 μm at $\dot{\varepsilon}_v=10^{-3}$ s^{-1} (Figure 25.10).

Diagrams of σ^{st} versus d calculated from Eqs. (25.4), (25.5), (25.6) for different strain rates are shown in Figure 25.11. For small grain sizes $d<d_1$, the flow stresses are independent of d (see Eq.(25.4)); for grain sizes $d>d_1$, a linear dependence of σ^{st} on d (see Eq.(25.5)) is limited by the value of σ_m. The first inflections in the lines $\sigma^{st}(d)$ correspond to grain sizes d_1, which depend on the strain rate: the higher $\dot{\varepsilon}_v$ is, the lower d_1 is. The second inflections in curves $\sigma^{st}(d)$ correspond to grain sizes d_2 at the ultimate stress σ_m. As seen from Eq. (25.6) and Figure 25.11, the higher $\dot{\varepsilon}_v$ is, the lower the value of d_2 is.

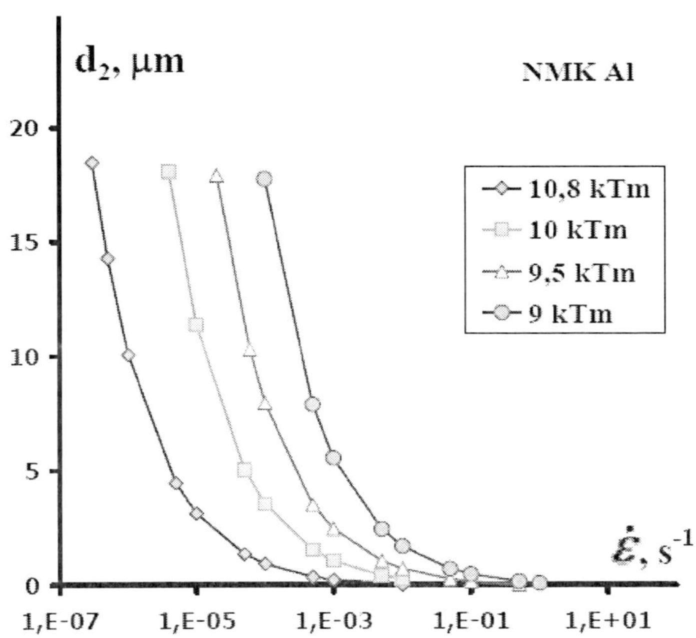

Figure 25.10 Critical grain size d_2 vs. strain rate for NMC aluminum at room temperature ($D_{b0}=8.74\cdot10^{-5}$ m^2/s, $T_m/T=3.18$, $\sigma_m/G=10^{-1}$) at different values of the activation energy of grain boundary diffusion Q_b^* Q_b^*.

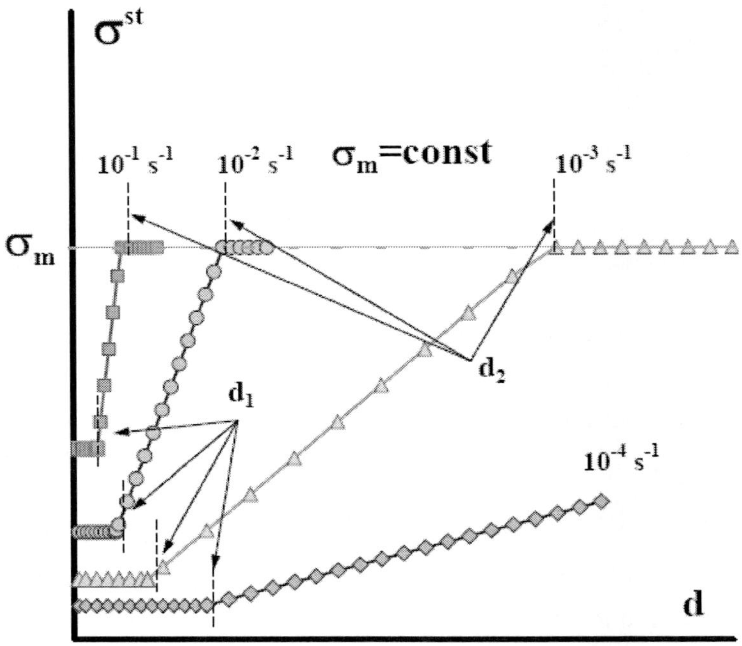

Figure 25.11. Flow stress σ^{st} vs. grain size at different strain rates.

Now let us consider the factors promoting an increase in σ^{st}. The first one is temperature. On the one hand, according to the Arrhenius law, the higher the homological temperature is, the higher the diffusion coefficient and, respectively, the lower the value of σ^{st} should be. On the other hand, increasing temperature intensifies the recovery processes in grain boundaries, and, consequently, the activation energy for grain boundary self-diffusion Q_b^* Q_b^* approaches its equilibrium value Q_b, which results in reduction of D_b^* D_b^*. Thus, the influence of deformation temperature on σ^{st} and ε^{st} is ambiguous and requires a special analysis in each particular case. The thermal prehistory before a strength test (temperature of ECAE, quenching conditions, aging time, etc.) plays a significant role due to the preceding recovery.

The second factor is the strain rate $\dot{\varepsilon}_v$ during testing. The dependence of σ^{st} on $\dot{\varepsilon}_v$ is also non-trivial because of a connection between D_b^* D_b^* and $\dot{\varepsilon}_v$. To evaluate this dependence, the approach described in Chapter 18 can be used. A schematic chart of $D_b^*(\dot{\varepsilon}_v)$ is shown in Figure 18.9 (see Chapter 18). At small strain rates, the value of D_b^* D_b^* insignificantly depends on $\dot{\varepsilon}_v$. At very high strain rates, when D_b^* D_b^* ~D_L, it is also almost independent of $\dot{\varepsilon}_v$. In the range of intermediate strain rates, the curve $D_b^*(\dot{\varepsilon}_v)$ can be described by a power-law $D_b^* \sim \exp(\dot{\varepsilon}_v^k)$, where $k = k_p=0.25$ for $d<d_1$ and $k=k_w=0.5$ for $d>d_1$. With the account of Figure 18.9, the dependence of σ^{st} on strain rate can be presented in the form shown in Figure 25.12. At small strain rates, when the grain boundary diffusion coefficient $D_b^* \sim D_b$ does not depend on $\dot{\varepsilon}_v$, the flow stress is $\sigma^{st} \sim (\dot{\varepsilon}_V)^k$. A similar behavior is observed at high strain rates, when the grain boundary diffusion coefficient $D_b^* \sim D_L$ is also independent of $\dot{\varepsilon}_v$. In the region of intermediate strain rates when $D_b^* \sim \exp\{\dot{\varepsilon}_v^k\}$, the stationary stress is

$$\sigma^{st} \sim \dot{\varepsilon}_v^k / \exp(g \cdot \dot{\varepsilon}_v^k), \tag{25.7}$$

and a maximum can appear in the diagram $\sigma^{st}(\dot{\varepsilon}_v)$ (Figure 25.12) at strain rate $(\dot{\varepsilon}_v)_{max}$ corresponding to a condition $\partial\sigma^{st} / \partial\dot{\varepsilon}_v|_{max} = 0$. In the case when $\rho_b^{st}\Delta b >> w_t^{st}$, k=1/4, $(\dot{\varepsilon}_v)_{max}$ is

$$(\dot{\varepsilon}_v^\rho)_{max} = (\alpha_B w_0)^4 \frac{1}{A_1\xi_1} \frac{G\Omega}{kT} \frac{\delta D_b^*}{b^3}. \tag{25.8}$$

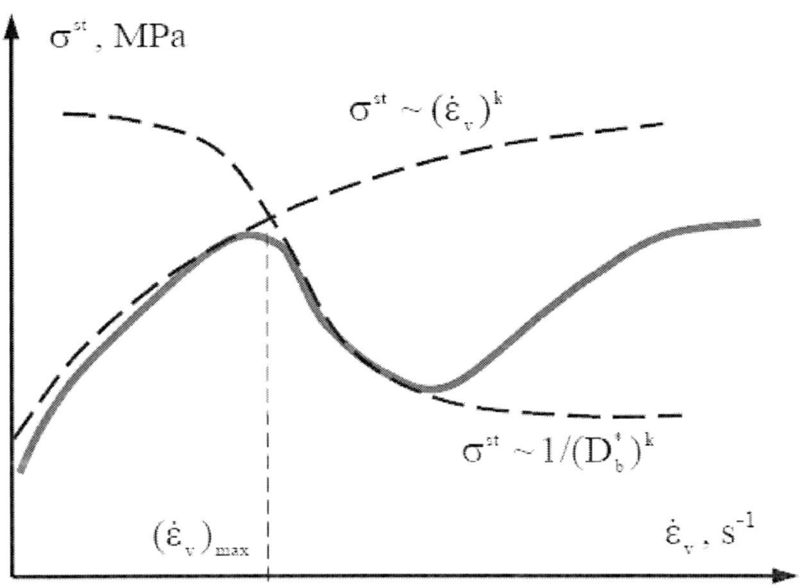

Figure 25.12. Stationary value of flow stress vs. strain rate.

As an example, calculations from Eq.(25.8) for typical parameters of NMC magnesium $Q_b=9kT_m$, $D_b=3.67\cdot10^{-15}$ m^2/s and d=2 μm give the optimum value of strain rate $(\dot{\varepsilon}_v^\rho)_{max}=1.2\cdot10^{-3}$ s^{-1}. For NMC aluminum with parameter values $Q_b=7kT_m$, $D_b=1.6\cdot10^{-14}$ m^2/s and d=1 μm, the optimum strain rate is $(\dot{\varepsilon}_v^\rho)_{max}=6.0\cdot10^{-3}$ s^{-1}. When $w_t^{st} >> \rho_b^{st}\Delta b$, k=1/2, formula for $(\dot{\varepsilon}_v)_{max}$ can be presented as:

$$(\dot{\varepsilon}_v^w)_{max} = (\alpha_B w_o)^2 \frac{C_1}{\xi_2}\left(\frac{b}{d}\right)^2 \frac{G\Omega}{kT}\frac{\delta D_b^*}{b^3}. \qquad (25.9)$$

In this case, the optimum strain rate for NMC magnesium is $(\dot{\varepsilon}_v^w)_{max}$ =5.8·10^{-3} s^{-1}. That is in a good agreement with the experimental value of about 10^{-3} s^{-1} (Perez-Prado et al. 2005) (Figure 25.6). For NMC aluminum, Eq. (25.9) gives $(\dot{\varepsilon}_v^w)_{max}$ =7.4·10^{-2} s^{-1}.

Before proceeding to further analysis, it is necessary to estimate the value of the grain boundary diffusion coefficient. From Eqs. (25.8) and (25.9) we obtain

$$D_{b(\rho)}^* = \xi_1 A_1 \dot{\varepsilon}_v \left(\frac{G}{\varphi_1 \sigma_\rho^{st}}\right)^4 \frac{kT}{G\Omega}\frac{\Omega}{\delta}, \qquad (25.10)$$

$$D_{b(w)}^* = \frac{\xi_2 \dot{\varepsilon}_v}{C_1}\left(\frac{G}{\varphi_1 \sigma_w^{st}}\right)^2 \left(\frac{d}{b}\right)^2 \frac{kT}{G\Omega}\frac{\Omega}{\delta}. \qquad (25.11)$$

The level of stresses achieved in a NMC material during deformation at room temperature is $\sigma^* \sim 10^{-2}G$. Inserting this value into Eqs. (25.10) and (25.11) instead of σ_ρ^{st} and σ_w^{st}, respectively, and using typical values of $\dot{\varepsilon}_v$ =10^{-4} s^{-1}, δ=2b, A_1=10, C_1=50, $\xi_1=\xi_2$=0.01, $\varphi_1=\varphi_2$=1 along with other parameters presented in Table 25.1, one can determine D_b^* D_b^*. For aluminum and copper at the room temperature and d=1 μm, that gives D_b^* $D_b^* \sim 10^{-19}$ m^2/s (see Table 25.1). Corresponding activation energies for grain boundary diffusion (see Eq. (15.147) and Table 25.1) are Q_b^* $Q_b^* \approx 6.5kT_m$ for NMC copper and Q_b^* $Q_b^* \approx 6.5kT_m$ for NMC aluminum. These values are by 12% for aluminum and 30% for copper lower than the equilibrium state. Probably, such small values of Q_b^* Q_b^* are possible if the excess free volume accumulated in grain boundaries after ECAE retained during the tensile test. As noted previously (see Chapter 19), the value of Q_b^* Q_b^* for ECAE can be as low as (3.5÷4)kT_m. This corresponds to a high excess free volume $\Delta\alpha$ induced by dislocations embedded in grain boundaries. If a low recovery rate of defects after ECAE maintaines during a subsequent tensile test, then a sufficiently low value of Q_b^* Q_b^* can be ensured.

Table 25.1 Calculation of grain boundary diffusion parameters for NMC materials at the stage of stationary flow

	T_m/T	$b\cdot 10^{10}$, m	$G\Omega/kT$	$\rho_b^{st}\Delta b >> w_t^{st}$			$w_t^{st} >> \rho_b^{st}\Delta b$		
				$D^*_{b(\rho)}$, m^2/s	D^*_{b0}, m^2/s	Q^*_b, kT_m	$D^*_{b(w)}$, m^2/s	D^*_{b0}, m^2/s	Q^*_b, kT_m
Al	3.18	2.86	102	$4.0\cdot 10^{-19}$	$7.2\cdot 10^{-6}$	9.6	$9.8\cdot 10^{-19}$	$6.3\cdot 10^{-6}$	9.3
Cu	4.63	2.56	231	$1.4\cdot 10^{-19}$	$2.4\cdot 10^{-6}$	6.6	$4.3\cdot 10^{-19}$	$1.8\cdot 10^{-6}$	6.2

25.3. Ultimate Strain Model

Determine strains that corresponds to the stationary stresses σ^{st}. From Chapter 23, ε_ρ^{st} and ε_w^{st} are:

$$\varepsilon_\rho^{st} = \dot{\varepsilon} t_1 \tag{25.12}$$

$$\varepsilon_w^{st} = \dot{\varepsilon} t_2 \tag{25.13}$$

Here t_1 and t_2 are described by Eqs. (15.154) and (15.155), respectively. When $\dot{\varepsilon} = \dot{\varepsilon}_v$ at room temperature, Eqs. (25.12), (25.13), (25.4), (25.5) give

$$\varepsilon_\rho^{st} = \overline{\sigma}_\rho^{st} / \xi_1, \tag{25.14}$$

$$\varepsilon_w^{st} = \overline{\sigma}_w^{st} / \xi_2, \tag{25.15}$$

where ξ_1 and ξ_2 are parameters related to the geometry of boundaries and dislocation flow. Eqs. (25.14) and (25.15) show that the values of ε_ρ^{st} and ε_w^{st} are proportional to $\overline{\sigma}_\rho^{st}$ and $\overline{\sigma}_w^{st}$, respectively. Therefore, factors that increase stress σ^* will also increase strain ε^*. It is evident that strain ε^{st} at the beginning of stationary flow under stress σ^{st} is not the ultimate strain during tensile tests of NMC materials. To evaluate the ultimate strain, it is necessary to take into account the failure mechanism. The failure of materials during superplastic deformation depends on the accumulation of defects at triple grain junctions and at dispersed second-phase particles inside grains (Perevezentsev et al. 1983, 1986).

Consider the failure mechanism of pure metals due to nucleation and growth of defects at triple grain junctions. Disclination-type defects appear at triple grain junctions as a result of interaction between lattice dislocations and grain boundaries and delocalization of dislocations embedded in the grain boundaries (Chapter 19). In the first approximation, such defects can be presented as biaxial disclination dipoles of power ω with arm d. The accumulation kinetics of power ω of such dipoles was described in Chapter 19. From Eq. (19.3) it can be seen that a stationary value of the dipole power ω^{st} depends on strain rate, grain size and D_b^*:

$$\omega^{st} = \xi_3 A_1 \frac{d^3}{\delta D_b^*} \dot{\varepsilon}_v \left(\frac{kT}{G\Omega} \right). \tag{25.16}$$

For small strains, i.e. far from the stationary flow, the expression for ω has a form:

$$\omega = \xi_3 \dot{\varepsilon}_v t \sim \xi_3 \varepsilon_v . \tag{25.17}$$

The proportionality factor ξ_3 between ω and the intragranular strain rate $\dot{\varepsilon}_v$ ε_v depends on geometric characteristics of grain boundaries, triple junctions and dislocation flows in adjacent grains; its typical value is ξ_3=0.01.

A disclination dipole generated during plastic deformation accumulates substantial elastic energy E_ω, which is determined by the expression (Likhachev et al. 1975):

$$E_\omega = \frac{Gd^2\omega^2}{4\pi(1-\nu)} \ln\left(\frac{R_\omega}{d} \right), \tag{25.18}$$

where R_ω is the effective radius of the elastic field around dipole. During deformation, the dipole power increases proportionally to strain ε (see Eq. (25.17)),)); while therefore, the dipole elastic energy E_ω grows proportionally to ε^2. At a certain strain value ε^*, this energy becomes sufficient for a wedge crack nucleation in a grain boundary vicinity in the region of affected by tensile stresses of dipoles. The related energy is

$$E_\gamma = (2\gamma_s - \gamma_b) L_c , \tag{25.19}$$

where γ_s is the surface energy, γ_b is the grain boundary energy and L_c=d/2 is the wedge crack length. Equating E_ω to E_γ, we can find the dipole power necessary for the nucleation of a crack of length d/2:

$$\omega_f^* = \sqrt{\frac{(2\gamma_s - \gamma_b)}{dG} \frac{2\pi(1-\nu)}{\ln(R_\omega/d)}} . \tag{25.20}$$

At typical parameters γ_s=Gb/8 (Missol, 1978), γ_b=Gb/24 (Orlov et al. 1981), R_ω=10d (Likhachev et al. 1975) and ν=0.3, Eq. (25.20) becomes

$$\omega_f^* = \sqrt{\frac{b}{2d}}. \tag{25.21}$$

A schematic chart of ω versus the deformation time for some value of ω_f^* is presented in Figure 25.13. It is seen that two patterns of the material behavior are possible within the developed model. In the first case (curve A in Figure 25.13), a stationary value of the disclination dipole power ω^{st} is below the critical value ω_f^*, and crack nucleation is impossible. Equating ω^{st} to ω_f^*, one can determine a maximum strain rate in this case:

$$(\dot{\varepsilon}_v)^{max} = \frac{\delta D_b^*}{\xi_3 A_1 d^3}\left(\frac{G\Omega}{kT}\right)\left(\frac{b}{2d}\right)^{1/2}. \tag{25.22}$$

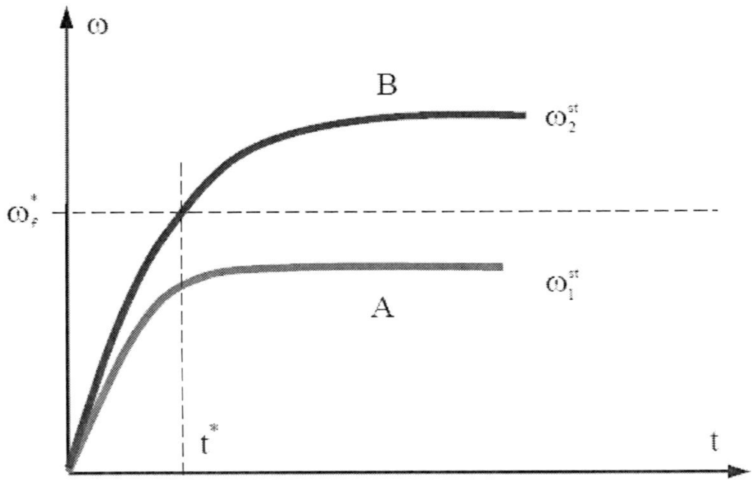

Figure 25.13. Disclination dipole power vs. deformation time.

Such condition $\dot{\varepsilon}_v < (\dot{\varepsilon}_v)^{max}$ can be met for high-temperature ($T>0.7T_m$) superplasticity with low strain rates $\dot{\varepsilon}_v \sim 10^{-3}\ s^{-1}$ when grain boundary diffusion is sufficiently largeintensive. In the second case (see curve B in Figure 25.13), a stationary value of the dipole power ω^{st} exceeds the critical value ω_f^*. As soon as strain reaches ε_m, microcracks are nucleated at triple grain junctions. This situation usually occurs at relatively low temperatures. The strain corresponding to the critical value ω_f^* (see Eq. 25.21) is

$$\varepsilon_m = \sqrt{b/2d}\,/\,\xi_3. \tag{25.23}$$

At d~1 μm and ξ_3=0.01, ε_{max} is close to 100%. As seen from Eq. (25.23), the elongation to failure depends only on the grain size d and decreases with an increase of d.

The above expressions describe the strain that can bring about the nucleation of wedge cracks at triple grain junctions. In a rough approximation, this value can also be used for estimating the ultimate deformation of NMC materials ε_{max}. A more thorough analysis requires a description of parameter ξ_3 and its spatial distribution along with the distribution of grain sizes, studying parameters γ_s and γ_b, and detailed examination of crack propagation and coalescence. However, for preliminary estimates we will use a simple approximation

$$\varepsilon_{max} \approx \varepsilon_m. \tag{25.24}$$

Now consider the relationship between ε_ρ^{st}, ε_w^{st} and ε_{max}. It is evident that in order to attain maximal elongation, the relation $\varepsilon^{st} < \varepsilon_{max}$ should be satisfied. We shall determine the necessary conditions for this relation. Figure 25.14, presents the values of ε^{st} plotted versus d according to Eqs. (25.14)-(25.15) and (25.23) along with the dependence of ε_m on d. At $d=d_1$ the line $\varepsilon^{st}(d)$, similarly to line $\sigma^{st}(d)$ in Figure 25.11, exhibits an inflection; at $d \geq d_1$ the value of ε^{st} increases linearly with d. With a further increase in d, the curve $\varepsilon^{st}(d)$ crosses the curve $\varepsilon_m(d)$ and reaches its maximum at the grain size d_m. A still further increase in grain size d brings about a decrease in ultimate strain. Thus, the dependence of ε^{st} on d is a curve with two singularities at d_1 and d_m whose positions depend on strain rate and temperature. The value of d_m at $d > d_1$ is calculated from the equality $\varepsilon_w^{st} = \varepsilon_m(d)$:

$$\frac{d_m}{b} = \left[\frac{1}{\sqrt{2}\xi_3} \frac{G\Omega}{kT} \frac{D_b^* \delta}{b^3} \frac{C_1}{\xi_2} \frac{1}{\dot{\varepsilon}_v} \right]^{2/5}. \tag{25.25}$$

Using typical values $D_b^* \sim 10^{-19}$ m^2/s, $b = 2.86 \cdot 10^{-10}$ m, $\xi_3 = 0.05$, $\xi_2 = 0.01$, $C_1 = 50$ and $G\Omega/kT_m = 32$ (Frost et al. 1982) for aluminum at room temperature and strain rate $\dot{\varepsilon}_v = 10^{-2}$ s^{-1}, from Eq. (25.25) we obtain $d_m \approx 1.4$ µm.

At small grain sizes $d < d_1$, the value of ε^{st} is determined by Eq. (25.14). Equating ε_ρ^{st} to ε_m, the strain rate below which high elongations are possible is

$$(\dot{\varepsilon}_v^\rho)_{max} = \frac{1}{\xi_3} \sqrt{\frac{b}{2d}} \left(\frac{G\Omega}{kT} \right) \frac{\delta D_b^*}{b^3} \frac{1}{\xi_1 A_1}, \tag{25.26}$$

For typical values $d \sim 1$ µm, $D_b \sim 10^{-19}$ m^2/s, Eq. (25.26) gives $(\dot{\varepsilon}_v^\rho)_{opt} \sim 10^{-3}$ s^{-1}.

Similarly, at $d > d_1$ the maximum strain rate is determined as

$$(\dot{\varepsilon}_v^w)_{max} = \frac{1}{\sqrt{2}\xi_3} \left(\frac{b}{d} \right)^{5/2} \left(\frac{G\Omega}{kT} \right) \frac{\delta D_b^*}{b^3} \frac{C_1}{\xi_2}. \tag{25.27}$$

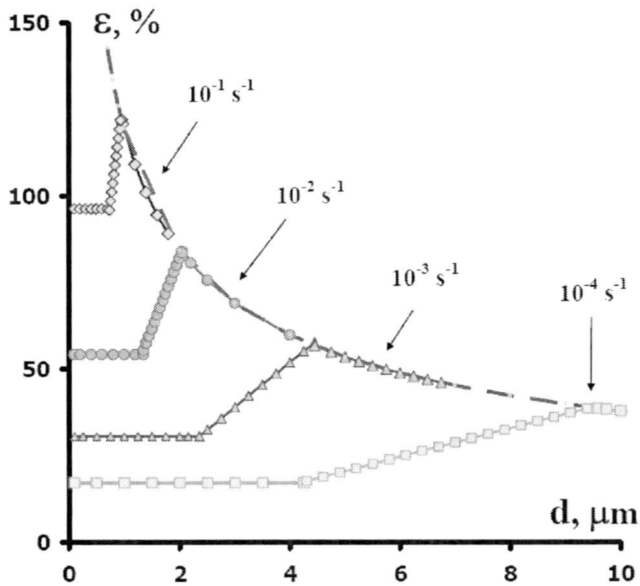

Figure 25.14. Deformation to failure vs. grain size at a room temperature.

For the parameter values describved above, ??? $\dot{\varepsilon}_v^w$ is equal to $10^{-2}s^{-1}$. Eq. (25.25) shows that the value of d_m substantially depends on strain rate and the grain boundary diffusion coefficient. Thus, for typical parameters at room temperature and at low strain rates, high strength and high plasticity can be simultaneously attained in NMC metals with the optimal grain size of slightly above one micron.

25.4. Comparison with Experimental Data

If a simultaneous increase of strength and plasticity in NMC materials is possible, it would be a great interest to realize these conditions in practice (Chuvil'deev et al. 2008b). Related experiments were performed on commercial alloy AMg6 produced by Kamensk-Uralsky Metallurgical Works whose chemical composition is presented in Table 25.2.

Table 25.2. Chemical composition of alloy AMg6

Chemical composition, wt. %							
Al	Mg	Mn	Fe	Si	Ti	Zn	Cu
92.88	6.0	0.71	0.24	0.1	0.04	0.02	0.01

To attain high σ^{st} and ε_m in this alloy at room temperature, it is necessary to (i) determine the optimum grain size d_{opt}^{RT}, (ii) obtain produce such structure by ECAE, and (iii) process the material at a strain rate not exceeding the critical value defined by Eqs. (25.26) and (25.27). In their turn, estimation of d_{opt}^{RT} and $(\dot{\varepsilon}_v)_{opt}$ require the grain boundary diffusion coefficient at room temperature. As noted previously, the activation energy Q_b^* for grain-boundary self-diffusion in aluminum alloys rises with increasing the magnesium content; in alloys with 6% Mg like AMg6, Q_b=12.3kT_m, D_{b0}=8.74·10^{-5} m^2/s and the grain boundary diffusion coefficient at room temperature is about 10^{-21} m^2/s. Using this value in Eq. (25.25) together with the parameters of tables 5.1 and 6.5, we find $d_{opt}^{RT}\approx 0.4$ μm at strain rate $\dot{\varepsilon}=10^{-3}$ s^{-1}. The temperature of ECAE necessary for obtaining this grain size (see Chapter 19) is about 523 K. The experimentally measured grain size in AMg6 alloy after 6 passes ECAE via route D at this temperature was 0.5 μm.

The tensile tests of NMC AMg6 alloy were performed at room temperature with the optimum strain rate $(\dot{\varepsilon}_v)_{opt}=3\cdot 10^{-3}$ s^{-1}. The experimental data for different structures of AMG6 AMg6 are presented in Table 25.3. It is seen that in the NMC material deformed under optimum conditions both the ultimate strength (σ_S = 430 MPa) and ductility (ε_{max}= 25%) are more than twice higher than in the cast condition (σ_S=210 MPa, ε_{max}=11%). This result demonstrates large potentials of ECAE for developing of unique properties in NMC materials.

Table 25.3. Tensile properties of AMg6 alloy after different processing

Condition	σ_S, MPa	ε_{max}, %
ECAE, 4 passes (Chuvil'deev et al. 2008b)	440	22
ECAE, 6 passes (Chuvil'deev et al.2008b)	430	25
ECAE, 6 passes + annealing (Chuvil'deev et al. 2008b)	350	42
ECAE, 6 passes + rolling (Chuvil'deev et al. 2008b)	540	14
Casting (Ed.Miklyaev, 1994)	210	11
Forging (Ed. Fridlyander, 2001)	330	19
Rolling (Ed. Fridlyander, 2001)	340	19

Similar results for commercial magnesium alloys (Chuvil'deev et al. 2004c, 2005) are presented in Table 25.4.

Table 25.4. Mechanical properties of NMC magnesium alloys at room temperature. The strain rate is 10^{-2} s^{-1} (Chuvil'deev et al. 2004c, 2005)

Alloy	Condition	σ_S, MPa	ε_{max}, %
AZ31	Extrusion	290	17
	ECAE (T_{ECAE}=473 K)	260	60
ZK60	Extrusion	270	20
	ECAE (T_{ECAE}=523 K)	230	45
AZ91	Extrusion	290	15
	ECAE (T_{ECAE}=473 K)	260	30

25.5. Optimization Generalization of Processing Results

The general requirement for providing optimal combinations of material ductility and strength is a high value of diffusion coefficient in grain boundaries at room temperature. That imposes certain conditions on characteristics of ECAE and post deformation heat treatment:

1 *Low processing temperature* reduces the rate of recovery and increases the density of defects embedded in grain boundaries. The resulting high level of nonequilibrium statenonequilibricity of grain boundaries retains in the material after ECAE.
2 *High strain rate during processing* promes for forming of nonequilibrium grain boundaries also retained after deformation.
3 *High cooling rate after ECAE* suppresses recovery and preserves grain boundary nonequilibricity induced during processing.
4 Short time and low temperature of post deformation heat treatment (aging, recovery annealing) prevent grain bounjdary transition to more equilibrium states.
5 *Grain size* also affects processes of recovery in NMC materials. For grain sizes $d>d_1$, a higher level of nonequilibrium can be attained in grain boundaries.

Under these conditions, the model described above can give adequate predictions properties of NMC materials. In particular, it provides a scientific basis to optimize the material strength and ductyility depending on processing/ loading conditions and thermodynamical/diffusion characteristics.

In some sence, mechanical behavior of NMC materials resembles superplasticity. The common in hardening of ordinary metals under structural superplasticity and hardening of NMC materials at room temperature is that in both cases this phenomenon relates to intragranular dislocation slip and conditioned by the accumulation of defects in grain boundaries. The difference is that due to a substantial temperature dependence of the rate of accommodation processes, the stationary density of defects in grain boundaries under structural superplasticity is substantially lower than under deformation at room temperature. As a result, effective stresses at superplastic deformation are significantly lower than at low-temperature deformation. Also, accommodation processes under superplasticity inhibit the accumulation of triple junction disclinations and the nucleation of microcracks. This is the reason for a profound difference in the plasticity of superplastic materials and NMC materials. At room temperature, diffusion processes do not have enough time to withdraw triple junction disclinations and the failure occurs relatively fast. The present model explains the occurrence of the stationary stage or plateau in the plastic flow curve and the existence of an optimum strain rate.

Substantial dependence of grain boundary diffusion on strain rate at room temperature allows one to understand why the value of rate sensitivity coefficient m at room temperature is substantially lower than under high-temperature superplastic deformation. The role of threshold stresses could be a possible reason for a low value of m at room temperature. Also, it should be noted that the model considers only the intragranular dislocation slip and interaction of lattice dislocations with grain boundaries and does not include grain boundary sliding which is not necessary to explain the observed plasticity increase in NMC materials.

25.6. SUMMARY

A model for evaluating the ultimate strength and maximal plasticity of NMC materials at homogeneous deformation has been developed. The model is based on the concept that the mechanical properties of NMC materials are determined mainly by the defects accumulated in grain boundaries during intragranular deformation. The conditions at which simultaneous high strength and high plasticity are possible under tension at room temperature are examined. It is shown that these conditions are satisfied when a highly nonequilibrium statenonequilibricity of grain boundaries after ECAE is maintained.

Nomenclature

No.	Symbol	Physical meaning
1.	A_1	Numerical coefficient in the expression for the delocalization time
2.	A' $_{A'}$	Numerical coefficient in the expression for superplastic flow stress
3.	A_R, B_R	Numerical coefficients in the Liefschitz-Wagner-Slezov equation
4.	A^T, B^T	Coefficients in the expression for temperature dependence of the activation volume for GB diffusion
5.	A_p	Numerical parameter depending on thermodynamic properties of material in the expression for grain boundary enthalpy under hydrostatic pressure
6.	A_v	Dorn constant
7.	A_Σ	The product of ultimate strength by maximum elongation
8.	A_ω	Numerical parameter in the expression for stationary power of triple junction disclinations generated during migration
9.	a	Lattice period
10.	a_p, b_a	Coefficients in the equation for the dependence of grain-boundary free volume on the applied hydrostatic pressure
11.	a_ρ	Characteristic distance between orientational misfit dislocations
12.	b	Burgers vector
13.	Δb_n, Δb_t	Normal and tangential components of Burgers vector of orientational misfit dislocations
14.	C_1	Numerical coefficient in the expression for the characteristic withdrawal time of tangential components of delocalized dislocationstangential components of Burgers vector of delocalized dislocations
15.	C_b	Concentration of alloying elements in the grain boundary
16.	C_{bp}	Ultimate solubility of alloying elements in the grain boundary at a given temperature
17.	C_v	Volume concentration of a alloying elements
18.	C_{vac}	Vacancy concentration inside the grain
19.	C_{vp}	Ultimate solubility of alloying elements in the volume at a given temperature
20.	$\dot{C}^+_{vac}$	Generation rate of excess vacancies
21.	C^{st}_{vac}	Stationary vacancy concentration
22.	ΔC_p	Difference between isobaric heat capacity of solid and liquid phases

No.	Symbol	Physical meaning
23.	C^*_v	Solubility limit of alloying elements in solid state
24.	D	Route D of ECAE
25.	D_b	Diffusion coefficient for equilibrium grain boundaries
26.	D_{b0}	Preexponential factor in the expression for diffusion coefficient for equilibrium grain boundaries
27.	D_c	Diffusion coefficient in the core of lattice dislocations
28.	D_{c0}	Preexponential factor in the expression for diffusion coefficient in the core of lattice dislocations
29.	D_m	Effective vacancy diffusion coefficient
30.	D_v	Volume self-diffusion coefficient
31.	D_{v0}	Preexponential factor in the expression for volume self-diffusion coefficient
32.	D^e_{b0}	Experimental value of preexponential factor for grain boundary self-diffusion coefficient
33.	D^p_b	Grain boundary self-diffusion coefficient under applied hydrostatic pressure
34.	D^p_{b0}	Preexponential factor in the expression for grain boundary self-diffusion coefficient under applied hydrostatic pressure
35.	D^*_b	Effective diffusion coefficient for nonequilibrium grain boundaries
36.	D^*_{b0}	Preexponential factor in the expression for diffusion coefficient for nonequilibrium grain boundaries
37.	D^*_{bS}, D^*_{bd}, $D^*_{b\ell}$	Grain boundary self-diffusion coefficient in different areas of a delocalized dislocation core
38.	D^*_{br}	Grain boundary self-diffusion coefficient in the area of delocalized dislocation core
39.	$D^*_{b(\rho)}$, $D^*_{b(w)}$	Value of grain boundary diffusion coefficient corresponding to the optimum strain rate at which the effect of simultaneous increase of strength and plasticity is realized, in case of domination of orientational misfit dislocations and tangential components of delocalized dislocationstangential components of Burgers vector of delocalized dislocations in grain boundaries
40.	$D^{(1)}_v$	Volume self-diffusion coefficient at stationary concentration of vacancies, which is determined by the rate of their absorption by dislocations
41.	$D^{(2)}_v$	Volume self-diffusion coefficient at stationary concentration of vacancies, which is determined by the rate of their absorption by migrating grain boundaries
42.	d	Grain size
43.	d_0	Initial grain size
44.	d_1	Critical grain size, at which the density of OMD and tangential components of delocalized dislocationstangential components of Burgers vector of delocalized dislocations are equal
45.	d_2	Grain size, at which the level of internal stresses σ_i reaches its critical value σ_m
46.	d_c	Average size of coarse (abnormal) grains at abnormal grain growth
47.	d_f	Average size of small grains at abnormal grain growth
48.	d_L	Grain size, at the attainment of which the L-mode of grain boundary

No.	Symbol	Physical meaning
		diffusion is realized
49.	d_m	Grain size, at which the homogeneous deformation at a tensile strength test at room temperature reaches its critical value $\varepsilon_m = \varepsilon_{max}$
50.	d_{min}	Minimum grain size, at which generation of dislocation pile-up in fine-grained material is possible
51.	d_R	Size of the recrystallization nucleus, the scale of diffusion mass transfer at recrystallization
52.	d_{sp}	Grain size after the superplasticity test (at strain-enhanced grain growth)
53.	d_o^a	Grain size before the superplasticity test
54.	$\overline{d}$	Average grain size
55.	d_{opt}^{RT}	Optimum grain size, at which the effect of simultaneous increase of strength and plasticity at room temperature occurs in NMC material
56.	d_{opt}^{sp}	Optimum grain size for superplasticity
57.	d^*	Grain refinement limit at ECAE
58.	E	Young modulus
59.	E_ω	Elastic energy of disclination dipole
60.	E_γ	Energy required for the crack development
61.	F	Free energy of critically-sized island nucleation
62.	F_b	Fraction of grain boundary area occupied by the precipitating phase
63.	F_m	Free energy required for melting of the grain-boundary S-phase area surrounding the initial island of L-phase
64.	F_γ	Free energy required for the generation of the solid/liquid interface around a new critically-sized island of L-phase
65.	F_1, F_2	Numerical coefficients in the expression for free volume of grain boundary depending on thermodynamic parameters of material
66.	ΔF	Free energy of overcoming point obstacles by a dislocation
67.	f_0	Normalization factor
68.	f_f	Volume fraction of small grains at abnormal growth
69.	f_f^*	Critical volume fraction of small grains, at which the average grain size at recrystallization is determined by the kinetics of coarse (abnormal) grain growth
70.	f_{fr}	Volume fraction of fragmented material
71.	f_v	Volume fraction of second-phase particles
72.	f_v^R	Volume fraction of recrystallized material
73.	f_v^*	Maximum volume fraction of second-phase particles
74.	G	Shear modulus
75.	G_1, G_2	Numerical coefficients depending on thermodynamic parameters of material in the expression for stationary flow stress
76.	g_1	Numerical coefficient in an expression for the dependence of OMD stationary density on strain rate
77.	g_2	Numerical coefficient in an expression for the dependence of stationary density of tangential components of Burgers vector of delocalized dislocations on strain rate
78.	$g(\dot{\varepsilon})$	Parameter in the law of strain-enhanced grain growth
79.	H	Metal hardness

No.	Symbol	Physical meaning
80.	H_c	Coercitive force
81.	I_v	Flow of lattice dislocations bombarding the grain boundary during intragranular deformation or GB migration
82.	I_v^+	Generation rate of lattice dislocation
83.	I_v^-	Intensity of withdrawal of lattice dislocations to sinks
84.	K	Grain boundary strengthening coefficient in the Hall-Petch equation
85.	K_0	Hall-Petch coefficient for a coarse-grained material
86.	K_1	Grain boundary strengthening coefficient in the equation for yield stress of a NMC metal in the case of no dislocation pile-up
87.	K_2	Grain boundary strengthening coefficient in the expression for yield stress of a NMC metal in the case of dislocation pile-up generation and OMD domination in grain boundaries
88.	K_3	Grain boundary strengthening coefficient in the expression for yield stress of a NMC metal in the case of dislocation pile-up generation and domination of tangential components of delocalized dislocationstangential components of Burgers vector of delocalized dislocations in grain boundaries
89.	K_c	Distribution coefficient for alloying elements
90.	K_c^{Sc}, K_c^{Zr}	Distribution coefficient of scandium and zirconium in aluminum
91.	K_p	Compression modulus
92.	K^ε	Numerical parameter equal to $\partial \lg \varepsilon^* / \partial \lg \dot{\varepsilon}$
93.	K^σ	Numerical parameter equal to $\partial \lg \sigma^* / \partial \lg \dot{\varepsilon}$
94.	k	Boltzmann constant
95.	k_ρ, k_w	Power exponents in the dependence of grain boundary diffusion coefficient on strain rate
96.	L	Length of dislocation path in Orowan equation
97.	L_c	Crack length
98.	$L_\perp$	Length of dislocation pile-up
99.	M	Full grain boundary mobility
100.	M_b	Equilibrium grain boundary mobility
101.	M_ω	Mobility of triple junction disclinations
102.	M_ρ	Mobility of orientational misfit dislocations
103.	M^*	Nonequilibrium grain boundary mobility
104.	M_l	Diffusion mobility of atoms in the area of fluctuations (L-phase)
105.	m	Coefficient characterizing stress sensitivity to strain-rate change in the expression for creep rate
106.	N	Number of ECAE passes
107.	N^{st}	Stationary number of dislocations in a grain
108.	N^*	Critical number of dislocations in grains, which corresponds to the critical level of dislocation density $\rho_v = \rho_v^*$
109.	n	Coefficient of deformation strengthening
110.	n_c	The fraction of alloying atoms in a precipitated particle
111.	n_v	Power exponent connecting stress and strain rate in the expression for creep rate

No.	Symbol	Physical meaning
112.	$n_{\perp}$	Number of dislocations in a pile-up
113.	P	Hydrostatic pressure
114.	P_m	Driving force for migration
115.	P_z	Zener force
116.	P_{ρ}	Driving force for migration connected with the reduction of the system's free energy at absorption of lattice dislocations by boundaries
117.	p	Power exponent in the expression for the dependence of grain boundary superplastic strain rate on grain size
118.	Q_l	Enthalpy of the L-phase critical island generation process
119.	Q_b	Activation energy for diffusion in equilibrium grain boundaries
120.	Q_b^*	Activation energy for diffusion in nonequilibrium grain boundaries
121.	Q_b^e	Experimental value of the activation energy for grain boundary diffusion
122.	Q_b^p	Activation energy for grain boundary self-diffusion under hydrostatic pressure
123.	Q_b^t	Theoretical value of the activation energy for grain boundary diffusion
124.	Q_c	Activation energy for diffusion along lattice dislocation cores
125.	Q^{eff}	Effective activation energy of superplastic flow
126.	Q_g	Energy of vacancy generation
127.	Q_m	Activation energy for vacancy migration
128.	Q_v	Activation energy for volume self-diffusion
129.	R	Radius of a grain-boundary R-cell
130.	R_l	Critical radius of a particle corresponding to the boundary breakaway moment from dispersed second-phase particle
131.	R_v	Radius of a second-phase particle
132.	R_x	Average distance between lattice dislocations
133.	R_{ω}	Screening radius of a disclination dipole
134.	r_l	Atomic radius
135.	r_A	Size of the dislocation core area, in which boundary diffusion parameters correspond to diffusion in a (supercooled) melt
136.	r_B	Radius of diffusion influence area of dislocation
137.	r_L	Radius of grain-boundary L-island
138.	r_L^p	Size of the L-phase island in the grain boundary under hydrostatic pressure
139.	r_L^S	Size of L-phase island in nonequilibrium grain boundary
140.	r_d	Dislocation core radiusRadius of dislocation core delocalization in a grain boundary
141.	r_d^0	Initial dislocation core radius
142.	$r_{\perp}$	Average distance between dislocations
143.	r_* r_*	Critical size of the L-phase island
144.	S_0	Dislocation core radius
145.	S_l	Nucleation entropy of a critical L-phase island
146.	S_{12}	Shear at the tip of the L-zone in the grain boundary

No.	Symbol	Physical meaning
147.	$S_{S/L}$	Entropy of the solid/liquid interface
148.	S_b	Grain boundary entropy
149.	S_d	Dislocation core width
150.	S_m	Entropy of vacancy migration
151.	S_b^c	Configurational entropy of grain boundary per unit area
152.	S_b^p	Grain boundary entropy under hydrostatic pressure
153.	S_b^+	Entropy of non-equilibrium grain boundary
154.	s_1	Lateral surface area of a cylindrical L-phase island
155.	s_2	Upper (lower) surface area of cylindrical L-phase island
156.	s_3	Area of the grain boundary S-phase
157.	s_S^0	Area of the initial (cylindrical) island of L-phase
158.	s_S^*	Area of critically-sized L-phase island
159.	T	Temperature
160.	T_1	Critical temperature of recrystallization in pure metals (initial grain boundary migration temperature in alloys)
161.	T_1^{min}	Minimum theoretical value of recrystallization critical temperature
162.	T_1^{max}	Maximum theoretical value of the recrystallization critical temperature
163.	T_2	Temperature at which abnormal grain growth is finished (critical initial temperature of collective recrystallization)
164.	T_b^0	Temperature at which the grain boundary energy was measured
165.	T_h	Effective heating temperature before recrystallization
166.	T_m	Melting point
167.	T_r	Recrystallization temperature
168.	T_v	Critical temperature corresponding to a transition from the strain rate limited by dislocation slip to the strain rate limited by dislocation climb
169.	T^*	Critical temperature at which the density of OMD and the density of tangential components of delocalized dislocationstangential components of Burgers vector of delocalized dislocations are equal
170.	ΔT_b	Difference between the melting temperature and the temperature at which the grain boundary energy is measured ($\Delta T_b = T_m - T_b^0$)
171.	t_1	Characteristic time of the OMD core decomposition into normal and tangential delocalized dislocations
172.	t_2	Characteristic time of withdrawal of tangential components of delocalized dislocationstangential components of Burgers vector of delocalized dislocations from grain boundaries
173.	t_{2B}	Characteristic climb time of effective dislocation with Burgers vector Δb_t along adjacent grain boundaries to the place of annihilation with similar dislocations of the opposite sign from other boundaries
174.	t_3	Time of diffusion withdrawal of grain junction disclination defects connected with normal components of delocalized dislocations from grain junctions
175.	t_4	Characteristic time of defect accommodation in grain boundaries at lattice dislocations embedding in migrating grain boundaries
176.	t_3^{exp}	Experimentally measured time of triple junction disclination diffusion

No.	Symbol	Physical meaning
		accommodation
177.	$t_3^{th(max)}$	Maximum theoretical time of triple junction disclination diffusion accommodation
178.	$t_3^{th(min)}$	Minimum theoretical time of triple junction disclination diffusion accommodation
179.	t_R	Incubation period of recrystallization
180.	t_a	Annealing time
181.	t_d	Characteristic time of delocalization of a lattice dislocation embedded in a random grain boundary (experimental)
182.	t_{diff}	Time of dislocation influence on the diffusion properties of a grain boundary (diffusion influence time)
183.	t_g	Characteristic time of dislocation emission by the source
184.	t_r	Characteristic time of absorption of dislocations by grain boundaries
185.	t_{rA}	Characteristic time of deformation at which stationary OMD density reaches the value of $(\rho_b^{st}\Delta b)_L$
186.	t_{rB}	Characteristic time of deformation at which stationary OMD density reaches the value of $(\rho_b^{st}\Delta b)_S$
187.	t_{rC}	Characteristic deformation time for which the stationary density of tangential components reaches the value corresponding to the L-mode diffusion
188.	t_{rD}	Characteristic deformation time for which stationary OMD density reaches the value of $\mathbf{W}_t^{st}$ at the S-mode diffusion
189.	t_{2A}	Characteristic time of withdrawal of tangential components of delocalized dislocationstangential components of Burgers vector of delocalized dislocations from grain boundary connected with grain boundary sliding
190.	t_{2B}	Characteristic time of purely diffusion withdrawal of tangential components of delocalized dislocationstangential components of Burgers vector of delocalized dislocations from grain boundary
191.	t_s	Time of dislocation slip movement through the grain of size d from the source at one grain boundary to the sink at another boundary (characteristic time of dislocation path through a grain)
192.	t_v	Characteristic time of withdrawal of excess vacancies to sinks
193.	t_{v1}	Characteristic time of excess-vacancy absorption by dislocations
194.	t_{v2}	Characteristic time of excess-vacancy absorption by migrating grain boundaries
195.	t_w	Time when the flow stress reaches its stationary value at the domination of tangential components of delocalized dislocationstangential components of Burgers vector of delocalized dislocations in grain boundaries
196.	t_ε	Characteristic deformation time
197.	t_ρ	Time when the flow stress reaches its stationary value at the domination of orientational misfit dislocations in grain boundaries
198.	U_0	The work of vacancy generation
199.	U_1	Excess energy of L-phase
200.	U_{12}	Shear at the tip of the L-zone in the grain boundary
201.	U_L	Shear rate at the L-zone of a boundary

No.	Symbol	Physical meaning
202.	U_s	Shear rate at the S-zone of a boundary
203.	V_o	Characteristic rate of dislocation movement in a lattice
204.	V_0	Excess free volume connected with lattice dislocation cores
205.	V^0	Volume of initial (cylindrical) island of L-phase
206.	V_1	Volume of the cylindrical island of L-phase
207.	V_b	Free volume of equilibrium grain boundary
208.	V^b_L	Free volume of L-phase
209.	V^b_S	Free volume of S-phase
210.	V^0_b	Structural Geometrical component of free volume
211.	V^T_b	Thermal component of free volume
212.	V^P_b	Free volume of a grain boundary in the material under hydrostatic pressure
213.	V^*_{b1}, V^*_{b2}	Numerical value of excess free volume used in calculations
214.	V^S	Full free volume of a nonequilibrium grain boundary
215.	$V_\perp$	Characteristic rate of dislocation movement in a latticeThe rate of dislocation slip
216.	V^+	Excess free volume of a nonequilibrium grain boundary
217.	V^*	Volume of critically-sized island of L-phase
218.	V_h	Heating rate
219.	V_m	The rate of grain boundary migration
220.	V^*_m	Critical value of migration rate, at which the grain boundary diffusion coefficient increases
221.	V^*_n	Free volume connected with normal components of delocalized dislocations
222.	V_p	Activation volume for the grain boundary self-diffusion
223.	V^+_r	Local eExcess free volume of grain boundary connected with delocalized dislocations
224.	V^S_r	Total free volume of an random boundary in the area of OMD core
225.	ΔV	Difference between atomic volumes of an alloying element and the matrix
226.	ΔV_m	Volume jump at the melting point
227.	v	Density of free volume connected with lattice dislocation cores
228.	W_1, W_2	Numerical coefficients in the expression for the activation energy of grain boundary self-diffusion, which depend on thermodynamic parameters of material
229.	w	Numerical parameter, which is equal to a difference of squared radii of critical and initial islands of L-phase
230.	w^0	Burgers vector density of a lattice dislocation embedded in grain boundary
231.	w_0	Numerical parameter proportional to the volume jump at the melting point
232.	w_n	Burgers vector density connected with normal components of delocalized dislocations
233.	w_t	Burgers vector density connected with tangential components of delocalized dislocations

No.	Symbol	Physical meaning
234.	w^0_n w_{t0}	Initial Burgers vector density of tangential normal components of delocalized dislocations
235.	w^0_t, W^0_n	Initial Burgers vector density density of tangential components of delocalized dislocationsBurgers vector density of tangential and normal components of delocalized dislocations in a grain boundary
236.	w^{st}_t	Stationary Burgers vector density of tangential components of delocalized dislocations
237.	w^{Σ}	Full density of defects in a nonequilibrium grain boundary
238.	x	Characteristic scale of diffusion mass transfer in the expression for the diffusion coefficient
239.	x_V	Characteristic distance passed by vacancies to the sink
240.	Z_1, Z_2	Numerical coefficients in the expression for a pre-exponent of grain boundary self-diffusion coefficient, which depends on thermodynamic parameters of material
241.	Z_d	Numerical coefficient in the dependence of intergranular strain rate on grain size
242.	Z_{opt}	Numerical coefficient in the expression for optimum grain size at superplasticity
243.	Z_v	Numerical parameter in the expression for temperature corresponding to a transition from strain rate limited by dislocation slip to the strain rate limited by climb
244.	z	Power exponent in the law for deformation-stimulated grain growth
245.	α	Relative free volume of grain boundary
246.	α^0	Relative fraction of a structural geometrical component of free volume
247.	α_A	Difference in the values of critical equilibrium free volume and current free volume
248.	α_B	Numerical parameter depending on thermodynamic properties of material in the expression for grain boundary self-diffusion coefficient
249.	α_L	Relative free volume of L-phase
250.	α_S	Relative free volume of S-phase
251.	α^S	Relative free volume for a nonequilibrium grain boundary
252.	α^P	Relative free volume of grain boundary in a material under hydrostatic pressure
253.	α^S_r	Total relative free volume of an random boundary in the area of OMD core
254.	α_ρ	Numerical parameter in the expression for contribution of lattice dislocations to the macroelastic limit
255.	α^*	Critical free volume of grain boundary
256.	$\alpha_\perp$	Numerical parameter (α=10) in the expression for capture cross-section
257.	$\alpha(\dot{\varepsilon})$	Parameter in the law of strain-enhanced grain growth
258.	$\Delta\alpha_1$, $\Delta\alpha_2$	Numerical value of excess relative free volume used in calculations
259.	$\Delta\alpha(P)$	Change in the relative free volume of grain boundary induced by hydrostatic pressure
260.	$\Delta\alpha_r$	Excess relative free volume of an random grain boundary in the area of OMD core
261.	β_1	Numerical coefficient in the expression for the driving force of migration
262.	β_L	Thermal coefficient of volume expansion expansion coefficient of a

No.	Symbol	Physical meaning
		melt
263.	β_S	Thermal coefficient of volume expansion expansion coefficient of a solid phase
264.	β_v	The fraction of plastic deformation work expended for the generation of one vacancy
265.	$\Delta\beta$	Difference between thermal expansion coefficients of a melt and a solid phase
266.	γ_0	Free energy of S-phase areas of the grain boundary
267.	γ_L	Surface energy of the liquid metal
268.	γ_b	Grain boundary energy
269.	γ_b^*	Nonequilibrium grain boundary energy
270.	γ_b^L	Grain boundary energy at a melting point
271.	γ_b^0	Grain boundary enthalpy
272.	γ_b^{0p}	Grain boundary enthalpy under hydrostatic pressure
273.	γ_b^p	Free energy of a grain boundary under hydrostatic pressure
274.	γ_b^{0+}	Enthalpy of nonequilibrium grain boundary
275.	γ_s	Surface free energy of the solid metal
276.	$\gamma_{S/L}$	Free energy per unit area of the soild/liquid interface
277.	$\gamma_{S/L}^0$	Enthalpy per unit area of solid/liquid interface
278.	$\Delta\gamma_b$	Excess energy of a grain boundary
279.	δ	Grain boundary width
280.	ε^*	Experimental deformation value at the stage of stationary flow of material
281.	ε_1	Critical deformation value corresponding to onset of the recrystallization nuclei generation
282.	ε_2	Critical deformation at which the total volume of material is subject to fragmentation
283.	ε_m	Critical deformation value at which microcracks are nucleated at grain junctions
284.	ε_{max}	Maximum elongation to failure
285.	ε_m^0	Ultimate level of homogeneous deformation at which internal stress σ_i reaches its critical value σ_m
286.	ε^{st}	Theoretical deformation value at the stage of stationary flow of material
287.	ε_w^{st}	The value of deformation corresponding to the homogeneous stage of stationary flow under superplasticity in the case of domination of tangential components of delocalized dislocationstangential components of Burgers vector of delocalized dislocations in grain boundaries
288.	ε_ρ^{st}	The value of deformation corresponding to the homogeneous stage of stationary flow under superplasticity in the case of OMD domination in grain boundaries
289.	$\Delta\varepsilon$	The value of microdistortions determined by X-ray analysis
290.	$\dot{\varepsilon}$	«Effective» strain rate
291.	$\dot{\varepsilon}^{opt}$	Optimum strain rate of superplastic deformation
292.	$\dot{\varepsilon}_b$	Intergranular strain rate

No.	Symbol	Physical meaning
293.	$\dot{\varepsilon}_v$	Intragranular strain rate
294.	$\dot{\varepsilon}_{vA}$	Intragranular strain rate at which the L-mode diffusion connected with OMD accumulation in the grain boundary is realized
295.	$\dot{\varepsilon}_{vB}$	Intragranular strain rate at which the S-mode diffusion in the grain boundary is realized, in case of OMD domination
296.	$\dot{\varepsilon}_{vC}$	Intragranular strain rate at which the L-mode diffusion is realized in the grain boundary, in case of w_t domination
297.	$\dot{\varepsilon}_{vD}$	Intragranular strain rate at which the S-mode diffusion is realized in the grain boundary, in case of w_t domination
298.	$\dot{\varepsilon}_{vC}$	Intragranular strain rate at which the L-mode diffusion is realized in the grain boundary
299.	$\dot{\varepsilon}_v^{I}$	Intragranular strain rate limited by the dislocation movement in the field of uniformly distributed point obstacles
300.	$\dot{\varepsilon}_v^{II}$	Intragranular strain rate connected with dislocation climb
301.	$(\dot{\varepsilon}_v)_{max}$	Optimum strain rate at which the effect of strength increase at room temperature is observed
302.	$(\dot{\varepsilon}_v)^{max}$	Maximum rate of intragranular deformation at which triple junction crack is not formed
303.	$(\dot{\varepsilon}_v)_{opt}$	Optimum rate of intragranular deformation at which the effect of simultaneous increase of strength and plasticity at room temperature is realized in NMC materials
304.	$(\dot{\varepsilon}_v^{w})_{max}$	Critical strain rate below which high elongations are possible at room temperature in the case of domination of tangential components of delocalized dislocationstangential components of Burgers vector of delocalized dislocations in grain boundary
305.	$(\dot{\varepsilon}_v^{\rho})_{max}$	Critical strain rate below which high elongations are possible at room temperature in the case of OMD domination in grain boundary
306.	$\dot{\varepsilon}^{\varepsilon}$	Critical strain rate corresponding to a sharp change of the tilting angle in dependence $K^{\varepsilon}(\dot{\varepsilon})$
307.	$\dot{\varepsilon}^{\sigma}$	Critical strain rate corresponding to a sharp change of the tilting angle in dependence $K^{\sigma}(\dot{\varepsilon})$
308.	ζ_1, ζ_2	Numerical coefficients depending on thermodynamic parameters of a material in the expression for the dependence of grain boundary diffusion coefficient on annealing time
309.	η	Acceleration parameter of grain boundary diffusion
310.	η_1	Coefficient of proportionality between the heating temperature rate and the recrystallization temperature
311.	θ	Grain boundary misorientation angle
312.	λ	Specific heat of fusion
313.	μ	Numerical coefficient in an expression for the dependence of grain refinement limit on the ECAE temperature
314.	μ_1	Numerical coefficient in an expression for temperature dependence of the grain refinement limit in a fcc metal
315.	μ_2	Numerical coefficient in the expression for temperature dependence of the grain refinement limit in a bcc metal
316.	ν	Poisson ratio
317.	ν_D	Debye frequency
318.	$\xi_{1,2}$	Numerical coefficients characterizing the homogeneity degree of

No.	Symbol	Physical meaning
		intragranular plastic deformation for OMD and tangential components, respectively
319.	ξ_3	Numerical coefficient characterizing the homogeneity degree of intragranular plastic deformation for triple junction disclinations
320.	ξ_m	The fraction of like-sign dislocations embedded in a migrating boundary
321.	ξ_v	Numerical coefficient characterizing the homogeneity degree of intragranular plastic deformation
322.	ρ□ρ	Density of metal
323.	ρ_b	OMD density of dislocations embedded in grain boundary
324.	ρ_{b0}	Initial OMD density
325.	ρ_b^m	Ultimate OMD density at which lattice dislocations start to accumulate in the grain volume during deformation
326.	ρ_b^{st}	Stationary OMD density
327.	$\rho_b^{st}\ \Delta b$	Stationary OMD density at which the S-mode diffusion is realized in a grain boundary
328.	$(\rho_b^{st}\ \Delta b)_L$	Stationary OMD density at which the L-mode diffusion is realized in a grain boundary
329.	ρ_g	Linear density of dislocational sources in grain boundaries
330.	ρ_v	Density of lattice dislocations
331.	ρ_v^*	Critical value of the lattice dislocation density at which the dislocation emission rate is equal to the rate of their withdrawal from grain boundaries
332.	ρ_v^{st}	Stationary density of lattice dislocations in a grain
333.	ρ_v^+	Excess density of like-sign lattice dislocations
334.	$\Delta\rho$	Change in electrical resistivity
335.	σ_{PN}	Peierls-Nabarro stress
336.	σ^o	Stress necessary for overcoming obstacles at zero temperature
337.	σ_1	Stress of intergranular plastic deformation initiation inside grains
338.	σ_1^*	Stress in the ECAE deformation zone
339.	σ^*	Critical flow stress in a grain generated by a dislocation pile-up in an adjacent grain
340.	σ_c	The value of elastic stress determined by X-ray analysis
341.	σ_d	Stress connected with dislocation strengthening
342.	σ_i	Internal stress
343.	$\sigma_i^{(b)}$	Grain boundary resistance to plastic deformation
344.	σ_i^*	Stress at the tip of dislocation pile-up
345.	σ_{int}	Stress field at the tip of the L-zone in a grain boundary
346.	σ_i^ρ	Contribution of orientational misfit dislocations to the macroelastic limit for NMC metal in the case of dislocation pile-up generation and OMD domination in grain boundaries
347.	σ_i^w	Contribution of tangential components to macroelastic limitthe grain boundary resistance of a NMC metal to plastic deformation in the case of dislocation pile-up generation and domination of tangential components in grain boundaries
348.	σ_m	Ultimate level of internal stress corresponding to the critical OMD

No.	Symbol	Physical meaning
		density ρ_b^m
349.	σ_o	Macroelasticity stress in the Hall-Petch equation
350.	$\sigma_o^{(1)}$	Macroelastic limit of NMC metal in the case of no dislocation pile-up
351.	σ^S	Flow stress at superplasticity (experimental value)
352.	σ_0^{sp}	Threshold stress of superplastic flow
353.	$\sigma_o^{кк}$	Macroelastic limit for coarse-grained material
354.	σ_p	Stress connected with strengthening by second-phase dispersed particle
355.	σ_s	Ultimate strength at tensile strength test
356.	σ_{ss}	Stress connected with solid-solution strengthening
357.	$\sigma_т^{(1)}$	Yield stress of NMC metal in the case of no dislocation pile-up
358.	$\sigma_т^{(2)}$	Yield stress of NMC metal in the case of dislocation pile-up generation and OMD domination in grain boundaries
359.	$\sigma_т^{(3)}$	Yield stress of NMC metal the case of dislocation pile-up generation and domination of tangential components of delocalized dislocationstangential components of Burgers vector of delocalized dislocations in grain boundaries
360.	σ_y	Yield stress
361.	σ_ρ	Contribution of the orientational misfit dislocations to flow stress
362.	σ_w	Contribution of the tangential components of delocalized dislocationstangential components of Burgers vector of delocalized dislocations to flow stress
363.	$\overline{\sigma}$	Relative stress (normalized to shear modulus)
364.	$\overline{\sigma}_\rho^{st}$	Contribution of the orientational misfit dislocations to relative (normalized to shear modulus) flow stress
365.	$\overline{\sigma}_w^{st}$	Contribution of the tangential components of delocalized dislocationstangential components of Burgers vector of delocalized dislocations to relative (normalized to shear modulus) flow stress
366.	τ	The level of external tangential stresses
367.	τ_{def}	Characteristic time of external action which is inversely proportional to the strain rate: $\tau_{def} \sim 1/\dot{\varepsilon}$ plastic deformation
368.	τ_y	Flow stress in the expression for continuous shear at the tip of L-zone in a grain boundary
369.	φ	Numerical parameter in the expression for grain boundary diffusion coefficient
370.	φ_1, φ_2	Numerical coefficients of the order of unity in the expression for internal stress σ_{iI} connected with OMD
371.	φ_2	Numerical coefficients in the expression for internal stress σ_I connected with w_t
372.	φ_3	Numerical coefficient in an expression for the internal stress field connected with triple junction disclinations
373.	φ_{1m}, φ_{2m}	Numerical coefficients depending on thermodynamic parameters of a material in an expression relating the OMD density and tangential components with the grain boundary migration rate
374.	φ_{d1}, φ_{d2}	Numerical coefficients in the dependence of grain boundary self-diffusion coefficient on grain size
375.	φ_{p1}, φ_{p2}, φ_{p3}	Numerical coefficients in the expression for migration driving force

No.	Symbol	Physical meaning
376.	χ	Capture cross-section
377.	χ_1, χ_2	Numerical coefficients in an expression for dependence of grain boundary diffusion coefficient on strain rate
378.	χ_ω	Geometrical coefficient in the expression for critical power of triple junction disclinations
379.	ψ	Geometrical factor in the expression for excess free volume connected with a lattice dislocation core
380.	ψ_1	Numerical coefficient in the expression for excess free volume connected with the delocalization of lattice dislocations
381.	ψ_2	Proportionality coefficient of between yield stress and initial plastic deformation stress in grains
382.	ω	Power of triple junction disclinations
383.	ω_0	Initial (before heating) power of triple junction disclinations
384.	ω^*	Critical power of triple junction disclinations at which broken wall dislocations are emitted from triple junctions
385.	ω_R	Critical power of triple junction disclinations at which fast grain boundary migration becomes possible
386.	ω_d	Additional GB misorientation connected with normal components of delocalized dislocations
387.	ω'_f	Critical power of triple junction disclinations at which triple junction cracks are formed
388.	ω^{st}	Stationary power of triple junction disclination
389.	ω^{st}_m	Stationary power of triple junction disclinations generated during migration
390.	Ω	Atomic volume

ABBREVIATIONS

№	Abbreviation	Interpretation
1.	AFM	atomic force microscopy
2.	bcc	body-centered cube
3.	CG	coarse-grained (metal)
4.	CTR	critical temperature of recrystallization
5.	DIGM	diffusion-induced grain boundary migration
6.	DSC	differential scanning calorimetry
7.	ECAE	equal-channel angular extrusion
8.	fcc	face-centered cube
9.	GB	grain boundary
10.	GBS	grain boundary sliding
11.	HAGB	high-angle grain boundaries
12.	hcp	hexagonal close-packed
13.	HPT	high-pressure torsion
14.	LD	lattice dislocation
15.	M0b	oxygen-free pure copper grade
16.	M1	technically pure copper grade
17.	MC	microcrystalline (metal)
18.	NC	nanocrystalline (metal)
19.	NMC	nano- and microcrystalline (metal)
20.	OMD	orientational misfit dislocations
21.	r.h.s.	right-hand side
22.	SP	superplasticity
23.	SPD	severe plastic deformation
24.	TEM	transmission electron microscopy
25.	TNGB	theory of nonequilibrium grain boundaries
26.	XRD	X-ray diffraction analysis

References

Aaron, H.B.; Bolling, G.F., *Surface Science*, 1972, 31, 27-49.

Akhmadeev, N.A.; Valiev, R.Z.; Kopylov, V.I.; Mulykov, R.R.,. *Izvesitya AS USSR. Metally (Russian Metallurgy)*, 1992, 5, 96-101. (in Russian

Amirkhanov, N.M.; Islamgaliev, R.K.; Valiev, R.Z., *The Physics of Metals and Metallography*, 1998, 86, 296-301.

Amirkhanov, N.M., Islamgaliev, R.K., Valiev, R.Z., *The Physics of Metals and Metallography*, 2001, 92, 518-525.

Amouyal, Y.; Divinski, S.V.; Estrin, Y.; Rabkin, E. *Acta Materialia*, 2007, 55, 5968-5979.

Amouyal, Y.; Divinski, S.V.; Klinger, L.; Rabkin, E., *Acta Materialia*, 2008, 55, 5500-5513.

Andrievskii, R.A.; Glezer, A.M., *The Physics of Metals and Metallography*, 2000, 89, 83-102.

Ashby, M.F.; Verrall, R.A., *Acta Metallurgica*, 1973, 21, 149-163.

Ashby, M.F.; Spaepen, F.A.; Williams, S., *Acta Metallurgica*, 1978, 26, 1647-1663.

Ballufi, R.W.; Kwok, T.; Bristowe, P.D.; Brokman, A.; Ho, P.S.; Yip, S., *Scripta Metallurgica*, 1981, 15, 951-956.

Barnet, M.R.; Keshavarz, Z.; Beer, A.G.; Atwell, D., *Acta Materialia*, 2004, 52, 5093-5103.

Benchabane, G.; Boumerzoug, Z.; Thibon, I.; Gloriant, T., *Materials Characterization*, 2008, 59, 1425-1428.

Berta, M.; Apps, P.J.; Prangnell, P.B., *Materials Science and Engineering A*, 2005, 410-411, 381-385.

Bida, G.V.; Gorkunov, Ed.S.; Shevnin, V.M. *Magnetic inspection of mechanical properties of rolled products*; Institute of Metal Physics of Ural branch of the Russian Academy of Science: Ekaterinburg, 2002; 253 p.

Bilby, B.A.; Cotrell, A.H.; Swinden, K.H., *Proc.R.Soc.*, 1963, A272, 304-312.

Bokstein B.S., Kopetskiy I.V., Shvindlerman L.S. *Thermodynamics and kinetics of grain boundaries in metals*; Metallurgy: Moscow, 1986; 224 p. (in Russian)

Bokstein, S.Z.; Gingburg, S.S.; Kishkin, S.T., *Autoradiography of interface surfaces and structural stability of alloys,* Metallurgy, Moscow, 1987, 272 p (in Russian)

Borisov, V.T.; Golikov, V.M.; Sherbedinsky, G.V., *The Physics of Metals and Metallography*, 1964, 17, 881-885. (in Russian).

Bowen, J.R.; Mishin, O.V.; Prangnell, P.B.; Juul Jensen, D., *Scripta Materialia*, 2002, 47, 289-294.

Brodova, I.G.; Shorokhov, E.V.; Shirinkina, I.G.; Zhgilev, I.N.; Yablonskikh, T.I., Astaf'ev, V.V.; Antonova, O.V., *The Physics of Metals and Metallography*, 2008, 105, 594-601.

Cabibbo, M.; Evangelista, E.; Scalabroni, C., *Micron*, 2005, 36, 410-414.

Cahn, R.W. *The Coming of Materials Science*; Pergamon Mayerials Series, England; University of Cambridge: Cambridge, 2001; 568 p.

Cao, W.Q.; Gu, C.F.; Pereloma, E.V.; Davies, C.H.J., *Materials Science and Engineering A*, 2008, 492, 74-79.

Chalmers, H.; Gleiter, H.A., *Philosophical Magazine*, 1971, 23, 1541-1546.

Chen, B.; Lin, D.-L.; Jin, L.; Zeng, X.-Q.; Lu, Ch., *Materials Science and Engineering A*, 2008, 483, 113-116.

Chen, H.-H.; Wang, J.-Y.; Lee, J.; Lee, S., *Journal of Alloys and Compounds*, 2008, 460, 305-308.

Chen, Y.C.; Huang, Y.Y.; Chang, C.P.; Kao, P.W., *Acta Materialia*, 2003, 51, 2005-2015.

Cheng, Y.Q.; Chen, Z.H.; Xia, W.J., *Materials Characterization*, 2007, 58, 617-622.

Chowdhury, S.G.; Mondal, A.; Gubicza, J.; Krallics G.; Fodor A., *Materials Science and Engineering A*, 2008, 490, 335-342.

Chuvil'deev, V.N.; Perevezentsev, V.N., In: *Physics of Defects of Material Layer Surface*, Leningrad Physical-Technical Institute, Leningrad, 1989, 83-89.

Chuvil'deev, V.N., *Melts*, 1996a, 2, 9-19. (in Russian)

Chuvil'deev, V.N., *Melts*, 1996b, 3, 3-15. (in Russian)

Chuvil'deev, V.N., *The Physics of Metals and Metallography*, 1996c, vol.81, No.5, pp.463-468; vol.81, No.6, pp.583-588; vol.82, No.1, pp.1-9. (in Russian).

Chuvil'deev, V.N.; Smirnova, E.S., *The Physics of Metasls and Metallography,* 1999, 88,67-72.

Chuvil'deev, V.N.; Pirozhnikova, O.E.; Petryaev, A.V., *The Physics of Metals and Metallography,* 2001a,92, 540-546.

Chuvil'deev, V.N.; Smirnova, E.S., *The Physics of Metals and Metallography*, 2001b, 92, 117-122.

Chuvil'deev, V.N.; Kopylov, V.I.; Zeiger,W., *Ann. Cnim. Sci. Mat.,* 2002, 27(3), 55-64.

Chuvil'deev, V.N.; Kopylov, V.I.; Makarov, I.M., *The Physics of Metals and Metallography,* 2003a, 96, 607-614.

Chuvil'deev, V.N.; Kopylov, V.I.; Nokhrin, A.V.; Makarov, I.M.; Malashenko, L.M.; Kukareko, V.A., *Materials Transactions (Materialovedenie)*, 2003b, 4, 9-18.

Chuvil'deev, V.N.; Nokhrin, A.V.; Kopylov, V.N., *Russian Metallurgy (Metals),* 2003c, 3, 248-257.

Chuvil'deev, V.N. *Non-equilibrium grain boundary in metals. Theory and applications*; Fizmatlit; Moscow, 2004; 304 p. (in Russian)

Chuvil'deev, V.N.; Kopylov, V.I.; Nokhrin, A.V.; Makarov, I.M.; Gryaznov, M.Yu., *The Physics of Metals and Metallography,* 2004a, 97, 1-6.

Chuvil'deev, V.N.; Kopylov, V.I.; Nokhrin, A.V.; Makarov, i.m.; Lopatin, Yu.G., *Doklady Physics*, 2004b, 49, 296-302.

Chuvil'deev, V.N.; Nieh, T.G.; Gryaznov, M. Yu.; Sysoev, A.N.; Kopylov, V.I., *Scripta Materialia,* 2004c, 50, 861-865.

Chuvil'deev, V.N.; Nieh, T.G.; Gryaznov, M.Yu.; Sysoev, A.N.; Kopylov, V.I., *Journal of Alloys and Compaunds,*2004d, 378, 253-257.

Chuvil'deev, V.N.; Nieh, T.G.; Gryaznov, M.Yu.; Sysoev, A.N.; Kopylov, V.I., In: *Severe Plastic Deformation: Toward Bulk Production of Nanostructured Materials,* Eds. B. Altan, G. Purcek, I. Miskiogly, Nova Science Publishers, New York, 2005, 204-211.

Chuvil'deev, V.N.; Gryaznov, M.Yu.; Sysoev, A.N.; Lopatin, Yu.G.; Kopylov, V.I., *Doklady Physics*, 2008a, 419, 189-192.

Chuvil'deev, V.N.; Kopylov, V.I.; Gryaznov, M.Yu.; Sysoev, A.N.; Ovsyannikov, B.V.; Flyagin, A.A., *Doklady Physics*, 2008b, 423, 334-339.

Chuvil'deev, V.N.; Nokhrin, A.V.; Makarov, I.M.; Lopatin; Yu.G.; Kopylov, V.I., *The Physics of Metals and Metallography*, 2009 (in press).

Ciccotti, G.; Guillope, M.; Pontikis, V. in: *Proc. Conf. Diffusion in Metals and Alloys*; Ed. Kednes, F.J.; Kossuth Univ.: Debrecen, Hungary, 1990.

Conrad, H., *Materials Science Engineering A*, 2003, 341, 216-228.

Conrad, H.; Jung K., *Scripta Materialia*, 2005, 53, 581-584.

Cottrell, A.H. *Theoretical structural metallurgy*; Edward Arnold (Publishers) Ltd.: London, 1959; 288 p.

Degtyarev, M.V.; Voronova, L.M.; Gubernatorov, V.V.; Chashchukhina, T.I., *Doklady Physics,* 2002, 47, 647-650.

Degtyarev, M.V. *Phasic development of ultradispersed structure in iron and structural steels under high-pressure deformation*; Institute of Metal Physics of Ural branch of the Russian Academy of Science: Ekaterinburg, 2005, 43 p.

Degtyarev, M.V.; Voronova, L.M.; Chashchuhina, T.I., *The Physics of Metals and Metallography,* 2005, 99, 276-285.

Degtyarev, M.V.; Chashchukhina, T.I.; Voronova, L.M.; Patselov, A.M.; Pilygin, V.P., *Acta Materialia*, 2007, 55, 6039-6050.

Deryagin, A.I.; Zavalishin, V.V.; Suropyatova, Yu.V. In: «*Problems of nanocrystalline materials*»; Editor Noskova, N.I.; Ustinov V.V.; Institute Of Metal Physics Ural branch of the Russian Academy of Science: Ekaterinburg, 2002; pp.433-438.

del Valle, J.A.; Carreno, F.; Ruano, O.A., *Acta Materialia*, 2006, 54, 4247-4259.

del Valle, J.A.; Penalba F.; Ruano, O.A., *Materials Science and Engineering A*, 2007, 467, 165-171.

Phase diagrams of binary metallic systems; Ed. Lyakishev, N.P.; Mashinostroenie: Moscow, 1996; 922 p. (in Russian).

Dudarev, E.F. *Microplastic deformation and yield stress of polycrystals*; Tomsk State University: Tomsk, 1988; 256 p. (in Russian)

Dugdale, D.S., *Journal of the Mechanics and Physics of Solids*, 1960, 8, 100-104.

Eddahbi, M.; del Valle, J.A.; Perez-Prado, M.T.; Ruano, O.A., *Materials Science and Engineering A*, 2005, 410-411, 308-311.

Edington, J.W., *Progress in Materials Scicence*, 1976, 21, 61-158.

El-Danaf, E.A., *Materials Science and Engineering A*, 2008, 492, 141-152.

Elagin, V.I. *Alloying of strained aluminum alloys by transition metals*; Metallurgy: Moscow, 1975; 248 p.

Elagin, V.I.; Zakharov, V.V.; Rostova, T.D. In: *Problems of metallurgy of light and special alloys*; All-Russian Institute of Light Alloys: Moscow, 1991.

Elsukov, E.P.; Lomaeva, S.F.; Konygin, G.N.; Dorofeev, G.A.; Povstugar, V.I.; Mikhailova, S.S.; Zagainov, A.V.; Maratkanova, A.N., *The Physics of Metals and Metallography*, 1999, 87, 114-119.

Efros, A.L. *Physics and geometry of disorder*. Nauka: Moscow, 1982, 176.

Estrin, Y.; Hellmig, R., *Metal Science and Heat Treatment*, 2006, 48, 504-207.

Estrin, Y.; Kim, H.S.; Nabarro, F.R.N., *Acta Materialia*, 2007, 55, 6401-6407.

Erdelyi, G.; Lojkowski, W.; Beke, D.L.; Godeny, I.; Kedves, F.J., *Philosophical Magazine A*, 1987, 56, 673-680.

Ewing, R.H.; Chalmers, B., *Surface Science*, 1972, 31, 161-171.

Farber, V.M., *Metal Science and Heat Treatment*, 2002, 44, 317-323.

Ferreasse, S.; Segal, V.M.; Alford, F.; Kardokus, J.; Strothers, S., *Materials Science and Engineering A*, 2008, 493, 130-140.

Figueiredo, R.B.; Langdon, T.G., *Materials Science and Engineering A,* 206, 430, 151-156.

Figueiredo, R.B.; Kawasaki, M.; Xu, Ch.; Langdon, T.G., *Materials Science and Engineering A*, 2008, 493, 104-110

Fionova, L.K., *The Physics of Metals and Metallography*, 1992, 4, 8-13. (in Russian)

Firstov, S.A.; Podrezov, Yu.N.; Kopylov, V.I.; Danilenko, N.I., *Russian Metallurgy (Metally)*, 2004, 1, 83-87.

Fridel, J. *Dislocations*, Pergamon Press: London, 1964; 626 p.

Friesel, M.; Lojkowski, W.; Gust, W.; Predel, B., *Defect and Diffusion Forum*, 1991, 75, 229-238.

Frost, H.J.; Ashby, M.F.; Spraepen, F., *Scripta Metallurgica,* 1980, 14, 1051-1054.

Frost, H.J.; Ashby, M.F. *Deformation-mechanism maps*; Pergamon Press: London, 1982; 328 p.

Fujita, T.; Horita, Z.; Langdon, T.G., *Materials Science and Engineering A*, 2004, 371, 241-250.

Furukawa, M.; Horita, Z.; Nemoto, M.; Valiev, R.Z., Langdon, T.G., *Acta Materialia*, 1996, 44, 4619-4629.

Furukawa, M.; Iwahashi, Y.; Horita, Z.; Nemoto, M.; Tsenev, N.K.; Valiev, R.Z.; Langdon, T.G., *Acta Materialia*, 1997, 45, 4751-4757.

Furukawa, M.; Utsunomiya, A.; Matsubara, K.; Horita, Z.; Langdon, T.G., *Acta Materialia*, 2001, 49, 3829-3838.

Furukawa, M.; Horita, Z.; Nemoto, MJ.; Langdon, T.G., *Materials Science and Engineering A,* 2002, 324, 82-89.

Gagen-Torn, K.V. *Influence of additives dopants on properties of unalloyed copper*. TsNIIEIMTsM: Moscow, 1979; 28 p. (in Russian)

Garcia-Infanta, J.M.; Zhilyaev, A.P.; Sharafutdinov, A.; Ruano, O.A., Carreno, F., *Journal of Alloys and Compounds*, 2009, 473 (1), p.163-166.

Geng, H.B.; Kang, S.B.; Min, B.K., *Materials Science and Engineering A*, 2004, 373, 229-238.

Gifkins, R.C., *Materials Science and Engineering*, 1967, 2, 181-185.

Gleiter, H.; Chalmers, B., *High-angle Grain Boundaries,* Pergamon Press, Oxford, 1972, 303.

Gleiter, H., *Acta Metallurgica*, 1979, 27, 1749-1754.

Gleiter, H.; Mahajan, S.; Bachmann, K.J., *Acta Metallurgica*, 1980, 28, 1603-1610.

Gleiter, H., *Progress in Materials Science,* 1981, 25, 125-183.

Gleiter, H, *Acta Materialia*, 2000, 48, 1-29.

Goloborodko, A.; Sitdikov, O.; Kaibyshev, R.; Miura, H.; Sakai T., *Materials Science and Engineering A*, 2004, 381, 121-128.

Golovin, S.A.; Pushkar, A. *Microplasticity and fatigue of metals*. Metallurgy: Moscow, 1980; 239 p. (in Russian).

Gol'dshteyn, M.I., Litvinov, V.S., Bronfin, B.M. *Metal Physics of high-strength alloys*; Metallurgy: Moscow, 1986; 312 p. (in Russian).

Gorelik, S.S.; Rastorguev, S.P.; Skakov, Yu A., *Radiographic and electron diffraction analysis,* Metallurgy, Moscow, 2002, 576 (in Russian).

Gorelik, S.S.; Dobatkin, S.V.; Kaputkina, L.M. *Recrystallization of metals and alloys*; Moscow Institute of Steel and Alloys: Moscow, 2005; 432 p. (in Russian)

Goto, M.; Han, S.Z.; Yakushiji, T.; Kim, S.S.; Lim, C.Y., *International Journal of Fatigue*, 2008, 30, 1333-1344.

Gottschalk, W.; Smidoda, K.; Gleiter, H., *Acta Metallurgica*, 1980, 28, 1653-1656.

Grabovetskaya, G.P. *Grain boundary diffusion and creep of submicrocrystalline metallic materials produced by severe plastic deformation*; Tomsk State University: Tomsk, 2008, 32 p.

Grabovetskaya, G.P.; Mishin, I.P.; Ratochka, I.V.; Psakhie, S.G.; Kolobov, Yu.R., *Technical Physics Letters*, 2008, 34, 136-138.

Grabski, M.W.; Korski, R., *Philosophical Magazine*, 1970, 22, 707-715.

Grain Boundary Structure and Properties. Eds. Chadwick G.A., Smith D.A.; L., N-Y, San Francisco, Academic Press, 1976, 380 p.

Grain Boundaries. Their Microstructure and Chemistry; Eds. Flewitt, P.E.J.; Wild, R.K.; John Wiley & Sons, Ltd: New York, 2001, 326 p.

Gubicza, J.; Schiller, I.; Chinh, N.Q.; Illy, J.; Horita, Z.; Langdon, T.G., *Materials Science and Engineering A*, 2007, 460-461, 77-85.

Gusev, A.I. *Nanomaterials, nanostructures, nanotechnologies*; Fizmatlit: Moscow, 2007; 416 p. (in Russian).

Gutierrez-Urrutia, I.; Munoz-Morris, M.A.; Morris, D.G., *Acta Materialia*, 2007, 55, 1319-1330.

Hasson, G.C.; Goux, C., *Scripta Metallurgica*, 1971, 5, 889-894.

Hebesberger, T.; Stuwe, H.P.; Vorhauer, A.; Wetscher, F.; Pippan, R., *Acta Materialia*, 2005, 53, 393-402.

Higashi, K.; Nieh, T.G.; Wadsworth, J., *Acta Metallurgica*, 1995, 43, 3275-3282.

Hillert, M.; Purdy, G.R., *Acta Metallurgica*, 1978, 26, 333-340.

Hirth, J.P.; Lothe, J. *Theory of dislocations*; McGraw Hill Book Company: N.-Y., 1969; 599 p.

Horita, Z.; Furukawa, M.; Nemoto, M.; Barnes, A.J.; Langdon, T.G., *Acta Materialia*, 2000, 48, 3633-3640.

Huang, C.X.; Yang, G.; Gao, Y.L.; Wu, S.D.; Zhang, Z.F., *Materials Science and Engineering A*, 2008, 485, 571-580.

Huang, J.C.; Fu, H.C.; Lou, B.Y.; Lee, H.L., *Scripta Materialia*, 1998, 39, 95-102.

Huang, W.H.; Yu, C.Y.; Kao, P.W.; Chang, C.P., *Materials Science and Engineering A*, 2004, 366, 221-228.

Internal friction method in metal science studies; Eds. Blanter, M.S.; Piguzov, Yu.V.; Metallurgy; Moscow, 1991; 248 p.

Islamgaliev, R.K.; Pyshmintsev, I.Yu.; Khotinov, V.A.; Korznikov, A.V., Valiev, R.Z., *The Physics of Metals and Metallography*, 1998, 86, 408-413.

Islamgaliev, R.K.; Yunusova, N.F.; Valiev, R.Z.; Tsenev, N.K., Perevezentsev, V.N.; Langdon, T.G., *Scripta Materialia*, 2003, 49, 467-472.

Islamgaliev, R.K.; Yunusova, N.F.; Valiev, R.Z. In: *Nanostructured materials by high-pressure severe plastic deformation*, Springer, 2006; pp.299-304.

Ivanisenko, Yu.V.; Sirenko, A.A.; Korznikov, A.V., *The Physics of Metals and Metallography*, 1999, 87, 329-334.

Ivanov, K.V. *Regularities of structure formation and creep mechanisms of submicrocrystalline Ni, Cu and Cu-*Al_2O_3, Institute of Physics of Advanced Materials, Siberian branch of the Russian Academy of Science: Tomsk, 2001; 23 p.

Jiang, Z.Z.; Yu, S.H.; Chun, Y.B.; Shin D.H.; Hwang S.K., *Materials Science and Engineering A*, 2008, 479, 285-292.

Jin, L.; Lin, D.; Mao, D.; Zeng X.; Ding W., *Journal of Alloys and Compounds*, 2006, 426, 148-154.

Johannesson, T.; Tholen, A., *Materials Science and Engineering*, 1972, 6, 189-195.

Kai, M.; Horita, Z.; Langdon, T.G., *Materials Science and Engineering A*, 2008, 488, 117-124.

Kaibyshev, O.A. *Superplasticity of commercial alloys*; Metallurgy: Moscow, 1987; 264 p. (in Russian)

Kaibyshev, O.A.; Valiev, R.Z. *Grain boundary and properties of metals*. Metallurgy: Moscow, 1987; 213 p. (in Russian).

Kaibyshev, R.; Kazakulov, I.; Gromov, D.; Musin, F.; Lesuer, D.R.; Nieh, T.G., *Scripta Materialia*, 2001, 44, 2411-2477.

Kaibyshev, R.; Sakai, T.; Musin, F.; Nikulin, I.; Miura, H., *Scripta Materialia*, 2000, 45, 1373-1380.

Kaibyshev, R.; Shipilova, K.; Musin, F.; Motohashi, Y., *Materials Science and Engineering A*, 2005, 396, 341-351.

Kang, H.G.; Lee, J.P.; Huh, M.Y.; Engler, O., *Materials Science and Engineering A*, 2008, 486, 470-480.

Kang, S.H.; Lee, Y.S.; Lee, J.H., *Journal of Materials Processing Technology*, 2008, 201, 436-440.

Kashyap, B.P.; Tangri, K., *Acta Materialia*, 1997, 45, 2383-2395.

Kaur, I.; Gust, W. *Fundamentals of grain and interphase boundary diffusion*. Ziegler Press: Stuttgarr, 1989; 448 p.

Kê, T.S., *Physics Review,* 1947, 71, 533-546.

Kê, T.S., *Journal of Applied Physics*, 1949, 20, 274-280.

Kedves, F.S.; Erdelyi, G., *Defect and Diffusion Forum*, 1989, 66-69, 175-188.

Khomskaya, I.V.; Zel'dovich, V.I.; Shorokhov, E.V.; Frolova, N.Yu., Zhgilev, I.N., Kheifets, A.E., *The Physics of Metals and Metallography*, 2008, 105, 586-593.

Kim, H.-K., *Journal of Materials Science*, 2004, 39, 7107-7109.

Kim, H.K.; Kim, W.J., *Materials Science and Engineering A*, 2004, 385, 300-308.

Kim, H.-K.; Lee, Y.-I.; Chung, C.-S., *Scripta Materialia*, 2005, 52, 473-477.

Kim, Y.G.; Hwang, B.; Lee, S.; Kim, W.G.; Shin, D.H., *Metallurgical and Materials Transactions A*, 2005, 36A, 2947-2955.

Kim, W.J.; An, C.W.; Kim, Y.S.; Hong, S.I., *Scripta Materialia*, 2002, 47, 39-44.

Kim, W.J.; Hong, S.I.; Kim, Y.S.; Min, S.H.; Jeong, H.T.; Lee, J.D., *Acta Materialia*, 2003, 51, 3293-3307.

Kittel, C. *Elementary solid state physics*; J.Wiley & Sons, Inc.: New York, 1962.
Knizhnik, G.S., *Surface*, 1982, 5, 50-56.
Ko, Y.G.; Kim, W.G.; Lee, Ch.S.; Shin, D.H., *Materials Science and Engineering A*, 2005, 410-411, 156-159.
Kolobov, Yu.R. *Diffusion-controlled processes at grain boundaries and plasticity of metallic polycrystals*; Nauka: Novosibirsk, 1998; 184 p. (in Russian).
Kolobov, Yu.R.; Grabovetskaya, G.P.; Ivanov, M.B.; Zhilyaev A.P.; Valiev R.Z., *Scripta Materialia*, 2001a, 44, 873-878.
Kolobov, Yu.R.; Grabovetskaya, G.P.; Ivanov, K.V.; Girsova, N.V., *The Physics of Metals and Metallography*, 2001b, 91, p.532-538.
Kolobov, Yu.R.; Valiev, R.Z.; Grabovetskaya, G.P.; Zhilyaev A.P.; Dudarev E.F.; Ivanov K.V.; Ivanov M.B.; Kashin O.A.; Naydenkin E.V. *Grain boundary diffusion and properties of nanostructured materials*; Nauka: Novosibirsk, 2001c; 232 p. (in Russian).
Komura, S.; Berbon, P.B.; Furukawa, M.; Horita, Z.; Nemoto, M.; Langdon, T.G., *Scripta Materialia*, 1998, 38, 1851-1856.
Kopetskiy Ch.V., Orlov A.N., Fionova, L.K. *Grain boundaries in pure materials*; Nauka: Moscow, 1987; 158 p. (in Russian).
Kopylov, V.I.; Makarov, I.M.; Nesterova, E.V.; Rybin, V.V., *Issues of Material Science*, 2002, 29, 273-278.
Kopylov, V.I.; Chuvil'deev, V.N., *Russian Metallurgy (Metally),* 2004, 1, 16-27.
Korznikov, A.V.; Safarov, I.M.; Laptenok, D.V.; Abdulin, B.F.; Valiev, R.Z., *Russian Metallurgy (Metally)*, 1993, 4, 131-136. (in Russian).
Korznikov, A.V.; Idrisova, S.; Noskova, N.I., *The Physics of Metals and Metallography*, 1998, 85, 327-331.
Kosevich, V.M., Ievlev, V.M., Palatnik, L.S., Fedorenko, A.I. *Structure of intercrystalline and interphase boundaries*; Metallurgy: Moscow, 1980; 256 p. (in Russian)
Krasil'nikov, N.A., *The Physics of Metals and Metallography*, 2001, 91, 293-298.
Kristallisation aus Schmelzen. Ed. Hein, K. and Buhrig, E.; VEB Deutscher Verlag für Grundstoffindustrie, Leipzig, 1983
Kubota, K.; Mabuchi, M.; Higashi, K., *Journal of Materials Science*, 1999, 34, 2255-2262.
Kuzhidlovsky, K.; Vuzhilovski, E.; Garbach, G., *The Physics of Metals and Metallography*, 1988, 65, 385-392. (in Russian).
Kwiecinski, J.; Wyrzykowski, J., *Acta Metallurgica*, 1989, 37, 1503-1507.
Kwiecinski, J.; Wyrzykowski, J., *Acta Metallurgica*, 1991, 39, 1953-1958.
Lapovok, R.; Thomson, P.F.; Cottam, R.; Estrin, Y., *Materials Science and Engineering A*, 2005, 410-411, 390-393.
Larin, S.A.; Perevezentsev, V.N.; Chuvil'deev, V.N., *The Physics of Metals and Metallography*, 1990, 1, 180-188. (in Russian)
Larikov, L.N., Yurchenko, Yu.F., *Thermal properties of metals and alloys*; Naukova dumka: Kiev, 1985; 437 p. (in Russian)
Handbook of diffusion in metals and alloys; Eds. Larikov, L. N., Isaichev V. I., Kiev, 1987; 510 p. (in Russian)
Lartigue, S.; Priester, L., *Acta Metallurgica*, 1983, 31, 1809-1819.
Lattice defects in quenched metals. Ed. Cotterill, R.M.J.; Academic Press: N.Y.-London; 1965.
Lee, S.; Furukawa, M.; Horita, Z.; Langdon, T.G., *Acta Materialia*, 2002, 50, 553-564.

Lee, S.; Furukawa, M.; Horita, Z.; Langdon, T.G., *Materials Science and Engineering A,* 2003, 342, 294-301.

Li, J.C.M., *Surface Science*, 1972, 31, 12-27.

Li, Y.; Langdon, T.G., *Acta Materialia*, 1999, 47, 3395-3403.

Likhachev, V.A.; Hayrov, R.Yu. *Introduction to disclination theory*. Leningrad State University Press: Leningrad, 1975; 183 p.

Likhachev, V.A.; Sen'kov, O.N., *Problems of strength*, 1987, 4, 31-37. (in Russian)

Liu, T.; Wang, Y.D.; Wu, S.D.; Lin Peng, R.; Huang, C.X.; Jiang, C.B.; Li, S.X., *Scripta Materialia*, 2004, 51, 1057-1061.

Lojkowski, W.; Grabsky, M.W., *Metals and Materials Science*, 1981, 329-333.

Lojkowski, W.; Wyrzykowski, J., *Scripta Metallurgica,* 1988, 22, 17-20.

Lojkowski, W., *Acta Metallurgica,* 1991, 39, 1891-1899.

Lojkowski, W.; Soderval, U.; Swiderski, J.; Mayer, S.; Gust, W.; Predel, B.; Lodding, A. In: *Procedings os IUMRS-ICAM-93*; Sunshine City, Ikateburo, Tokyo, Japan, 1993.

Lotkov, A.I.; Baturin, A.A.; Grishkov, V.N.; Kopylov, V.I., *Physical Mesomechanics*, 2007, 10, 179-189.

Makarov, I.M., *Materials Transactions (Materialovedenie, in Russian)*, 1999, 9, 47-53.

Mao, J.; Kang, S.B.; Park, J.O., *Journal of Materials Processing Technology*, 2005, 159, 314-320.

Martin, G.; Blackburn, D.A.; Adda, Y., *Physics State Solids*, 1967, 23, 223-228.

Martin, J. W. *Micromechanisms in particle-hardened alloys*. Cambridge Univ. Press: Cambridge, 1980; 167 p.

Martin J., Doerty R. *Stability of microstructure of metallic systems*; Cambridge Univ. Press: Cambridge, 1976; 248 p.

Machine Engineering. Encyclopedia. volume 3. Nonferrous metals and alloys. Composite metallic materials; Ed. Frindlyander, I.N.; Machine Engineering: Moscow, 2001; 880 p. (in Russian)

Matsubara, K.; Miyahara, Y.; Horita, Z.; Langdon, T.G., *Acta Materialia*, 2003, 51, 3073-3084.

May, J.; Hoppel, H.W.; Goken, M., *Scripta Materialia*, 2005, 53, 189-194.

Mazilkin, A.A.; Straumal, B.B.; Protasova, S.G.; Kogtenkova, O.A.; Valiev, R.Z., *Physics of Solid State*, 2007, 49, 868-873.

McLean, D. *Grain boundaries in metals*; Clarendon Press: Oxford, 1957; 322 p.

Mecking, H.; Estrin, Y., *Scripta Metallugica*, 1980, 14, 815-819.

Messemaeker, J. De; Verlinden, B.; Van Humbeeck, J., *Materials Letters*, 2004, 58, 3782-3786.

Meyers, M.A.; Andrade, U.R.; Chokshi, A.H., *Metallurgical and Materials Transactions A*, 1995, 26A, 144-149.

Meyers, M.A.; Mishra, A.; Benson, D.J., *Progress in Materials Science*, 2006, 51, 427-556.

Microplasticity;. Ed. McMahon, C.D.; Inrescience publishers: N.-Y., 1967; 340.

Miklyaev, P.G. *Mechanical properties of light alloys at pressure-processing temperatures and rates*; Metallurgy: Moscow, 1994; 280 p. (in Russian)

Minger, B.; Kulyasova, O.B.; Islamgaliev, R.K.; Gorb, G.; Karnthaler, H.P.; Zehetbauer, M.J., *Journal of Materials Science*, 2007, 42, 1477-1482.

Mishin, O.V.; Juul Jensen, D.; Hansen, N., *Materials Science and Engineering A*, 2003, 342, 320-328.

Mishra, A.; Kad, B.K.; Gregori, F.; Meyers, M.A., *Acta Materialia*, 2007, 55, 13-28.

Mishra, R.S.; Mahoney, M.W.; Grimes, R., *Scripta Materialia*, 1999, 42, 163-168.

Missol, W. *Surface energy of phase interface in metals*; Slask: Katoroice, 1978; 176 p.

Molodova, X.; Gottstein, G.; Winning, M.; Hellmig, R.J., *Materials Science and Engineering A*, 2007, 460-461, 204-213.

Morris, D.G.; Munoz-Morris, M.A., *Acta Materialia*, 2002, 50, 4047-4060.

Mott, N.F., *Proc. Phys. Soc.*, 1948, 60, 391-394.

Mughrabi, H.; Hoppel, H.W.; Kautz, M.; Valiev, R.Z., *Zeitschrift fur Metallikunde*, 2003, 10, 1079-1083.

Mukherjee, A.K., *Materials Science and Engineering A*, 2002, 322, 1-22.

Mulyokov, Kh.Ya.; Korznikova, G.F.; Abdulov, R.Z.; Valiev, R.Z., *Journal of Magnetism and Magnetic Materials*, 1990, 89, 207-213.

Munoz-Morris, M.A.; Garcia Oca, C.; Morris, D.G., *Scripta Materialia*, 2003, 48, 213-218.

Musin, F.; Kaibyshev, R.; Motohashi, Y.; Itoh, G., *Scripta Materialia*, 2004, 50, 511-516.

Neishi, K.; Horita, Z.; Langdon, T.G., *Materials Science and Engineering A*, 2003, 352, 129-135.

Neishi, K.; Horita, Z.; Langdon, T.G., *Materials Science and Engineering A*, 2002, 325, 54-58.

Nieh, T.G.; Wadsworth, D; Sherby, O.D. *Superplasticity in Metals and Ceramics*; Cambridge Univ. Press: Cambridge, 1997.

Nokhrin, A.V.; Chuvil'deev, V.N.; Smirnova, E.S.; Makarov, I.M., Lopatin, Yu.G., Kopylov, V.I., *Russian Metallurgy (Metally)*, 2004, 2, 139-152.

Nokhrin, A.V.; Smirnova, E.S.; Chuvil'deev, V.N.; Kopylov, V.I., *Russian Metallurgy (Metally)*, 2003, 3, 211-220.

Noskova, N.I.; Mulyukov, R.R. *Submicrocrystalline and nanocrystalline metals and alloys*; Institute Of Metal Physics of Ural branch of the Russian Academy of Science: Ekaterinburg, 2003; 279 p. (in Russian).

Novikov, I.I.; Portnoy, V.K. *Superplasticity alloys with ultrafine grain*; Metallurgy: Moscow, 1981; 168 p. (in Russian)

Orlov A.N., Perevezentsev V.N., Rybin V.V. *Grain boundaries in metals*; Metallurgy: Moscow, 1980; 156 p. (in Russian).

Orlov, A.N.; Perevezentsev, V.N.; Rybin, V.V., *Surface*, 1982, 6, 134-142. (in Russian).

Orlov, A.N.; Trushin, Yu.V. *Energy of point defects in crystals*; Atomizdat: Moscow, 1983 (in Russian).

Padmanabhan, K.A., *Journal of Materials Science*, 1981, 16, 531-533.

Padmanabhan, K.A., *Journal of Materials Science Letters*, 1983, 2, 45-46.

Panin, V.E.; Grynaev, Yu.V.; Danilov, V.I.; Zuev, L.B.; Egorushkin, V.E.; Elsukova, T.F.; Koneva, E.V.; Poletika, T.M.; Kul'kov, S.N.; Psah'e, S.G.; Korostelev, S.Yu.; Chertova, N.V. *Structural levels of plastic deformation and failure*; Nauka. Sibir branch: Novosibirsk, 1990; 255 p. (in Russian)

Park, K.-T.; Hwang, D.-Y.; Lee, Y.-K.; Kim, Y.-K., Shin, D.H., *Materials Science and Engineering A*, 2003, 341, 273-281.

Perevezentsev, V.N.; Rybin, V.V.; Chuvil'deev,, V.N., *Surface*, 1983, 10, 108-115 (in Russian).

Perevezentsev, V.N.; Rybin, V.V.; Chuvil'deev, V.N., *Surface*, 1985a, 4, 113-120 (in Russian).

Perevezentsev, V.N.; Rybin, V.V.; Chuvil'deev, V.N., *Surface*, 1985b, 11, 101-108 (in Russian).

Perevezentsev, V.N.; Rybin, V.V.; Chuvil'deev, V.N., *Surface*, 1986, 11, 130-139 (in Russian).

Perevezentsev, V.N. in: *Issues of the theory of defects in crystals*; Physical-Technical Institute: Leningrad, 1987; pp. 85-101 (in Russian).

Perevezentsev, V.N.; Pirozhnikova, O.E.; Chuvil'deev, V.N., *The Physics of Metals and Metallography,* 1991, 4, 34-41 (in Russian).

Perevezentsev, V.N., Rybin, V.V.; Chuvil'deev, V.N., *Acta Metallurgica et Materialia*, 1992, 40, 887-924.

Perevezentsev, V.N.; Pirozhnikova, O.E.; Chuvil'deev, V.N., *Inorganic Materials (Neorganicheskie materialy),* 1993, 29, 421-425 (in Russian).

Perevezentsev, V.N.; Chuvil'deev, V.N.; Kopylov, V.I.; Sysoev, A.N.; Langdon, T.G., *Annales de Chimie Science des Materiaux*, 2002a, 27, 36-44.

Perevezentsev, V.N.; Chuvil'deev, V.N.; Sysoev, A.N.; Kopylov, V.I.; Langdon, T.G., *The Physics of Metals and Metallography*, 2002b, 94, S45-S51.

Perevezentsev, V.N.; Sherban', M.Yu.; Murashkin, M.Yu.; Valiev, R.Z., *Technical Physics Letters*, 2007, 33, 648-651.

Perez-Prado, M.T.; del Valle, J.A.; Ruano, O.A., *Materials Letters*, 2005, 59, 3299-3303.

Petryaev, A.V.; Chuvil'deev, V.N., *The Physics of Metals and Metallography*, 2000, 89, 122-126.

Physical Metallurgy. Part II. Eds. Cahn, R.W.; Haasen, P.; North-Holland Physics Publishing: Amsterdam – Oxford – New York – Tokyo, 1983; 640 p.

Piatti, G., *Journal of Materials Science*, 1983, 18, 2471-2483.

Pilyugin, V.P.; Gapontseva, T.M.; Chashchukhina, T.I.; Voronova, L.M.; Shchinova, L.I.; Degryarev, M.V., *The Physics of Metals and Metallography*, 2008, 105, 409-420.

Pozdnyakov, V.A., *The Physics of Metals and Metallography*, 2003, 96, 105-120.

Prangnell, P.B.; Bowen, J.R.; Apps, P.J., *Materials Science and Engineering A*, 2004, 375-377, 178-185.

Prowan, J.W.; Bamiro, O.A., *Acta Metallurgica*, 1977, 25, 309-319.

Pumphrey, P.H.; Gleiter, H., *Philosophical Magazine*, 1974, 30, 593-602.

Pumphrey, P.H.; Gleiter, H., *Philosophical Magazine*, 1975, 32, 881-885.

Pyshmintsev, I.Yu.; Valiev, R.Z.; Aleksandrov, I.V.; Khotinov, V.A., Kil'mametov, A.R.; Popov, A.A., *The Physics of Metals and Metallography*, 2001, 92, 92-98.

Recovery and recrystallization of metals. Ed. Himmel. L.; New York – London: Interscience Publishers, 1964.

Recrystallization of metallic materials; Ed. Haessner, F.; Dr. Riederer Verlag GmbH: Stuttgard, 1978; 352 p.

Rhee, K.; Lapovok, R.; Thomson, P.F., *Journal of Minerals, Metals and Materials Society*, 2005, 57, 62-66.

Ridley, N.; Bate, P.S.; Zhang, B., *Materials Science and Engineering A*, 2007, 463, 224-230.

Ruoff, A.L., *Journal of Applied Physics*, 1967, 38, 3999-4003.

Ruoff, A.I.; Balluffi, R.W., *Journal of Applied Physics*, 1963, 34, 2862-2872.

Rybin, V.V.; Zisman, A.A; Zolotarevskiy, N. Yu., *Physics Solid State,* 1985, 27, 181-186 (in Russian).

Rybin, V.V. *Severe plastic deformation and metal failure*; Metalurgiya: Moscow, 1986; 224 p. (in Russian).

Rybin, V.V., *Material Science Issues*, 2002, 1, 11-34. (in Russian).

Sabirov, I.; Estrin, Y.; Barnett, M.R.; Timokhina, I.; Hodgson, P.D., *Scripta Materialia*, 2008a, 58, 163-166.

Sabirov, I.; Estrin, Y.; Barnett, M.R.; Timokhina, I.; Hodgson, P.D., *Acta Materialia*, 2008b, 56, 2223-2230.

Schafler, E.; Steiner, G.; Korznikova, E.; Kerber, M.; Zehetbauer, M.J., *Materials Science and Engineering A*, 2005, 410-411, 169-173.

Scripov, V.P. *Spontaneous crystallization of overcooling liquids*; Nauka: Moscow, 1984, 232 p. (in Russian)

Seeger, A.; Schottky, G., *Acta Metallurgica*, 1959, 7, 495-503.

Segal, V.M.; Reznikov, V.I.; Drobushevskiy, A.E.; Kopylov, V.I., *Izvestiya AS USSR. Metally (Russian Metallurgy)*, 1981, 1, 115-119. (in Russian)

Segal, V.M.; Reznikov, V.I.; Kopylov, V.I.; Pavlik, D.A.; Malyshev, M.F. *Processes of plastic metal structure formation*; Nauka i tekhnika: Minsk, 1994; 232 p. (in Russian).

Semiatin, S.L.; Berbon, P.B.; Langdon, T.G., *Scripta Materialia*, 2001, 44, 135-140.

Sergueeva, A.V.; Mara, N.A.; Valiev, R.Z.; Mukherjee, A.K., *Materials Sciecne and Engineering A*, 2005, 410-411, 413-416.

Sergueeva, A.N.; Stolyarov, V.V.; Valiev, R.Z.; Mukherjee, A.K., *Materials Science and Engineering A*, 2002, 323, 318-325.

Sergueeva, A.V.; Mara, N.A.; Valiev, R.Z.; Mukherjee, A.K., *Materials Science and Engineering A*, 2005, 45, 747-752.

Setman, D.; Schafler, E.; Korznikova, E.; Zehetbauer, M.J., *Materials Science and Engineering A*, 2008, 493, 116-122.

Shafler, E.; Pippan, R., *Materials Science and Engineering A*, 2004, 387-389, 799-804.

Shafler, E.; Kerber, M.B., *Materials Science and Engineering A*, 2007, 462, 139-143.

Shammazov, A.M.; Tsenev, N.K.; Valiev, R.Z.; Myshlyaev, M.M.; Bikbulatov, M.M.; Lebedich, S.P. *The Physics of Metals and Metallography*, 2000, 89, 314-318.

Shmatko, O.A.; Usov, Yu.V. *Structure and properties of metals and alloys. Electrical and magnetic properties of metals*; Naukova dumka, Kiev, 1987; 325 p. (in Russian)

Shtremel, M.A. *Strength of alloys. Part I. Lattice defects*; Moscow Institute od Steel and Alloys: Moscow, 1999; 384 p. (in Russian).

Shtremel', M.A., *Metal Science and Heat Treatment*, 2002, 44, 324-327.

Siegel, R.W., *Materials Science and Engineering A*, 1993, 168, 189-197.

Sitdikov, O.; Sakai, T.; Avtokratova, E.; Kaibyshev R.; Miura, H.; Sakai, T., *Materials Science and Engineering A*, 2007, 444, 18-30.

Sivaraman, A.; Chakkingal, U. *Materials Science & Engineering A*, 2008, 487 (1), 264-270.

Smagorinskiy, M.E.; Bulyanda, A.A.; Kudryashov, S.V. *Reference book on thermomechanical and thermocyclic metal working*. Politehnika: St. Petersburg, 1992; 416 p. (in Russian)

Smidoda, K.; Gottschalk; Gleiter, H., *Acta Metallurgica*, 1978, 26, 1833-1836.

Smirnov, B.I. *Dislocation structure and hardening of crystals*; Nauka: Leningrad, 1981; 236 p. (in Russian)

Metals Reference Book. Ed. Smithells, C.I.; Fifth Edition, Butterworths and Co. (Publishers) Ltd., London & Boston, 1976

Song, R.; Ponge, D.; Raabe, D.; Speer, J.G.; Matlock, D.K., *Materials Science and Engineering A*, 2006, 441, 1-17.

Structure and properties of internal surfaces of interface in metals. Ed. Bokstein, B.S; Nauka: Moscow, 1988; 272 p. (in Russian)

Stolyarov, V.V.; Soshnikova, E.P.; Brodova, I.G.; Bashlykov, D.V.; Kil'mametov, A.R., *The Physics of Metalls and Metallography*, 2002, 93, 567-574.

Straumal, B.B.; Klinger, L.M.; Shvinlerman, L.S., *Acta Metallurgica,* 1984, 32, 1355-1364.

Straumal, B.B.; Baretzky, B.; Mazilkin, A.A.; Phillipp, F.; Kognetkova, O.A.; Volkov, M.N.; Valiev, R.Z., *Acta Materialia*, 2004, 52, 4469-4478.

Styczynska, M.; Lojkowski, W., *Scripta Metallurgica*, 1985, 19, 1409-1413.

Su, C.W.; Chua, B.W.; Lu, L.; Lai, M.O., *Materials Science and Engineering A*, 2005, 402, 163-169.

Sun, P.L.; Kao, P.W.; Chang, C.P., *Scripta Materialia*, 2004, 51, 565-570.

Sutton, A.P.; Vitek, V., *Phil. Trans. R. Soc. Lond.*, 1983, A309, 1-36.

Suwas, S.; Gottstein, G.; Kumar, R., *Material Science and Engineering A*, 2007, 471, 1-14.

Suzdalev, I.P. *Nanotechnology: physics and chemistry of nanoclusters, nanostructure and nanomaterials*; KomKniga: Moscow, 2006; 592 p. (in Russian).

Swiatnicki, W.A.; Lojkowski, W.; Grabsky, M.W., *Acta Metallurgica*, 1986, 34, 599-605.

Handbook of Physical Data; Eds. Grigor'ev, I.S.; Meylikhov, E.Z.; Energoatomizdatr 1991, 1232 p. (in Russian)

Todd, R.I., *Metallurgical Science Forum*, 1997, 243-245, 99-108.

Tsuji, N.; Ito, Y.; Saito, Y.; Minamino, Y., *Scripta Materialia*, 2002, 47, 893-899.

Tikhonov, A.V.; Kononenko, V.A.; Prokopenko, V.A.; Rafalovskiy, V.A. *Mechanical properties of metals and alloys*; Naukova dumka: Kiev, 1986; 567 p.

Ultrafine-grain metals. Eds Burke, J.J.; Weiss, V.; Syracuse Univ. Press: Siracuse, 1973; 384 p.

Utsunomiya, H.; Hatsuda, K.; Sakai, T.; Saito, Y., *Materials Science and Engineering A*, 2004, 372, 199-206.

Valiev R.Z.; Gertsman, V.Yu.; Kaibyshev O.A., *Physics State Solids (a)*, 1980a, 61, 95-99.

Valiev, R.Z.; Kaibyshev, O.A.; Sergeev, V.I., *The Physics of Metals and Metallography,* 1980b, 49, 1291-1298 (in Russian).

Valiev, R.Z.; Gertsman, V.Yu.; Kaibyshev, O.A.; Sergeev, V.I., *Metallofizika (in Russian)*, 1983a, 5, 94-100.

Valiev, R.Z.; Gertsman, V.Yu.; Kaibyshev, O.A., *Physics State Solids (a)*, 1983b, 77, 97-105.

Valiev, R.Z.; Gertsman, V. Yu.; Kaibyshev, O.A., *Scripta Matellurgica,* 1983c, 17, 853-856.

Valiev, R.Z.; Kaibyshev, O.A., *Acta Metallurgica,* 1983d, 31, 2121-2128.

Valiev, R.Z.; Gertsman, V.Yu.; Kaiybushev, O.A., *Solid State Physics*, 1990a, 22, 2213-2216.

Valiev, R.Z.; Vladimirov, V.I.; Gertsman, V. Yu.; Nazarov, A.A.; Romanov, A.E., *The Physics of Metals and Metallography*, 1990b, 3, 31-38 (in Russian).

Valiev, R.Z.; Korznikov, A.V.; Mulykov, R.R, The Physics of Metals and Metallography, 1992, 4, 760-86 (in Russian).

Valiev, R.Z.; Ivanisenko, Yu.V; Rauch, E.F.; Baudelet, B., *Acta Materialia,* 1996, 44, 4705-4712.

Valiev, R.Z.; Salimonenko, D.A.: Tsenev, N.K.; Berbon, P.B.; Langdon, T.G., *Scripta Materialia*, 1997, 37, 1945-1950.

Valiev, R.Z.; Islamgaliev, R.K., *The Physics of Metals and Metallography*, 1998, 85, 367-378.

Valiev, R.Z.; Aleksandrov, I.V., Nanostructural Materials Produced by Severe Plastic Deformation, Logos, Moscow, 2000, 272 (in Russian).

Valiev, R.Z.; Aleksandrov, *Doklady Physics*, 2001, 46, 633-635.

Valiev, R.Z., In: *Problems of Nanocrystalline Materials,* Institute of Metals Physics of the Russian Academy of Sciences, Ekaterinburg, 2002, 17-30 (in Russian).

Valiev, R.Z.; Sergueeva, A.V.; Mukherjee, A.K., *Scripta Materialia,* 2003, 49, 669-674.

Valiev, R.Z.,: Langdon, T.G., *Progress in Materials Science,* 2006, 51, 881-981.

Valiev, R.Z.; Islamgaliev, R.K.; Semenova, I.P., *Materials Science and Engineering A*, 2007, 463, 2-7.

Varin, R.A., *Physics State Solids (a)*, 1979, 52, 347-356.

Varin, R.A., *Scripta Metallurgica,* 1980, 14, 337-340.

Varin, R.A., Z. Metallkunde, 1982, 10, 654-658.

Varin, R.A.; Tangri, K., *Z. Metallikunde*, 1982a, B73, 144-148.

Varin, R.A.; Tangri, K., *Zeitschrift fur Metallikunde*, 1982b, B73, 654-658.

Varin, R.A.; Kurzudlowski, K.J.; Tangri, K., *Materials Science and Engineering A*, 1986, 80, 11-14.

Vasil'ev, L.S.; Lomaev, I.L.; Elsukov, E.P., *The Physics of Metalls and Metallography*, 2006, 102, 186-197.

Vieregge, K.; Herzing, Chn.; Lojkowski, W., *Scripta Metallurgica*, 1991, 25, 1707-1712.

Wang, J.; Iwahashi, Y.; Horita, Z.; Furukawa, M.; Nemoto, M.; Valiev, R.Z., Langdon, T.G., *Acta Materialia*, 1996, 44; 2973-2982.

Wang, M.; Shan, A., *Journal of Alloys and Compounds*, 2008, 455, 10-14.

Wang, Y.; Chen, M.; Zhou, F.; Ma, E., *Nature*, 2002, 419, 912-915.

Wang, Y.Y.; Sun, P.L.; Kao, P.W.; Chang, C.P., *Scripta Materialia*, 2004, 50, 613-617.

Waryoba, D.R.; Kalu, P.N.; Crooks, R., *Materials Science and Engineering A*, 2008, 494, 47-51.

Watanabe, T. K., *Materials Science Forum*, 1997, 243-245, 21-30.

Watanabe, H.; Mukai, T., *Scripta Materialia*, 1999a, 40, 477-484.

Watanabe, H.; Mukai, T.; Mabuchi, M.; Higashi, K., *Scripta Materialia*, 1999b, 41, 209-213.

Wei, W.; Chen, G.; Wang, J.; Chen, G., *Rare Metals*, 2006, 25, 697-703.

Wei, W.; Wei, K.X.; Fan, G.F., *Acta Materialia*, 2008, 56, 4771-4779.

Wilkinson, D.C.; Caceres, C.H., *Journal of Materials Science Letters*, 1984, 3, 395-399.

Wolf, D., *Scripta Metallurgica*, 1989a, 23, 1913-1918.

Wolf, D., *Philosophical Magazine*, 1989b, 59, 667-680.

Wolf, D., *Acta Metallurgica*, 1990, 38, 781-790.

Xia, K.; Wang, J.T.; Wu, X.; Chen, G.; Gurvan, M., *Materials Science and Engineering A*, 2005, 410-411, 324-327.

Xu, Ch.; Dixon, W.; Furukawa, M.; Horita, Z.; Langdon, T.G., *Materials Letters*, 2003, 57, 3588-3592.

Yamashita, A.; Horita, Z.; Langdon, T.G., *Materials Science and Engineering A*, 2001, 300, 142-147.

Yunusova, N.F.; Islamgaliev, R.K.; Bardinova, M.A.; Kil'mametov, A.R.; Valiev, R.Z., *Metal Science and Heat Treatment*, 2007, 49, 135-140.

Ziman, J. *Models of disorder*; Cambridge Univ. Press: Cambride, 1979; 592.

Zel'dovich, V.I.; Shorokhov, E.V.; Frolova, N.Yu.; Zhgilev, I.N., Kheifets, A.E.; Khomskaya, I.V., Gundyrev, V.M., *The Physics of Metals and Metallography*, 2008, 105, 402-408.

Zhilyaev, A.P.; Gimazov, A.A.; Soshnikova, E.P.; Rezesz, A.; Langdon, T.G., *Materials Science and Engineering A*, 2008, 489, 207-212.

Zinov'ev, V.E. *Thermal properties of metals at high temperatures*; Metallurgy: Moscow, 1989; 384 p. (in Russian)

Zuberova, Z.; Kunz, L.; Lamark, T.T.; Estrin, Y.; Janecek, M., *Metallurgical and Materials Transactions A*, 2007, 38A, 1934-1940.

INDEX

A

B

C

D

E

F

K

L

M

N

O

P

Q

R

S

T

U

V

W

X

Z